Properties and Applications of Diamond

Properties and Applications of Diamond

John Wilks MA, DPhil, DSc
Formerly Lecturer, Clarendon Laboratory, University of Oxford; Emeritus Fellow,
Pembroke College, Oxford, UK

Eileen Wilks BSc, PhD
Formerly Research Officer, Clarendon Laboratory, University of Oxford, UK

Butterworth-Heinemann Ltd
Linacre House, Jordan Hill, Oxford OX2 8DP

  A member of the Reed Elsevier group

OXFORD   LONDON   BOSTON   MUNICH
NEW DELHI   SINGAPORE   SYDNEY
TOKYO   TORONTO   WELLINGTON

First published 1991
Paperback edition 1994

© Butterworth-Heinemann Ltd 1991

British Library Cataloguing in Publication Data
Wilks, John
  Properties and applications of diamond.
  1. Industrial diamonds. Applications
  I. Title II. Wilks, Eileen,
  621.9

ISBN 0 7506 1915 5

Library of Congress Cataloguing in Publication Data
Wilks, J. (John). *1922–*
  Properties and applications of diamond/John Wilks,
  Eileen Wilks.
  p. cm.
  Includes bibliographical references and index.
  ISBN 0 7506 1915 5
  1. Diamonds. Industrial. I. Wilks, Eileen. II. Title.
  TJ1193.W55 1991
  620.1'98–dc20   90–12794
                             CIP

Composition by Scribe Design, Gillingham, Kent
Printed and bound by Antony Rowe Ltd, Eastbourne

Preface

Although we have worked on diamonds for many years we continue to be struck by the great variety of form which diamond exhibits and the considerable variations in its properties. We have also been much impressed by the way in which scientific colleagues wishing to make use of a diamond are often at a loss to answer the question 'What sort of a diamond?' Diamonds mined from the earth have always shown a very varied form and this variety is increasing with the opening up of new mines. Today man-made diamonds are being grown to an ever increasing size, and have taken over from natural diamonds in many applications. Diamond powder can be sintered to form relatively large polycrystalline blocks of great strength and toughness. The growth of diamond films from a low pressure vapour has opened up a large new field of technology. The use of diamond is steadily expanding into new applications and technologies. This book is an attempt to set out the properties and applications of diamond both to the newcomer and to workers in these fields.

Acknowledgements

We are much indebted to many friends and colleagues for advice and discussions over a long period of time. We are particularly grateful to Paul Daniel, the editor of the *Industrial Diamond Review*, who is a veritable mine of information on diamond and its applications. We thank Dr Alan Collins for advice and for helpful comments on the manuscript for Chapters 3 and 4. We also thank the staff of the Radcliffe Science Library for their invaluable help in finding our way through the very scattered literature of diamond.

We are particularly grateful to the Governing Body of Pembroke College Oxford for a Research Fellowship for J.W. which gave the opportunity to produce this book. We thank De Beers Industrial Diamond Division and the Diamond Research Laboratory, particularly Dr R.J. Caveney, for various support and assistance over a long period of time.

We thank the many authors and publishers who have given us permission to make use of their diagrams and photographs. We are particularly grateful to those authors who have supplied us with copies of their photographs. The source of all this material is indicated in the captions to the figures. Figures 1.7, 1.8, 15.33 and 15.34 are by courtesy of *Nature* published by Macmillan. Figures 1.15, 1.16, 1.19, 7.3, 7.4, 7.14, 7.15, 7.17, 9.13, 11.7, 11.8, 11.9, 11.10, 11.11, 12.2, 12.3, 12.6, 12.8, 12.12, 14.6, 14.7, 14.8, 15.24, 15.25 and 16.36 are by courtesy of the *Journal of Materials Science* published by Chapman and Hall. Figure 3.16 is by courtesy of *Solid State Communications*, Figures 8.11 and 8.12 by courtesy of *Materials Research Bulletin*, and Table 7.4 by courtesy of *Scripta Metallurgica*, all published by Pergamon Press. Figure 5.32 is by courtesy of the Mineralogical Society of America. Colour plates 1 and 2 are by courtesy of De Beers Consolidated Mines and Anco Industrial Diamond Corp., and the picture on the cover by courtesy of Diamant Boart Industrial.

Contents

Part III: Applications and wear of diamonds

Chapter 1

Diamonds today

Diamonds have been valued as gemstones and noted for their hardness for at least two millennia but have only been available in relatively large numbers since about 1870 when extensive deposits were discovered in South Africa. Their use as jewellery is well known. Because of their extreme strength and hardness they are used in industrial processes to cut or work other materials and in a variety of other technological applications. In fact about half the diamonds which are mined today are used in industry and technology, together with a considerably greater volume produced by synthesis from other forms of carbon. This chapter describes the various sources of diamond now available.

1.1 Diamonds from the earth

We first summarize the principal sources of natural diamonds and how they are mined. We also stress the considerable variety of form exhibited by these diamonds. Further details of the mining and extraction of diamond are given in a book by Bruton (1981) which also gives a good general account of the gem industry.

1.1.a Genesis of diamond

One of the first known references to diamond is by the Roman writer Pliny (*circa* AD 70) who describes crystals apparently diamond, coming from India, presumably from alluvial river deposits like those which are still worked today. One of the earliest descriptions of these deposits was given by the French traveller and merchant Tavernier (1684) who observed very large numbers of diggers, men, women and children in the Kollur gorge of the Krishna river. No sources of diamond outside India were known until the early eighteenth century when alluvial deposits were found in Brazil. The next stage in the diamond story came in 1867 when diamonds were found in alluvial deposits in South Africa near the Orange river.

The discovery of the African diamonds set off a great urge of exploration and by 1870 perhaps 20000 diggers were working along the Vaal and Orange rivers. This activity soon showed that diamonds could sometimes be found by digging straight down into the earth into a rock which was clearly not alluvial material. After clearing away some feet of alluvium or overburden the diggers arrived at a yellow coloured rock containing diamond which they described as yellow ground.

However, it soon became clear that the yellow colouration was due to weathering and that lower down the diamond bearing rock had a bluish colour and was therefore called blue ground.

Extensive excavations particularly near Kimberley showed that the diamond bearing blue ground formed a *pipe* going down vertically into the earth. By the end of the century it was clear that these pipes were in fact the vents of extinct volcanoes whose cones had been worn away by terrestrial denudation, see for example Bonney (1899). The life history of such a volcano is shown schematically in Figure 1.1. Initially a mass of molten rock or magma in the depths of the earth at very high temperature and under very high pressure pushes out veins and dykes into the surrounding rock (Figure 1.1(a)). Eventually one of these dykes breaks through to the surface and the great pressure produces a volcanic eruption of molten material high into the atmosphere (Figure 1.1(b)). Much of this debris falls around the vent to give the well known conical form of a volcano (Figure 1.1(c)).

Eventually all activity ceases, the volcano becomes extinct, and the cone is left to the destructive weathering action of the atmosphere. Finally, as the surface of the land is worn down in the course of geological time, the cone wears down to the level of the surrounding countryside (Figure 1.1(d)). There is now almost no trace of the previous activity except that a change of vegetation on the different soil may reveal the pipe to an alert prospector. For a recent review of current

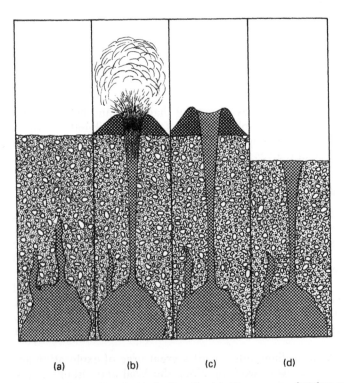

(a)   (b)   (c)   (d)

Figure 1.1 Formation of a kimberlite pipe (a), the pressure of molten magma begins to form dykes in the surrounding rock; (b), a dyke reaches the surface: the volcano erupts and begins to form a cone; (c), eruption is complete; (d), erosion has removed the cone (after Linari-Linholm, 1973)

opinion on the formation of diamond and of the properties of the blue ground now known as kimberlite, see Gurney (1989).

We see from the above accounts that diamonds have been formed at great depths from carbonaceous material under high pressures and temperatures, and subsequently ejected with the parent rock in volcanic eruptions. It follows that the former cones of the volcanoes probably consisted mainly of diamondiferous material. Hence the subsequent denudation of the cone led to diamond bearing material being washed down the rivers, and the formation of alluvial deposits. In fact, surveys around the mouth of the Orange river which drains Kimberley and other areas have revealed massive deposits of diamond buried in the beaches and on the sea bed, extending along the coast for about 50 miles on either side of the river.

Diamonds have also been found in meteorites although only as crystals of diameter 10 μm or less. Although not a significant source of supply these crystals have aroused interest and speculation about their origin, see for example Wentorf and Bovenkerk (1961) and Bibby (1982). More recently Lewis *et al.* (1987) claim to have observed extremely small meteoritic diamonds no more than 5 nm in size but the exact form of these crystals is still under discussion, see for example Blake *et al.* (1988). Evidence is also emerging that micron size crystals of diamond may be present in rocks other than kimberlite. Thus Sobolev and Shatsky (1990) describe cubo-octahedron diamonds about 12 μm in size found in garnets in metamorphic rocks.

1.1.b Mining and extraction

It is useful to give a brief outline of these processes as background information for the users of diamond. Nearly all major production today involves the mining of pipes like those described in the previous section (Figure 1.1). These pipes are roughly circular or elliptical in cross-section with surface areas ranging from perhaps $0.1 \, km^2$ to $1 \, km^2$. To obtain the diamonds, the pipe material has first to be dug out and then processed to extract the diamonds from a vastly greater volume of other material.

The uppermost material of a pipe is removed by standard open-cast methods until the mine becomes inconveniently deep. At this stage it is necessary to drill a shaft and proceed by standard methods such as block caving shown schematically in Figure 1.2. The diagram shows a pipe of kimberlite contained between harder and more resistant surrounding rock. Initially the uppermost part of the pipe where the surrounding rock has been cut back was worked by open-cast methods. Eventually, however, a series of concrete lined tunnels or levels perhaps 15 m apart were driven through the pipe perhaps 200 m below the surface. The kimberlite is then encouraged to fall through holes or drawpoints in the concrete tunnels into the levels where it is transported away for processing. Hence, as the mining proceeds, the whole mass of kimberlite and the fallen debris on top move slowly down leaving a vast hole like the well known 'big hole' at Kimberley.

In the large scale alluvial deposits on the south-west coast of Africa the diamonds are found in a layer of alluvial gravel perhaps a metre thick buried under sand and other material up to 12 m deep. The mining techniques are relatively straightforward but involve moving vast quantities of the overburden. Attempts to recover more diamonds from the sea bed, see for example Nesbitt (1967), have proved difficult because of the roughness of the sea.

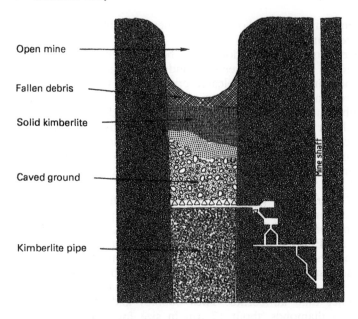

Open mine

Fallen debris

Solid kimberlite

Caved ground

Kimberlite pipe

Figure 1.2 Section of a kimberlite pipe mined by block caving (De Beers Consolidated Mines)

Once the diamond bearing material has been removed from the earth it is necessary to extract the diamonds. This is a formidable task as the diamonds are present only in concentrations of the order of 1 part in 10 million, that is about 1 ct per tonne. (The carat, ct, is the standard unit of mass for the diamond and is equal to 0.2 g.) Very large processing plants are needed to crush the ore to form a slurry which is passed through various stages of separation and floatation techniques to form a concentrate carrying the diamonds. The final extraction of the diamonds is principally by the use of grease tables or X-rays. In the older grease techniques the concentrated slurry runs down a sloping table covered with grease and the diamonds stick to the grease while the other material passes on. (It was therefore an unpleasing discovery when diamonds from a new mine did not stick to the grease. However, this misfortune was due to the presence of a surface film which could be removed by chemical treatment prior to the grease tables (Bruton, 1981).) Details of the X-ray techniques of extraction are given in Section 4.1. For further details of the mining and extraction processes, including the recovery of the smaller stones, see Bruton (1981) and Linari-Linholm (1973).

1.1.c Variety and classification

Diamonds recovered from the earth are much more varied in appearance than one might suppose from the clear colourless cut stones displayed by the jeweller. The diamonds recovered from a mine can be divided into three broad categories: gemstones, industrials and boart. The first category consists of all those diamonds which can be sold as gemstones. The industrials are diamonds somewhat similar to the first group but which cannot be used as gems because of their colour or the presence of inclusions or other defects. The lowest quality material is described as boart. This term is used somewhat loosely, but so far as mine production is concerned it is used to describe all the diamond material other than the gems and the industrials. This boart exhibits a wide range of forms from single crystals with cracks and inclusions to agglomerates of dirty coloured micro-crystals. The only use for boart is to be crushed down to provide grit and powder for grinding. The relative yields of gemstones, industrials and boart vary very considerably from mine to mine, but on average the yields of the three groups are probably quite comparable.

Uncut and unpolished diamonds are known in the trade as *rough* and Plate 1 shows a collection of this material to be used either as gem or industrial stones. We see that many of these diamonds have a pronounced colour, while an inspection with a low power eye lens shows a wide range of geometric forms. Further examples of the variety of diamond are given in Plate 2 which shows a number of industrial stones selected to show different shapes and colours.

The variations in diamond are also shown by a study of the production from different South African mines by Harris *et al.* (1975). These authors first sorted the diamonds into sieve sizes between +7 and +21 corresponding to stones between 0.14 ct and 5 ct, this range of sizes being chosen to include a major part of the production and to permit an examination of each stone with a ×5 or ×10 eye lens. One or two thousand stones of each size from each mine were classified according to shape, colour, transparency, number of inclusions, and presence of any 'coat' on the surface (Section 5.6.a). Table 1.1 summarizes some of their results, for two pipe mines (Finsch and Premier) and one alluvial mine (Dreyers Pan in Namaqualand). The table is largely self-explanatory, the figures give the percentage fraction of the diamonds with each characteristic. We discuss the various geometric shapes mentioned in the table in Chapter 5 and note here only their considerable variety. (The figures given for stones with and without inclusions do not total to 100% because inclusions could only be assessed in transparent stones. Similarly the figures for the types of coat do not total to 100% because most of the stones had no coat.)

Table 1.1 displays both the variety of diamonds from one mine and the appreciable differences between diamonds from different mines. For example, we see large differences between mines in the fractions of octahedrons, irregularly shapes stones, coloured stones, opaque stones, inclusions, and stones with transparent coats. In fact the variations from mine to mine are generally sufficiently marked that with a large sample of diamonds from one mine it is often possible to identify that mine by a visual inspection, see for example Cotty and Wilks (1971). It is of course, much more difficult to determine the provenance of one diamond on its own. Further differences observed in diamonds from South African mines have been analysed and discussed by Harris, Hawthorne and Oosterveld (1984).

Thus far we have described only the variety in diamond which is immediately apparent on a simple visual inspection. However, the diamonds used as high quality gemstones and in advanced technologies are subjected to much further scrutiny. Top quality gem diamonds are sorted by experts trained to distinguish small traces of colour and lack of clarity which reduce the brilliance of the final cut gemstones. Selection for technological applications may involve a range of measurements giving information on the impurity content and on imperfections in the crystal structure. Methods of detecting and estimating the concentrations of the various impurities in diamond by chemical analysis and by observing their optical spectra are described in Chapters 2, 3, and 4. Methods of assessing various geometrical

Table 1.1 Classification of diamonds from three mines according to size and other properties. (The sizes +7 to +21 correspond to a range from about 0.14ct to 5 ct.) (after Harris *et al.*, 1975)

	Finsch Mine									Premier Mine									Dreyers Pan Mine								
Size	+21	+19	+17	+15	+13	+12	+11	+9	+7	+21	+19	+17	+15	+13	+12	+11	+9	+7	+21	+19	+17	+15	+13	+12	+11	+9	+7
Form																											
Octahedra	18	20	18	15	14	11	12	10	7	4	5	5	8	9	6	6	6	5	–	–	38	34	30	–	21	16	9
Dodecs	22	24	25	27	26	29	31	33	32	10	13	18	15	15	17	16	21	18	–	–	25	34	41	–	55	65	62
Flattened dodecs	2	3	3	3	4	3	3	3	4	2	2	2	2	5	2	2	2	3	–	–	10	15	13	–	15	14	17
Cubo-octa	0	0	0	0	0	0	0	0	0	0	0	0	0	0	0	0	0	0	–	–	0	0	0	–	0	0	0
Cubo-dodec	0	0	0	0	0	0	0	0	0	0	0	0	0	0	0	0	0	0	–	–	0	0	0	–	0	0	0
Cubes	0	0	0	0	0	0	0	0	0	0	0	0	0	0	0	0	0	0	–	–	0	0	0	–	0	0	0
Macles	19	17	17	18	17	15	18	17	18	19	22	23	16	15	14	11	10	12	–	–	19	12	11	–	5	4	2
Spheres	0	0	0	0	0	0	0	0	0	0	0	1	1	1	0	1	0	0	–	–	0	0	0	–	0	0	0
Irregular	36	34	35	34	37	39	32	33	36	60	55	50	56	53	55	60	59	59	–	–	8	7	5	–	4	1	10
Transparency																											
Translucent	76	77	85	85	92	92	98	96	97	74	82	89	87	94	95	95	95	97	–	–	100	99	100	–	100	100	100
Opaque	24	23	15	14	8	8	2	4	3	26	18	11	13	5	5	5	5	3	–	–	0	1	0	–	0	0	0
Angularity																											
Planar	2	4	4	5	6	2	3	2	4	4	2	3	7	13	5	6	8	8	–	–	3	3	1	–	1	1	5
Rounded	94	91	93	94	91	95	89	93	94	95	98	97	91	81	92	92	89	91	–	–	87	82	79	–	84	84	78
Regularity																											
Regular	9	15	16	20	16	11	8	5	7	1	2	3	8	12	2	3	4	3	–	–	49	56	60	–	60	72	75
Distorted	33	32	30	25	28	32	44	41	38	14	18	23	18	17	24	31	25	23	–	–	24	26	32	–	31	24	13
Inclusions																											
None	9	14	19	22	26	23	27	22	25	8	7	10	11	16	15	20	26	25	–	–	64	71	76	–	75	76	81
Few	21	21	27	23	27	30	27	29	29	13	16	18	25	31	30	29	30	30	–	–	18	19	14	–	19	20	15
Many	46	43	39	41	39	39	44	45	43	52	59	61	50	47	49	46	38	42	–	–	18	10	10	–	6	4	3
Colour																											
Colourless	27	26	29	28	33	37	36	34	35	32	42	42	36	45	46	40	41	41	–	–	52	51	60	–	59	55	55
Yellow	23	18	20	18	15	15	9	9	8	17	8	10	13	6	5	5	4	2	–	–	10	11	11	–	10	17	21
Brown	15	24	26	27	33	32	40	44	49	19	24	30	31	37	41	46	48	47	–	–	3	5	4	–	3	6	14
Green	17	17	15	16	12	10	9	8	5	2	7	2	6	5	3	2	1	1	–	–	34	33	25	–	28	21	10
Black	2	2	1	1	1	0	1	0	0	5	3	4	3	1	1	1	1	1	–	–	0	0	0	–	0	0	0
Grey	16	13	9	9	6	5	5	5	4	24	13	12	11	6	4	7	5	7	–	–	0	1	0	–	0	0	0
Coats																											
Transparent	16	15	13	14	8	9	4	4	4	2	7	1	3	1	2	2	2	1	–	–	34	32	25	–	27	20	10
Opaque	0	0	0	0	0	0	0	0	0	0	0	0	0	0	0	0	0	0	–	–	0	0	0	–	0	0	0
Graphite	4	2	2	1	0	0	0	0	0	4	3	11	5	1	1	1	1	4	–	–	0	0	0	–	0	0	0

imperfections in the crystal structure are described in Chapter 6. (Variations in crystal structure are often so well marked and so characteristic that a record of this structure taken on an X-ray topograph (Section 2.3.b) may be sufficient to identify a stolen diamond even if all its faces have been repolished (Lang and Woods, 1976).

1.1.d Sources and production

Since the opening up of the mines round Kimberley there has been a continuous search for other pipes both in Africa and elsewhere, and diamonds have now been found in regions as far apart as Siberia and Western Australia. The latter source is of particular interest as the diamonds are found in a pipe rock which is not kimberlite but a somewhat different type of rock known as lamproite, for further details see Atkinson (1987) and Mitchell (1989). The Siberian mines are also noteworthy because they lie within the arctic circle and are therefore difficult to work because the ground is frozen for most of the year.

Table 1.2 presents a list of countries mining substantial quantities of diamond and their total production. We see that the main sources are located in Australia, the USSR, and the southern part of Africa. It is important to note that the production figures in Table 1.2 give no indication of the quality of the diamonds and this varies greatly. For example over 90% of the diamonds from the coastal deposits are said to be of gem quality, while the mines in Zaire and Australia give yields much below the general average, Suttill (1987) quotes the yield of gemstones at the Argyle mine in Australia as only about 5%. (The very high yield of gemstones in the coastal deposits may be due to diamonds of lesser quality being broken up preferentially by the gravel and boulders in the river bed en route to the sea, as was suggested by an experiment described by Linari-Linholm (1973).)

Table 1.2 World natural rough diamond production in millions of carats (Kennedy, 1989)

	1985	1986	1987	1988
Australia	7.06	29.2	30.0	35.0
Zaire	19.6	20.5	21.0	23.0
Botswana	12.6	13.0	13.0	15.0
USSR	12.0	12.0	12.0	12.0
South Africa	9.9	10.2	9.6	9.0
Namibia	0.91	1.0	1.0	0.9
South America	0.85	0.85	0.85	0.85
Ghana	0.6	0.55	0.44	0.3
CAR	0.5	0.6	0.6	0.45
Sierra Leone	0.4	0.4	0.36	0.3
Liberia	0.4	0.3	0.35	0.35
Tanzania	0.35	0.3	0.15	0.13
Angola	0.9	0.2	0.9	1.0
Other countries	0.47	0.5	0.45	0.45
World total	66.54	89.60	90.70	98.73

1.2 Man-made diamonds

The synthesis of diamond was first announced by the General Electric company in 1955. Since then both the size of synthetic diamonds and the volume of production have grown considerably. World production is now of order of 300 million ct, or about 3 times the production of diamonds mined from the earth. As the word synthetic is often used in a derogatory sense it must be noted that synthesized diamonds are just as real as those found naturally. In addition, they offer the prospect of an assured and unlimited supply of diamond, which can be designed to give a high degree of uniformity and controlled impurity content.

1.2.a Principles of synthesis

It has been known for a long time that diamond and graphite are different forms or *allotropes* of crystalline carbon. It was also known that transitions could be induced between different allotropes by changing the conditions of pressure and temperature, as in the transitions between grey and white tin and between the various forms of sulphur. By the beginning of the century geological evidence clearly implied that diamonds were formed deep in the earth where both the pressure and temperature were very high (Section 1.1.a).

In 1906 Nernst enunciated his heat theorem later known as the third law of thermodynamics which made possible the calculation of the equilibrium conditions in a chemical reaction from thermal data (Nernst, 1906), for details see Wilks (1961). The first estimate of the temperatures and pressures at which graphite should convert to diamond was made by Pollitzer (1912) but used only the very limited data then available. The full line in Figure 1.3 shows the transition line calculated by Berman and Simon (1955) using reliable data, and the lower line their extrapolation based on plausible assumptions about the behaviour of the thermal properties of diamond and graphite at higher temperatures. The upper line in Figure 1.3 shows a revised extrapolation taking into account later thermal data (Berman, 1965). For a discussion of progress in determining the curve more precisely in the region about 2000 K see Berman (1979).

It will be useful to note that in the literature, diamond pressures and temperatures are quoted in more than one set of units. Temperatures may be given either on the centigrade scale (°C) or the absolute scale (K), the relation between the two being:

$$T(K) = T(°C) + 273.14$$

or as a working approximation at high temperature

$$T(K) = T(°C) + 300$$

Pressures are often quoted in imperial units, as well as the SI unit the Pascal, so it is useful to remember that:

$$1 \text{ bar} = 10^5 \text{ Pa} \simeq 1 \text{ atmosphere}$$

The significance of the Berman–Simon curve in Figure 1.3 is that on the lower side of the line graphite is stable, while on the upper side diamond has less thermodynamic energy and is therefore the preferred allotrope. The curve therefore suggests that by applying high pressures to graphite at room temperature one might hope to obtain diamond. However, the curve also shows that diamonds are not thermodynamically stable at room temperature even though they appear quite stable and do not transform to graphite. This does not happen because the atoms in both diamond and graphite are held together by carbon–carbon bonds which are very strong and can only be broken by thermal energy at much higher

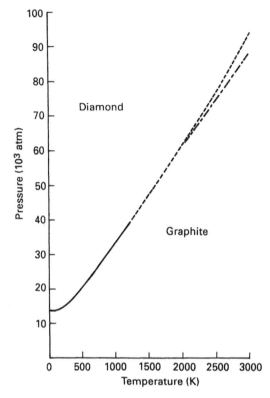

Figure 1.3 Diamond-graphite equilibrium curve: (——) calculated from reliable data by Berman and Simon (1955); (----) calculated using later data (Berman 1965); (—·—·—) linear extrapolation proposed by Berman and simon

temperatures. In fact diamond isolated from any other material only begins to convert to graphite at temperatures above about 1800 K (Section 13.3.a). This result implies that graphite can only be converted directly to diamond at temperatures above 1800 K, when the required pressure will be of the order of 60 kbar.

High pressures and high temperatures are obtained in a type of press developed by Bridgman who made a wide range of experiments at high pressures including several unsuccessful attempts to synthesize diamond (Bridgman, 1955). Details of a later press of essentially the same type are shown in Figure 1.4 and described by Wedlake (1979). The reaction chamber is surrounded by a cylindrical electric heater in turn surrounded by a ceramic cylinder which acts as thermal insulation. The chamber is compressed between tungsten carbide anvils and contained by an encircling annulus of tungsten carbide, both the anvils and the annulus being constrained by outer steel rings or 'belts'. A detailed account of the design of the initial 'belt' apparatus is given by Hall (1960a). Note that the gaskets shown in the figure act as pressure seals and play a vital role, see Hall (1958, 1960b).

Bovenkerk *et al.* (1959) have described various ways of synthesizing diamond explored by the General Electric company. Two main lines of approach were tried.

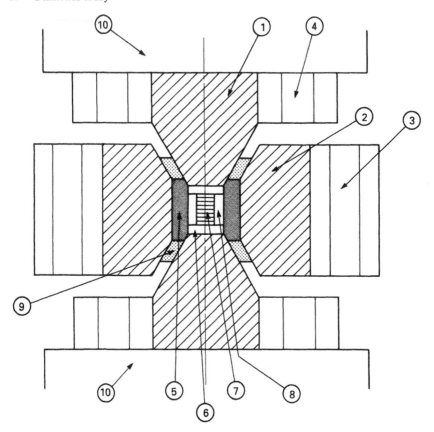

Figure 1.4 Schematic section through a typical belt type apparatus 1, tungsten carbide anvil; 2, tungsten carbide die; 3, steel support rings for the die; 4, steel support rings for the anvil; 5, ceramic tube; 6, electrical contacts; 7, reaction volume; 8, heater sleeve; 9, gaskets; 10, press plattens (Wedlake, 1979)

The first being the direct transformation of graphite to diamond. This method presents considerable difficulty because while increasing the pressure promotes the transition it also has the effect of reducing the reaction rate at which the transition takes place. Hence it was soon realized that extremely high temperatures would be needed to obtain a sufficient reaction rate and this would require even higher pressures. The second alternative was to set up some reaction in the diamond stable region shown in Figure 1.3 in the hope that under suitable conditions diamond would crystallize out.

Various reactions to obtain diamond were explored including the decomposition of carbides and other carboniferous matter but the only really successful results were obtained by dissolving carbonaceous material in molten iron or nickel under conditions such that some of the dissolved carbon crystallized out as diamond. The temperature required for such a process must obviously be greater than the melting point of the metal, or rather of the metal containing dissolved carbon. Figure 1.5 shows the melting curves of pure nickel and a eutectic mixture of carbon and nickel superposed on the Berman–Simon equilibrium line. The shading shows the region

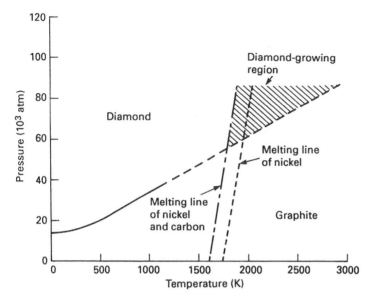

Figure 1.5 The diamond–graphite equilibrium line and the melting line of nickel and the nickel–carbon eutectic (Bovenkerk *et al.*, 1959)

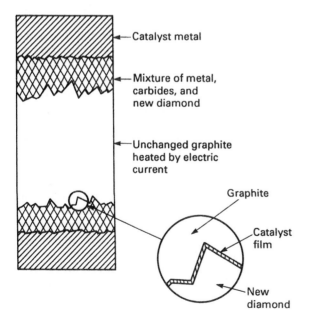

Figure 1.6 Sketch of reaction chamber for making diamonds (after Bovenkerk *et al.*, 1959)

where diamond growth was obtained at pressures over 60 kbar and temperatures of over 1800 K.

Figure 1.6 shows a sketch given by Bovenkerk *et al.* (1959) of the contents of the reaction chamber during synthesis. Initially the chamber (initially no more than about 10 mm in linear dimensions) was filled with a block of graphite between two metal cylinders. On heating under pressure the graphite in contact with the metal reacts to form carbides and diamonds. Then as the pressure and temperature is maintained the mixture of metal, carbides and diamond grows at the expense of the original carbon. The authors stress the importance of the thin film of metal observed at the interface between the graphite and the new diamond (shown in the inset in Figure 1.6). The conversion of graphite to diamond appears to be brought about by this film, only about 0.1 mm thick, so the authors describe it as a catalytic film and the metal as a solvent/catalyst. The diamonds were obtained at the end of the experiment by breaking open the capsule and dissolving away the metal in acid.

1.2.b Development of high pressure synthesis

The announcement in 1955 of the successful General Electric synthesis was followed by a period of considerable activity. Hall (1958) devised a convenient new type of tetrahedral press to avoid the General Electric patent on the belt press which he had designed. Within 5 or 6 years several groups announced successful syntheses, De Beers in 1959, Giardini, Tydings and Levin (1960), and Pugh, Lees and Bland (1961). The Swedish firm ASEA gave details of successful synthesis achieved but not reported in 1953 (Liander and Lundblad, 1960). However, although it was now possible for almost any laboratory to make diamond there was a long way to go to obtain sizeable crystals of good quality.

The first synthetic crystals were small and generally black or green in colour because of impurities coming from the bath of molten metal. In fact the details of how to keep the diamonds free of inclusions are still highly confidential. For some general remarks on these inclusions, see Wedlake (1979). Even when free of inclusions synthetic diamonds commonly have a strong yellow colouration because of impurity atoms of nitrogen which are always present in the materials of the chamber. This colouration is generally of no consequence in diamonds for industrial purposes but various experiments have been made to produce colourless stones by doping the starting material with elements to remove the nitrogen (see Wedlake, 1979). Other possibilities include doping with boron to produce semiconducting diamonds (Section 3.4) as described by Wentorf and Bovenkerk (1962), see also Wedlake (1979).

In any crystal growing operation careful and accurate control of the growth conditions is necessary to obtain crystals of good quality and morphology. This requirement was not easy to fulfill in the case of diamond. Techniques had to be devised, first to obtain accurate measurements of the pressures and temperatures throughout the chamber, and then to control them during the synthesis. In addition it was necessary to obtain a greater understanding of the chemistry and thermodynamics of the processes involved. Virtually all these developments have been commercial ventures so much essential detail has been left unpublished but the main lines of progress are quite well documented.

Information on the diamond–graphite transition is given in three papers by the General Electric team. Bundy *et al.* (1961) describe a series of syntheses similar to

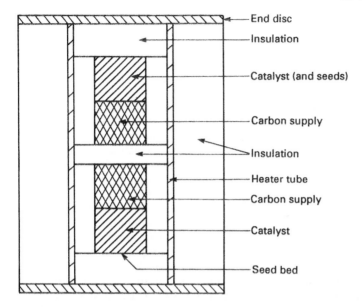

- End disc
- Insulation
- Catalyst (and seeds)
- Carbon supply
- Insulation
- Heater tube
- Carbon supply
- Catalyst
- Seed bed

Figure 1.7 Schematic diagram of reaction chamber for growing diamond by the reconstitution method (Wentorf, 1971)

those described above but at higher temperatures and pressures which confirmed the Berman–Simon line up to 2800 K. Bundy (1963a) describes a technique of transient electrical heating to obtain still higher pressures and temperatures. Energy was released by the discharge of a condenser so that the graphite in the chamber was heated and had begun to cool before the pulse of heat reached the walls of the chamber and caused a relaxation of the pressure or even melting. Using this technique Bundy (1963a) was able to measure the melting curve of graphite up to 4600 K at pressures up to 130 kbar. Then, using the same technique, he produced a direct transition of diamond to graphite with a pressure of about 125 kbar at 3000 K (Bundy, 1963b). Subsequently Wentorf (1965) showed that diamonds could be produced in this way from a range of carbonaceous material (including peanuts).

To obtain single-crystal diamonds of good quality and appreciable size it is necessary to maintain steady conditions of pressure and temperature, to keep the molten bath chemically clean, and to avoid secondary nucleations of diamond (Bundy, Strong and Wentorf, 1973). Figure 1.7 given by Wentorf (1971) shows a reaction chamber designed to obtain improved growth conditions. The arrangements in this chamber differ in two ways from those described above. First, diamond grit is used as the source of carbon rather than graphite. This avoids the large volume change when graphite converts to diamond and facilitates the control of the pressure and temperature. In particular, the driving force for the crystallization arises only from the differences in the solubilities of the carbon in the metal due to temperature gradients in the bath, an effect uncomplicated by the presence of graphite. The second difference is that small diamonds are introduced at the cooler upper and lower ends of the chamber to act as seed crystals.

Details of this technique often referred to as the reconstitution method are described and discussed by Wentorf (1971) and Strong and Chrenko (1971). Using this technique General Electric were able to produce diamonds up to 0.3 ct in size on an experimental basis, either almost colourless or with pleasant shades of pale blue, deep blue and intense yellow. As described by Crowningshield (1971) they were of moderate quality when viewed as gemstones because of metallic inclusions and other defects. Nevertheless, they showed the possibility of producing large gem quality diamonds albeit at considerable expense.

The main developments were of course centred on increasing the size of diamonds which could be made available commercially. By 1981 De Beers were offering nicely formed crystals of size 100 to the ct, that is of the order of 1 mm size. For a general account of the synthesis process and production technology see Wedlake (1979). Details of the phase diagram and the thermodynamics of the metal-carbon solutions are discussed by Strong and Hanneman (1967). Korsunskaya *et al.* (1973), Muncke (1979) and Laptev *et al.* (1985).

Finally we mention that the work on diamond synthesis at General Electric led to the discovery of two new materials one of considerable technological importance. In experiments to investigate the transition of graphite to diamond Bundy and Kasper (1967) compressed crystallites of graphite at temperatures greater than 1000°C at about 130 kbar (with the c axes of the crystallites parallel to the direction of compression). These conditions led to the formation of a polycrystalline material with crystallites of the order of 0.1 μm in size. Analyses and X-ray diffraction studies showed this material to be essentially crystalline carbon but with the atoms linked together not as in diamond but as the atoms in the mineral wurtzite which has a hexagonal symmetry as against the cubic symmetry of the diamond lattice (Section 1.5). The possibility of such a carbon structure had already been predicted by Ergun and Alexander (1962). The material is now known by the not altogether appropriate name of hexagonal diamond, and has also been produced by shock synthesis (Section 1.2.d). It has also been found in meteorites but has not so far been produced as large crystals. Not much information is available on its properties but some further comments are given by Wedlake (1974, 1979).

The other high pressure development concerns work on boron nitride. The usual form of boron nitride has a structure similar to that of graphite but on applying pressures and temperatures of the order of 85 kbar and 2100 K it undergoes a transition somewhat similar to the graphite–diamond transition. (Wentorf, 1957, 1961). The product, cubic boron nitride, so called because of its crystal symmetry, is an extremely strong and hard material second only to diamond. In addition, because of its different chemical constitution it is more resistant to wear by ferrous and similar metals. Hence it has various industrial applications (see Hibbs and Wentorf, 1974; Metzger, 1986; Krar and Ratterman, 1990) and is now manufactured in considerable quantities.

Finally we note that increasing control is being maintained on the materials used in the synthesis process in order to produce diamond with specific properties and impurity content. For example, diamond generally consists of the normal carbon isotope ^{12}C with about 1.1% of the heavier but chemically identical isotope ^{13}C, however, it is now possible to grow crystals consisting mainly of ^{13}C; Chrenko (1988) reports measurements of the Raman scattering in diamonds containing up to 89% ^{13}C and Collins *et al.* (1988) report measurements of the optical absorption and luminescence of diamonds containing 99% ^{13}C. It is also possible to obtain diamonds in which the common isotope ^{14}N is replaced by the isotope ^{15}N (with effects on the optical properties which are mentioned in Section 3.7.c). As described in Section 1.4 the technique of growth by vapour deposition has recently been used to manufacture diamond grit with a 99.9% concentration of ^{13}C for use as the starting material for growing larger crystals.

1.2.c Large diamonds

As described in the previous section the reconstitution method offers the possibility of growing large diamonds of good quality but until recently the cost of production was much greater than that of natural diamonds. Even with improvements in the control of growth conditions to obtain crystals free of impurities and with a good morphology a basic problem still remains. As in all crystal growing operations good crystals are generally only obtained at relatively slow rates of growth. For example, Yazu, Tsuji and Yoshida (1985) mention growth times of 60 h to produce 0.3 ct size diamonds. Clearly the cost of maintaining high pressures and temperatures for such long periods to produce perhaps only one diamond is quite uneconomic. However, recent developments make possible the production of larger numbers in each growth cycle thus greatly reducing the cost of each diamond.

To obtain a greater number of diamonds Yazu, Tsuji and Yoshida (1985) used a reaction chamber design as shown schematically in Figure 1.8. Instead of one bath of molten metal there are now two or more independent baths arranged one above the other, in each of which carbon is transported down to a layer of seed

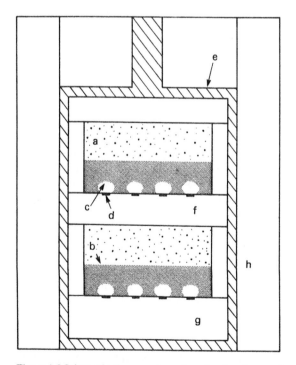

Figure 1.8 Schematic diagram of reaction chamber for growing several diamonds at two levels. a, carbon source material (diamond powder); b, metal alloy flux; c, growing diamond crystal; d, diamond seed crystal; e, heating unit; f, partition walls; g, bottom wall; h, insulating and pressure medium (Shigley *et al.*, 1987)

crystals at the base of the bath. Hence by setting several seed crystals in each bath, and with two or more baths in the chamber, one may produce perhaps 30 or more diamonds in a single run.

The above design includes a particularly ingenious feature. The driving force moving the carbon from the source down to the seeds is provided by the temperature gradient between the hotter top end of each bath and the somewhat cooler base. Therefore to obtain the optimum growth conditions the temperature of each bath must be appropriate both to the value of the pressure in the capsule and to the value of the temperature gradient. However, because the temperature gradient extends over the whole chamber, each bath is at a different temperature, so at first sight it seems that optimum conditions can only be obtained for one bath. This difficulty is overcome by choosing different metal alloys with appropriately different phase diagrams as the solvents in each bath (Yazu, Tsuji and Yoshida, 1985). Note also that the seed crystals are placed at the bottom rather than the top of the bath because impurities tend to rise to the top and act as unwanted centres of crystallization.

The above reconstitution methods are now capable of producing large crystals of quite good quality. Diamonds of up to 1.2 ct in weight made by the Japanese firm Sumito are described by Shigley *et al.* (1986) and diamonds up to 5 ct made by De Beers are described by Shigley *et al.* (1987). Like most synthetic stones they generally show a pronounced yellow colouration which may vary in intensity between different growth zones in the diamond (Section 5.4.b). Although at first sight the above stones appear to be of good quality closer inspection generally shows numbers of small inclusions and other defects. However, much development work is continuing to improve both the size and the quality of the crystals, for example Collins (1989) refers to a diamond as large as 11.1 ct that has been produced by De Beers.

1.2.d Shock synthesis

The shock synthesis method uses the fact that extremely high pressures and temperatures accompany shock waves generated by explosive charges. For example, an experiment in which a specimen of quartz was struck by an explosively driven plate generated a pressure exceeding 600 kbar (600000 atm) and a temperature of over 1400 K (DeCarli and Jamieson, 1959). In the first shock synthesis of diamond a specimen of graphite was subjected to a pressure estimated as 300 kbar for 1 μs. An inspection of the shocked product showed the presence of micron size particles which were identified as diamond by their X-ray diffraction patterns (DeCarli and Jamieson, 1961).

In other experiments Du Pont produced diamonds by shocking pieces of cast-iron containing approximately 3% graphite. Using pressures of up to 1000 kbar this method produced crystals of both diamond and the so-called hexagonal diamond mentioned in Section 1.2.b, for details see Trueb (1968). Subsequently crystals of graphite were compacted with copper powder and shocked to pressures between 250 kbar and 450 kbar for 10–30 μs at temperatures of the order of 1400 K. This process produced normal cubic type diamond in the form of polycrystalline aggregates up to about 5 μm in size made up from much smaller crystallites Trueb (1971). Although not useful for the production of large diamonds, shock synthesis has been developed by Du Pont to produce a friable type of diamond grit which is used for fine polishing processes, see Bergmann, Bailey and Coverly (1982) and Section 15.5.

The mechanism of the graphite–diamond transition under the conditions of shock synthesis is not well understood but experiments have been carried out to study the reaction and the conditions which determine whether it yields normal or hexagonal diamond, see for example Sekine et al. (1987), Kurdyumov, Ostrovskaya and Pilyankevich (1988) and Simonsen et al. (1989). Finally, diamond has recently been observed in the solid material known as detonation soot produced by chemical explosives (Greiner et al., 1988). These authors identified the diamond by X-ray diffraction methods and state that it is in the form of crystallites 4 nm to 7 nm in size, and that it makes up 25% by weight of the soot.

1.3 Sintered polycrystalline diamond (PCD)

Natural diamond exists in various polycrystalline forms described in Section 5.8 but these forms are quite rare. Here we are concerned with a man-made type of polycrystalline diamond generally known as PCD which is formed by sintering together a mass of fine diamond crystallites. This material is almost as strong and hard as single crystal diamond and has two considerable advantages. In spite of its strength, single crystal diamond fractures relatively easily along certain cleavage planes (Section 7.4). However, in polycrystalline material the crystallites are oriented in random directions, so the passage of a crack tends to be held up when passing from one crystallite to another. Hence PCD is a much tougher material than diamond (see Section 12.2.c). The second advantage of PCD is that it can be produced in relatively large blocks because the sintering process avoids the problems involved in growing large single crystals. In fact, the limit to the size of PCD material is set by the dimensions of the reaction chambers in the presses.

The principles of the sintering process are described by Wentorf, DeVries and Bundy (1980). As in all sintering processes heat and pressure is applied until the particles fuse into each other. For most pure materials it is found that sintering proceeds reasonably rapidly at absolute temperatures about 70% of the absolute melting temperature. As discussed in Chapter 8 diamond has an effective melting temperature of about 3300 K (Muncke, 1974, 1979), so crystallites must probably be raised to a temperature of the order of 2300 K, and a pressure of the order of 70 kbar applied to maintain the stability of the diamond. However, if diamond powder is compressed, as in a typical sintering process, the maximum pressures are produced at the points of contact of the grains and not in the voids between the grains where the pressure will be much less. Therefore at the high temperatures involved the diamond adjacent to the voids will transform to graphite, so the standard technique of sintering by heat and pressure cannot be used successfully with diamond.

The methods now used for the production of PCD material are outlined by Wentorf, DeVries and Bundy (1980) and Tomlinson and Wedlake (1983). A schematic arrangement of the reaction chamber prior to sintering is shown in Figure 1.9. On heating under pressure the cobalt melts and infiltrates into the voids and thus maintains the pressure on the diamond surfaces and prevents graphitization during the sintering process. Various combinations of pressure and temperature may be used, Yazu et al. (1983) mention 50 kbar at about 1600K, and Davey, Evans and Robertson (1984) 99 kbar at 1200K.

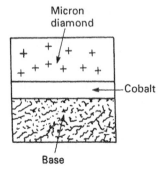

Micron
diamond

—Cobalt

Base

Figure 1.9 Typical arrangement of the starting material in a reaction chamber for producing PCD (Tomlinson and Wedlake, 1983)

Although the interstices between the diamond crystallites in PCD are generally filled with cobalt metal, the strength of the material does not arise primarily from this metal acting in the same way as the binder in many composite materials. The strength of good quality PCD depends on the strength and extent of the diamond–diamond bonds produced by the crystallites growing together around their areas of contact. Figure 1.10(a) shows an optical micrograph of a polished surface of typical PCD material, and Figure 1.10(b) an SEM micrograph of a typical surface after etching in a bath of aqua regia to remove the cobalt. We see that the initial diamond powder has been converted by intergrowth into a single matrix of diamond. The crystallites are so firmly bonded together in well made material that if any fracture occurs it is generally transgranular, that is, it passes through the grains rather than around them (Figure 1.11). The structure of PCD material is further discussed in Section 12.1.

50 μm

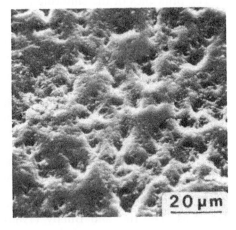

20 μm

Figure 1.10 (a), optical micrograph of polished surface of PCD; (b), SEM micrograph of surface of PCD after etch to remove cobalt

Plate 1 Collection of rough (unpolished) gem and industrial diamonds (De Beers Consolidated Mines)

Plate 2 Industrial diamonds selected to show a range of shape and colour (Anco Industrial Diamond Corp.)

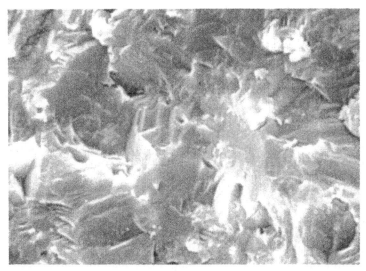

Figure 1.11 SEM micrograph of PCD material showing transgranular fracture through the grains (Lammer, 1988)

PCD materials are marketed as proprietary products rather than as materials with specified forms and structures. The principal manufacturers each make a range of products designed for particular operations, with the sizes of the crystallites ranging from perhaps 2 μm to 25 μm. Therefore when discussing the properties of PCD it is necessary to remember that the values of, for example, the strength and toughness, will vary to some extent between different products. The best known proprietary names are probably General Electric's Compax© and the De Beers' Syndite© together with variants of these names. Typical samples of PCD material for use as cutting tools are shown in Figure 1.12. The PCD has the

Figure 1.12 PCD blocks on a tungsten–carbide base for use as a tools (De Beers Industrial Diamond Division)

form of a thin layer which is generally produced integrally on a thicker base of tungsten carbide; the PCD provides the cutting edge while the less costly carbide adds considerable extra strength and toughness. In a typical product the PCD layer may be about 0.7 mm in depth and the carbide three or four times thicker. These blocks are presently available in various shapes and sizes up to 50 mm diameter. Solid blocks of PCD alone are available up to about 5 mm in size.

Today most commercial production of PCD material is based on the use of cobalt, as described above, but there is also the possibility of obtaining improved products by using other sintering agents and different conditions of pressure and temperature. Two important variants already available are designed to avoid the PCD losing strength at high temperatures when the presence of cobalt encourages the graphitization of the diamond. General Electric market a product in which the cobalt is leached out after manufacture. De Beers produce a material in which the cobalt is replaced by silicon which apparently acts as a binder holding the crystallites together. Although this material is less strong it retains its strength to higher temperatures than normal PCD and can be made in volumes that would be difficult to leach out. These two materials are further described in Section 12.2.e.

Various other experimental methods of forming PCD material have been reported. Nakai and Yazu (1987) used starting mixtures containing between 93% and 99.5% diamond plus cobalt or nickel and approximately 0.2% tungsten or other carbide and thus obtained PCD with volume concentrations of diamond of up to 99%. Other authors have produced PCD by the shock compression of diamond powder (Akashi and Sawaoka, 1987) and of diamond powder plus graphite (Potter and Ahrens, 1988). Naka *et al.* (1987) describe experiments starting not with diamond powder but with purified graphite mixed with about 10% titanium and up to 40% iron by weight. Sintering for 15 min at 7 GPa and 1700°C produced compact masses of diamond and metal carbide. At present there is not much information on the structure and strength of these different products.

1.4 Diamond films

It is now possible to grow diamond by the deposition of carbon from a carbonaceous gas at less than atmospheric pressure and to obtain a thin film of polycrystalline diamond on substrates such as silicon, glass, molybdenum and tungsten. These films offer considerable possibilities as a protective coating on other materials, as a means of improving the performance of cutting tools, and perhaps as semiconductor materials. At present, however, the material and its application are still primarily in the development stage.

1.4.a The growth process

The principal efforts to synthesize diamond have hitherto been on the lines described in Section 1.2 but there has long been speculation on other possible methods. The first report of growing diamond from a vapour source appears to be that of Eversole (1962), who introduced methane at low pressure (0.15 torr) over fine diamond powder at temperatures around 1300 K. At that temperature some of the methane decomposes into carbon and hydrogen, and some of the carbon is deposited as diamond on the surface of the powder. A particularly detailed set of experiments was made by Angus, Will and Stanko (1968) who passed methane at

0.3 torr over diamond powders of size 0–1 μm and 1.5 μm at a temperature of 1050°C and observed increases in the weight of the powder of up to 15%.

To demonstrate that the increase in weight was due to the formation of diamond Angus, Will and Stanko (1968) made chemical analyses which showed that the new material was virtually all carbon, suggesting that it was either graphite or diamond. Careful measurements of the density of the powder showed that in at least two cases the values were very close to that for single crystal diamond and quite different from that for graphite. The absence of graphite was also shown by the resistance of the powder to a selective chemical etch, by the absence of any graphite lines in X-ray and electron diffraction patterns, and by the absence of any sign of electrical conductivity when placed in a high frequency microwave field. References to subsequent work on these lines are given by Wedlake (1979)

Although diamond had been grown from a vapour the above results were discouraging in at least two respects. First, the rate of growth was extremely slow and after some deposition of diamond the process tended to produce graphite. It was then necessary to stop the deposition and clean off the graphite by switching to an atmosphere of hydrogen before proceeding with further growth. Secondly, subsequent experiments showed that the new growth on the diamond was highly polycrystalline.

At about the same time as the above experiments, Derjaguin *et al.* (1968) reported that they had been able to grow filamentary crystals or whiskers of diamond on single crystal diamond by deposition from a carbonaceous gas, the whiskers being about 10–20 μm in diameter and perhaps 200 μm long. Not much experimental detail was given. These Russian workers continued to develop the method and a good review of this work is given by Spitsyn, Bouilov and Derjaguin (1981). Most of the experiments were based on the use of methane as before but with the addition of *atomic* hydrogen, generated either by a hot filament or electric discharge in the methane. The atomic hydrogen has two effects, it keeps the growing diamond free of graphite, and also appears to promote the growth process. Using this technique it was found possible to grow 'high-perfection single crystalline layers' on a diamond surface, with a thickness of the order of several microns, at growth rates of the order of 1 μm/h, very much higher than obtained in the previous experiments mentioned above.

Spitsyn, Bouilov and Derjaguin (1981) observed that their single crystal films became less perfect as their thickness increased, and they noted a gradual transition from a single crystal film to a polycrystalline film when the thickness was of the order of 5 μm. The same authors also showed that the temperature of the diamond substrate had a critical effect on determining whether the film was single or polycrystalline.

In addition to growing diamond films on diamond Spitsyn, Bouilov and Derjaguin (1981) were also able to grow single crystals of diamond on substrate materials other than diamond, such as copper, silicon and tungsten. They described the spontaneous formation of diamond nuclei on the surface of the substrates and their subsequent growth into small crystals of about 10 μm size in about 30 min. It was also observed that the rate of nucleation of the diamonds was between 10 and 10^2 times faster on elements forming carbides (silicon and tungsten), and about 10^2 times faster on polycrystalline than single crystal material. Also, in accord with this last observation, the rate of nucleation was greatly enhanced by scratching the substrate prior to desposition.

The above techniques were further developed by Matsumoto and co-workers in

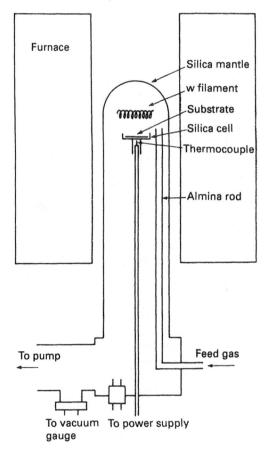

Figure 1.13 Schematic diagram of the arrangements for the vapour deposition of diamond (Matsumoto *et al.*, 1982)

a series of experiments using non-diamond substrates. Matsumoto *et al.* (1982) used an apparatus similar to that of previous workers which is shown schematically in Figure 1.13. As before, methane was used as the carbon source but the concentration of hydrogen was greatly increased by using a gaseous mixture of hydrogen and methane with a methane content of between 0.5% and 5%. The atomic hydrogen was produced from the molecular hydrogen by a tungsten filament run at about 2000°C. Then with a gas pressure of 10 torr to 100 torr it was possible to deposit 5 μm size crystals on silicon, molybdenum and silica substrates at 700°C to 1000°C in about 3 h. The authors also described how the morphology of the crystals depended on the pressure and temperature, and on the rate of flow and constitution of the gas mixture. Examples of their crystals are shown in Figure 1.14.

Following the above experiments there has been a steady development of technique to obtain continuous polycrystalline films of diamond on non-diamond substrates. These techniques include the use of a glow discharge to dissociate the hydrogen over a large area of substrate (Matsumoto, 1985). More recently plasma

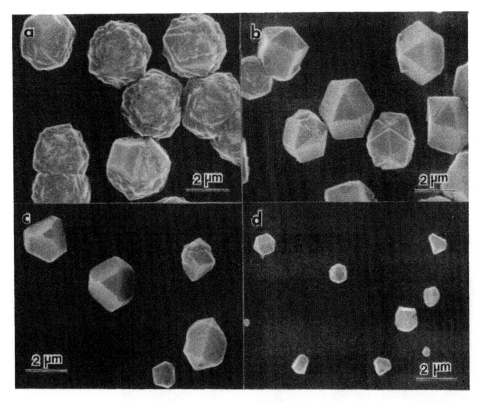

Figure 1.14 Diamond crystals grown by vapour deposition using different methane concentrations: (a), 2%; (b), 1%; (c), 0.67%; (d), 0.5%; (Matsumoto *et al.*, 1982)

arcs have been used to direct a plasma jet of gas on to the diamond as in Figure 1.15, using either an R.F. plasma (Matsumoto, Hino and Kobayashi, 1987) or a D.C. plasma (Kurihara *et al.*, 1988). As a result of these developments the rates of film growth in the latter two sets of experiments were of the order of 60 μm/h. Hirose and Terasawa (1986) have grown films from various organic vapours including alcohols and acetone. It has even been found possible to deposit a film by directing either an oxy-acetylene,or some other types of combustion flame, on to a water cooled substrate (Hirose, 1988; Snail *et al.*, 1988).

Parallel with the increase in the rates of deposition there have also been large improvements in the mechanical coherence of the film. In the initial experiments of Matsumoto *et al.* (1982) the diamond growth was in the form of isolated crystals, whereas the films now grown are well formed and coherent so that they can exist independently of the substrate. Figure 1.16 shows SEM micrographs of a surface of a typical polycrystalline film and of a section of the same film (De Beers Industrial Diamond Division). The films when viewed separately from the substrate are generally straw coloured (like other synthetic material) and appear matt on the relatively rough upper side (Figure 1.16(a)) but shiny on the lower side where they were in contact with the substrate. They are commonly 10–20 μm thick and are generally translucent but not transparent.

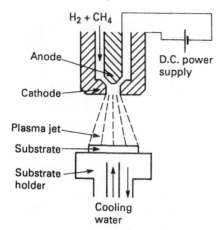

H₂ + CH₄

Anode

D.C. power
supply

Cathode

Plasma jet

Substrate

Substrate
holder

Cooling
water

Figure 1.15 Schematic diagram of the arrangements for plasma jet method of growing diamond films (Kurihara *et al.*, 1988)

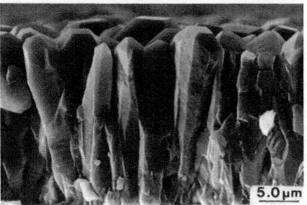

4.0 μm   (a)

5.0 μm   (b)

Figure 1.16 SEM micrographs of a diamond film grown by vapour deposition on a silicon substrate: (a), the surface; (b), a cross section view; (De Beers Industrial Diamond Division)

1.4.b Current developments

Now that methods of growing diamond films are well established there is considerable interest in the details of the techniques of growth and in the properties of the films. We now briefly indicate the main directions in which work is proceeding. Overall views of the present activity have been given in two conference reports (Japan New Diamond Forum, 1988; Chang, Nelson and Hiraki, 1989).

Naturally, one of the main points of interest is how best to assess the structure and quality of a film. The principal techniques of assessment make use of the electron microscope and the Raman effect. The SEM provides a detailed view of the surface which gives some idea of the degree of crystallinity while further information is obtained by sectioning the film (see Figure 1.16). Electron diffraction techniques in the transmission electron microscope can be used both to assess the lattice perfection of a single crystallite and the polycrystallinity of a wider area.

The Raman effect is the shift in the frequency of light scattered from a crystal as described in Section 3.1.c. It turns out to be a sensitive indicator of the structural perfection of a diamond crystal. In a typical experimental arrangement a beam of monochromatic light is passed through the film and the scattered light examined in a spectrometer to measure any change in frequency, the result being presented as a spectrum of the shifts observed. Figure 1.17 shows Raman spectra observed by Matsumoto *et al.* (1982) from films obtained on molybdenum substrates by deposition from mixtures of methane and hydrogen under three different conditions of pressure, temperature and gas concentration. Curve c is similar to that given by a good single crystal diamond, but the other curves are appreciably different, the broad peak around $1600\,\text{cm}^{-1}$ being associated with the presence of graphitic type deposits.

A variety of other techniques are also being used to study the structure of the film. Hu, Joshi and Nimmagadda (1989) report studies using both Auger and XPS techniques (Section 13.8.a), and Bruley *et al.* (1989) describe studies using energy loss electron spectroscopy (EELS). Various studies involving observations of the structure of X-ray absorption edges have been made by Capehart *et al.* (1989), and by Takata *et al.* (1989) who used this X-ray method to determine the length of the carbon–carbon bonds in the film. Studies of the luminescence of films in the SEM have been made by Kawarada *et al.* (1989). Collins, Kamo and Sato (1989) describe the cathodoluminescence of single crystal diamonds grown from a mixture of methane and hydrogen as a function of the methane content. The results suggest that the details of the emission spectra can give useful information on the state of perfection of the crystal lattice.

An increasing number of papers are concerned with obtaining basic information on the growth process. The kinetics of the dissociation of the carbonaceous gases and the growth of diamond are discussed by Derjaguin and Fedoseev (1977) and by Fedoseev *et al.* (1984), both these references are in Russian but are summarized in a review by Badzian and DeVries (1988). Experimental studies of the kinetics of the growth process have been made by Harris, Belton, Weiner *et al.* (1989) who placed a probe close to the substrate to obtain samples of the gas mixture for analysis by mass spectroscopy. Studies of the effect of the pressure and temperature of the gas are described by Badzian *et al.* (1988) and Meilunas *et al.* (1989). Somewhat similar experiments by Suzuki, Sawabe and Inuzuka (1988) also included measurements of the electric field in their D.C. plasma.

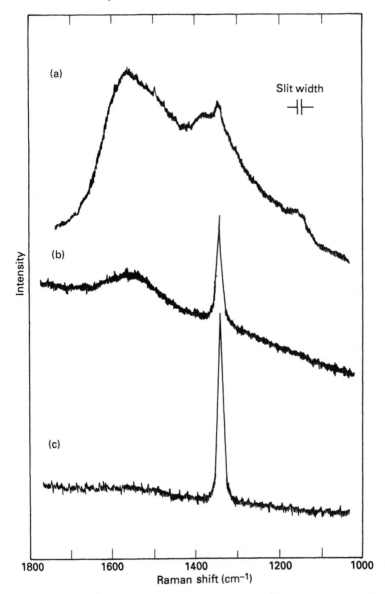

Figure 1.17 Raman spectra of different carbonaceous films deposited on molybdenum. The spectrum in curve (c) is similar to that of diamond, see text (Matsumoto *et al.*, 1982)

There is also considerable interest in the possibility of obtaining single crystal films, as well as polycrystalline films, on non-diamond substrates. However, the mismatch between the diamond lattice and that of the substrate has so far precluded growth as a single crystal. The difficulties are considerable because, as mentioned above, single crystal film even growing on diamond tends to become polycrystalline with increasing thickness. We note, however, that Sato, Haka and Kamo (1988) show a section of a film in which crystallites about 2 μm in diameter have grown up vertically from the substrate as columnar single crystals for about 15 μm. We ourselves have seen that using a high power microscope it is quite possible to look right through such crystallites to the substrate below.

Finally, there is now a considerable literature on films which are described as 'diamond like'. This somewhat imprecise term is used to refer to a range of films including both imperfect low quality diamond films and amorphous films of hydrocarbons which may contain between about 15 atomic % and 60 atomic % of hydrogen. Although the amorphous hydrocarbon films are not so hard as diamond they are relatively abrasion resistant and may be used in various applications as protective layers. Their amorphous nature results in a smooth surface which has a low friction and is said to be resistant to chemical corrosion. Moreover, because their mechanical and optical properties are determined both by the deposition conditions and the type of hydrocarbon there is the possibility of designing films for particular needs. Reviews of the properties of these films have been given by Dischler and Brandt (1985) and Angus and Hayman (1988); see also Chang, Nelson and Hiraki (1989) and Franks, Enke and Richardt (1990).

1.4.c Applications

At present the interest in diamond films centres on three different fields of application, their use as hard protective coatings on tools and other devices, the possibilty of constructing semiconducting electronic devices, and as a valuable adjunct to previous methods of growing diamond under controlled conditions. We consider each of these fields in turn.

The coating of cutting tools with thin films of various hard materials is becoming increasingly common but none of these materials have the hardness of diamond. Therefore the possibility of coating say a tungsten carbide tool with a diamond film is most attractive as the diamond is very hard and the carbide strong and tough. Therefore considerable attention is being paid to the problem of obtaining good adhesion between the film and the substrate. In general, the best adhesion between a film and its substrate is obtained when the film has grown epitaxially on the substrate, that is when the deposited atoms take sites correlated with the positions of the atoms in the substrate, and the lattice of the film has a particular orientation with respect to that of the substrate. For example, a diamond film will grow on the surface of a single crystal of diamond as a single crystal with the same orientation, without any loss of strength across the boundary. However, if the atoms in the substrate have arrangements and spacings much different from those in diamond it will be difficult to obtain epitaxial growth and the bond to the substrate will be correspondingly weaker. Even so, Yagi (1988) and Murakawa, Takeuchi and Hirose (1988) have reported considerable improvements in the performance of tungsten–carbide tools when coated with diamond films. Not much information is yet available on the mechanical properties of these films but see Section 12.6.

Besides the use of diamond films as protective coatings there has been interest in and speculation on their use as semi-conductor materials. This interest derives from the possibility of growing films doped with boron so as to produce semi-conducting diamond. However, it is by no means certain that any such diamond material will be more useful than other semi-conductor materials now being developed (see Section 17.7). In any case considerable effort will be required to obtain films with the high degree of crystal perfection necessary for the manufacture of useful devices.

Finally we note that the technique of growing diamond by vapour deposition offers the possibility of growing diamonds with more precise and controlled properties. A striking example of these possibilities was recently reported by the General Electric company who grew films from methane isotopically enriched to have a very low content of the isotope ^{13}C. The diamond sheet was then 'crushed and powdered' and used as the starting material in a normal high-pressure reconstitution process to produce single crystals of diamond about 1 ct in size with a low concentration of ^{13}C (Anthony *et al.*, 1990). Such diamonds have a very high thermal conductivity (Section 6.3.c) and will be valuable as heat sinks (Section 17.6.d). This use of the vapour deposition method to control the starting material for the growth of large crystals obviously opens up considerable possibilities.

1.5 The diamond lattice

As in any other crystalline solid the atoms in diamond are linked together in a simple and regular way. Each carbon atom shares one of its outer four electrons with each of four other carbon atoms, spaced symmetrically about it. This sharing of electrons results in a force between each pair of atoms which has the effect of a *bond* holding them tightly together. The net effect of all these carbon–carbon bonds is that the atoms take up the regular arrangement shown schematically in Figure 1.18.

Each ball in Figure 1.18 represents a carbon atom and the four rods connected to each ball represent the electron bonds, the distance between the centres of any pair of atoms is known from experiments on X-ray diffraction to be about 0.155 nm. Given that all four bonds are symmetric about a carbon atom this arrangement specifies the relative positions of all the atoms in the crystal. Note, however, that the figure is somewhat misleading in that it suggests that the atoms are spaced quite widely apart, whereas in fact a more realistic model would show the balls touching each other. However, it would then be impossible to see the geometric arrangement of the atoms in the model.

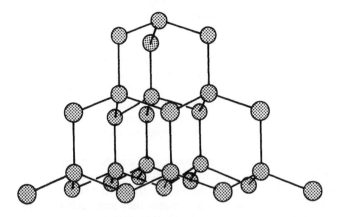

Figure 1.18 Model of the diamond lattice

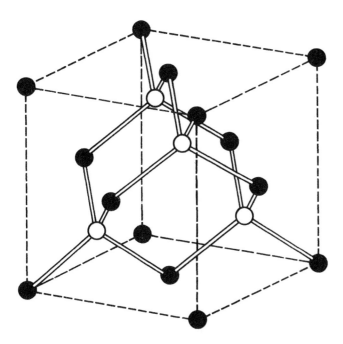

Figure 1.19 Model of Figure 1.18 viewed in another direction. (After Kittel, 1976)

At first sight the arrangement of the atoms in Figure 1.18 appears to give quite a simple structure but this is somewhat misleading. In particular, the lattice shown in the figure may look quite different when viewed in other directions. For example Figure 1.19 gives another view of the relative positions of a group of atoms in the diamond lattice using the same type of model as Figure 1.18. Eight of the atoms shown black are located at the corners of a cube, and six other black atoms are situated at the centres of the cube faces. Within the cube are four more carbon atoms shown as open circles each of which is linked by four bonds to one of the black atoms around it. This arrangement repeats itself all through the diamond, the black atoms forming a lattice of the type known as face-centred cubic. In addition, and this is not obvious from the small number of atoms shown in the figure, the atoms shown as open circles also lie on the points of another face-centred cubic lattice similar and parallel to the first, but somewhat displaced from it. Thus the complete diamond lattice may be described as consisting of two interpenetrating face-centred cubic lattices. These regular arrangements of the atoms in diamond result in certain symmetries in its properties, some of which are discussed further in Sections 5.1 and 5.2.

References

Akashi, T. and Sawaoka, A. B. (1987) *Journal of Materials Science*, **22**, 3276–3286
Angus, J. C. and Hayman, C. C. (1988) *Science*, **241**, 913–921
Angus, J. C., Will, H. A. and Stanko, W. S. (1968) *Journal of Applied Physics*, **39**, 2915–2922
Anthony, T.R., Banholzer, W.F., Fleischer, J.F., *et al.* (1990) *Physical Review B*, **42**, 1104–1111

Atkinson, W. J. (1987) *Industrial Diamond Review*, **47**, 1–8

Badzian, A. R., Badzian, T., Roy, R., *et al.* (1988) *Materials Research Bulletin*, **23**, 531–548

Badzian, A. R. and DeVries, R. C. (1988) *Materials Research Bulletin*, **23**, 385–400

Bergmann, O. R., Bailey, N. F. and Coverly, H. B. (1982) *Metallography*, **15**, 121–139

Berman, R. (1965) In *Physical Properties of Diamond* (ed. R. Berman), Clarendon Press, Oxford, pp. 371–393

Berman, R. (1979) In *The Properties of Diamond* (ed. J. E. Field), Academic Press, London, pp. 3–22

Berman, R. and Simon, F. E. (1955) *Zeitschrift für Elektrochemie*, **59**, 333–338

Bibby, D. M. (1982) In *Chemistry and Physics of Carbon*, Vol. 18 (ed. P. A. Thrower), Marcel Dekker, New York, pp.1–91

Blake, D. F., Freund, F., Krishnan, K. F. M., *et al.* (1988) *Nature*, **332**, 611–613

Bonney, T. G. (1899) *The Geological Magazine*, **6**, 309–321

Bovenkerk, H. P., Bundy, F. P., Hall, H. T., *et al.* (1959) *Nature*, **184**, 1094–1098

Bridgman, P. W. (1955) *Scientific American*, Nov, 42–46

Bruley, J., Cuomo, J. J., Guarnieri, R. C. and Whitehair, S.J. (1989) In *Extended Abstracts Technology Update on Diamond Films* (eds R. P. H. Chang, D. Nelson and A. Hiraki), Materials Research Society, Pittsburg, pp. 99–102

Bruton, E. (1981) *Diamonds*, (2nd edn.) N.A.G. Press, London

Bundy, F. P. (1963a) *Journal of Chemical Physics*, **38**, 618–630

Bundy, F. P. (1963b) *Journal of Chemical Physics*, **38**, 631–643

Bundy, F. P., Bovenkerk, H. P., Strong, H. M. and Wentorf, Jr, R. H. (1961) *Journal of Chemical Physics*, **35**, 383–391

Bundy, F. P. and Kasper, J. S. (1967) *Journal of Chemical Physics*, **46**, 3437–3446

Bundy, F. P., Strong, H. M. and Wentorf, Jr, R. H. (1973) In *Chemistry and Physics of Carbon, 10*, (eds P. L. Walker, and P. A. Thrower), Marcel Dekker, New York, pp. 213–263

Capehart, T. W., Perry, T. A., Beetz, C. B., *et al.* (1989) *Applied Physics Letters*, **55**, 957–959

Chang, R. P. H., Nelson, D. and Hiraki, A. (eds) (1989) *Extended Abstracts Technology Update on Diamond Films*. Materials Research Society, Pittsburg

Chrenko, R.M. (1988) *Journal of Applied Physics*, **63**, 5873–5875

Collins, A. T. (1989) *Industrial Diamond Review*, **49**, 24–27

Collins, A.T., Davies, G., Kanda, H. and Woods, G.S. (1988) *Journal of Physics C*, **21**, 1363–1376

Collins, A. T., Kamo, M. and Sato, Y. (1989) *Journal of Physics: Condensed Matter*, **1**, 4029–4033

Cotty, W. F. and Wilks, E. M. (1971) *Diamond Research 1971*, supplement to *Industrial Diamond Review*, pp.8–11

Crowningshield, R. (1971) *Gems and Gemology*, **13**, 302–314

Davey, S.T., Evans, T. and Robertson, S.H. (1984) *Journal of Materials Science Letters*, **3**, 1090–1092

DeCarli, P. S. and Jamieson, J. C. (1959) *Journal of Chemical Physics*, **31**, 1675–1676

DeCarli, P. S. and Jamieson, J. C. (1961) *Science*, **133**, 1821–1822

Derjaguin, B. V. and Fedoseev, D. V. (1977) *Growth of Diamond and Graphite from Gas Phase* (in Russian). Moscow: Nauka

Derjaguin, B. V., Fedoseev, D. V., Lukyanovich, V. M., *et al.* (1968) *Journal of Crystal Growth*, **2**, 380–384

Dischler, B. and Brandt, G. (1985) *Industrial Diamond Review*, **45**, 131–133

Ergun, S. and Alexander, L. E. (1962) *Nature*, **195**, 765–767

Eversole, W. G. (1962) *United States Patent Nos 3030187 and 3030188*

Fedoseev, D. V., Derjaguin, B. V., Varshavskaya, I. G. and Semenova-Tyan-Shanskaya, A. S. (1984) *Crystallisation of Diamond* (in Russian). Nauka, Moscow

Franks, J., Enke, K. and Richardt, A. (1990) *Metals and Materials*, **6**, pp. 695–700

Giardini, A. A., Tydings, J. E. and Levin, S. B. (1960) *American Mineralogist*, **45**, 217–221

Greiner, N. R., Phillips, D. S., Johnson, J. D. and Volk, F. (1988) *Nature*, **333**, 440–442

Gurney, J. J. (1989) In *Kimberlites and Related Rocks*, Vol.2, *Proceedings of the Fourth International Kimberlite Conference* (Perth, 1986), (ed. J. Ross), Blackwells Scientific, Melbourne, pp.935–965

Hall, H. T. (1958) *Review of Scientific Instruments*, **29**, 267–275

Hall, H. T. (1960a) *Review of Scientific Instruments*, **31**, 125–131

Hall, H. T. (1960b) In *Proceedings of an International Symposium on High Temperature Technology*, (California, 1959), McGraw-Hill, New York, pp.145–156

Harris, J. W., Hawthorne, J. B. and Oosterveld, M. M. (1984) *Annals Science University of Clermont-Ferand, II*, **74**, 1–13

Harris, J. W., Hawthorne, J. B., Oosterveld, M. M. and Wehmeyer, E. (1975) In *Physics and Chemistry of the Earth*, Vol.9. (eds L. H. Ahrens, J. B. Dawson, A. R. Duncan andA. J. Erlank). Pergamon Press, Oxford, pp.765–783

Harris, S. J., Belton, D. N., Weiner, A. M. and Schmieg, S. J. (1989) *Journal of Applied Physics*, **66**, 5353–5359

Hibbs, Jr, L. E. and Wentorf, Jr, R. H. (1974) *High Temperatures - High Pressures*, **6**, 409–413

Hirose, Y. (1988) *Program and Abstracts, First International Conference on the New Diamond Science and Technology*, Japan New Diamond Forum, Tokyo, pp 38–39.

Hirose, Y. and Terasawa, Y. (1986) *Japanese Journal of Applied Physics*, **25**, L519–L521

Hu, H. S., Joshi, A. and Nimmagadda, R. (1989) In *Extended Abstracts Technology Update on Diamond Films*, (eds R. H. P. Chang, D. Nelson and A. Hiraki), Materials Research Society, Pittsburg, pp.103–106

Japan New Diamond Forum (1988) *Program and Abstracts, First International Conference on the New Diamond Science and Technology*. Japan New Diamond Forum, Tokyo

Kawarada, H., Yokota, Y., Mori, Y. and Hiraki, A. (1989) In *Extended Abstracts Technology Update on Diamond Films*, (eds by R. P. H. Chang, D. Nelson and A. Hiraki), Materials Research Society, Pittsburg, pp.95–98

Kennedy, A. (1989) *Metals and Minerals Annual Review*, C27–C29

Kittel, C. (1976) *Introduction to Solid State Physics*, (5th edn.) Wiley, New York

Korsunskaya, I. A., Kamenetskaya, D. S. and Ershova, T. P. (1973) *Soviet Physics Doklady*, **18**, 346–348

Krar,S.F. and Ratterman,E. (1990) *Superabrasives: Grinding and Machining with CBN and Diamond*. New York: McGraw-Hill

Kurdyumov, A. V., Ostrovskaya, N. F. and Pilyankevich, A.N. (1988) *Soviet Powder Metallurgy*, January, 32–37

Kurihara, K., Sasaki, K., Kawarada, M. and Koshino, N. (1988) *Applied Physics Letters*, **52**, 437–438

Lammer, A. (1988) *Industrial Diamond Review*, **48**, 179–182

Lang, A. R. and Woods, G. S. (1976) *Industrial Diamond Review*, **36**, 96–103

Laptev, V.A., Pomchalov, A.V., Belimenko, L.D. and Samoilovich, M.I. (1985) *Inorganic Materials*, **21**, 1154–1157

Lewis, R. S., Ming, T., Wacker, J. F. *et al.* (1987) *Nature*, **326**, 160–162

Liander, H. and Lundblad, E. (1960) *Arkiv för Kemi*, **16**, 139–149

Linari-Linholm, A. A. (1973) *Occurrence, Mining and Recovery of Diamonds*. De Beers Consolidated Mines Ltd, London

Matsumoto, S. (1985) *Journal of Materials Science Letters*, **4**, 600–602

Matsumoto, S., Hino, M. and Kobayashi, T. (1987) *Applied Physics Letters* , **51**, 737–739

Matsumoto, S., Sato, Y., Tsutsumi, M. and Setaka, N. (1982) *Journal of Materials Science*, **17**, 3106–3112

Meilunas, R., Wong, M. S., Sheng, K. C. *et al.* (1989) *Applied Physics Letters*, **54**, 2204–2206

Metzger, J.L. (1986) *Superabrasive Grinding*. Butterworths, London

Mitchell, R. H. (1989) In *Kimberlites and Related Rocks*, Vol.1, Proceedings of the Fourth International Kimberlite Conference (Perth, 1986), (ed. J. Ross), Blackwells Scientific, Melbourne, pp. 7–45

Muncke, G. (1974) In *Diamond Research 1974*, supplement to *Industrial Diamond Review*, pp. 7–10

Muncke, G. (1979) In *The Properties of Diamond*, (ed. J. E. Field), Academic Press, London, pp.473–499

Murakawa, M., Takeuchi, S. and Hirose, Y. (1988) In *Program and Abstracts, First International Conference in the New Diamond Science and Technology*, Japan New Diamond Forum, Tokyo, pp. 208–209

Naka, S., Itoh, H. and Tsutsui, T. (1987) *Journal of Materials Science*, **22**, 1753–1757

Nakai, T. and Yazu, S. (1987) *United States Patent, Number 4, 636, 253.* 13 January 1987

Nernst, W. (1906) *Nachrichten von der Königlichen Gesellschaft der Wissenschaften zu Göttingen,* Mathematisch-physikalische Klasse, pp. 1–39

Nesbitt, A. C. (1967) In *Science and Technology of Industrial Diamonds*, Vol. Two: Technology, (ed. J. Burls), De Beers Industrial Diamond Division, London, pp 351–367

Pliny, G. P. (circa 70) *Historia Naturalis, Book 37*, (transl. D. E. Eichholz), 1962. Heinemann, London

Pollitzer, F. (1912) In *Chemischer und chemisch-technischer Voträge*, Begrundet von Prof. F.B. Ahrens, *Band XVII*, Verlag Ferdinand Enke, Stuttgart, pp. 333–501

Potter, D. K. and Ahrens, T. J. (1988) *Journal of Applied Physics*, **63**, 910–914

Pugh, H.LI.D., Lees, J. and Bland, J. A. (1961) *Nature*, **191**, 865–866

Sato, Y., Hata, C. and Kamo, M. (1988) In *Program and Abstracts, First International Conference on the New Diamond Science and Technology*, Japan New Diamond Forum, Tokyo, pp. 50–51

Sekine, T., Akaishi, M., Setaka, N. and Kondo, K-I. (1987) *Journal of Materials Science*, **22**, 3615–3619

Shigley, J. E., Fritsch, E., Stockton, C. M. *et al.* (1986) *Gems & Gemology*, Winter, 192–208

Shigley, J. E., Fritsch, E., Stockton, C. M. *et al.* (1987) *Gems & Gemology*, Winter, 187–206

Simonsen, I., Chevacharoenkul, S., Horiey, Y. *et al.* (1989) *Journal of Materials Science*, **24**, 1486–1490

Snail, K.A., Hanssen, L.M., Carrington, W.A. *et al.* (1988) In *Program and Abstracts, First International Conference on the New Diamond Science and Technology*, Japan New Diamond Forum, Tokyo pp. 142–143

Sobolev, N. V. and Shatsky, V. S. (1990) *Nature*, **343**, 742–746

Spitsyn, B. V., Bouilov, L. L. and Derjaguin, B. V. (1981) *Journal of Crystal Growth*, **52**, 219–226

Strong, H. M. and Chrenko, R. M. (1971) *Journal of Physical Chemistry,* **75**, 1838–1843

Strong, H. M. and Hanneman, R. E. (1967) *Journal of Chemical Physics*, **46**, 3668–3676

Suttill, K. R. (1987) *Engineering and Mining Journal*, **188**, Dec. 56–57

Suzuki, K., Sawabe, A. and Inuzuka, T. (1988) *Applied Physics Letters*, **53**, 1818–1819

Takata, Y., Edamatsu, K., Yokoyama, T. *et al* (1989) *Japanese Journal of Applied Physics*, **28**, L1282–L1285

Tavernier, J. B. (1684) *Six Travels through Turkey and Persia.* (Trans. V.Ball), London (1889)

Tomlinson, P. N. and Wedlake, R. J. (1983) In *Proceedings of the International Conference on Recent Developments in Speciality Steels and Hard Materials* (Pretoria, 1982), (ed. N. R. Comins and J. B. Clark), Pergamon Press, Oxford, pp. 173–184

Trueb, L. F. (1968) *Journal of Applied Physics*, **39**, 4707–4716

Trueb, L. F. (1971) *Journal of Applied Physics*, **42**, 503–510

Wedlake, R. J. (1974) In *Diamond Research 1974*, supplement to Industrial Diamond Review, pp.2–6

Wedlake, R. J. (1979) In *The Properties of Diamond*, (ed. J. E. Field), Academic Press, London, pp.501–535

Wentorf, Jr, R. H. (1957) *Journal of Chemical Physics*, **26**, 956

Wentorf, Jr, R. H. (1961) *Journal of Chemical Physics*, **34**, 809–812

Wentorf, Jr, R. H. (1965) *Journal of Physical Chemistry*, **69**, 3063–3069

Wentorf, Jr, R. H. (1971) *Journal of Physical Chemistry*, **75**, 1833–1837

Wentorf, Jr, R. H. and Bovenkerk, H. P. (1961) *Astrophysics Journal*, **134**, 995–1005

Wentorf, Jr, R. H. and Bovenkerk, H. P. (1962) *Journal of Chemical Physics*, **36**, 1987–1990

Wentorf, Jr, R. H., DeVries, R. C. and Bundy, F. P. (1980) *Science*, **208**, 873–880

Wilks, J. (1961) *The Third Law of Thermodynamics.* Oxford University Press, Oxford

Yagi, M. (1988) In *Program and Abstracts First InternationalConference on the New Diamond Science and Technology*, Japan New Diamond Forum, Tokyo, pp. 158–159

Yazu, S., Nishikawa, T., Nakai, T. and Doi, Y. (1983) In *Proceedings of the International Conference on Recent Developments in Speciality Steels and Hard materials* (Pretoria, 1982), (ed N.R.Comins and J.B.Clark), Pergamon Press, Oxford, pp.449–456

Yazu, S., Tsuji, K. and Yoshida, A. (1985) *European Patent B01J 3/06, C01B 3106* 22222

The structure of diamond

We consider the properties of diamond in Parts I and II and then discuss their applications in Part III. It is only relatively recently that largish synthetic diamonds have become readily available. Therefore much of our knowledge of diamond comes from studies of natural material, and the results set out in Parts I and II were obtained primarily with natural diamond. Most of these results are applicable to all types of diamonds, but we also note various points of difference arising in the man-made materials. We begin by describing the impurity content of diamond and particularly the rather varied content of impurities in natural diamonds.

Chapter 2

Impurities in diamond

An ideal diamond consists of a lattice of carbon atoms as described in Section 1.5, and nothing else, and is a clear colourless crystal. Real diamonds often exhibit some colouration and dark spots within the body of the crystal. The spots are *inclusions* which consists of foreign materials mainly introduced during the formation of the diamond. Colouration generally indicates that impurity *atoms* other than carbon have been incorporated into the diamond on an atomic scale. (Note, however, that the absence of colour does not mean that there are no impurities). There is also the posibility of gaps or *voids* between inclusions and the surrounding diamond, and of voids unconnected with inclusions, which may contain gases or condensed gases.

This chapter reviews experiments made to detect the various chemical elements which may be found in diamond and to assess their concentrations. It must be remembered that both inclusions and impurities may not be distributed homogeneously through the crystal. Therefore we may analyse the crystal as a whole and perhaps detect all the elements present but still not know where they are located. On the other hand, by inspecting a small volume near the surface, by some beam technique, we obtain a precise analysis of this volume but no information on the rest of the crystal. The various methods of analysis are, of course, equally applicable to natural and man-made diamonds but most of the results so far relate to natural diamonds, which generally contain a more varied range of impurities. Therefore the emphasis of this chapter is on the properties of natural diamonds.

2.1 Chemical elements in diamond

We now describe various methods used to determine what chemical elements are present in a diamond, and in what concentrations.

2.1.a Analysis by vaporization

Raal (1957) vaporized diamonds by burning and then photographed the optical spectra of the excited atoms. He thus identified about a dozen metallic elements at concentrations of the order of 1 ppm to 10 ppm. Today, vaporization is usually achieved by heating in a vacuum and the analysis performed with a mass spectrometer. This method was apparently first used on diamond by Kaiser and

Bond (1959) to assess the nitrogen content in the crystal lattice by working with gem quality diamond free of inclusions.

In all methods of analysis it is important to avoid misleading results arising from dirt and other contaminations both in the apparatus and on the surface of the specimen. This is particularly important when, as in diamond, many of the elements are present only at a level of 1 ppm or considerably less. Hence, surface contamination can produce quite incorrect results, and it is essential to adopt rigorous cleaning procedures before commencing the analysis, see for example Sellschop (1975). Particular attention must be paid to any contamination which has entered the diamond by cracks and fractures and can be very difficult to remove.

One method used to vaporize the diamond prior to analysis is to direct a beam of ions on to a small area of the specimen to sputter off the surface atoms. In this case one must remember that the distribution of impurities in a diamond is not necessarily homogeneous so the stone should be sampled at more than one point. Conversely, the method is well suited to studying spatial differences in the distribution of the impurities. As before, all the diamond analysed is destroyed, but this volume now constitutes only a small part of the whole crystal. The method is also able to detect the presence of surface contamination. When the ion beam first strikes a particular location its initial effect will be to sputter off any surface contamination, and then as the beam penetrates into the diamond the yield should change to a time independent value unaffected by contamination.

The method of vaporization followed by mass spectroscopy is an attractive one but care is needed to obtain the correct answer. A mass spectrometer measures the ratio of the electron charge on the sputtered ions to their mass, and this ratio is then related to a particular chemical element. For example, if nitrogen atoms are sputtered off one expects to observe a charge to mass ratio equal to e/m_N where e is the electronic charge and m_N the mass of a nitrogen atom, and one identifies the numerical value of this ratio with nitrogen. However, the sputtering may create double or mutiple electronic charges ne on the sputtered ions, and these ions may not be single atoms but so-called dimers consisting of a small number of say q atoms grouped together. The ratio of charge to mass now becomes ne/qm where n and q are integers, and it may be difficult to relate this ratio to the element responsible. This is a secondary effect but may be important when seeking to estimate a particular impurity. It seems probable that matters may be improved by new plasma sources which liberate the ions from the specimen by less violent processes.

2.1.b The microprobe analyser

A considerable disadvantage of the above methods of analysis is that they inevitably destroy the material analysed. Diamonds vary so much in constitution that it is very desirable to experiment on specimens with a known constitution, and this becomes impossible if the analysis destroys the diamond. However, the X-ray microprobe analyser gives a method of non-destructive analysis. This technique directs a focused beam of high energy electrons on to the surface of a specimen where they excite the surface atoms. The excited atoms then emit X-rays with well defined energies which are characteristic of the particular atom in question. Thus an analysis of the energy of the X-rays leads to a chemical analysis of the material close to the surface. The standard instrument is the microprobe analyser in which an electron beam is focused on to a small area of the specimen. The beam is scanned across the surface, as in a scanning electron microscope (Section 4.1), and the analyser presents a chart of the spatial chemical elements located within the order of 5 μm of the surface.

The microprobe analyser is at its best when analysing a material such as a steel which may contain say 80% atomic parts of iron and 10% of manganese, and readily measures the relative concentrations of the two elements. However, in analysing diamond we are interested in concentrations of impurities less than that of the carbon by factors of the order of 10^6 or more. Moreover some of the impurities of particular interest, hydrogen, boron, nitrogen and oxygen, have low atomic numbers and few electrons compared with iron and manganese, so the intensity of the X-ray signal is correspondingly reduced.

2.1.c Nuclear activation analysis

This method of analysis irradiates the diamond with some nuclear particles which react with the various nuclei in the diamond to produce new unstable nuclei which decay with the emission of nuclear particles, X-rays and gamma rays. It is then possible by observing the type and energy of these decay products to deduce the type of nucleus from which they come. Hence a measurement of the intensities of these emissions together with suitable calibrations can lead to estimates for the concentrations of the different parent nuclei, that is of the chemical constitution.

As an example of the activation technique we mention an estimation of the nitrogen content in diamond by irradiating the surface layer of the specimen with 8 MeV alpha particles (Sellschop *et al.*, 1974). The alpha particles may convert a nitrogen nucleus to an unstable isotope of fluorine which decays with a half life of 66 s (Figure.2.1), the end products being an isotope of oxygen and two 0.511 MeV gamma rays emitted simultaneously. Hence, the number of nitrogen atoms present can be derived by counting the number of pairs of coincident 0.511 MeV gamma rays, and then making a suitable calibration using a specimen with a known concentration of nitrogen.

The method described above works well if nitrogen is the only impurity, but if other impurities are present they may give rise to somewhat similar emissions and confuse the picture. For example, in the analysis for nitrogen just described Sellschop *et al.* calibrated the system by irradiating a specimen of boron nitride whose nitrogen content is known. However, this irradiation also converted the boron nuclei to an unstable form which also decayed with the emission of a pair of 0.511 MeV gamma rays! Fortunately, the latter nuclei decay more slowly than those derived from nitrogen, with a half life of about 10 min, and the two sets of gamma rays can be distinguished by observing the number of counts as a function of time. Figure 2.2 shows the results of such an experiment, note the double scales on the time axis. The initial fast decay is associated with the nitrogen nuclei and the final slower decay with the boron nuclei. Hence, the number of counts due to the nitrogen alone is readily obtained by a suitable analysis of the curves.

The above example underlines the importance of choosing the most suitable nuclear reaction for detecting a particular impurity. It is essential that the decay processes identify the parent nuclei unambiguously and distinguish them from any other impurity nuclei which may have been activated by the irradiation. Several methods have been used to meet these requirements. Lightowlers and Collins (1976) measured the boron content of semiconducting diamonds by irradiating with a high energy beam of protons and observing the pairs of alpha particles liberated

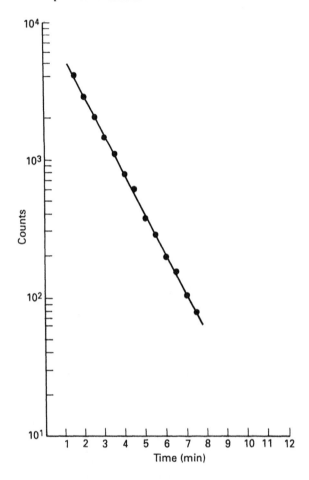

Figure 2.1 Decay of coincident gamma rays following the activation of nitrogen in diamond by alpha particle bombardment (Sellschop *et al.*, 1974)

immediately by the reaction. Sellschop *et al.* (1974) used irradiation by fast neutrons to determine the concentrations of oxygen and silicon in diamond by observing the intensity and decay of the resulting gamma emission.

   Activation analysis can also be carried out using low energy thermal neutrons. This is a useful technique because these slow neutrons penetrate the whole crystal and react very readily with other nuclei. They will therefore activate most of the foreign elements found in a diamond. An irradiation generally produces a large and complicated spectrum of gamma rays (Figure 2.3) in which particular elements are identified by making precise measurements of the energy of the gamma rays and of the decay rates or half lives. Thermal neutron activation analysis appears to be well suited for making a survey of the whole range of impurities in a diamond but the identification and analysis of the full emission spectra of all the elements which may be present will generally be a major task. A further complication is that

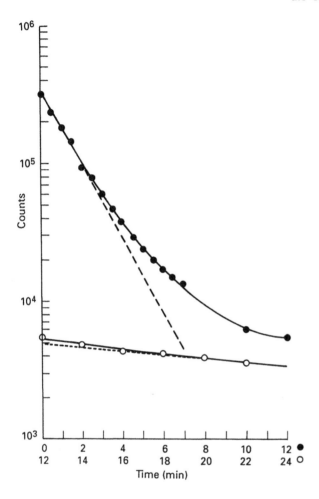

Figure 2.2 The decay of coincident gamma rays following the activation of nitrogen and boron in boron nitride (Sellschop *et al.*, 1974)

the neutrons react much more readily with some nuclei than others; the cross-sections of different nuclei for the capture of a neutron may vary by factors of up to 10^5 or more. Hence particular care must be taken to ensure that the emissions from one impurity species are not masked by those from another.

2.1.d Elements in diamond

We now collect together some of the results of the above methods. Most of these analyses do not indicate whether the impurity atoms form part of inclusions or are incorporated in some way in the diamond. As before, we remind the reader that different methods of analysis may sample the diamond in very different ways.

A comprehensive review of all the elements observed in natural diamonds is given by Bibby (1982) who sets out the methods employed, the number of

40

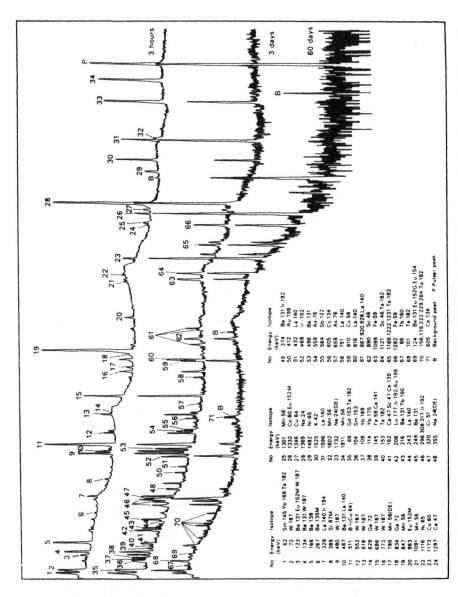

Figure 2.3 Typical energy spectrum of gamma radiation following the thermal neutron activation of a sample of diamonds (Sellschop *et al.*, 1974)

diamonds analysed, and the concentrations observed. One may perhaps distinguish three groups of chemical elements. The first includes the elements H, N, O and S which are generally present at concentrations between about 1 ppm and at least 1000 ppm by weight. At a lower level of concentration, say from about 0.1 ppm to 100 ppm, we find a range of elements principally B, Mg, Al, Si, S, Ca, V, Cr, Mn, Fe, Co, Ni, Cu, Ag and Ba. The third group consists of a large number of metallic elements often at very low levels ranging from 10^{-6} ppm to 10^{-1} ppm. (As the above concentrations are given in parts per million *by weight*, the atomic concentrations of the heavier elements are even smaller.)

Information on the second group of elements was obtained by Fesq *et al.* (1975) using neutron activation analysis, see also Sellschop (1975). These authors took about 1500 stones from each of three South African mines and sorted them into batches according to colour and the presence or absence of inclusions visible in a ×50 polarizing microscope. The authors note that many or most of the inclusions in diamond are formed from a small group of earth minerals built up from a dozen or so chemical elements (Section 2.2.b). Hence one might expect the above analysis to show correlations between the masses of these elements in a given diamond. Therefore Fesq *et al.* looked for correlations between the elements most commonly found in the minerals of the inclusions: O, Na, Mg, Al, K, Ca, Si, T, V, Cr, Mn and Fe. They found that for diamonds with visible inclusions there were strong correlations between these elements, for example Figure 2.4 shows the correlations between Mn, Al and Mg. Note that the scales in the diagram are logarithmic and that the range of measured concentrations extends over four orders of magnitude.

The various results described above show the considerable differences in constitution which exist between different diamonds. The results also underline the difficulty and complexity involved in achieving a complete analysis of even one particular diamond. An analysis for one element may be much complicated by the presence of other impurities. Also in order to obtain values of the low levels of concentration often observed, and the extremely low levels sometimes observed, great care must be taken to prepare samples free of surface contamination, and to avoid contamination of the equipment used for analysis.

2.1.e Hydrogen and oxygen

We have already noted in the previous section that appreciable concentrations of hydrogen and oxygen may be found in diamond. Atomic concentrations of up to 1000 ppm or more were reported by Hudson and Tsong (1977) using ion beam analysis. There is also some evidence that hydrogen may be incorporated in the lattice and responsible for two absorption lines in the infared at $1403 \, \text{cm}^{-1}$ and $2107 \, \text{cm}^{-1}$ as described in Section 3.9. In addition hydrogen has been observed when diamonds are crushed (see Section 2.2.b). However, it was not clear where all the hydrogen was located, in voids round inclusions, dissolved in the lattice, or incorporated in the lattice.

In experiments to study the location of the hydrogen Sellschop *et al.* (1979) bombarded diamond with fluorine ions which react with hydrogen nuclei to give characteristic triplets of high energy gamma rays. This reaction is highly resonant so the yield falls off rapidly if the energy of the fluorine atoms differs much from the optimum value of 16.48 MeV, and this behaviour was used as the basis of the depth-profiling technique. Suppose that the ion beam is adjusted so that the ions

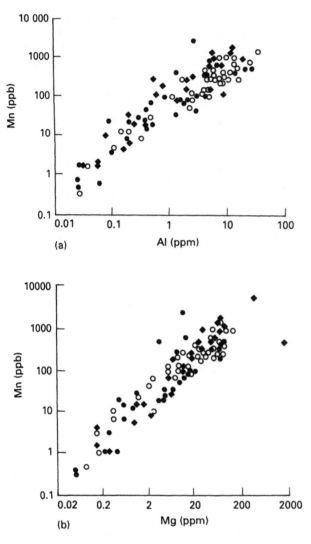

Figure 2.4 Plots showing the concentrations of manganese plotted against the concentration of aluminium and magnesium in diamonds from four South African mines. Each symbol indicates a different mine (Fesq *et al.*, 1975)

have the resonant energy as they strike the surface. In this case they produce the resonance reaction with any atoms of hydrogen on the surface. If the energy of the beam is now increased somewhat above the resonant value it will no longer activate these surface atoms, but as it penetrates into the diamond it loses energy, so at some depth the ions will have energies equal to the resonant value, and therefore activate hydrogen atoms at this depth. Thus the yield of gamma rays at a particular bombarding energy comes from nuclei at a particular depth. Therfore, as the energy loss of the high energy ions in the diamond is known from published data, a plot of gamma ray yield versus bombarding energy can be converted to give the concentration of hydrogen as a function of the depth below the surface.

Sellschop *et al.* (1979) observed that the yield of gamma rays decreased for a period of about 10 min after the ion beam was directed on the diamond. Therefore measurements were taken when the yield had reached an equilibrium value which was about half the intial value. Figure 2.5 shows that below a depth of 100 nm this equilibrium atomic concentration was almost constant, with a value of about 300 ppm, and the authors state that after an initial drop near the surface the concentration in all the specimens was constant down to about 400 nm.

In a further experiment to determine the location of this hydrogen a diamond was held in a high vacuum at 1200°C for 30 min in an attempt to bake out any dissolved gas. This treatment, in fact, had the effect of increasing the observed concentration of hydrogen by an order of magnitude. However, these measurements had been made on two faces which had been heavily bombarded by the ion

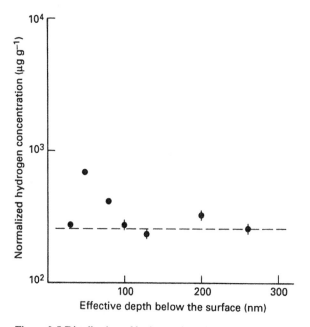

Figure 2.5 Distribution of hydrogen in a diamond as a function of depth below the surface, see text (Sellschop *et al.*, 1979)

beam during previous analyses, so the authors also measured the hydrogen concentration at two faces of the diamond not previously bombarded. They then observed concentrations of between 200 ppm and 400 ppm, much less than the values observed at the first two faces. The authors explain these results by proposing that the large concentrations of hydrogen produced by baking the first two faces had come from droplets of magma containing water incorporated into the diamond during growth. The baking at 1200°C dissociated this water and allowed the hydrogen atoms to diffuse through the diamond. Most of these atoms passed out through the surface but some were trapped in regions near the surface damaged by the earlier bombardments, where they were detected by the subsequent analysis.

No optical absorption in diamond has yet been associated with oxygen. Therefore there is no reason to suppose that any oxygen is incorporated in the lattice, although it is certainly present as part of the earth minerals forming inclusions. However, there has also been a suggestion (Sellschop *et al.*, 1974) that the measured oxygen content may be greater than would be expected from the concentrations of the earth minerals, suggesting that oxygen may be present in the gaseous form in various voids, or as part of water in magma droplets.

To obtain further information on the oxygen content Sellschop *et al.* (1979) bombarded a diamond with a beam of helium ions to transform oxygen nuclei to excited states of fluorine which decay with a yield of gamma rays. They worked with the same diamond which they had analysed for hydrogen and found an average atomic concentration of oxygen of about 80 ppm. After baking the diamond at 1200°C as previously described, they observed that this concentration had increased to an average of about 300 ppm. The authors suggest that this increase arose from the same dissociation of water in magma droplets that was postulated to explain increases in the concentration of hydrogen. The fact that the increase is smaller with oxygen is perhaps due to the fact that oxygen is less mobile in the lattice.

2.2 Visible inclusions

2.2.a Size and geometry

The inclusions in diamond may be studied by viewing them either through the natural faces of the diamond, or if these are opaque or unsuitably positioned, through flat surfaces polished to act as windows. The inclusions often exhibit a well developed crystal shape or habit enclosed by good diamond which has grown round them, the habit being either characteristic of the material in the inclusion or sometimes being imposed by the habit of the diamond. Many of these inclusions may have crystallized at the same time as the diamonds and such inclusions are referred to as *syngenetic*. There is also a possibility that the inclusions may be of earlier date and have been incorporated complete into the growing diamond as a *protogenetic* inclusion.

Besides inclusions of well marked habit diamonds may also contain inclusions with less well developed crystal forms. Descriptions of these various types of inclusions are given by Harris (1968) and Orlov (1977). Harris (1968) describes three forms, minerals in powder form generally near the periphery of the diamond, powders in sub-surface fractures around earlier syngenetic inclusions, and minerals in cleavage or fracture planes intersecting the surface of the diamond. These forms appear to be the result of foreign material entering the diamond from the surrounding environment at some time after its formation. They are termed *epigenetic* in contrast to the inclusions incorporated during growth.

The performance of a diamond as a gem stone is much influenced by the presence of inclusions which absorb the light in the stone, a dark inclusion being more troublesome than a lighter or transparent one. The effect produced by an inclusion also depends on its location in the gem stone relative to the paths of the light reflected internally around the diamond. For these reasons there are now grading schemes to assess the scale of inclusions, e.g., Bruton (1981). The basic criterion in the gem trade for a diamond to be deemed free of inclusions is that none are visible when the diamond is viewed with a ×10 eye lens. This is of course an arbitrary criterion as an inspection with a higher power microscope will probably reveal smaller inclusions. However, when selecting diamonds for tools where even small inclusions near the cutting edge may have a serious effect an inspection at magnifications of at least ×50 or ×100 is now quite usual. Nevertheless it is convenient to begin by considering inclusions with linear dimensions ranging from 0.1 mm upwards which can readily be seen with a ×10 lens.

There is not much detailed information available on how frequently inclusions are present in diamonds. Harris, Hawthorne and Oosterveld (1984) in the course of a comprehensive study of diamonds from four Kimberley mines, surveyed batches of about 1500 small stones of about 2 mm size from each mine, using a ×10 lens, and found that the numbers of diamonds with inclusions ranged from 89 to 320. We mention these figures only as an order of magnitude, as the inclusion content will vary from mine to mine and from one geographical region to another.

2.2.b Constitution of inclusions

Inclusions in diamond were first analysed by crushing the diamonds and then examining the inclusions by standard mineralogical techniques, principally with the polarizing optical microscope. In this way many of the inclusions were identified as fairly common minerals occurring in the earth. Subsequently more powerful techniques have become available. We referred to the X-ray microprobe analyser in the previous section and now briefly outline the technique of X-ray diffraction.

In X-ray diffraction methods a fine beam of X-rays is passed through or reflected from a crystal and gives rise to several scattered beams which leave the crystal at various angles. In the simple Laue technique the positions of these beams are recorded on a photographic film placed normal to the beam about 100 mm from the crystal. Figure 2.6 shows a typical set of Laue spots obtained from a good

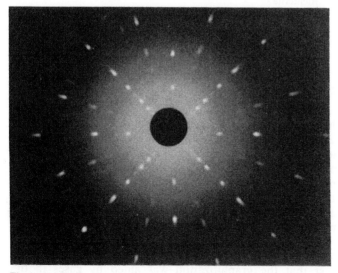

Figure 2.6 Reversal print of X-ray negative film showing a typical Laue pattern (from a cube face of diamond)

quality diamond and by measuring the positions of these spots it is possible to obtain both an accurate value of the spacing between the carbon atoms and information on their geometrical arrangement in the crystal. The pattern of Figure 2.6 is characteristic of diamond, but by repeating the experiment with the beam reflected from an inclusion a different pattern will be obtained which can be used to identify the inclusion. A full discussion of these and other X-ray techniques is given by Cullity (1978).

The considerable literature on the mineralogy of the inclusions in diamond has been reviewed by Orlov (1977), Harris and Gurney (1979) and Meyer (1987). The majority of the inclusions in natural diamond are formed of minerals found in the earth's crust and which appear as intrusions (xenoliths) in the kimberlite ore containing the diamonds. Table 2.1 illustrates the range of minerals commonly found in diamond (Meyer, 1987).

An inclusion generally consists of just one mineral. Not surprisingly the minerals in syngenetic and protogenetic inclusions are generally different from those in epigenetic inclusions. It seems likely that only relatively few inclusions are protogenetic, and that one of the more common protogenetic minerals is diamond itself, but such inclusions are quite rare. As might be expected the mineral content of the inclusions in diamonds from different localities varies considerably as described by Harris and Gurney (1979).

Table 2.1 Inclusions identified in diamond (after Meyer, 1987)

Protogenetic and/or syngenetic		Epigenetic	Uncertain
Forsterite	Omphacite	Serpentine	Phlogopite
Enstatite	Pyrope-almandine	Calcite	Biotite
Diopside	Kyanite	Graphite	Muscovite
Cr-pyrope	Sanidine	Haematite	Amphibole
Cr-spinel	Coesite (quartz)	Kaolinite	Magnetite
Mg-ilmenite	Rutile	Acmite	Apatite
Sulphides	Ruby	Richterite	
Zircon	Ilmenite	Perovskite	
Diamond	Chromite	Mn-ilmenite	
Native iron	Sulphides	Spinel	
	Diamond	Xenotime	
		Sellaite	
		Geothite	

The presence of graphite in diamond, indicated in Table 2.1, is of particular interest in view of the part played by graphite in the synthesis process. Graphite appears as black opaque material around mineral inclusions and as fine dark clouds. They have been identified and distinguished from rather similar dark inclusions of mineral sulphides by Harris (1972) and Giardini and Melton (1975a). The origin of the graphite is discussed by Harris (1972) who concludes that it arises by the internal graphitization of the diamond at some period of its history under conditions sufficient to produce graphitization without breaking down the whole diamond lattice. Harris and Vance (1972) produced similar graphatization around inclusions by heating diamonds in a vacuum to temperatures between 900°C and 1650°C. They also suggest that the graphitization was aided by gases in the diamond.

Man-made diamonds exhibit a quite different inclusion content, in particular there are no earth minerals. The principal inclusions come from the metal solvents, iron, cobalt and nickel used in the synthesis process, see for example Wentorf (1971). Naturally one of the main aims in developing the production of synthetic diamond is to reduce the level of inclusions as much as possible. An account of the initial identification of nickel in synthetic diamonds by X-ray diffraction experiments is given by Lonsdale, Milledge and Nave (1959), and a review of other work on metallic inclusions in synthetic material by Wedlake (1979).

As mentioned above, diamonds may also contain gaseous or condensed impurities in voids between mineral inclusions and the diamond and perhaps elsewhere. These gases have been analysed by crushing the diamonds and detecting the gases in a mass spectrometer (Melton, Salotti and Giardini, 1972; Melton and Giardini, 1974; Giardini and Melton, 1975b). Some results obtained in this way are

Table 2.2 Average volumes and % composition of gases released by crushing various natural and synthetic diamonds in high vacuum at 200°C (after Giardini and Melton, 1975b)

Sample	STP volume (cm^3) per carat $\times 10^5$	Ar	N_2	H_2	O_2	H_2O	CH_4	CO	CO_2	CH_3OH or CH_3-CH_2OH	Other hydro-carbons
21 Arkansas crystals	14.0	0.3	7.4	24.2		33.0	8.5	3.6	22.7	0.7	0.1
14 African crystals	1.5	0.3	9.8	24.7	0.1	30.2	6.5	11.5	15.8	0.6	0.4
1 Brazilian crystal	0.9	0.02	2.1	5.0	1.5	80.6	4.6		5.6	0.05	0.05
4 Synthetic crystals	7.5	0.6	5.1	26.2		25.8	24.8		17.4		

shown in Table 2.2. It must be appreciated that these results were obtained from only a few diamonds selected at random, that Arkansas diamonds have a somewhat unusual form, and that the synthetic stones were produced by the techniques of some years back. Even so the experiments gave a clear indication that in any diamond we must be prepared to find the gases H_2, O_2, H_2O, CH_4, CO and CO_2 with a total STP volume of perhaps 10^{-3} to 10^{-4} of the volume of the diamond. We also note that the synthetic diamonds gave quite similar yields of gaseous impurity, perhaps surprisingly.

2.2.c Diamonds and geology

The formation of diamonds in the earth has long been of interest, but the details of the processes involved are still not clear. For an account of current views see Richardson *et al.* (1984), Meyer (1985) and Haggerty (1986). Quite apart from the question of their formation diamonds are also of geological interest because of the inclusions they contain. These inclusions must have been formed when the diamond was growing in diamond stable conditions at pressures above 60 kbar and temperatures above 1500 K. These conditions are only found by going down into the earth's crust about 200 km. Then at some time after their formation the diamonds were transported to the surface of the crust by the movement of volcanic rocks. They must have been carried up very rapidly perhaps in less than hours because otherwise they would have come into equilibrium with the pressure and temperature of their surroundings and reverted to graphite. Hence, as emphasized by Frank (1967), inclusions present the geologist with samples of material from great depths which have been subsequently isolated and protected by the surrounding diamond.

Besides providing samples from the earth's crust the inclusions also give information relevant to the age of the diamonds. Because the inclusions have been isolated by the surrounding diamond, the techniques of radioactive dating may be used to make some estimates of the time that has elapsed since the inclusions were first incorporated in the diamond, that is of the age of the diamond. The values obtained for diamonds from different sources range from the order of 100 million years to 3000 million years. For references to this work see the reviews by Harris (1987) and Gurney (1989), also an article by Phillips, Onstott and Harris (1989).

Another point of interest arises because most chemical elements exist in the form of two or more stable isotopes which are virtually identical chemically but have somewhat different atomic masses. For example carbon exists as two stable isotopes ^{12}C and ^{13}C which have atomic masses of 12 and 13, and the carbon in the CO_2 of the atmosphere and in living matter consists mainly of ^{12}C with a fraction of ^{13}C which is generally quite close to a value of about 1.1%. However, these isotopic ratios may depend somewhat on the source of the material containing the element, and may therefore give potentially useful information about the source. In fact diamonds from different sources may show appreciable differences in the concentrations of ^{13}C (Orlov, 1977) often of the order of 10% and sometimes of 30% (Koval'skii and Cherskii, 1973). In addition the concentration of ^{13}C may sometimes vary within a single diamond particularly one with marked growth layers or sectors (Chapter 5). Hence the isotopic ratio ^{12}C/^{13}C gives some indications of the conditions prevailing during the growth of the diamond and is now a topic of increasing study, see for example Milledge et al. (1983) and Swart et al. (1983). In addition very similar studies have been made of the isotopic ratios exhibited by the nitrogen impurity atoms in diamond (Boyd et al., 1988) and by the helium and neon atoms (Honda et al., 1987). See also comments in reviews by Harris (1987) and Gurney (1989).

Another method of estimating the conditions under which the diamond was formed and held before being brought to the surface depends on the fact that diamond held at high temperatures and pressures undergoes annealing processes which result in the nitrogen atoms forming different small groups or aggregates according to the particular conditions involved (Section 6.4). Hence an assessment of the various forms of nitrogen aggregates in a diamond may give information on the conditions of formation, see Evans and Harris (1989).

Finally we note that a good overall view of current geological interest in diamond is given in the Proceedings of the Fourth Kimberlite Conference (Ross, 1989), and that a recent report of micron-size diamonds in garnets in metamorphic rocks has important implications for tectonic models of the earth's surface (Sobolev and Shatsky, 1990).

2.3 Detection of microscopic inclusions

An analysis of the chemical elements in diamond is not complete until we know whether the various elements are located in the diamond lattice or in visible or invisible inclusions. The inclusions discussed in Section 2.2 were sufficiently large that one could distinguish their individual shape and optical properties. However, it may well be that some diamonds contain a whole range of smaller and much · smaller inclusions, some quite invisible, others being responsible for the thin transparent veils or cloud-like sheets well known to diamond graders see for example (Bruton,1981; Lenzen,1983). We now describe techniques used to detect and assess these smaller inclusions.

2.3.a The polarizing microscope

The polarizing microscope is essentially an ordinary transmitted light microscope in which polarizing plates are set above and below the specimen. In the usual mode of operation these plates or polars are set with their optic axes mutually perpendicular, so that no light can traverse both plates, and the field of view in the microscope is quite dark. It is well known that certain transparent crystals exhibit *birefringence* or *double refraction*, a phenomenon which causes a beam of light entering a crystal to be split into two separate beams of different polarization. One effect of this behaviour is shown by placing a slab of birefringent material between the crossed polars of the microscope, the field is no longer dark as some light now passes through the crystal and the polars. For an explanation of this effect see for example Gay (1967).

Most pure crystals including diamond are not birefringent, and if inserted between crossed polars produce no effect. However, the presence of an inclusion in any crystal produces a certain amount of elastic strain or distortion in the surrounding lattice, and in some crystals including diamond these strains cause the crystal to become birefringent. Hence the crystal viewed between polars is seen to pass light through the regions of strain, the light being more intense the greater the strain. Using a white source of light, the transmitted light is white at low levels of birefringence, but with greater strains shows a spectrum of characteristic colours, see for example Gay (1967).

As an example of the use of the polarizing microscope Figure 2.7 shows a moderately large inclusion in a diamond viewed between crossed polars. Note that the birefringent region is much bigger than the size of the actual inclusion because the elastic strains fall off quite slowly as one moves away from the centre of the disturbance. Hence, because the microscope detects the relatively large strained region rather than the actual inclusion, it will have a superior performance over the normal microscope in detecting small inclusions at the limit of resolution, see for example Seal (1966). Although precise figures are not available this method can probably reveal the presence of inclusions of sizes down to 0.5 μm and perhaps somewhat less. Recently birefringence has been used to observe stacking faults in diamond (Jiang, Lang and Tanner, 1987) and dislocations (Van Enckevort and Seal, 1988), although only in specimens very free of other sources of birefringence.

In order to obtain good results from a polarizing microscope its optical components must be of high quality so that in the absence of a specimen there is an almost complete extinction of the light through the polars. Otherwise any transmitted light will tend to mask the light produced by weak strains and low

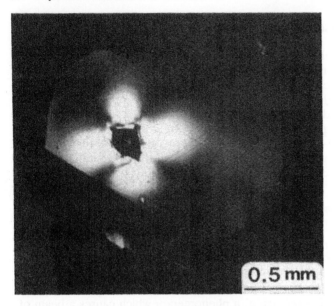

Figure 2.7 Birefringence pattern centred on a visible inclusion (Wilks and Wilks, 1971)

levels of birefringence. It is also important to avoid any scattering of light in the specimen which may result in extraneous light reaching the eyepiece. Therefore it is often necessary to prepare carefully polished flat windows on a diamond which allow the light to enter and leave the crystal without any scattering.

2.3.b X-ray topography

We have already mentioned the use of X-ray diffraction and Laue patterns to determine the lattice constants and structure of crystals. We now describe the diffraction process in more detail in order to explain another technique which gives valuable information on the internal structure of crystals.

The atoms in a perfect crystal form a regular array and may be seen as lying on a range of parallel sets of planes. Figure 2.8 is a schematic diagram showing the atoms in three adjacent planes of one parallel set. The diffraction process is such that a beam of X-rays of width AA travelling in the direction of the arrow will in

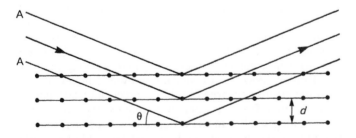

Figure 2.8 Sketch to illustrate a Bragg reflection of X-rays, see text

general pass through the crystal undeviated. However, at certain angles of incidence a diffracted beam is produced as if the incident beam were reflected by the atomic layers. These so-called Bragg reflections occur when the angle θ shown in the figure satisfies the Bragg relation

$$2d \sin \theta = n\lambda \tag{1}$$

where d is the spacing between the planes, n an integer, and λ the wavelength of the X-rays. Note that if a beam of given wavelength is allowed to enter a crystal set at an arbitrary orientation with respect to the beam, condition (1) will not in general be satisfied and no reflected beams are observed. (In order to obtain the Laue patterns mentioned in Section 2.2.b one must work with an X-ray beam made up of a continuous range of wavelengths, so that some value of λ in the range will satisfy the Bragg relation.)

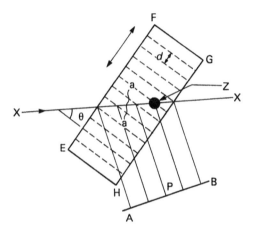

Figure 2.9 Schematic diagram of arrangement for taking X-ray topographs, see text

Consider now the schematic arrangement shown in Figure 2.9. EFGH is a crystal and the dashed lines indicate one particular set of atom planes lying perpendicular to the plane of the paper. A narrow beam XX of X-rays of a single wavelength λ lies in the plane of the paper, and the crystal is oriented so that the beam makes an angle θ with the lattice planes satisfying the condition for a Bragg reflection. In this case a fraction of the beam is reflected as it passes each plane of atoms and a set of parallel rays AB emerge from the crystal and strike a photographic plate P. It follows that each point on the plate between A and B corresponds to a particular point in the crystal, the intersection between the beam and one particular plane of atoms.

So far we have assumed that the crystal is perfect, that is the atoms all lie in absolutely regular rows, but this is not always the case. Therefore we now consider how the diffraction is modified if a small volume Z of the crystal is imperfect in one of two ways. The atoms in the volume may not be sited exactly at the points of the lattice, or they may be quite regularly arranged but the whole small volume may be somewhat misoriented with respect to the main body of the crystal. In the latter case the different orientation may not satisfy the Bragg condition and there is no reflected beam. On the other hand, some randomness in the positions of the individual atoms has a quite different effect and results in an increase in the intensity of the reflected beam. Thus, in either case, an irregularity in the lattice produces a change of intensity on the photographic plate. Hence this type of arrangement gives a method of detecting inclusions and other defects, the image produced being known as an X-ray topograph. (The complete treatment of X-ray diffraction and topography is quite involved. General accounts have been given by Lang (1970) and Tanner (1976).)

The usual experimental arrangement is to use a monochromatic X-ray beam collimated to the form of a thin ribbon of the order of 10 μm thick, with the width of the ribbon perpendicular to the plane of the paper in Figure 2.9. The photographic plate then records a picture or *section topograph* of a section of the crystal by the beam. The geometry of the figure implies that the size of the image will be somewhat less than the size of the crystal, but a careful choice of orientation and of the order of reflection permit magnifications not much below unity (Lang, 1970). In any case, it is particularly important to use special emulsions and other techniques to avoid loss of resolution in the photographic plate.

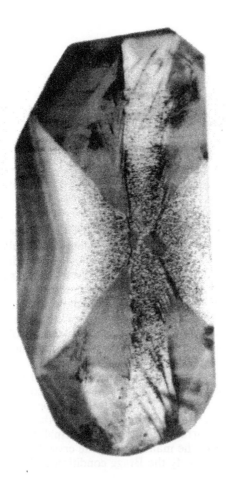

Figure 2.10 Section topograph of a natural centre-cross diamond as described in Section 5.5 showing large numbers of small inclusions in some sectors. The width of the ribbon-shaped X-ray beam cutting the crystal is 15 μm (Lang, 1974)

As an example of this technique, Figure 2.10 shows a section topograph of a type of diamond described in Section 5.5. This is not a typical diamond but the figure is a good example of a topograph revealing a large number of small inclusions which appear as black spots in parts of the light areas. The spots are more clearly seen in the original print and show a distribution of diameters down to less than 1 μm. Like birefringence techniques, X-ray topography may image both an inclusion and its strain field. Hence, even though the inclusion itself is beyond the resolution of the system the larger volume of the strain field may still be detected.

It is also useful to describe another method of X-ray topography to which we refer in subsequent chapters. The experimental arrangement is similar to that shown in Figure 2.9 except that the crystal and the plate P are both mounted on a precision slide which can be moved or oscillated parallel to the line EF so that the X-ray beam traverses the whole of the crystal. The resulting *projection topograph* on the plate is then a superposition of a range of section topographs covering the whole crystal, and gives a more complete picture. For example the line feature in the crystal shown as aa in Figure 2.9 will appear as a line in the projection topograph, but in a section topography only as a spot where it intersects the beam XX. Because it pictures the whole crystal projection topographs are often preferred to section topographs. However, since the projection topograph involves a superposition of images its ultimate resolving power is less than that of the section topograph, but by carefully optimizing the conditions when taking projection topographs good contrast and resolution may be obtained.

2.3.c Light scattering techniques

Inclusions may be readily observed in the optical microscope by using some form of dark ground illumination whereby the specimen is strongly illuminated by a beam of light perpendicular to the axis of the microscope, as shown schematically in Figure 2.11. If the specimen is a perfect crystal we expect the beam to pass through with virtually no scattering. Hence if the crystal is viewed through the microscope the beam should be invisible in its passage through the crystal. However, if a diamond contains any visible inclusions then the beam will be scattered intensely if it strikes any of them. In addition smaller inclusions not seen in the normal transmitted-light mode may also produce scattering and appear as bright diffraction spots.

Experimental arrangements to study light scattering in diamond have been described by several authors. Lang (1974) used a high pressure mercury lamp as a

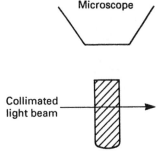

Figure 2.11 Schematic diagram of a crystal viewed by dark ground illumination

source, while Wilks and Wilks (1980) and Van Enckevort and Seal (1987) describe arrangements using a low power laser to produce the beam. The latter authors also used a technique of scanning the light beam to produce a microtomograph, as described by Vand, Vedam and Stein (1966) and Moriya and Okawa (1978, 1980). In all these experiments the crucial point is to reduce any extraneous scattered light in the background to an absolute minimum so as not to obscure the light scattered by defects in the diamond. Therefore it will generally be necessary to polish smooth faces on the diamond to permit the light beam to enter and leave without any scattering from the interface. It is also useful to blacken the other faces of the diamond to absorb any light reflected around the inside of the diamond, see for example Lang (1974) who coated the faces with a mixture of carbon black and sulphur.

As an example of the dark ground technique Figure 2.12 taken by Lang (1974) reveals a large number of inclusions which scatter light and appear as small bright spots. (Note that this concentration of inclusions is unusually high, and that this diamond, like that of Figure 2.10, is of a rather unusual type.) The diameters of the spots seen in the micrograph range down to the order of $1\,\mu m$ or less, so the smallest spots will be diffraction discs rather than geometric images. It is not clear which of the above three methods, birefringence, X-ray topography and light

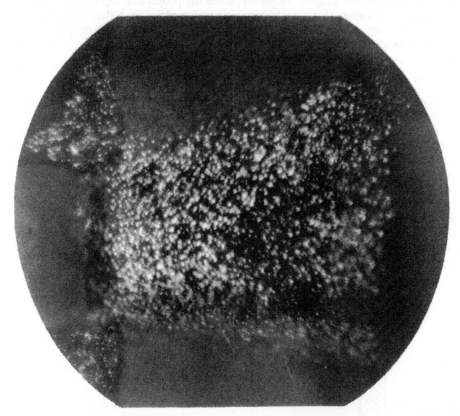

Figure 2.12 Inclusions in a natural diamond viewed by scattered light. Diameter of the field is about 1 mm (Lang, 1974)

scattering is most effective for the detection of fine inclusions, because there are no accounts of the same diamond being studied by the different methods. However, all three methods operating at their best can reveal the presence of particles or strain fields with a linear scale of the order of 1 μm or less.

2.3.d Neutron activation analysis

This method is much less direct than those just described but may offer the possibility of detecting the presence of much smaller inclusions. We described in Section 2.1.c the use of neutron activation analysis to determine the chemical elements in diamonds. In particular, Fesq *et al.* (1975) analysed a large number of diamonds containing visible inclusions and detected a group of elements whose concentrations correlated with each other and which appear to be associated with various minerals present in the inclusions. These authors also examined batches of diamonds from the same mines selected to be free of visible inclusions when viewed in a ×50 polarizing microscope. One might have expected that these diamonds would be free of the mineral elements observed with the inclusions, but all the same elements were observed although at concentrations reduced by the order of 10^2. Fesq *et al.* also observed correlations between these reduced concentrations which were similar, though not as good, to those observed in the diamonds with visible inclusions. Hence they concluded that at least some of the nominally inclusion-free diamonds contained mineral inclusions on a micron or sub-micron scale, the correlations being less good than for the larger inclusions at least partly because smaller concentrations lead to greater statistical variations.

2.4 Electron microscopy

The optical microscopy methods described above have a resolution limit of the order of a micron set by the wavelength of visible light. To examine a crystal for defects substantially less than a micron in scale it is necessary to turn to the transmission electron microscope (TEM) which has a resolution of almost atomic dimensions. We now give a brief account of the instrument and of the types of inclusions it has already revealed.

The transmission electron microscope is essentially a microscope using a beam of electrons instead of the light source used in an optical microscope. This electron beam is focused by magnetic lens to pass through the specimen and form an image on some form of fluorescent screen where it may be photographed, see for example Watt (1985). The high resolution of the TEM arises because the electrons have a much shorter wavelength than visible light. One considerable difficulty arises when studying crystals in the TEM because the electrons are both heavily absorbed and scattered by the specimen. It is therefore necessary to work with specimens often no more than 0.1 μm thick. Specimens of diamond are usually prepared from thin slivers perhaps only a micrometer thick by using some form of chemical etch (Evans and Phaal, 1962) or ion-beam polishing. Because of the need to prepare thin specimens, inspection in the TEM inevitably involves some destruction of the diamond. Also the areas inspected are determined somewhat arbitrarily by the progress of the etching processes.

To a rough approximation the TEM forms an image in rather the same way as an optical microscope, and many electron micrographs may be regarded as an image similar to those seen in the optical microscope. However, a detailed interpretation of the TEM image is by no means straightforward principally because the wavelength of the X-rays is of the same order as the spacing between the atoms. Therefore, as the atoms in a crystal form a regular array reminiscent of a diffraction grating, the final image may include marked features due to diffraction effects. In this case a quite involved interpretation may be needed to relate the detail in the micrograph to the geometry of the actual defects in the crystal, see for example Hirsch *et al.* (1977). We now briefly summarize the principal inclusion-like features which have been studied in the TEM.

2.4.a Inclusions

The TEM provides a powerful method of observing the presence of sub-micron size inclusions but very few studies have been carried out. Not only does the method require very thin specimens to be prepared but there is such a variety of inclusions in natural diamond that a survey of any number of diamonds becomes a daunting prospect. However, the TEM has been used to assess the size of sub-micron inclusions of the metal solvent/catalyst in synthetic diamonds (Wong *et al.*, 1985). These authors studied diamonds grown with iron, nickel or cobalt as the catalyst, and reported the presence of inclusions ranging in size from 20 nm to 700 nm with the most probable sizes ranging from 50 nm to 150 nm. The same authors also used measurements of extended X-ray absorption fine structure (EXAFS) to determine the crystal structure of the inclusions. They found, perhaps surprisingly, that the structure of all the inclusions whether of iron, nickel or colbalt was face centred cubic whereas normally iron is body centred cubic and cobalt close packed hexagonal (and only nickel is face centred cubic).

2.4.b Platelets

Figure 2.13 shows a micrograph of a thin section of diamond taken by Evans and Phaal (1962) in one of the first studies of diamond in the electron microscope. The micrograph shows a large number of conspicuous dark objects. Subsequent studies using suitable stages to tilt the specimen have shown that these defects have the shape of a thin disc and that the plane of the disc coincides with one of the {001} or cube planes of the crystal lattice (the terms cube and {001} are explained in sections 5.1 and 5.2). These so-called *platelets* are particularly well shown in the centre of Figure 4.10 which is a micrograph of a polished cube face of diamond obtained by another technique described in Sections 4.1 and 4.4. The bright lines in the centre of the figure arise at the edges of platelets lying in cube planes perpendicular to the surface, which are being viewed edge on. The sizes of platelets vary greatly in different diamonds with linear dimensions ranging from a few nanometers to a few micrometers (Evans and Phaal, 1962; Woods, 1976; Lang, 1979). The presence of these planar defects was originally predicted from studies of X-ray diffraction. The Laue patterns of some diamonds exhibit a number of characteristically shaped spots or spikes (Lonsdale, 1942), and the form of these spikes implies that they arise from planar defects lying in {001} planes (Frank, 1956).

Subsequent electron microscope studies have shown that the diamond lattice on either side of a platelet is distorted as if pushed apart by the intrusive presence of the platelet. Several sets of measurements have now established that the relative

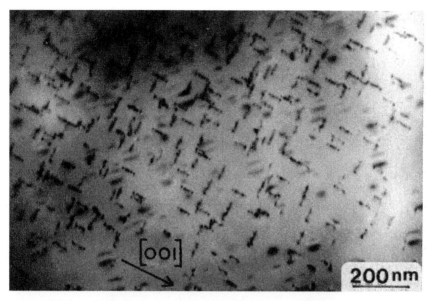

Figure 2.13 Transmission electron micrograph of a specimen of diamond showing platelets. The plane of the specimen is approximately (001) and the platelets lie in {010} and {100} planes. (Evans and Phaal, 1962)

displacement of the two sides of the diamond lattice by the lattice is apparently the same for all sized platelets. The most accurate assessment appears to be that of Bursill *et al.* (1981) who made use of Moiré fringes between slightly overlapping sections of a cracked diamond and quote a value between 0.124 nm and 0.129 nm. For references to other measurements see Barry, Bursill and Hutchison (1983). There has been considerable discussion of the constitution of the platelets. There is evidence to suggest that they consist of one or two planes of nitrogen atoms, but also some arguments to the contrary. We discuss this further in Section 6.4.b after describing some experiments on the annealing of diamond, which provide relevant information on this point.

2.4.c Voidites

These defects were first observed by Stephenson (1978) and as yet have only been observed in a small number of diamonds. Figure 2.14 shows one such voidite observed in the TEM by Hirsch *et al.* (1986). These voidites are octahedral in form, being bounded by {111} planes, with diameters ranging from 1 nm to 10 nm (Anstis and Hutchison, 1982; Barry, 1986), and have been investigated in detail by Hirsch, Hutchison and Titchmarsh (1986) and Barry *et al.* (1987). Phase contrast electron microscopy shows that the scattering density of the voidites is between one-half and a third that of the surrounding diamond.

The array of spots seen within the image of the voidite in Figure 2.14 appear to be Moiré fringes arising from interference between the diamond and a crystalline solid within the voidite (Hirsch *et al.*, 1986; Barry, 1986). The experiments of Hirsch *et al.* also detected the presence of nitrogen in a voidite by analysing the

Figure 2.14 Transmission electron micrograph of a voidite in diamond (Hirsch *et al.*, 1986)

energy spectrum of the X-rays emitted under the electron beam. However, both Hirsch *et al.* and Barry state that the observed fringe structures are not consistent with the voidites being full of ordinary solid nitrogen. According to Barry the fringes are consistent with the interior containing solid ammonia (NH_3) under high pressure, but the infrared spectrum shows no sign of any absorption which might be expected ffrom the stretching of N–H bonds (Evans and Woods, 1987); Hirsch *et al.* suggest the possibility of a new phase of solid nitrogen.

2.4.d Planar defects on octahedral planes

A further type of defect has been observed recently by Walmsley *et al.* (1987) in a natural diamond which had grown in the relatively uncommon cubic habit. As described in Section 5.5, cuboid growth is often accompanied by a high concentration of intrusive particles of the order of microns or less in size, some probably being mineral inclusions as described above. The diamond studied by Walmsley *et al.* contained visibly cloudy zones presumably caused by the presence of fine particles. An examination in the TEM revealed up to 140 planar disc-like defects randomly distributed on octahedral {111} planes. These defects were very nearly circular with diameters mainly between $0.5\,\mu m$ and $1.5\,\mu m$, and appeared to be lens-like in shape and less dense than the surrounding diamond. Interference fringes in the micrographs suggest that the maximum thickness of the defect is about 3 nm. About 20 of the 140 defects were surrounded by stacking faults (Section 6.1.a) as in the TEM micrograph of Figure 2.15 which shows a section of diamond with a disc enclosed by a stacking fault, each giving rise to a characteristic

Figure 2.15 Transmission electron micrograph of a planar feature in a diamond, width of field 2.4 μm; see text. (Walmsley *et al.*, 1987)

set of interference fringes. Walmsley *et al.* estimate that the pressure in the disc distorting the diamond must be of the order of 1.6 GPa, but make no suggestion on the constitution of the defect.

References

Anstis, G. R. and Hutchison, J. J. (1982) In *Proceedings of the 10th International Congress on Electron Microscopy* Deutsche Gesellschaft für Elektronenmikoskopie, Frankfurt, **2**, 93–94

Barry, J.C. (1986) *Ultramicroscopy*, **20**, 169–176

Barry, J. C., Bursill, L. A. and Hutchison, J. L. (1983) *Philosophical Magazine A*, **48**, 109–121

Barry, J, C., Bursill, L. A., Hutchison, J. L. *et al.* (1987) *Philosophical Transactions of the Royal Society*, **A321**, 361–401

Bibby, D. M. (1982) In *Chemistry and Physics of Carbon*, Vol 18, (ed. P.A.Thrower), Marcel Dekker, New York, pp.1-91

Boyd,S.R., Pillinger,C.T., Milledge,H.J. *et al.* (1988) *Nature*, **331**, 604–607

Bruton, E. (1981) *Diamonds*, (2nd edn). N.A.G. Press, London

Bursill, L. A., Hutchison, J. L, Lang, A. R. and Sumida, N. (1981) *Nature*, **292**, 518–520

Cullity, B. D. (1978) *Elements of X-ray Diffraction*, (2nd edn.) Addison Wesley, Reading, Mass

Evans,T. and Harris,J.W. (1989) In *Kimberlites and Related Rocks*, Vol.2, *Proceedings of the Fourth International Kimberlite Conference* (Perth, 1986), (ed. J.Ross), Blackwells Scientific, Melbourne, pp. 1001–1011

Evans, T. and Phaal, C. (1962) *Proceedings of the Royal Society*, **A270**, 538–552

Evans,T. and Woods,G.S. (1987) *Philosophical Magazine*, **55**, 295–299

Fesq, H.W., Bibby, D. M., Erasmus, C. S. *et al.* (1975) In *Physics and Chemistry of the Earth*, Vol.9. (eds L.H. Ahrens, J.B. Dawson, A.R. Duncan, and A.J. Erlank), Pergamon Press, Oxford, pp. 817–836

Frank, F. C. (1956) *Proceedings of the Royal Society*, **A237**, 168–174

Frank, F. C. (1967) In *Science and Technology of Industrial Diamonds*, Vol. 1, Science, (ed. J. Burls), De Beers Industrial Diamond Division, London, pp. 119–135

Gay, P. (1967) *An Introduction to Crystal Optics*. Longmans, London

Giardini, A. A. and Melton, C. E. (1975a) *American Mineralogist*, **60**, 934–936

Giardini,A.A. and Melton,C.E. (1975b) *Fortschritte der Mineralogie*, Special Issue to Vol.52, 455–464

Gurney,J.J. (1989) In *Kimberlites and Related Rocks*, Vol.2, *Proceedings of the Fourth International Kimberlite Conference* (Perth, 1986), (ed. J.Ross), Blackwells Scientific, Melbourne, pp. 935–965

Haggerty, S. E. (1986) *Nature*, **320**, 34–38

Harris, J. W. (1968) *Industrial Diamond Review*, **28**, 402–410, 458–461

Harris, J. W. (1972) *Contributions to Mineralogy and Petrology*, **35**, 22–33

Harris,J.W. (1987) In *Mantle Xenoliths*, (ed. P.H.Nixon), John Wiley, Chichester, pp. 477–500

Harris, J. W. and Gurney, J. J. (1979) In *The Properties of Diamond*, (ed. J.E. Field), Academic Press, London, pp. 555–591

Harris, J. W., Hawthorne, J. B. and Oosterveld, M. M. (1984) *Annalles Science, University of Clermont-Ferand, II*, **74** pp.1–13

Harris, J. W. and Vance, E. R. (1972) *Contributions toMineralogy and Petrology*, **35**, 227-234

Hirsch, P.B., Howie, A., Nicholson, R. B., *et al.* (1977) *Electron Microscopy of Thin Crystals* Robert E. Krieger, Malabar, Florida

Hirsch, P. B., Hutchison, J. L. and Titchmarsh, J. (1986) *Philosophical Magazine Letters*, **A54**, L49–L54

Honda,M., Reynolds,J.H., Roedder,E. and Epstein,S. (1987) *Journal of Geophysical Research*, **92**, 12, 507–12, 521

Hudson, P. R. W. and Tsong, I. S. T. (1977) *Journal of Materials Science*, **12**, 2389–2395

Jiang, S.S., Lang, A.R. and Tanner, B.K. (1987) *Philosophical Magazine*, **56**, 367–375

Kaiser, W. and Bond, W. L. (1959) *Physical Review*, **115**, 857–863

Koval'skii,V.V. and Cherskii, N.V. (1973) *Industrial Diamond Review*, **33**, 54–56

Lang, A. R. (1970) In *Modern Diffraction and Imaging Techniques in Material Science*, (eds S. Amelinckx, R. Gevers, G,Remaut and I. van Landuyt), North-Holland, Amsterdam, pp. 407–479

Lang, A. R. (1974) *Proceedings of the Royal Society*, **A340**, 233–248

Lang,A.R. (1979) In *The Properties of Diamond*, (ed. J.E.Field), Academic Press, London, pp. 425–469

Lenzen, G. (1983) *Diamonds and Diamond Grading*, Butterworths, London, p. 207

Lightowlers, E. C. and Collins, A. T. (1976) *Diamond Research 1976*, supplement to *Industrial Diamond Review*, pp. 14–21

Lonsdale, K. (1942) *Proceedings of the Royal Society*, **A179**, 315–320

Lonsale, K., Milledge, H. J. and Nave, E. (1959) *Mineralogical Magazine*, **32**, 185–201

Melton, C. E. and Giardini, A. A. (1974) *American Mineralogist*, **59**, 775–782

Melton, C. E., Salotti, C. A. and Giardini, A. A. (1972) *American Mineralogist*, **57**, 1518–1523

Meyer, H. O. A. (1985) *American Mineralogist*, **70**, 344–355

Meyer,H.O.A. (1987) In *Mantle Xenoliths*, (ed. P.H. Nixon), John Wiley, Chichester, pp 501–522

Milledge,H.J, Mendlessohn,M.J., Seal,M. *et al.* (1983) *Nature*, **303**, 791–792

Moriya,K. and Okawa,T. (1978) *Journal of Crystal Growth*, **44** 53–60

Moriya,K. and Okawa,T. (1980) *Philosophical Magazine A*, **41**, 191–200

Orlov, Yu. L. (1977) *The Mineralogy of the Diamond*. John Wiley, New York

Phillips,D., Onstott,T.C. and Harris,J.W. (1989) *Nature*, **340**, 460–462

Raal, F. A. (1957) *American Mineralogist*, **42**, 354–361

Ross, J (ed.) (1989) *Kimberlite and Related Rocks, Vol 2 Proceedings of the Fourth International Kimberlite Conference (Perth 1986)* Blackwells Scientific, Melbourne

Richardson,S.H., Gurney,J.J. Erlank,A.J. and Harris,J.W. (1984) *Nature*, **310**, 198–202

Seal, M. (1966) *Nature*, **212**, 1528–1531

Sellschop, J. P. F. (1975) *Diamond Research 1975*, supplement to *Industrial Diamond Review*, pp. 35–41

Sellschop, J. P. F., Madiba, C. C. P., Annegarn, H. J. and Shongwe, S. (1979) *Diamond Research 1979*, supplement to *Industrial Diamond Review*, pp. 24–30

Sellschop, J. P. F., Mingay,D.W., Bibby, D. M. and Erasmus, C. S. (1974) *Diamond Research 1974*, supplement to *Industrial Diamond Review*, pp. 43– 50

Sobolev,N.V. and Shatsky,V.S. (1990) *Nature*, **343**, 742–746

Stephenson,R.F. (1978) *Ph.D.Thesis*, University of Reading

Swart,P.K., Pillinger,C.T., Milledge,H.J. and Seal,M. (1983) *Nature*, **303**, 793–795

Tanner, B. K. (1976) *X-ray Diffraction Topography*. Pergamon Press, Oxford

Vand,V., Vedam,K. and Stein,R. (1966) *Journal of Applied Physics*, **37**, 2551–2557

Van Enckevort,W.J.P. and Seal,M. (1987) *Philosophical Magazine A*, **55**, 631–642

Van Enckevort,W.J.P. and Seal,M. (1988) *Philosophical Magazine A*, **57**, 939–954

Walmsley, J. C., Lang, A. R., Rooney, M-L. T. and Welbourn, C. M. (1987) *Philosophical Magazine Letters*, **55**, 209–213

Watt, I. M. (1985) *The Principles and Practice of Electron Microscopy*. Cambridge University Press, Cambridge

Wedlake, R. J. (1979) In *The Properties of Diamond*, (ed. J.E. Field), Academic Press, London, pp. 501–535

Wentorf, R. H. (1971) *Journal of Physical Chemistry*, **75**, 1833–1837

Wilks,E.M. and Wilks,J. (1971) *Industrial Diamond Review*, **31**, 238–242

Wilks, E. M. and Wilks, J. (1980) *Industrial Diamond Review*, **40**, 8–13

Woods,G.S. (1976) *Philosophical Magazine*, **34**, 993–1012

Wong,J., Koch,E.F., Hejna,C.I. and Garbauskas,M.F. (1985) *Journal of Applied Physics*, **58**, 3388–3393

Chapter 3

Optical absorption and colour

When a beam of monochromatic radiation traverses a transparent medium its intensity falls according to the relationship

$$I = I_0 \, e^{-\alpha x}$$

where I_0 is the initial intensity, x the path length in the medium, and α the coefficient of absorption which is a function of the wavelength of the light. The coefficient α is best determined by placing a parallel sided block of the medium in the beam of an absorption spectrometer. This instrument measures the decrease in intensity of the beam in passing through the medium and produces a spectrum of the coefficient of absorption as a function of the wavelength.

Measurements of the absorption of infrared, visible and ultraviolet radiation in diamond provide an important method of assessing the presence of impurity atoms and other defects. However, the absorption spectra of diamonds often show a considerable amount of complex structure, not all of which is understood. Comprehensive reviews of the absorption are given by Davies (1977a) and Walker (1979). Here we are concerned only with the main features of the spectra in sufficient detail to serve as background information when discussing other properties of diamond which are modified by the presence of the impurities and defects.

The spectra in Figures 3.2 to 3.9 which we discuss below were conveniently obtained with two commercial spectrometers, a Perkin Elmer 325 spectrometer for the infrared and a Perkin Elmer Lambda 9 spectrometer for the visible and ultraviolet spectra, the range of wavelengths being chosen to include the principal features of the spectra. To obtain an exact value of the coefficient of absorption it is necessary to take account of the energy loss due to reflection at the entry surface and the effect of multiple reflections within the crystal, as well as to allow for instrumental errors and any limiting of the beam by the size of the specimen. However, for our present survey it is sufficient to allow for the reflection loss by zeroing the spectrometer at wavelengths where the absorption is known to be very small. (One should also note that the measured absorption may be increased above its true value if the light is scattered by inclusions in the diamond.)

Diamonds generally contain more than one type of impurity or defect, and each will give rise to different optical centres making separate contributions to the absorption. As a result absorption spectra are often complex and not easy to interpret. Therefore we begin with examples of the spectra of diamonds chosen so that the absorption from impurities comes primarily from only one type of optical centre. The area under an absorption peak or band, the so-called *integrated absorption*, is then proportional to the concentration of the optical centres responsible, see for example Ditchburn (1952).

In order to present as complete a picture as possible of the impurity content of a diamond, values of the absolute magnitude of the coefficient of absorption should be given whenever possible. Values presented in arbitrary units are much less useful, as are values of the transmittance through a crystal of unspecified thickness. Measurements made with a diffusive reflectance attachment will certainly give distorted spectra.

Finally we note that three alternative ways of indicating the wavelength of radiation are commonly used in the literature. Perhaps the most fundamental characteristic of monochromatic radiation is its frequency f but this frequency is hardly ever quoted. The light is characterized in one of three ways by its wavelength in air λ, its wavenumber $v = 1/\lambda$ in units of cm^{-1}, or its energy hf in units of electron volts (h being Planck's constant 6.626×10^{-34} in SI units). As different authors use different conventions it will be useful to note the relationships

$$\text{energy (eV)} = 1240/\lambda(nm) = 1.240\,v\,(cm^{-1})/10^4$$

3.1 Perfect diamond

We start by describing the absorption spectra of an ideal perfect diamond free of all impurities and lattice defects, and then describe various additional components of absorption observed in real diamonds and the nature of the optical centres giving rise to them.

3.1.a Diamond as an insulator

Any discussion of the optical properties of diamond requires some understanding of the nature of dielectric material, so we begin with a brief sketch of some characteristics of diamond as an insulator. As explained in Section 1.5 the outer electrons of the carbon atoms in diamond are localized in bonds between adjacent atoms. At first sight this explains why diamonds are generally electrical insulators, the electrons are held in the bonds and therefore cannot move through the crystal when an electric field is applied. In fact the position is much more complex, as set out in texts on semiconductors, see for example Rosenberg (1988). In the rest of this section we summarize some of the main features of these treatments as applied to diamond.

According to the electron theory of metals and semiconductors each electron in a diamond must have an energy different from that of all the other electrons (the so-called exclusion principle). Also, certain values of energy are forbidden to all electrons. Thus Figure 3.1 shows the three ranges of energy values in diamond. The lower range is known as the *valence band* and the upper range as the *conduction band*, the region in between is the *band gap* forbidden to all electrons. In diamond at room temperature all the electrons have energies in the lower valence band, moreover the number of these electrons is just the number needed to completely fill the band.

Let us suppose that we attempt to produce a flow of current by applying an electric field across the diamond. An electric field causes a current in a conductor

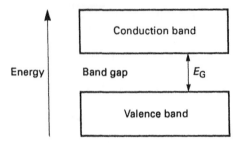

Figure 3.1. Schematic diagram of the band gap in diamond

by accelerating the electrons and this involves increasing their energy. However, in diamond the only values of energy not already taken up by other electrons are in the conduction band above the energy gap. The gap has a width of approximately 5.5 eV while the mean available thermal energy is only of the order of 0.025 eV, so no electrons can cross the gap and the diamond is an excellent insulator. Rather similar considerations determine the absorption of light. Absorption will only occur if energy can be transferred from the light to the electrons in the diamond, but these electrons can only take up energy by moving across the band gap, and this requires a minimum energy of about 5.5 eV. Therefore as the photons of visible light have energies in the range from about 1.7 eV (730 nm) to 3.1 eV (400 nm) they pass through pure diamond without any absorption.

3.1.b Optical absorption

Although a perfect diamond produces no absorption of visible light, it exhibits considerable absorption in both the infrared and ultraviolet regions. For example Figure 3.2 shows the visible/ultraviolet and infrared spectra of a diamond selected to have a low level of impurities and defects. The rapid onset of absorption in the ultraviolet (UV) region above about 5.5 eV occurs because the light then has enough energy to transfer electrons across the band gap. In the infrared (IR) region the energy of the light is only of the order of 100 meV, so the absorption cannot be due to electrons being excited across the gap. It arises because of processes in which photons of radiation produce additional vibrations of the atoms in the crystal at their characteristic frequencies of vibration, the energy of each photon being transferred to the lattice vibrations as a phonon or quantum of vibrational energy. As we see below, the infrared absorption spectrum contains much information about the diamond.

The mechanism of the absorption in the infrared is best understood by considering first the absorption of light in alkali halide crystals in which adjacent atoms carry electric charges of opposite sign. As in all crystals, the thermal vibrations of the atoms can be resolved into modes of different frequencies and the modes of highest energy (the so-called optic modes) are such that adjacent atoms move in opposite directions. The presence of an electric field will tend to change the relative spacing of positively and negatively charged atoms. Hence radiation with a frequency similar to that of one of the vibrational modes will induce a resonance which results in absorption. A complete quantum treatment

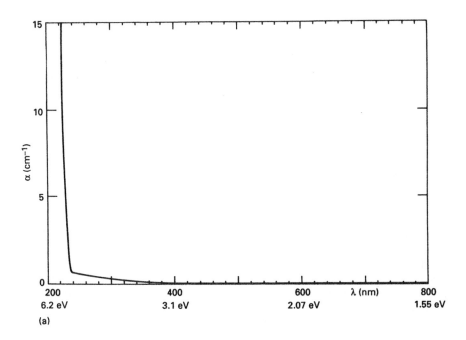

(a)

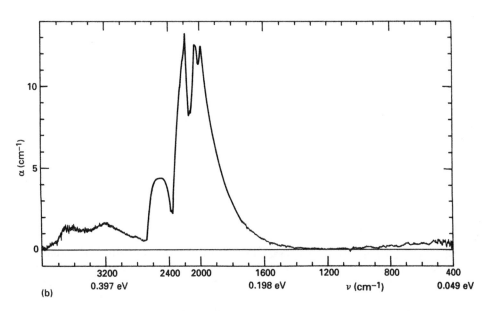

(b)

Figure 3.2 Absorption spectra of a diamond with a low level of impurities (a), visible and ultraviolet; (b), infrared

shows that these alkali halides will absorb radiation with a range of energies in the infrared corresponding to a range of lattice vibrations.

The mechanism of absorption just described is not applicable to diamond because the atoms are not charged. Another possibility is that the electric field associated with the light wave may polarize the crystal (producing a relative displacement of the negative electrons and the positive nuclei) and then interact with this polarization. However, a complete calculation (Lax and Burstein, 1955) shows that this process is forbidden by the symmetry of the diamond. In fact the absorption in diamond arises from processes involving two phonons, one of which can be regarded as inducing polarizing charges on the atoms while the second phonon causes a vibration of these charges which couples to the radiation field of the light (see for example Reissland, 1973). The only allowed interactions are those involving photons and phonons satisfying the relationship

$$hf = \epsilon_1 \pm \epsilon_2 \tag{1}$$

where hf, ϵ_2 and ϵ_2 are the energies of the photon and of the two phonons. There is also the restriction that the two phonons must be of different polarization. Further, in the case of diamond the energies of the photons are much greater than the thermal energy at room temperature, so the probability of the light exciting the difference term in Equation (1) is very low, and it makes no contribution to the absorption.

The maximum vibrational frequency of the lattice, the *Raman frequency*, is known both from neutron scattering experiments (Warren *et.al.*, 1967) and from experiments described in the following section in which light in the visible region is scattered from the diamond with an appreciable change in wavelength (the Raman effect). The Raman frequency for diamond corresponds to an energy of 0.165 eV, so Equation (1) can only be satisfied if $hf < 0.33$ eV, that is for wavenumbers less than 2665 cm^{-1}. In fact Figure 3.2 shows that the absorption falls to a low value at about this wavenumber, the small residual absorption at higher wavenumbers arising from processes involving three phonons with a corresponding cut off at 0.495 eV or 4000 cm^{-1}. Experiments have now established a fairly complete correlation between the form of the IR absorption spectra and the lattice vibrations as deduced from neutron scattering and Raman scattering, see for example Solin and Ramdas (1970).

The thermal vibrations of the lattice also manifest themselves in the UV absorption spectra. An examination of the steep rise in the absorption curve in Figure 3.2(a) at high resolution reveals considerable structure. Also, the position of the rise shifts to slightly lower energies at high temperature. The vibrational energy of the phonons can assist a photon to lift electrons across the gap and Clark, Dean and Harris (1964) have shown that the structure in the absorption curve correlates reasonably well with the dominant frequencies in the phonon spectrum.

3.1.c Raman scattering

In the infrared absorption processes described above a quantum of radiation is generally only absorbed if $hf < 0.33$ eV and this condition restricts the absorption to the infrared part of the spectrum. Visible light cannot be absorbed in this way. However, we now describe another way in which visible light interacts with the thermal vibrations of a solid, known as the Raman effect.

It is often observed that when a beam of monochromatic visible light passes through a crystal some scattered light is produced with a well defined wavelength appreciably different from that of the incident light. In one of the first experiments on diamond Ramaswamy (1930) observed that when the two mercury lines of wavelength 404.7 nm and 435.8 nm were passed through a diamond they each gave rise to a scattered line with a single wavelength shifted by an amount corresponding closely to a reduction of energy of 0.165 eV. This occurs because some of the energy of the light has been transferred to the lattice vibrations.

Because the phonons in a crystal have a range of energies from zero to the maximum value it might seem that the incident radiation would create phonons with this range of energies. However, we would then expect the scattered light to exhibit a spread of energies whereas only one sharp line is observed, corresponding to the maximum phonon energy. This happens because the mechanism of the interaction imposes certain restrictions on the process (conservation of wave vector) see for example Reissland (1973). Therefore the above effect, the so called first order Raman effect, gives almost no information on the form of the vibration spectrum. However, there is also a second order effect in which the visible light gives up energy by creating two phonons and the restrictions for this process turn out to be less severe. Hence the spectrum of the light scattered in this way is no longer a sharp line but exhibits a spread of frequencies, and measurements of this frequency spectrum are usually displayed in terms of the frequency *shifts* produced by the scattering. Observations of the second order Raman spectrum in diamond (Solin and Ramdas, 1970) correlate well with studies of the lattice dynamics based on observations of the scattering of neutrons by the lattice (Warren *et al.*, 1967).

Finally we note that the single sharp line of the first-order Raman effect in diamond with its characteristic frequency shift of 0.165 eV presents a method for the identification of diamond. Moreover changes in lattice structure may be quite sensitively reflected in the spectrum. For example measurements of the first order Raman effect give a convenient way to inspect the form of diamond and diamond-like films produced by vapour deposition (Section 1.4). Thus the spectra in Figure 1.17 readily reveal the differences between films of pure diamond and those with only a diamond-like structure.

3.2 Optical centres due to nitrogen

Impurity atoms in insulating solids can act as optical centres which create optical absorption. Consider the effect on the electronic structure of a diamond when a carbon atom is replaced by a nitrogen atom. Nitrogen is adjacent to carbon in the periodic table of the elements so nitrogen atoms are rather similar in size to carbon but have an outer ring of five electrons, one more than carbon. Four of these outer electrons will bond with the four nearest carbon neighbours leaving one electron over. According to the theory of semiconductors the energy of this extra electron is not restricted to values in the valence and conduction bands (Figure 3.1) but may take up values in the band gap. Hence there is the possibility of electronic transitions to and from a level in the band gap which absorb light of less energy than the band gap. These optical centres may consist of either a single nitrogen atom or a small group of such atoms.

We now outline the principal optical centres associated with nitrogen atoms in the diamond lattice and the form of the absorption associated with them. The absorption spectra of most diamonds are actually a superposition of contributions from several different centres. Nevertheless it is possible to select diamonds in which the additional absorption arises mainly from just one of the principal centres.

3.2.a The A centre

Figure 3.3 shows absorption curves where the additional absorption due to impurity atoms arises primarily from the so-called A centre. At wave numbers greater than $1400\,\mathrm{cm}^{-1}$ the IR spectrum in Figure 3.3(b) is similar to that in Figure 3.2(b) and is associated with the vibrations of the lattice which are not much affected by the impurities. (Because this spectrum due to the lattice vibrations is similar in most diamonds it provides a convenient reference level for zeroing the spectrometer when making absorption measurements in other parts of the infrared spectrum.) Figure 3.3(b) shows in addition a substantial and characteristic infrared absorption in the range $1000\,\mathrm{cm}^{-1}$ to $1400\,\mathrm{cm}^{-1}$. The A centre also produces an additional absorption in the UV so the steep rise in the absorption commences at appreciably lower energies, cf. Figures 3.3(a) and 3.2(a).

Kaiser and Bond (1959) made measurements on several diamonds with spectra of the form shown in Figure 3.3 and observed a range of values for the absorption in the region $1000\,\mathrm{cm}^{-1}$ to $1500\,\mathrm{cm}^{-1}$. By analysing the nitrogen content of the diamond by vacuum fusion gas analysis, they showed that the extra absorption in both the infrared and ultraviolet were proportional to the concentration of nitrogen. Kaiser and Bond also measured the lattice constant using X-rays and observed a small change in the constant proportional to the concentration of nitrogen. From the magnitude of this change they inferred that the nitrogen atoms must be sited substitutionally on carbon sites rather than interstitially.

The additional absorption in the ultraviolet due to the A centre has been studied in a range of experiments which imply that the centre consists of two nitrogen atoms replacing two adjacent carbon atoms. The two extra electron spins associated with each centre pair off and the crystal shows no paramagnetism. Davies (1976) has observed the response of the ultraviolet absorption spectrum to a uniaxial stress applied in different directions and hence deduced the crystal symmetry of the defect. He was thus able to argue that the only arrangement of 2, 4 or 6 neighbouring nitrogen atoms which can account for the observed results is that of two adjacent atoms.

At first sight it might seem that the A centre increases the ultraviolet absorption because it provides an electronic level in the band gap and thus permits electrons to be brought to the top of the gap by radiation of less than the band gap energy. In fact the position is probably more complicated, as shown by Davies and Nazaré (1979) who observed the effect of a uniaxial stress on the structure of the additional ultraviolet absorption (including the sharp lines designated N5, N6, N7 and N8 by Clark, Ditchburn and Dyer (1956a)). These results suggest that the absorption arises from transitions localized at the A defect and which are broadened by the strain fields created by the presence of the A defects.

The additional absorption produced by the A centre is only possible because the presence of the lattice defect changes the symmetry of the lattice and therefore the conditions governing the absorption of radiation. Hence a photon of radiation can now be absorbed in a so-called one phonon process in which only one phonon is created (Lax and Burstein, 1955). In fact Figure 3.3(b) shows that the maximum energy of the infrared light absorbed in the defect region is quite close to the value

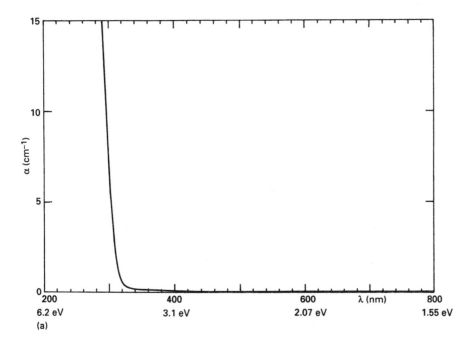

(a)

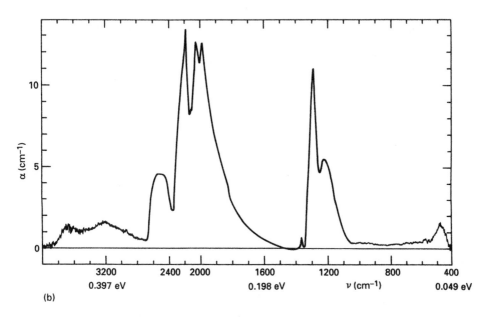

(b)

Figure 3.3 Absorption spectra of a diamond in which A centres are the predominant impurities (a), visible and ultraviolet; (b), infrared

Table 3.1 Optical centres in diamond. E_{max} gives the energy of the radiation at the position of maximum absorption due to the centre or (in the case of the N3 centre) the position of the zero-phonon line

Centre	E_{max}			Colouration
A	0.159 eV	1282 cm^{-1}	7.8 μm	none
B	0.147 eV	1185 cm^{-1}	8.4 μm	none
Single N	0.140 eV	1130 cm^{-1}	8.85 μm	strong yellow
N3	2.985 eV		415 nm	pale yellow
Platelets	0.169 eV	1370 cm^{-1}	7.3 μm	none
Boron	0.305 eV	2460 cm^{-1}		blue
	0.347 eV	2790 cm^{-1}		
Dislocations				brown

of 0.165 eV (1332 cm^{-1}), the maximum energy of a single phonon (Section 3.1). We expect the shape of the spectrum to depend both on the vibrational spectra of the crystal and on the particular nature of the defect but no detailed calculations have yet been made.

The various optical centres in diamond are often characterized by quoting the energy, the wavenumber, or wavelength of some feature in their spectrum, either of their maximum intensity or in the case of vibronic bands (Section 3.2.d) of the zero-phonon line. Because of the form of the output of most spectrometers the wavenumber description tends to be used in the infrared and the wavelength description in the visible and ultraviolet. These various values for the A centre and other optical centres in diamond are given in Table 3.1.

3.2.b The B centre

Figure 3.4 shows the spectra of a diamond which exhibits considerable additional absorption in the infrared with a spectrum different from that produced by the A defect. In fact the spectra of Figure 3.4 are typical of a diamond containing only the so-called B centre. Note also that a comparison of Figures 3.4(a) and 3.3(a) shows that the B centres produce appreciably less absorption in the ultraviolet than the A centres. Sobolev and Lisoivan (1972) using a group of such diamonds demonstrated that the magnitude of the additional absorption in both the infrared and ultraviolet is proportional to the nitrogen concentration measured by a nuclear activation analysis technique.

The B centre has not been investigated as thoroughly as the A centre. However, it produces no paramagnetism and is therefore probably a small group or aggregate consisting of an even number of substitional nitrogen atoms, perhaps 4 to 6. Further work is still required to establish the exact nature of the B centre, but perhaps the most likely possibility is that it consists of four adjacent nitrogen atoms plus a vacancy (Bursill and Glaisher, 1985); see also Section 6.4.a. (As described in Section 3.8 both A and B centres transform with the addition of a vacancy to form, respectively, the so-called H3 and H4 centres.)

3.2.c Single substitutional nitrogen

Figure 3.5(b) shows an infrared absorption spectrum which is typical of many or most synthetic diamonds but which is only observed in a relatively few natural

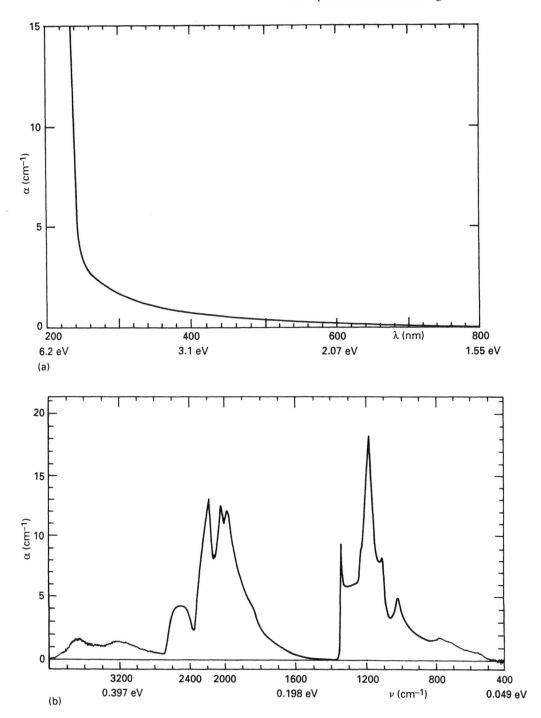

Figure 3.4 Absorption spectra of a diamond in which B centres are the predominant impurities (a), visible and ultraviolet; (b), infrared

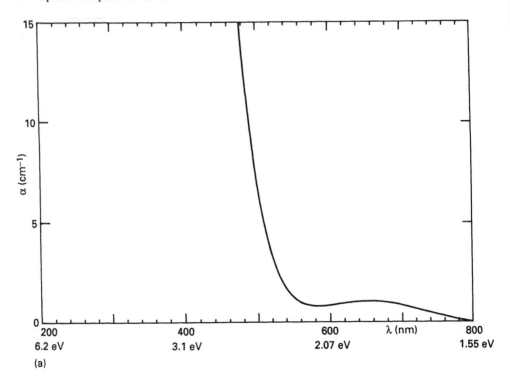

(a)

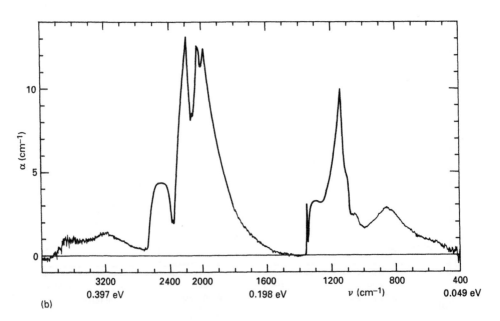

(b)

Figure 3.5 Absorption spectra of a diamond in which single nitrogen atoms are the predominant impurities (a), visible and ultraviolet; (b), infrared

stones. The absorption again shows a broad band between about $1000\,cm^{-1}$ and $1500\,cm^{-1}$, but with a shape quite different from that of the A or the B spectrum. This infrared absorption is accompanied by additional absorption in the ultraviolet which differs from that produced by the A and B centres. Note that Figure 3.5(a) is somewhat misleading as it suggests that the ultraviolet absorption rises steadily to an extremely high value as is the case for the A and B type absorption. In fact the absorption shows a peak at about 270 nm and then falls at higher energies before rising again, as is observed in diamonds with a lower concentration of centres, see for example Dyer et al. (1965).

Diamonds with the absorption spectra of Figure 3.5 are paramagnetic and give rise to an electron spin resonance signal (Smith et al., 1959). A short account of electron spin resonance in diamond is given by Bleaney and Owen (1965) and an extensive review by Loubser and Van Wyk (1978). It is found that the additional optical absorptions shown in Figure 3.5 in both the infrared and ultraviolet are proportional to the strength of the resonance signal (Dyer et al., 1965; Sobolev et al., 1969; Chrenko, Strong and Tuft, 1971). Hence the optical and resonance signals must arise from the same centre, the magnitude of the absorption in the ultraviolet being considerably greater than in the infrared. In addition, detailed measurements of the hyperfine structure of the resonance signal show that the centre has a nuclear spin of $I = 1$, a value which is not at all common and is characteristic of nitrogen (Loubser and Du Preez, 1965). This varied information implies that the centre is a single nitrogen atom replacing a carbon atom, the paramagnetism arising from the additional electron provided by the nitrogen atom. Note that as it is a relatively straightforward operation to calibrate the absolute sensitivity of a spin resonance detector, this technique gives a reliable method of determining the concentration of the single nitrogen atoms. Finally we note that the complete spectrum of this system, like that of other similar systems, may be quite complex. See for example a discussion by Collins and Woods (1982) of a subsidiary peak at $1344\,cm^{-1}$ associated with the main absorption at $1130\,cm^{-1}$.

3.2.d The N3 centre

The spectrum in Figure 3.6 is that of a diamond which contains a relatively high concentration of the so-called N3 centres. The absorption due to the centre is characterized by a peak at about 415 nm together with a band extending to shorter wavelengths, this absorption appears superposed on a monotonically increasing background of absorption arising from other centres. There is also some small additional absorption between 450 nm and 480 nm superposed on the background. This latter component is sometimes referred to as the N2 set of lines (Clark, Ditchburn and Dyer, 1956a) but in fact forms part of the N3 system (Davies, Welbourn and Loubser, 1978).

Absorption of the form shown in Figure 3.6 is accompanied by a paramagnetism which gives rise to the so-called P2 electron spin resonance signal, the strength of this signal being proportional to the strength of the optical absorption. Measurements show that the resonance arises from a centre with electron spin 1/2 and a nuclear spin 1, values characteristic of the single nitrogen centre. However, detailed measurements of the hyperfine structure (Davies, Welbourn and Loubser, 1978) suggested that the centre consists of three adjacent substitutional nitrogen atoms lying in a {111} plane and bonded to one particular carbon atom, while Bursill and Glaisher (1985) have proposed a model with three nitrogen atoms

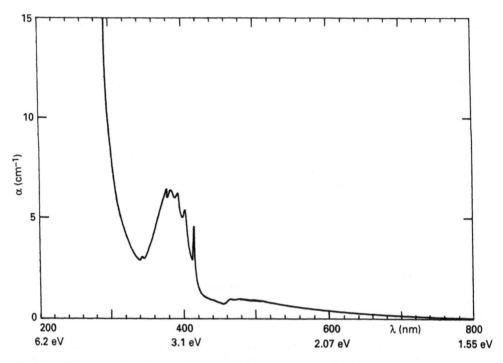

Figure 3.6 Visible and ultraviolet absorption spectrum of a diamond in which N3 centres contribute to the absorption

surrounding a vacancy. The absorption in Figure 3.6 is a typical example of a *vibronic* spectrum. The sharp peak or *zero-phonon* line corresponds to an absorption process in which only the centre is excited by the radiation, while the broad band is caused by the absorption of higher energy radiation which excites the centre and at the same time creates phonons in the lattice; for further details see Fitchen (1968) and Davies (1970, 1981a).

The N3 is a quite different and distinct centre from the A and B centres but appreciable N3 absorption bands are only observed in diamonds with a pronounced A or B type absorption. Davies and Summersgill (1973) noted that N3 spectra tend to be more pronounced in diamonds with appreciable B type absorption, and Woods (1986) observed a direct proportionality between B type absorption and N3 type absorption in the so-called 'regular' diamonds described in Section 6.4.b. The N3 centre makes no perceptible contribution to the infrared absorption as transitions in the infrared are much less readily excited than the electronic transitions giving rise to the vibronic band absorption (Davies and Summersgill, 1973).

3.2.e Platelets

Some diamonds exhibit an absorption peak in the infrared at about 170 meV (1370 cm^{-1}) superposed on a background of absorption arising from other centres (Figure 3.7). The magnitude of this peak, sometimes referred to as the B' peak,

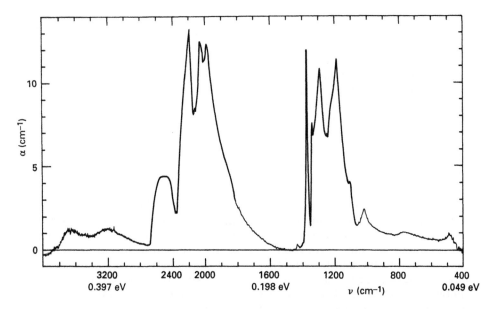

Figure 3.7 Infrared spectrum including a large narrow peak at 1370 cm^{-1} due to the presence of platelets

correlates with the presence of the platelets described in Section 2.4.a. Thus the integrated strength of the 1370 cm^{-1} absorption peak is closely proportional to the intensity of the extra spikes seen in Laue diffraction patterns of diamonds containing platelets (Sobolev, Lisoivan and Lenskaya, 1968; Evans and Rainey, 1975). The area of the platelets has also been observed to be proportional to the intensity of the spikes (Evans, 1973).

As described in Section 2.4.c, the platelets vary in diameter from a few nanometers to a few micrometers and appear to consist of one or two extra atomic layers inserted in the diamond lattice. There is still considerable discussion on the form of the platelets, both on the nature of the extra atoms and their structural arrangement. The fact that the peak is only observed in diamonds containing an appreciable amount of nitrogen suggests that the platelets contain nitrogen, and probably consist either partly or entirely of nitrogen. However, the results of various studies of the platelets are quite complicated (see Section 6.4.b).

3.3 Type I and Type II diamonds

In the previous section we described five different types of optical centres due to nitrogen. Hence, *a priori*, we may expect diamonds selected at random to contain all five centres in a wide range of possible concentrations. The observed absorption will then be a complex combination of the spectra due to the different centres. Even so, it is quite possible to classify diamonds according to the principal features of their spectra. Although not ideal this system is much used and should be understood.

About 2% of all diamonds show little absorption in the infrared in the range $800\,\text{cm}^{-1}$ to $1400\,\text{cm}^{-1}$. All the rest contain enough nitrogen to produce contributions to the absorption which are clearly visible in spectra on the scale of Figures 3.2 to 3.9. These diamonds are known as Type I, and contain nitrogen in some or all of the forms described above. Those containing principally the non-magnetic A and B centres are known as Type Ia, those containing primarily the paramagnetic single atom centres are known as Type Ib. The Type Ia group is further divided into Type IaA and Type IaB according to whether A or B centres make the principal contribution to the absorption.

The diamonds containing little or no nitrogen are defined as Type II. A very few of these show some electrical conductivity which, as we describe in Section 3.4, is due to the presence of boron impurities. To distinguish between these two forms, the non-conducting majority are described as Type IIa and the conducting specimens as Type IIb. It is useful to note that although Type II diamond is now defined by reference to its infrared spectrum it has in the past also been defined by reference to its ultraviolet spectrum, in particular by its transmission of radiation down to about 220 nm. However, this criterion is unsatisfactory because a Type IaB diamond produces little additional absorption in the ultraviolet but has a very characteristic infrared absorption. (It seems likely that the term *intermediate* diamond used in the past referred essentially to diamonds containing primarily B centres, but it was ambiguous and is now obsolete.)

The above classifications are useful but have their limitations. Most natural diamonds are Type Ia and contain both A and B centres. Most synthetic diamonds contain nitrogen primarily as single atoms and are good examples of Type Ib. However, a small number of natural diamonds also contain the single nitrogen centre and and are often called Type Ib even if they contain an equal or greater concentration of A centres. There is also a further complication if the impurity atoms are not distributed uniformly in the crystal, see for example Chapter 5. Even so, it is now possible to make useful estimates of the concentrations of these three centres from the infrared spectra.

To determine the concentration of the optical centres it is necessary to resolve the region of the infrared spectrum between $1000\,\text{cm}^{-1}$ and $1500\,\text{cm}^{-1}$ due to the various optical centres into its component parts. Davies (1981a) has described a relatively simple method of decomposing the spectrum into the absorptions due to the A and B components when only these are present. However, in some diamonds the same spectral region may contain other components and a more elaborate analysis is required (Clark and Davey, 1984). The latter authors observed two additional components of the spectra the so-called C spectrum which appears to be part of the single nitrogen spectrum and the D spectrum which is proportional to the platelet absorption (Woods, 1986). Once a spectrum has been resolved into these components, we may deduce the concentrations of the nitrogen in the A and B centres from relationships determined by the analysis of diamonds with essentially A or B type spectra (Woods, Purser *et al.*, 1990).

$$N_A = (17.5 \pm 0.4)\,\alpha_A\,(1282) \tag{3.1}$$

$$N_B = (103.8 \pm 2.5)\,\alpha_B\,(1282) \tag{3.2}$$

where α_A (1282) and α_B (1282) are the coefficients of absorption in units of cm^{-1} due to the A and B components of the spectra measured at $1282\,\text{cm}^{-1}$, and the concentrations of nitrogen N_A and N_B are given as the atomic concentrations in parts per million. A corresponding expression for a spectrum associated only with single nitrogen atoms is given by Woods, Van Wyk and Collins (1990) as

$$N_{\mathrm{N}} = (22.0 \pm 1.1)\, \alpha_{\mathrm{N}}\, (1130) \tag{3.3}$$

The figures quoted in Equations (3.1) to (3.3) replace values obtained in earlier experiments which are less reliable. Problems have arisen in the past because of the complexities of the situation. It was not always fully appreciated in earlier work that the nitrogen in a diamond may take several forms and that it was necessary to analyse diamonds in which all the nitrogen was associated with essentially just one type of optical centre. The fact that diamonds are often inhomogenous still presents a major difficulty. It must also be born in mind that in any analysis of diamond it is difficult to obtain accurate values for the *absolute* values of small concentrations of impurities, particularly those of low atomic number.

At the start of the section we described Type II diamonds as those which show little or no absorption in the infrared between $800\,\mathrm{cm}^{-1}$ and $1400\,\mathrm{cm}^{-1}$. In fact, the concentrations of nitrogen in diamonds exhibit a continuous range of values so the criterion of what constitutes a Type II diamond is essentially arbitrary. Nevertheless we may obtain a working definition in line with general practice by describing Type II diamonds as those which show no appreciable absorption peaks in the range $800\,\mathrm{cm}^{-1}$ to $1400\,\mathrm{cm}^{-1}$ when their spectra are plotted to give a maximum reading of $15\,\mathrm{cm}^{-1}$ as for example in Figure 3.2. On this scale the limit of detection is of the order of $0.1\,\mathrm{cm}^{-1}$, which according to equation (3.1) corresponds to a concentration of A form nitrogen of about 1 ppm. A rather similar figure for the single nitrogen concentration follows from Equation (3.3) of about 2 ppm. These figures appear to be in line with those quoted by other workers, see for example Chrenko and Strong (1975) who quote values of 1 ppm to 5 ppm.

3.4 Optical centres due to boron

Figure 3.8 shows an infrared spectrum with a small background absorption between $1000\,\mathrm{cm}^{-1}$ and $1500\,\mathrm{cm}^{-1}$ but no sign of any contribution from A, B or single nitrogen features. The spectrum is thus somewhat similar to that of a Type IIa diamond (Figure 3.2) except that there are pronounced peaks at $2460\,\mathrm{cm}^{-1}$ (0.307 eV) and $2790\,\mathrm{cm}^{-1}$ (0.347 eV), and at higher energies a continuous band of absorption. The latter extends towards the red end of the visible spectrum (Figure 3.8a), and if in sufficient strength produces a blue colouration in the diamond. Diamonds with this type of spectrum are described as Type IIb. They contain boron and are of particular interest in that they are semiconductors with an appreciable electrical conductivity.

3.4.a Diamond as a semiconductor

We begin by explaining how the presence of boron leads to an electrical conductivity in diamond. Consider a diamond in which a small number of carbon atoms have been replaced substitutionally by atoms of boron. Boron lies next to carbon in the periodic table of the elements so that the atom is of quite similar size but has only 3 electrons to share with the 4 surrounding carbon atoms. There is thus the possibility of electrons from the full valence band moving to the boron sites to bond with adjacent carbon atoms left unbound. The boron is therefore said

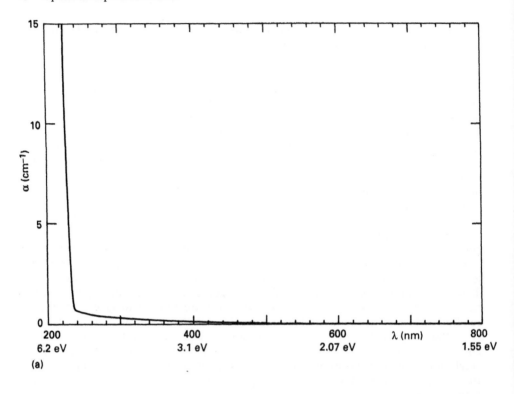

(a)

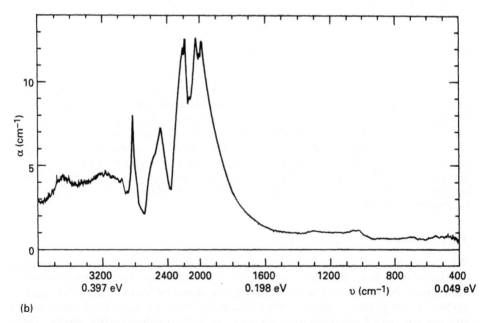

(b)

Figure 3.8 Absorption spectra of a diamond in which boron atoms are the predominant impurities (a), visible and ultraviolet; (b), infrared

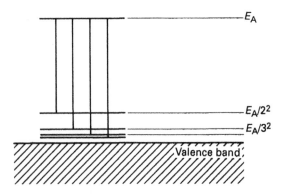

Figure 3.9 Sketch of energy levels associated with a boron atom

to act as an *acceptor* site, the energy of the extra electron being somewhat higher than the energy of the top of the valence band. Once electrons have moved up to the acceptor levels the valence band is no longer completely full. Therefore the electrons in the band can be accelerated without leaving the band, and carry a current which is proportional to the number of missing electrons or *holes*. Note that the diamond will only conduct if electrons in the valence band have enough energy to move up to the higher levels associated with the boron atoms.

We now consider a simple model of how the boron atoms are responsible for the additional peaks in the infrared spectrum. The model assumes that the hole associated with the boron atom acts in a rather similar way to the electron in a hydrogen atom. In this case if the ionization energy of the boron atom is E_A we expect the atom to have a range of possible states with energy $E_A/1^2$, $E_A/2^2$, $E_A/3^2$ etc which because we are dealing with a hole are measured from the top of the valence band as in Figure 3.9. Hence we now have the possibility of transitions between the ground state with energy E_A and the other states as shown by the vertical lines in the figure.

The magnitude of E_A can be found from experiments because its value determines the probability that electrons will be thermally excited from the valence band to the boron ground state thus leaving holes in the valence band and producing electrical conductivity. In fact measurements of the electrical conductivity as a function of temperature imply that the ionization energy is about 0.368 eV (Collins and Lightowlers, 1979). (Measurements of the effect of a magnetic field on a current in the diamond, the so-called Hall effect show that the current is carried by holes in the valence band rather than by electrons in the conduction band (Austin and Wolfe, 1956).)

Hence taking E_A as 0.368 eV the simple hydrogen-like model predicts that the boron states will stand above the valence band by energies of

0.368, 0.092, 0.041, 0.023, 0.015, . . . eV

Therefore we expect the absorption spectrum to show peaks coreesponding to transitions from the ground state (0.368 eV) to the other states, with energies corresponding to the energy difference between the states. Hence the model predicts absorption lines at energies

0.276, 0.327, 0.345, 0.355, . . . eV

The line at 0.276 eV is forbidden by an additional transition rule but the lines at 0.327 and 0.345 eV may be identified with the observed lines at 0.307 and 0.347 eV. Other lines in the same series have been reported by Smith and Taylor (1962).

3.4.b Compensation

It might be thought that if the 0.347 eV peak is due to boron acceptor centres, then the absorption should be proportional to the concentration of boron, at least for small concentrations. However, this is generally not so because of the presence of other impurities. There is usually some nitrogen present in a diamond besides boron and this cannot be ignored.

As described in Section 3.2 the extra electron associated with a nitrogen atom has an energy which lies in the band gap somewhat below the bottom of the conduction band. Thus in principle the nitrogen may act as a *donor* level, with these electrons being thermally excited to the conduction band where they carry a current. However, the absorption spectra of nitrogen in the visible region (Section 3.2) show that none of the nitrogen centres affect the absorption of light with energy less than 2 eV. This result implies that the nitrogen levels lie at least 2 eV below the conduction band, an unusually large distance but one in agreement with a calculation by Watkins and Mesmer (1970). The thermal energy at room temperature is therefore insufficient to permit the electrons to reach the conduction band so the nitrogen does not produce any conductivity.

Consider now a diamond containing small numbers n_B and n_N of substitutional boron and nitrogen atoms. In this case the empty levels associated with the boron atoms may be filled by electrons falling in from the higher nitrogen levels. Indeed if $n_N > n_B$ all the boron levels will be filled or *compensated*, no electrons can move up from the valence band, and the diamond is a complete insulator. However, if $n_N < n_B$ there will remain a number $\eta = n_B - n_N$ of uncompensated boron levels which will permit electrons to move up from the valence band and the diamond will show a conductivity proportional to η. Thus the presence of the nitrogen reduces the number of boron atoms available to produce absorption from n_B to $\eta = n_B - n_N$, hence we expect the absorption to be proportional to η and not to n_B. (There is also the possibility of other impurities, but this is a complication we ignore as nitrogen and boron are certainly the most significant.)

The concentrations of acceptor and donor atoms in semiconducting diamond may be found by observing the resistance and the Hall effect as functions of temperature (Collins and Williams, 1971) or perhaps more accurately from measurements of the electrical capacity as a function of temperature (Lightowlers and Collins, 1976). These experiments, like those of Austin and Wolfe (1956), show that the strength of the infrared absorption at 0.347 eV is proportional to the number of effective acceptors $\eta = n_B - n_N$. (Note that suitable procedures must be used when measuring the electrical conductivity in order to obtain good ohmic contacts to the diamond, for details see Collins, Lightowlers and Williams (1970).)

Acceptor atoms in synthetic semiconducting diamonds were identified as boron by Chrenko (1973) and Sellschop *et al.* (1977). Using nuclear activation techniques the latter authors measured the concentrations of boron and showed that they agreed quite well with the values of the acceptor concentrations, n_B derived from electrical measurements. Earlier reports identifying the acceptor atoms as aluminium were erroneous, see for example Collins and Williams (1971). It is now clear from experiments on the synthesis of diamond that the presence of aluminium acts as a getter which removes nitrogen from the diamond and can thus result in low levels of boron remaining uncompensated (Chrenko, 1973). Values of the boron concentration in some natural diamonds are given by Lightowlers and Collins (1976).

Finally we note that, although diamond containing enough nitrogen to compensate all the boron centres shows no optical absorption and no electrical conductivity asociated with the boron atoms, the boron atoms may play a part in luminescence, see Section 4.2.b.

3.4.c Impurity levels in Type II diamond

It is sometimes said that Type IIb diamonds are the purest form of diamonds. In this connection it must be noted that the absorption due to boron is due to a process different from those responsible for the A, B and single nitrogen absorption between $1000 \, cm^{-1}$ and $1400 \, cm^{-1}$. The boron absorption arises by purely electronic transitions whereas the nitrogen absorption is due to one-phonon processes (Section 3.2.a) involving the lattice vibrations of the carbon atoms, and the probability of a one-phonon transition is much less than that of an electronic transition. Hence, although the extra absorption due to boron in Figure 3.8 is not greatly less than that produced by the nitrogen (Figures 3.4–3.7) the concentration of boron is much less than the concentrations of the nitrogen centres.

The concentration of boron in diamond has been measured relatively directly by nuclear activation techniques using high energy protons. Analyses of five diamonds by Lightowler and Collins (1976) and of five by Sellschop et al. (1977) gave atomic concentrations of boron of only 0.02 ppm to 0.25 ppm. We also know that in a conducting diamond the nitrogen concentration must be less than the boron concentration, for if not, the boron centres would be completely compensated by the nitrogen. Electrical measurements on six diamonds by Wedepohl (1957) and on five diamonds by Lightowlers and Collins (1976) gave values for the concentration of donor atoms, presumably nitrogen, of the order of 0.02 ppm. Thus there is clear evidence that some Type IIb diamonds have a very low impurity content. According to Chrenko (1973) a concentration of no more than 0.5 ppm of uncompensated boron is sufficient to produce a bluish colouration in the diamond. Hence if we select a conducting diamond which is also colourless we expect it to have a very low level of both boron and nitrogen impurities.

3.5 Optical centres due to plastic deformation

Diamonds with a brown colouration, sometimes tinged with mauve, often given ultraviolet and visible absorption spectra of the form shown in Figure 3.10. The absorption rises steadily as the energy of the light increases with an appreciable value in the visible region giving rise to the brown colouration. Moreover in a Type II diamond this absorption in the ultraviolet and visible is not accompanied by any corresponding absorption in the infrared strong enough to be visible in spectra plotted on the scales used above.

The optical centres responsible for the brown type spectrum are quite distinct from the centres already described. There is no correlation whatever between the

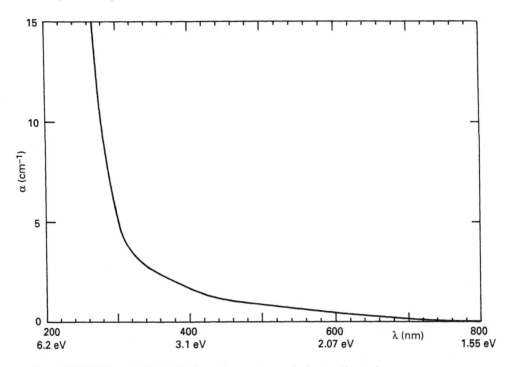

Figure 3.10 Visible and ultraviolet absorption spectrum of a brown diamond

strength of this brown absorption and that arising from any of the nitrogen or boron centres. Indeed a brown colouration is often seen in Type II diamonds which have very low concentrations of nitrogen. The common feature of all these brown diamonds appears to be that they have suffered plastic deformation at some period in their history. This plastic deformation has produced slip bands and dislocations (Chapter 6) as described by Orlov (1977), Hanley, Kiflawi and Lang (1977) and Wilks and Wilks (1987). The latter authors inspected 27 diamonds selected only for their brown colouration and observed clear signs of plastic deformation in all of them.

The implication of the above observations is that the optical centres responsible for the brown colouration are lattice defects associated with dislocations. Other experiments described in Section 6.2.b show that dislocations luminesce when stimulated by an electron beam. It therefore appears probable that the optical centres responsible for the luminescence also produce the absorption giving the brown colouration. It has been suggested in the past that the brown colouration arises not from a true absorption but from the scattering of the light by microscopic particles, perhaps of graphite. However, measurements using a photometer and an integrating sphere show that there is an actual loss of luminous energy indicating that the process is essentially one of absorption (Bastin, Mitchell and Whitehouse, 1959; Wilks and Wilks, 1987).

3.6 The colour of diamonds

The perception of colour depends on the relatively small spectral range of radiation detected by the eye, a range lying between 400 nm and 700 nm, as indicated in Figure 3.11. Even quite low values of the absorbance in this spectral range may produce an appreciable colouration. If a diamond is viewed in white light, a greater absorption at the short wave blue end of the spectrum will give rise to a yellow tinge, while a greater absorption at the long wave yellow end will produce a bluish

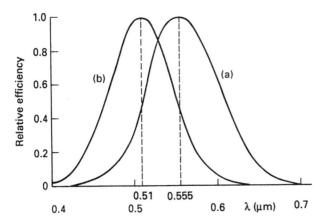

Figure 3.11 Relative luminous efficiency of the eye at different wavelengths for (a), high; (b), low levels of intensity (Longhurst, 1973)

tinge. Besides the colour centres described above which may be observed in natural or synthetic diamonds, there are also other centres which are usually only observed after irradiating the diamond with high energy electrons or other particles. These centres are described in Sections 3.7 and 3.8. However, in this section we are concerned only with the colour of natural and synthetic diamonds which have not been treated by irradiation.

The colour of a diamond is determined primarily by the form of its absorption spectra, but it is important to realize that several other factors affect the apparent colour, even in high quality and highly transparent stones. First, the apparent colour of a diamond depends on its size. Colour is produced by the preferential absorption of some components of the white light, so the longer the path of the light in the diamond, the more light is absorbed, and the more coloured the diamond. That is, even if two diamonds have identical absorption spectra the larger stone will appear more strongly coloured. The appearance of a stone also depends on the nature of the lighting, and for the commercial sorting of diamonds this is quite closely specified, see for example Bruton (1981) and Pagel-Theisen (1980). Even in a given light the actual colour seen by an observer depends on the colour senstivity of his eyes, and this sensitivity varies from person to person, sometimes considerably.

It is also important to note when discussing colour that many diamonds fluoresce. As we discuss in the next chapter, these diamonds absorb light at the ultraviolet end of the spectrum and then give out light of lower energy. This emitted light often has a bluish colouration and therefore changes the apparent colour of the diamond. For example, we have observed two diamonds with quite similar absorption spectra each with similar N3 components, and yet one appeared pale yellow as would be expected and the other almost colourless. An inspection under ultraviolet light showed that the latter diamond (and only this diamond) emitted a strong blue fluorescence, apparently sufficient to offset the effect of the N3 system. Therefore the type of illumination used to display diamonds may unduly affect their selection and marketing.

It must be stressed that in order to obtain a true appreciation of the colours of diamond it is necessary to view a large number of stones, to view different stones side by side, and if possible to have available a set of diamonds exhibiting a range of colours as standards for comparison. Sometimes the colours are quite pronounced, sometimes quite delicate. In high quality gemstones, which are almost colourless, very fine shades of colour are apparent to the expert, and greatly influence the value of the diamond. However, we are concerned here with the much greater range of colour exhibited by diamonds in general.

We now consider the colour of diamonds in which particular optical centres either predominate or can be readily detected, see Table 3.1. Diamonds containing only nitrogen in the A or B form show no absorption in the visible region and will be colourless. The N3 centre produces an absorption which extends sufficiently into the visible region to absorb some blue and produce a pale yellow colour. The single N centre produces strong absorption at the blue end of the spectrum and a pronounced yellow colouration. The nitrogen platelets produce no absorption in the visible region and no colour. Boron centres absorb at the red end of the spectrum and may produce a blue colouration. Dislocations appear to be responsible for an absorption increasing steadily towards the short wave end of the spectrum which gives rise to a brownish appearance. Of course the colour achieved depends greatly on the concentration of the centres. For example, Chrenko and Strong (1975) give examples of the colours produced by diamonds synthesized to contain only either single nitrogen centres or boron, see Table 3.2. In addition to the above centres there are a number of others which may on occasion be observed in some diamonds. For example an absorption band in the region of 2.2 eV, arising from an unidentified centre gives rise to a pink colouration (Raal, 1958). Finally we mention that a few natural diamonds show localized greenish colourations probably caused by some adjacent natural radioactivity in their former geological environment (Section 3.8). Also the occasional diamond is sometimes found with

Table 3.2 Colouration due to single nitrogen and boron atoms (Chrenko and Strong, 1975)

Centre	Concentration	Colour
Single nitrogen	5–10 ppm	Pale yellow
	50–100 ppm	Golden yellow
	150 ppm	Golden tinged green
	300–400 ppm	Green
Boron	1 ppm	Light blue
	10 ppm	Deep blue

a uniform bottle green colour not yet related to a particular centre (Collins, 1982a; Orlov, 1977).

Finally we mention that some diamonds are almost colourless but nevertheless have a somewhat grey appearance. An inspection of such diamonds generally reveals the presence of defects which reduce the so-called clarity of the stone, see for example Pagel-Theisen (1980). These defects may be inclusions and internal cracks on a relatively large scale which appear black when the stone is viewed in transmitted light. Sometimes the inclusions are on a finer scale which appear only as a faint cloud perhaps giving an appearance of milkiness to the diamond. In addition if the surface of a natural diamond is rough and irregular this roughness will greatly impede the transmission and reflection of light and the stone will appear greyish when compared with a similar diamond with polished faces. In fact the clarity of the interior of a diamond can only be assessed by polishing flat surfaces on opposite sides of the stone to act as windows.

3.7 Optical centres produced by irradiation

We now consider some other optical centres which may be created by irradiating diamond with high energy electrons typically of 1 MeV to 2 MeV, gamma rays, or neutrons. It is well known that all three types of irradiation will displace some atoms away from their lattice sites to interstitial positions between the sites, thus producing two separate types of defect, an empty lattice site or *vacancy* and an *interstial* atom elsewhere in the crystal. Neutrons will also produce more extensive damage, but if electrons or gamma rays are used the defects are primarily vacancies and interstitials. (We note in connection with the following spectra that while irradiations are often made at room temperature the resulting spectra are usually observed at liquid nitrogen temperature to bring out detail otherwise obscured by the thermal vibrations of the atoms.)

3.7.a The GR1 and ND1 centres

Figure 3.12 shows the absorption spectrum of a Type IIa diamond before and after irradiation with 1.0 MeV electrons (Clark, Ditchburn and Dyer. 1956a). The additional absorption following the irradiation has several components. The lines between 1.6 eV and 2.4 eV appear to be a typical vibronic band with a zero phonon line at 1.673 eV (Davies, 1977a), and is known as the *GR1 band* (GR for general radiation). Various peaks between 2.8 eV and 3.0 eV are identified as GR2 to GR8 and are generally seen to have the same relative strength with respect to the GR1 line (Walker, Vermeulen and Clark, 1974). In addition, at higher energies, there is a second vibronic system known as the ND1 system, with a zero phonon line at 3.150 eV (Dyer and Du Preez, 1965). These two systems are also produced by gamma and neutron irradiation, albeit with some differences in detail (Clark, Ditchburn and Dyer, 1956a). In particular, neutron irradiation produces more damage to the lattice and the vibronic systems are superposed on a background absorption which increases towards the ultraviolet (Collins, 1982a).

As already mentioned, one expects an irradiation to produce interstitial atoms and vacancies. Detailed optical studies of the GR1 line, particularly experiments to observe the effect of a uniaxial stress, show that the GR1 centre behaves in the same way as one would expect for a neutral vacancy, that is a vacant lattice site

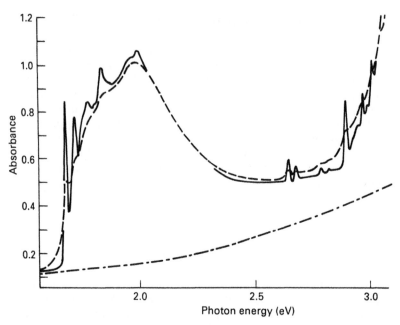

Figure 3.12 Dot-dash line shows the spectrum of the total absorbance of a Type IIa diamond at room temperature. The dashed line shows the room temperature spectrum after 1.0 Mev electron irradiation, and the full line this spectrum recorded at 80 K (Clark, Ditchburn and Dyer, 1956a)

which carries no electric charge (Clark and Walker, 1973). Annealing experiments by Chrenko and Strong (1975) also suggest that this vacancy is immobile up to 600°C. The position regarding the ND1 centre is less clear. It seems at present that the centre is probably a vacancy which carries a negative charge (Davies, 1977a; Lowther,1984).

3.7.b Interstitial atoms

The formation of vacancies and interstitials has been studied by electron irradiations of diamond at low temperatures of about 15 K or 20 K. By varying the energy of the bombarding electrons Bourgoin and Massarani (1976) estimated the threshold energy needed to produce damage as 35 ± 5 eV in agreement with theoretical estimates of the energy of formation of a vacancy–interstitial pair. In another set of experiments Lomer and Wild (1971) observed that an irradiation produced an electron spin resonance signal which they thought was probably associated with the interstitials.

Massarini and Bourgoin (1976) also observed various changes in the electrical resistance of Type II diamonds when specimens irradiated by electrons at about 15 K were warmed to room temperature, including a recovery stage at around 260 K. The authors deduced an activation energy for this stage of 1.3 eV and suggested that it was associated with the migration of interstitial atoms to nearby vacancies. On the other hand Flint and Lomer (1983) suggest that the interstitial may be mobile at 50 K. It must also be remembered that an interstitial may be involved in other annealing processes. For example, an annealing of the electron spin resonance signal at about 50 K (Lomer and Marriot, 1979; Flint and Lomer, 1983) may be due to the thermal release of carriers from interstitial atoms acting as donors (Bourgoin, Massarini and Visocekas, 1978). Finally, two other annealing stages in the region of 300°C and 500°C were observed by Lomer and Welbourn (1977) who suggested that they are associated with the migration of interstitials from or to other defects and impurities but further experiments are needed to make the position clear. A serious difficulty with such annealing experiments is that the results appear to depend considerably on the presence of any other defects in the diamond, and the number and nature of these defects are generally not well known (Clark, Mitchell and Parsons, 1979).

3.7.c Annealing effects

If an irradiated diamond is annealed at a temperature above about 600°C substantial changes are produced in the absorption spectra. Figure 3.13 shows that the absorption due to the GR and ND1 centres in a Type Ia diamond are much reduced while at the same time a marked absorption band appears in the region of 2.6 eV. More detailed measurements show that this band consists of two vibronic systems H3 and H4 with zero phonon lines at 2.463 eV and 2.498 eV respectively (Figure 3.14). Moreover, the ratio of the strengths of these zero phonon lines H3/H4 in a range of diamonds subjected to different amount of irradiation is proportional to the relative strengths of the A and B centres (Davies and

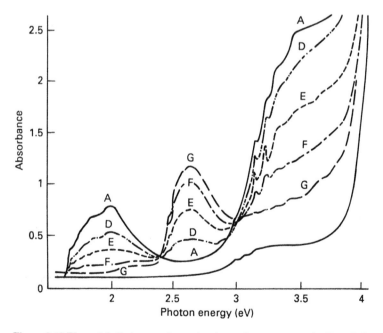

Figure 3.13 The unlabelled curve shows the absorption spectrum of a Type I diamond and curve A the spectrum after 1.0 Mev electron irradiation. Curves D,E,F and G show the spectrum after successive anneals at temperatures up to 930°C. All spectra recorded at 290 K (Clarke, Ditchburn and Dyer, 1956b)

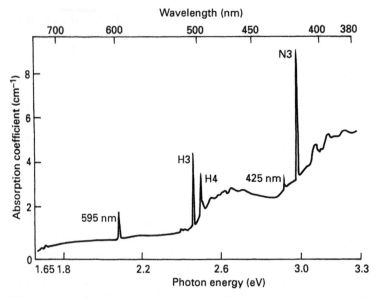

Figure 3.14 Absorption spectrum of an irradiated and annealed Type Ia diamond recorded at about 100 K (Collins, 1982b)

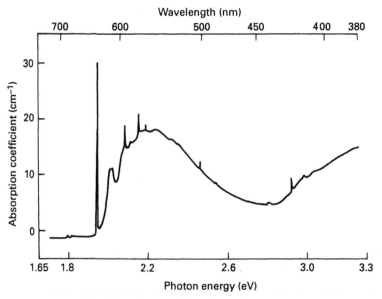

Figure 3.15 Absorption spectrum of an irradiated and annealed Type Ib diamond recorded at about 100 K (Collins, 1982b)

Summersgill, 1973). The implication is that the H3 and H4 centres are formed by the migration of vacancies to the A and B centres. Figure 3.14 also shows a line at 2.086 eV (595 nm) produced by the irradiation. This has been described by Davies and Summersgill (1973) and by Collins (1982b).

Figure 3.15 shows the spectrum of an irradiated and annealed diamond of Type Ib which contains only small concentrations of A or B nitrogen but an appreciable amount of single atom nitrogen. The treatment has again produced a typical vibronic absorption system but with the zero phonon line at an energy of 1.945 eV. (A small zero phonon H3 peak at 2.463 eV is presumably due to a small concentration of A centres.) Annealing studies show that there is a correlation between the growth of the 1.945 eV system and the decrease in the absorption at 1.673 eV and 3.150 eV associated with the vacancies. These results together with studies of the effect of uniaxial strain suggest that the 1.945 eV centre is a single nitrogen atom plus a vacancy (Davies and Hamer, 1976; Davies, 1977a, b). Table 3.3 gives a summary of the principal optical centres produced by irradiation, but it must be remembered that the spectra also show other less pronounced structure.

Table 3.3 Optical centres produced by irradiation
(V = vacancy)

	Zero-phonon line		Vibronic	Structure
GR1	1.673 eV	741 nm	$\checkmark$	V
ND1	3.150 eV	393 nm	$\checkmark$	V⁻
After subsequent annealing				
H3	2.463 eV	503 nm	$\checkmark$	A + V
H4	2.498 eV	496 nm	$\checkmark$	B + V
1.945	1.945 eV	637 nm		Single N + V
2.086	2.086 eV	595 nm		?

For example, Davies (1977b, 1981b) discusses this structure in terms of a range of defects with different charge and symmetry states. Collins, Davies and Woods (1986) describe uniaxial stress measurements on the so-called H1b and H1c systems originally reported by Clark, Ditchburn and Dyer (1956b), and deduce that the centres responsible are formed by the migration of 2.086 eV centres to A and B centres respectively.

Most of the experiments on irradiated diamond have been concerned with effects produced in the visible and ultraviolet spectra, but changes also occur in the infrared. Smith and Hardy (1960) observed an absorption band with peaks of 0.139 eV and 0.148 eV after neutron irradiation, the bands being particularly obvious in a Type II diamond which showed no absorption in this region prior to irradiation. More recently Woods and Collins (1982) and Woods (1984) have studied the infrared spectrum of the H1a and H1b systems. Collins, Stanley and Woods (1987) have observed the spectra of diamonds synthesized to contain equal amounts of normal nitrogen ^{14}N and of the heavier isotope ^{15}N. The effect of the ^{15}N was to introduce a new line at 1426 cm^{-1} of similar height and width to the usual H1a line at 1450 cm^{-1}, the implication being that the H1a centre contains nitrogen. As the presence of the second isotope produced only one more line at a different frequency it follows that each centre contains only one nitrogen atom.

3.8 Colour changes produced by irradiation

Several of the optical centres described in the previous section produce absorption in the visible region which gives colour to the diamond. The final colour produced by a treatment will depend, of course, on the initial colour and constitution of the diamond and on the type and energy of the irradiating particles. The intensity of the colour change increases with the dose or time of irradiation, indeed sufficient absorption will cause a diamond to appear black. Because of the number of variables involved any short summary of the colour changes must be treated with caution. However, the changes produced by irradiation without any subsequent annealing are relatively simple. The ND1 system lies entirely in the ultraviolet, so any colouration is produced by the low energy end of the GR1 spectrum which absorbs in the red. Hence the electron irradiation of a previously colourless stone produces a bluish tinge, while the irradiation of a pale yellow Ia diamond produces a blue-green colour. As already mentioned neutron irradiation also produces an increased background absorption particularly at shorter wavelengths and this results in a distinctive green colour.

More varied colours are observed when an irradiation is followed by heat treatments at temperatures in the range 600°C–800°C. For example the heat treatment of a Type Ia diamond after irradiation with either electrons or neutrons produces H3 and H4 centres which give rise to a yellow or amber colour. On the other hand, when a Type Ib diamond is treated, the principal new centre is the 1.945 eV system which gives rise to a different range of colours described variously as pink to mauve (Collins, 1982a) and brownish red to brownish purple (Lenzen, 1983).

The results of treating Type II diamonds are not so well documented. According to Bruton (1981) and Lenzen (1983) Type II diamonds generally turn brown, but of course many Type II stones show a brownish tinge even in their natural state. A complete account of the effects of irradiation and annealing should therefore specify the initial condition of the diamond. It has also been observed that when Type IIb diamonds are irradiated the amount of additional absorption may not be proportional to the flux of irradiation. It appears that in a IIb diamond the first centres created by the irradiation may lose electrons to the low lying boron acceptor centres. It is only after the boron centres are fully compensated that the irradiation begins to create new optically active centres (Dyer and Ferdinando, 1966).

We conclude this brief review of colour changes produced by irradiation by noting that changes may also be produced by natural processes. Some natural diamonds show localized greenish spots or a greenish coating on their surface, which are ascribed to an irradiation of the diamond by radioactive material adjacent to it in the depths of the earth (Custers, 1957; Vance, Harris and Milledge, 1973; Orlov, 1977; Hanley, Kiflawi and Lang, 1977).

The various colour changes produced by treatments of diamond present commercial possibilities. To appreciate these we recall that the characteristic appearance of a diamond cut as a gemstone is due to light which is internally reflected round the diamond and refracted at the entrance and exit surfaces. The reflections give the diamond its sparkle and the refraction produces its fire or colour range. The greatest brilliance is obtained in quite colourless diamond free of all visible absorption and these diamonds are generally the most valuable. On the other hand if there is sufficient absorption to produce a pale yellow or brown tinge the brilliance is reduced and the diamond is much less valuable. Therefore some process which would remove this colour would be of considerable interest, but no practical method is yet available. However, a small number of diamonds show pronounced and pleasing yellow or pink colours and these are highly valued. Therefore much work has been done to produce this type of colouration in less valuable off-white stones.

To produce colouration in a diamond the stone may be irradiated by electrons, gamma rays or neutrons. An electron beam is a useful and easily controllable source, but the irradiation will not be uniform because the beam loses strength as it passes through the diamond. However, by using 2 MeV electrons the penetration depth is as much as 2 mm, sufficient to give satisfactory results. Gamma radiation irradiates the diamond uniformly but only acts very slowly. Neutrons penetrate the whole specimen, act fairly rapidly, and are readily available in nuclear reactors. After the irradiations the diamonds are annealed at temperatures in the range 600°C to 800°C, sufficient to allow the vacancies to become mobile. (If the diamonds are heated to higher temperatures special precautions must be taken against damage by oxidation or graphitization.)

The diamonds most commonly selected for treatment are probably Type Ia stones in which N3 centres give a yellow tinge which reduces their value. An irradiation of these stones with either electrons or neutrons followed by a heat treatment produces H3 and H4 centres which give a more pleasing yellow or golden amber colour. The exact result obtained is of course influenced by the centres already present. For example, diamonds which do not luminesce are said to give a more attractive result than luminescing stones (Collins, 1982a). In fact a wide range of possibilities has given rise to many different attempts to produce attractive colours.

There is some general feeling against 'artificial' materials of all kinds. Therefore if two similar diamonds have an equal appearance and one is natural and the other coloured by treatment then the natural will command a greater price. Hence it is standard practice in the gem trade that any specification of a diamond must include a statement of any treatment the stone has undergone (Lenzen, 1983). Methods of identifying diamonds which have been treated are discussed by Collins (1982a) and Woods and Collins (1986). It appears that a diamond showing no absorption peaks at 595 nm, 1936 nm and 2024 nm has almost certainly not been treated.

3.9 Optical centres due to hydrogen and nickel

As discussed in the previous chapter there is considerable evidence that *hydrogen* is a common constituent of diamond with an atomic concentration which in some diamonds may be as much as 0.1%. Chrenko, McDonald and Darrow (1967) observed two lines in the infrared spectra in some so-called coated diamonds at 1430 cm^{-1} and 3107 cm^{-1} (Figure 3.16), and proposed that they arose from the vibration of hydrogen atoms bonded to carbon atoms. Subsequently Runciman and Carter (1971) observed that the ratio of the intensities of these two lines was the same in several diamonds. They also measured the concentration of hydrogen by vacuum fusion analysis, and estimated that the magnitude of the absorption is consistent with it arising from an interstitial hydrogen atom bonded to a carbon (or nitrogen) atom. A more complete discussion has been given by Woods and Collins (1983) who assign both lines to a carbon–hydrogen bond vibrating in a

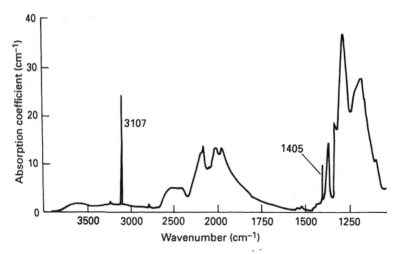

Figure 3.16 Infrared absorption spectrum of a diamond showing two lines at 3107 cm^{-1} and 1405 cm^{-1} associated with the presence of hydrogen (Davies, Collins and Spear, 1984)

stretching mode. Other lines in the infrared which may be associated with hydrogen have been described by Davies, Collins and Spear (1984).

Nickel also appears to give rise to optical centres. Synthetic diamonds grown using nickel as a catalyst-solvent are often greenish in colour particularly near inclusions of nickel (Wedlake, 1979). Absorption spectra of diamonds synthesized with nickel show sharp lines in the visible region at 1.883 eV and 2.51 eV, and a sharp line in the infrared at 1332 cm^{-1} quite distinct from the 1344 cm^{-1} line characteristic of Type Ib diamonds (Collins and Spear, 1982). Besides these sharp lines the absorption in the visible region increases steadily towards low energies, giving rise to the greenish colour mentioned above. The presence of optical centres associated with nickel is also implied by cathodoluminescent studies described in Section 4.4 and by an electron spin resonance observed by Loubser and Van Ryneveld (1966).

References

Austin, I.G. and Wolfe,R. (1956) *Proceedings of the Physical Society B*, **69**, 329–338

Bastin,J.A., Mitchell, E.W.J. and Whitehouse, J. (1959) *British Journal of Applied Physics*, **10**, 412–416

Bleaney,B and Owen,J. (1965) *Diamond Research 1965*, supplement to Industrial Review, pp.15–23

Bourgoin, J.C. and Massarani B. (1976) *Physical Review B*, **14**, 3690–3694

Bourgoin,J., Massarani,B. and Visocekas,R. (1978) *Physical Review B*, **18**, 786–793

Bruton, E. (1981) *Diamonds*, (2nd edn.) N.A.G. Press, London

Bursill,L.A. and Glaisher,R.W. (1985) *American Mineralogist*, **70**, 608–618

Chrenko,R.M. (1973) *Physical Review B*, **7**, 4560–4567

Chrenko, R.M., McDonald, R.S. and Darrow, K.A. (1967) *Nature*, **213**, 474–476

Chrenko, R.M.and Strong, H.M. (1975) *Physical Properties of Diamond*, Report No.75CRDO89. General Electric Company, Schenectady, New York

Chrenko, R.M., Strong, H.M. and Tuft, R.E. (1971) *Philosophical Magazine*, **23**, 313–318

Clark,C.D. and Davey, S.T. (1984) *Journal of Physics C*, **17**, 1127–1140; *Journal of Physics C*, **17**, L399–L403

Clark, C.D., Dean, P.J. and Harris, P.V. (1964) *Proceedings of the Royal Society*, **A277**, 312–329

Clark, C.D., Ditchburn, R.W. and Dyer, H.B. (1956a) *Proceedings of the Royal Society*, **A234**, 363–381

Clark, C.D., Ditchburn, R.W. and Dyer, H.B. (1956b) *Proceedings of the Royal Society*, **A237**, 75–89

Clark, C.D., Mitchell, E.W.J. and Parsons, B.J. (1979) In *The Properties of Diamond*, (ed J.E.Field), Academic Press, London, ch 2,p.38

Clark, C.D. and Walker, J. (1973) *Proceedings of the Royal Society*, **A334**, 241–257

Collins, A.T. (1982a) *Journal of Gemmology*, **18**, 37–75

Collins, A.T. (1982b) *Journal of Physics D*, **15**, 1431–1438

Collins, A.T., Davies, G. and Woods, G.S. (1986) *Journal of Physics C*, **19**, 3933–3944

Collins, A.T. and Lightowlers, E.C. (1979) In *The Properties of Diamond*, (ed J.E. Field), Academic Press, London, pp. 79–105

Collins,A.T., Lightowlers,E.C. and Williams,A.W.S. (1970) *Diamond Research 1970*, supplement to *Industrial Diamond Review*, pp.19–22

Collins, A.T. and Spear, P.M. (1982) *Journal of Physics D*, **15**, L183–L187

Collins, A.T., Stanley, M. and Woods, G.S. (1987) *Journal of Physics D*, **20**, 969–974

Collins, A.T. and Williams, A.W.S. (1971) *Journal of Physics C*, **4**, 1789–1800

Collins, A.T. and Woods G.S. (1982) *Philosophical Magazine B*, **46**, 77–83

Custers, J.F.H. (1957) *Gems and Gemmology*. Winter 1957–58, 111–114

Davies, G. (1970) *Journal of Physics C*, **3**, 2474–2486

Davies, G. (1976) *Journal of Physics C*, **9**, L537–L542

Davies, G. (1977a) In *Chemistry and Physics of Carbon, Vol. 13*, (ed P.L. Walker and P.A.Thrower), Marcel Dekker, New York, pp. 1–143

Davies, G. (1977b) *Diamond Research 1977*, supplement to *Industrial Diamond Review*, pp. 15–24

Davies, G. (1981a) *Nature*, **290**, 40–41

Danies, G. (1981b) *Reports on Progress in Physics*, **44**, 787–830

Davies, G., Collins,A.T. and Spear, P. (1984) *Solid State Communications*, **49**, 433–436

Davies, G. and Hamer, M.F. (1976) *Proceedings of the Royal Society*, **A348**, 285–298

Davies, G. and Nazaré, M.H. (1979) *Proceedings of the Royal Society*, **A365**, 75–94

Davies, G. and Summersgill, I. (1973) *Diamond Research 1973*, supplement *to Industrial Diamond Review*, pp. 6–15

Davies, G., Welbourn, C.M. and Loubser, J.H.N. (1978) *Diamond Research 1978*, supplement to *Industrial Diamond Review*, pp.23–30

Ditchburn, R.W. (1952) *Light*, Blackie, London, p.460

Dyer, H.B. and Du Preez, L. (1965) *Journal of Chemical Physics*, **42**, 1898–1906

Dyer, H.B. and Ferdinando, P. (1966) *British Journal of Applied Physics*, **17**, 419–420

Dyer,H.B., Raal,F.A., Du Preez,L. and Loubser,J.H.N. (1965) *Philosophical Magazine*, **11**, 763–774

Evans, T. (1973) *Diamond Research 1973*, supplement to *Industrial Diamond Review*, pp 2–5

Evans,T. and Rainey,P. (1975) *Proceedings of the Royal Society*, **A344**, 111–130

Fitchen, D.B. (1968) In *Physics of Colour Centres*, (ed W.B.Fowler), Academic Press, New York, pp.293–350

Flint,I.T.and Lomer,J.N. (1983) *Physica*, **116B**, 183–186

Hanley, P.L., Kiflawi, I. and Lang, A.R. (1977) *Philosophical Transactions of the Royal Society*, **284**, 329–368

Kaiser, W. and Bond, W.L. (1959) *Physical Review*, **115**, 857–863

Lax, M. and Burstein, E. (1955) *Physical Review*, **97**, 39–52

Lenzen, G. (1983) *Diamonds and Diamond Grading*. Butterworths, London

Lightowlers, E.C. and Collins, A.T. (1976) *Diamond Research 1976*, supplement to *Industrial Diamond Review*, pp.14–21

Lomer, J.N. and Welbourn, C.M. (1977) *Diamond Research 1977*, supplement to *Industrial Diamond Review*, pp. 5–10

Lomer, J.N. and Marriott,D. (1979) In *Proceedings of the International Conference on Defects and Radiation Effects in Semiconductors*, (Nice,1978) (ed J.H.Albany), Institute of Physics, Bristol, pp.341–346

Lomer, J.N. and Wild,A.M.A. (1971) *Philosophical Magazine*, **24**, 273–278

Longhurst, R. S. (1973) *Geometrical and Physical Optics*, (3rd edn.) Longman, London, pp.424–444

Loubser, J.H.N. and Du Preez, L. (1965) *British Journal of Applied Physics*, **16**, 457–462

Loubser, J.H.N. and Van Ryneveld, W.P. (1966) *Nature*, **211**, 517

Loubser, J.H.N. and Van Wyk, J.A. (1978) *Reports on Progress in Physics*, **41**, 1202–1248

Lowther, J.E. (1984) *Journal of Physics and Chemistry of Solids*, **45**, 127–131

Massarani, B. and Bourgoin, J.C. (1976) *Physical Review B*, **14**, 3682–3689

Orlov, Yu.L. (1977) *The Mineralogy of the Diamond*. John Wiley, New York

Pagel-Theisen, V. (1980) *Diamond Grading ABC*. Rubin and Son, Antwerp

Raal, F.A. (1958) *Proceedings of the Physical Society*, **71**, 846–847

Ramaswamy, C. (1930) *Nature*, **125**, 704

Reissland, J.A. (1973) *The Physics of Phonons*. Wiley, London

Rosenberg, H.M. (1988) *The Solid State*. (3rd edn). Clarendon Press, Oxford

Runciman, W.A. and Carter, T. (1971) *Solid State Communications*, **9**, 315–317

Sellschop, J.P.F., Renan, M.J., Keddy, R.J., Mingay, D.W. *et al.* (1977) *International Journal of Applied Radiation and Isotopes*, **28**, 277–279

Smith,S.D. and Hardy,J.R. (1960) *Philosophical Magazine*, **5**, 1311–1314

Smith, S.D. and Taylor, W. (1962) *Proceedings of the Physical Society*, **79**, 1142–1153

Smith, W.V., Sorokin, P.P., Gelles, I.L. and Lasher, G.J. (1959) *Physical Review*, **115**, 1546–1552

Sobolev, E.V. and Lisoivan, I. (1972) *Soviet Physics- Doklady*, **17**, 425–427

Sobolev, E.V., Lisoivan, V.I. and Lenskaya, S.V. (1968) *Soviet Physics- Doklady*, **12**, 665–668

Sobolev, E.V., Litvin, Yu, A., Samsonenko, N.D. *et al.* (1969) *Soviet Physics Solid State*, **10**, 1789–1790

Solin, S.A. and Ramdas, A.K. (1970) *Physical Review B*, **1**, 1687–1698

Vance, E.R., Harris, J.W. and Milledge, H.J. (1973) *Mineralogical Magazine*, **39**, 349–360

Walker, J. (1979) *Reports on Progress in Physics*, **42**, 1605–1659

Walker, J., Vermeulen,L.A. and Clark, C.D. (1974) *Proceedings of the Royal Society*, **A341**, 253–266

Warren,J.L., Yarnell,J.L.,Dolling,G. and Cowley,R.A. (1967) *Physical Review*, **158**, 805–808

Watkins, G.D. and Messmer, R.P. (1970) In *Proceedings of the 10th International Conference on the Physics of Semiconductors*, (Cambridge, Mass:1970), (eds S.P.Keller,J.C.Hensel and F.Stern), pp. 623–629. Division of Technical Information; United States Atomic Energy Commission

Wedepohl, P.T. (1957) *Proceedings of the Physical Society*, **70**, 177–185

Wedlake, R.J. (1979) In *The Properties of Diamond*, (ed. J.E.Field), Academic Press, London, pp. 501–535

Wilks, E.M. and Wilks, J. (1987) *Wear*, **118**, 161–184

Woods, G.S. (1984) *Philosophical Magazine B*, **50**, 673–688

Woods, G.S. (1986) *Proceedings of the Royal Society*, **A407**, 219–238

Woods, G.S. and Collins, A.T. (1982) *Journal of Physics C*, **15**, L949–L952

Woods, G.S. and Collins, A.T. (1983) *Journal of the Physics and Chemistry of Solids*, **44**, 471–475

Woods, G.S. and Collins, A.T. (1986) *Journal of Gemmology*, **20**, 75–82

Woods, G.S., Purser, G C., Mtimkulu, A.S.S. and Collins,A.T. (1990) *Journal of the Physics and Chemistry of Solids*, **51**, 1191–1197

Woods, G.S., Van Wyk, J.A. and Collins, A.T. (1990) *Philosophical Magazine*, **62**, 589–595

Chapter 4

Luminescence

Most diamonds exhibit luminescence when they are irradiated by ultraviolet light, X-rays, or beams of high energy electrons. That is, they absorb energy from the incoming beam, and then some of this energy is emitted as radiation of longer wavelength in either the infrared, visible, or ultraviolet parts of the spectrum. Luminescence is of practical importance in at least three ways. As mentioned in Section 3.6 it may affect the apparent colour of a diamond, it is used extensively in the mines to sort diamonds from the base rock material, and provides an important technique for studying the structure of diamond in the scanning electron microscope. There is now an extensive literature on luminescence in diamond including studies of the polarization of the radiation (Clark, Maycraft and Mitchell, 1962; Clark and Norris, 1970), of luminescence from defects produced by irradiation (Collins, 1981; O'Donnell and Davies, 1981), and of the effects of uniaxial stress (Mohammed, Davies and Collins, 1982). Here we confine our discussion to the main principles of the luminescence particularly those underlying its use in the scanning electron microscope (SEM).

4.1 Luminescence and the SEM

Luminescence in diamond is most easily observed by illuminating a group of diamonds in a darkened box or room by say the 365 nm wavelength radiation from an ultraviolet lamp. Under these circumstances many diamonds will emit a glow of visible light which is often quite bright, although the 365 nm radiation lies quite outside the visible region. The colour of this luminescent light is most generally blue, but yellow luminescence is also common, sometimes tinged with green. (Details of ultraviolet viewing boxes which screen the observer from the direct effect of the harmful ultraviolet radiation are given by Bruton (1981).)

The luminescence of diamond arises from various optical centres such as those described in the previous chapter. The incoming radiation excites electrons in the optical centres to higher energy states, and they then fall back to their initial state emitting light of characteristic wavelengths. Although the colour of the luminescing diamonds gives some indication on the nature of the centres which are emitting the light, much more specific information is obtained by taking emission spectra of the luminescence. The emitted light is passed through a spectrometer which analyses its various components and presents them as an emission spectrum, analogous to the absorption spectra of the last chapter. These measurements are generally made with the diamonds cooled to liquid nitrogen temperature (77 K) in order to obtain greater detail in the spectra. Most of the measurements to date have been principally concerned with fluorescence in the visible region but experiments have also been made to observe emission in the infrared and the ultraviolet regions.

It should be noted that observations of fluorescence can provide a particularly sensitive method of detecting small concentrations of optical centres, much more so than measurements of absorption because a small concentration of centres may make only a very small diminution in the intensity of light passing through a crystal. However, a fluorescent signal may be enhanced by using a more intense irradiation and a more sensitive detector, the limit of sensitivity at a particular wavelength being set only by the general background fluorescence at that part of the spectrum. Although some diamonds do not fluoresce when observed in a simple viewing box, it is probable that most will show some fluorescence when observed with a sufficiently sensitive system (Sellschop, 1979).

Luminescence is often referred to as photoluminescence or cathodoluminescence according to the mode of excitation. Photoluminescence is generated by visible or ultraviolet light and by X-rays, cathodoluminescence by electron beams. These various modes of excitation all produce somewhat similar types of emission spectra, though with differences in detail arising in part from the different energies of the incident radiation. Thus the higher energy X-rays and electrons suffer multiple collisions in the diamond and therefore excite the electrons to a range of levels. Each method has its own characteristics and advantages. For example, excitation by photons in the visible or ultraviolet is simple and convenient to produce, and offers the possibility of discriminating between different optical centres. Thus if a diamond contains two different emission systems, a suitable choice of the wavelength of the exciting light may permit the excitation of one system but not the other, this being particularly useful if the spectra of the two systems overlap, see for example Collins (1974).

Photoluminescence excited by X-rays is not commonly used in laboratory experiments but has become important as a method of sorting diamonds from the ore in which they are found, see for example Suttill (1987). The crushed ore is allowed to fall through a beam of X-rays which causes fluorescence in the diamond but not in the rest of the ore. The fluorescence is picked up by a suitable detector and a jet of high pressure air separates the diamond from the other material. (The successful operation of this device depends on a powerful source of excitation which is readily obtained with X-rays. High energy electron beams can also be produced quite easily but the electrons only travel an appreciable distance in a high vacuum and are therefore unsuitable for this application.)

The electron beam technique, or cathodoluminescence, is particularly useful when used with the scanning electron microscope. We recall that in the normal mode of the microscope an electron beam is focused on the specimen and scanned across it in a television-style raster. At each point of the specimen the beam is scattered by the surface to an extent depending on the details of the surface, including its surface topography and its chemical composition. A detector collects some fraction of the scattered electrons and the resulting electric current is amplified and used to produce a television type display whose intensity is modulated according to the electron current. The system has a much greater depth of focus than either the optical or transmission electron microscope and is thus well suited for the examination of three dimensional structures.

When the SEM is used in the cathodoluminescence (CL) mode the electron collector is replaced by an optical system which focuses luminescence generated by the beam on to a photo-electric detector, usually sensitive mainly to the visible spectrum. The output of the detector is then used to modulate the intensity of an image in the SEM. The electron beam in the SEM has an energy of the order of 30 keV and penetrates into the diamond for a distance of the order of 5 μm (Hanley *et al.*, 1977; Davies, 1979). Hence the micrograph shows the pattern of the luminescence intensity from this surface layer, see for example Figure 5.11. This is obviously an important technique because differences in luminescence indicate differences in the concentrations of defects and impurities. The rest of this chapter therefore gives a general outline of how the luminescence in diamond is related to the various optical centres which may be present. For a more general account of SEM and CL techniques, see for example Goldstein and Yakowitz (1975). (We also note that Van Enckevort and Lochs (1988) describe a technique of using a scanned laser beam to reveal detail by photoluminescence.)

4.2 Form of the emission spectra

In discussing absorption spectra in the previous chapter we stressed the need to measure the absolute values of the coefficient of absorption because these can be directly related to the concentration of impurities in a diamond. It would be very useful to make absolute measurements of the luminous intensity of emission but this is more difficult because the intensity depends on the type and intensity of the excitation and on the sensitivity and spectral range of the detector. In fact, emission spectra are generally presented with the luminous intensities in arbitrary units, so for the most part we are able to discuss only the form of luminescence spectra and not their absolute magnitudes.

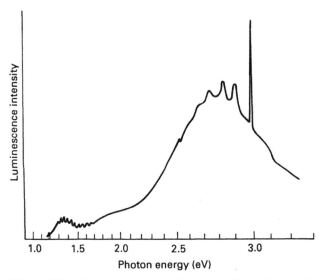

Figure 4.1 Luminescence spectrum of a natural diamond excited by an electron beam (Wight *et al.*, 1971)

Studies of the emission spectra of diamond have revealed much detailed structure. Comprehensive reviews of this work, not all of which is well understood, have been given by Davies (1977) and Walker (1979). For our present purposes, however, we need only appreciate some of the main features of the spectra. We begin by considering Figure 4.1 which shows a spectrum excited by an electron beam. This spectrum exhibits two main types of feature, a broad band of emission spreading over the whole visible region together with a prominent set of peaks which appear superposed on the band. Both types of feature are commonly observed in emission spectra and arise in different ways. The set of peaks is an example of a vibronic band related to a particular vibronic band in the absorption spectrum. The broad band emission is less well understood and does not arise from a single centre. Both systems are briefly described.

4.2.a Vibronic emission bands

A typical example of the vibronic type of emission is given in Figure 4.2(a) which shows a photoluminescence spectrum excited by the 365 nm line from a mercury lamp. Figure 4.2(b) shows a component in the absorption spectrum of the same diamond which we recognize as the N3 nitrogen system. We see also that the

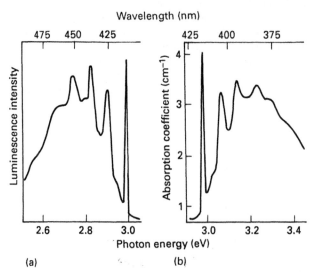

Figure 4.2 (a) Photoluminescence and (b) absorption spectra due to the N3 system (Dyer and Matthews, 1957)

sharpest peak in the luminescence spectrum occurs at virtually the same energy (2.985 eV) as the sharpest peak in the absorption spectrum, the zero-phonon line of the N3 system. There is an obvious similarity between the two systems. (As the scale of the emitted intensity is arbitrary, the similarity of the absolute magnitudes of the two curves has no significance.) Roughly speaking one set of bands appears to be the mirror image of the other, reflected about the position of the zero-phonon line. In fact, many vibronic systems observed in an absorption spectrum can be excited to give emission spectra of this general form. The most common vibronic system in natural diamond is that due to the N3 centre (Section 3.2.d) which absorbs at the extreme blue end of the spectrum and so gives the diamond a yellow colouration; most of the emission is between 400 nm and 475 nm, resulting in a blue colouration. Note that this blue colour should not be confused with that due to the broad band fluorescence described below (which is only observed with electron-beam and X-ray excitation).

The details of the mechanisms of emission are rather complicated but the general form of the vibronic spectra may be understood from quite simple considerations.

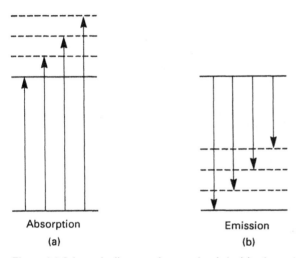

Absorption

(a)

Emission

(b)

Figure 4.3 Schematic diagram of energy levels in (a), absorption; (b), emission; see text

The full horizontal lines in Figure 4.3(a) show the two electronic levels associated with the zero-phonon transition of some optical centre in diamond. When light is absorbed in this transition an electron is transfered from the lower ground state to the upper state as described in Section 3.2.d. However, it is also possible that the absorbed light, besides exciting an electron, may produce lattice vibrations or phonons. This possibility is represented in the figure by drawing in the levels shown by dashed lines to indicate states of the diamond in which an additional phonon is excited besides the electron excitation. Hence a transition between the ground state and one of these upper levels will absorb radiation of greater energy than the zero-phonon line. There are many phonon states which might be excited with a spread of energies in a spectrum showing various maxima, and these are responsible for the subsidiary peaks beyond the zero-phonon line in Figure 4.2(b).

The mechanism of emission is somewhat different and is illustrated in Figure 4.3(b). Following absorption, the upper level is populated with electrons which may then relax to the ground state. Therefore the luminescence spectrum shows a strong emission with the same energy as the zero-phonon absorption line. There is also the further possibility that the electrons in the upper level of Figure 4.3(b) may give up their extra energy by both emitting radiation and creating lattice vibrations. In this case the possible states of the diamond after the emission of the radiation are shown schematically in Figure 4.3(b) by the dashed lines, each corresponding to the electronic ground state together with an excited phonon state. Representing allowed transitions by vertical arrows as in Figures 4.3(a) we see that the emission spectra will include a range of frequencies corresponding to energies less than that of the zero-phonon line. As the shape of the vibronic band is determined by the spectrum of the lattice vibrations this simple picture leads us to expect that the emission and absorption spectra will show the mirror image relationship to each other as seen in Figure 4.2. In fact systems such as the N3 system give good examples of this mirror effect, but with other systems, such as the H3, the emission spectrum may only approximate to a mirror image of the absorption spectrum (for details see Walker (1979)). Finally we note that the above discussion refers to the photo-excitation of luminescence but that similar considerations apply to spectra excited by cathodoluminescence as in the SEM.

4.2.b Broad band luminescence

Nearly all diamonds show some broad band cathodoluminescence in the visible region, which is known as band A emission. Although observed in most diamonds, the position of the band and the details of its shape depend on the constitution of the diamond and the form of the excitation. For example, Figure 4.4 based on summaries by Dean (1965) and Collins (1974) indicates the differences in emission

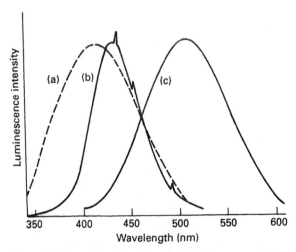

Figure 4.4 Examples of band A cathodoluminescence at 77 K: (a), Type Ia diamond; (b), Type IIb diamond; (c), synthetic Type Ib diamond (Collins, 1974)

spectra which may be observed from Type Ia, Type IIb and synthetic Type Ib diamonds. The typical band from a Type Ia diamond shown in the figure is responsible for the very common blue fluorescence; in the other two types of diamond the band is shifted towards lower energies. The luminescence in the Type IIb shows a less saturated blue colour with a suggestion of green, while the Type Ib shows a distinctly greenish colour. (Note that the figure shows only the shapes and positions of the spectra, the height of the maximum intensity being taken arbitrarily to be the same for each diamond.)

Band A luminescence is generally believed to arise in the same way as the rather similar bands produced in some semiconductors by a process known as donor–acceptor pair combination (Thomas, Gershenzon and Trumbore, 1964). We now give a simple summary of this mechanism, but it must be remembered that the actual situation in diamonds containing a range of optical centres may be very complicated. Consider a semiconductor with a band gap of width E_G, and donor and acceptor centres with ionization energies E_D and E_A as in Figure 4.5.

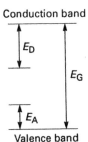

Conduction band

E_D

E_G

E_A

Valence band   **Figure 4.5** Energy levels of neutral donor and acceptor centres, see text

Assuming a more or less random distribution of these centres we expect that some donor centres will be near to acceptor centres, and it is these donor–acceptor pairs which are held responsible for the band A luminescence.

When the donor and acceptor centres in a pair are close together it is assumed that an electron from the donor centre will occupy the hole provided by the acceptor, so that both centres of the pair are ionized. If the crystal is now irradiated this will create pairs of electrons and holes, and some of these pairs will be captured by donor–acceptor pairs. In this capture the electron goes to the donor centre and the hole to the acceptor. Both centres are now electrically neutral with the ionization energies E_D and E_A shown in Figure 4.5. From this condition an electron from the donor state may recombine with the hole in the acceptor leaving the donor–acceptor pair back in its ionized state. The energy released in this transition is equal to:

$$E(r) = E_G - (E_A + E_D) + e^2/4\pi\epsilon r \tag{4.1}$$

where e is the electronic charge, ϵ the absolute permittivity and r the distance between donor and acceptor. This energy is available to appear as luminescence or lattice vibrations, and if the donor and acceptor are close together, its magnitude is determined to a significant extent by the last term in Equation (4.1) which arises from the electrostatic interaction. Hence as r can only take discrete values determined by the crystal lattice, we might expect to see a series of lines each arising from donor–acceptor pairs with a certain spacing, as is observed in gallium phosphide (Thomas, Gershenzon and Trumbore, 1964).

Although a discrete series of lines is not observed in diamond, Dean (1965) proposed that the band A luminescence also arises from donor–acceptor pairs but that the detailed line structure is masked by the presence of more than one type of donor–acceptor pair, as well as by other factors. Dean accounts for the different position of the band in different types of diamond (Figure 4.4) by postulating that different conditions of genesis and synthesis give rise to different separations between donors and acceptors, the mean separation being greatest in synthetic material. It appears that this treatment may account for the general features of the broad band emission but it is difficult to make detailed calculations because little or nothing is known of either the details of the donor–acceptor pairs, particularly the values of r, or of the influence of other defects in the crystal. For a further discussion of these points see Collins (1974), Collins and Lightowlers (1979) and Walker (1979).

One further point follows from the above discussion. If we assume that band A luminescence arises from donor–acceptor pairs, the fact that it is generally observed in diamond implies that donor–acceptor pairs are commonly present in diamond. Dean (1965) assumed that the donor and acceptor atoms were nitrogen and aluminium respectively but the main acceptor is now known to be boron (Section 3.4). This suggests that many diamonds may contain boron, although it is usually compensated (Section 3.4.b) and therefore does not affect either the optical absorption or the electrical conductivity.

4.3 Lifetimes and quenching

One might expect that the intensity of the vibronic emission bands associated with optical centres in diamonds would be proportional to the number of centres in each diamond. However, we now describe experiments which show that the position is more complex because the behaviour of a particular optical centre may be modified by the presence of other centres and defects in the diamond.

4.3.a Line widths and decay times

Although we are concerned primarily with luminescence it is informative to consider first an experiment on the *absorption* due to similar vibronic systems in different diamonds. Davies (1970) irradiated and annealed sixteen diamonds in a similar way in order to produce the same concentration of H3 centres in each stone. He then observed that the positions of the resulting zero-phonon lines in the absorption spectra varied from diamond to diamond, from about 503.2 nm to 503.4 nm, and that the width of the line increased over this range from about 0.2 to 0.8 nm (Figure 4.6). The same author also observed that the magnitude of these changes was greater in diamonds with higher values of the infrared absorption at the A nitrogen peak at $1282 \, \text{cm}^{-1}$. These results imply that the absorption due to the H3 centres may be modified by the presence of A type nitrogen.

The effect of A nitrogen was also shown in other experiments by Thomaz and Davies (1978) on the emission of light by excited optical centres. After a centre has been excited it does not emit radiation instantaneously, but like a radioactive nucleus has a characteristic decay time. Hence, if a diamond is excited by a short pulse of radiation we expect the subsequent intensity of the emission to fall exponentially with a half life equal to the decay time. Figure 4.7 shows the decay times for the N3 system in seven different diamonds plotted against the absorption due to the presence of A nitrogen (Thomaz and Davies, 1978). We see that the decay time steadily decreases from about 40 ns to 20 ns with increasing A nitrogen absorption. The authors also stated that the decease in decay time did not correlate

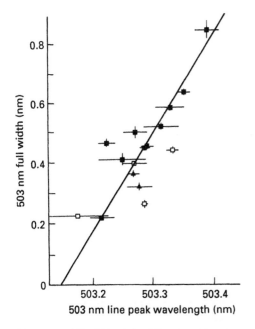

Figure 4.6 Full width of the 503 nm emission line in 16 diamonds plotted against the value of the peak wavelength (Davies, 1970)

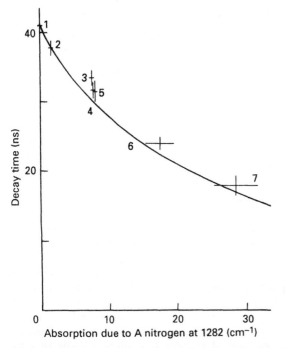

Figure 4.7 Luminescent decay time of the N3 system in 7 diamonds plotted against the absorption at 1282 cm^{-1} due to the presence of A nitrogen (Thomaz and Davies, 1978)

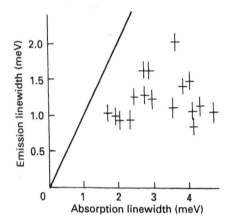

Figure 4.8 Full width of the H3 zero-phonon line measured in absorption compared with the width measured from photo-luminescence spectra for 17 diamonds. If the widths were equal, the points would lie on the straight line (Davies and Crossfield, 1973)

with the concentration of the N3 centres, or with the concentrations of B nitrogen centres, platelets, or naturally occurring H3 centres.

In another experiment on both emission and absorption spectra Davies and Crossfield (1973) studied some seventeen natural diamonds, all of similar size, which had been treated to produce about the same concentration of H3 centres in each diamond. They observed both the emission and absorption spectra of these centres and found that the width of the zero-phonon peak in emission was consistently less than the width in absorption, by up to a factor of 4 (Figure 4.8). (As stressed by the authors it is important in experiments of this type to use diamonds which are relatively homogeneous, or at least to observe the emission and absorption from the same volume of the specimen.)

4.3.b Quenching

Before discussing the above results we describe some other experiments described by Davies and Crossfield (1973). These authors observed that when the set of seventeen diamonds mentioned above with similar H3 systems was excited by 365 nm ultraviolet light the intensity of the luminescence varied markedly from diamond to diamond. Also, the intensity appeared to correlate with the concentration of A nitrogen centres measured by the strength of the infrared absorption at 1282 cm^{-1}, the fluorescent intensity being less at larger concentrations of A centres. Rather similar observations have been made by Sobolev and Dubov (1975).

The above phenomena has been discussed by Davies and Crossfield (1973) and Crossfield *et al.* (1974). These authors assume that the A nitrogen centres produce elastic strain fields which interact with the H3 and other vibronic centres. This interaction, probably in the form of an electric dipole–quadrupole coupling, provides a mechanism whereby an excited electronic state can decay and get rid of its excess energy without the emission of radiation. Estimates show that the decay times for these transitions are much shorter than the decay times for radiative emission. Hence an excited centre may prefer to decay by the non-radiative process, with the result that the luminescent emission may be much reduced.

The treatment by Crossfield *et al.* also explains why the spectral lines are narrower when observed in emission rather than absorption as noted in the previous Section. To follow the argument we must first note that the width of a spectral line produced by a transition between an excited state and the ground state increases as the lifetime of the excited state decreases (see for example Eisberg, 1961). When radiation is absorbed, some of the H3 centres excited will be perturbed by the strain fields and have relatively short lives, so the absorption lines will be correspondingly broader. However, these perturbed states tend not to contribute to the luminesence as they probably return to the ground state by a non-radiative transition. Therefore the emission comes primarily from only weakly perturbed or unperturbed states so that the emisssion lines are narrower than those observed in absorption.

We have discussed the above experiments in some detail because it is necessary to appreciate that the presence of impurities such as A nitrogen reduces the intensity of luminescence, and that this may be a large effect. Hence, the observation of, say, an H3 emission spectrum certainly indicates the presence of H3 centres but does not give entirely reliable information about the concentration of these centres.

4.4 Infrared, thermo and tribo luminescence

The luminescent emission spectra of diamonds may extend into the near infrared region, but only a limited number of studies have so far been made. Figure 4.9 shows the spectrum of a Type II diamond described by Wight *et al.* (1971) including the band A luminescence described above and two other bands, the so called band B with a maximum at about 1.8 eV, and band C with a maximum at about 1.3 eV.

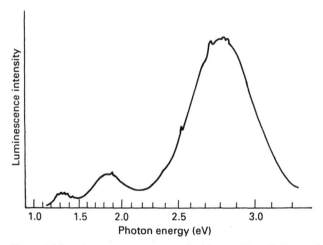

Figure 4.9 Luminescence spectrum excited from a Type II diamond showing emission in the near infrared (Wight *et al.*, 1971)

Figure 4.10 SEM micrograph showing growth layering and platelets on an {001} face revealed by luminescence in the near infrared. Field width 1.07 mm. (Lang and Makepeace, 1977).

The same authors also described an emission band in Type Ib diamonds with a maximum at 1.40 eV. This band is particularly strong in synthetic diamonds grown with a nickel catalyst (Collins and Spear, 1983), and is also observed in natural diamonds implanted with nickel (Vavilov *et al.*, 1982).

Another spectrum observed by Wight *et al.* shows a quite broad peak with a maximum at about 1.25 eV, and which correlates strongly with the strength of the 170 meV (1360 cm^{-1}) absorption peak due to the nitrogen platelets. These bands and the mechanisms which may be responsible for them are discussed by Wight *et al.* but further investigations are required to present a complete picture.

We also note that although cathodoluminescent techniques in the scanning electron microscope generally rely on luminescence in the visible or ultraviolet, the platelets described in Section 2.4.b as well as other features may be observed in the SEM by detecting their emission in the infrared (Figure 4.10). The relevant technique which depends primarily on the use of a sensitive infrared detector is described by Lang and Makepeace (1977).

We finally note that it has been known since the time of Boyle, *circa* 1670, that luminescence in diamond may also be stimulated by heating (thermoluminescence) or by rubbing (triboluminescence), see Mellor (1924). Some more recent experiments on thermoluminescence in diamond are described by Nahum and Halperin (1963). Observations that diamonds sometimes emit a blue or green glow during polishing presumably arise from either or both of these processes.

References

Bruton,E. (1981) *Diamonds*, (2nd edn). N.A.G.Press, London

Clark, C. D. and Norris, C. A. (1970) *Journal of Physics C*, **3**, 651–658

Clark, C. D., Maycraft,G.W. and Mitchell, E. W. J. (1962) *Journal of Applied Physics* supplement to Vol. **33** 378–382

Collins, A. T. (1974) *Industrial Diamond Review*, **34**, 131–137

Collins, A. T. (1981) *Journal of Physics C*, **14**, 289–294

Collins,A.T. and Lightowlers,E.C. (1979) In *The Properties of Diamond*, (ed. J.E.Field), Academic Press, London, pp. 79–105

Collins, A. T. and Spear, P. M. (1983) *Journal of Physics C*, **16**, 963–973

Crossfield, M. D., Davies, G., Collins, A. T. and Lightowlers, E.C. (1974) *Journal of Physics C*, **7**, 1909–1917

Davies, G. (1970) *Journal of Physics C*, **3**, 2474–2486

Davies, G. (1975) *Diamond Research 1975*, supplement to *Industrial Diamond Review*, pp. 13–17

Davies, G. (1979) In *The Properties of Diamond*, (ed. J.E. Field), Academic Press, London, pp.165–181

Davies, G. and Crossfield, M. (1973) *Journal of Physics C*, **6**, L104–L108

Dean, P. J. (1965) *Physical Review*, **139**, A588–A602

Dyer, H. B. and Matthews, I. G. (1957) *Proceedings of the Royal Society*, **A243**, 320–335

Eisberg,R.M. (1961) *Fundamentals of Modern Physics*, Wiley, New York

Goldstein,J,I. and Yakowitz,H. (1975) *Practical Scanning Electron Microscopy*. Plenum Press, New York

Hanley, P.L., Kiflawi, I. and Lang, A.R. (1977) *Philosophical Transactions of the Royal Society*, **284**, 329–368

Lang, A. R. and Makepeace, A. P. W. (1977) *Journal of Physics E*, **10**, 1292–1296

Mellor, J.W. (1924) *Inorganic and Theoretical Chemistry*, Volume V, Longmans Green, London, p.768

Mohammed, K., Davies, G. and Collins, A. T. (1982) *Journal of Physics C*, **15**, 2779–2788, 2789–2800

Nahum, J. and Halperin, A. (1963) *Journal of the Physics and Chemistry of Solids*, **24**, 823–834

O'Donnell, K. P. and Davies, G. (1981) *Journal of Luminescence*, **26**, 177–188

Sellschop,J.P.F. (1979) In *The Properties of Diamond*, (ed J.E.Field), Academic Press, London, pp. 107–163

Sobolev, E. V. and Dubov, Yu.I. (1975) *Soviet Physics Solid State*, **17**, 726–727

Suttill,K.R. (1987) *Engineering and Mining Journal*, **188**, December 56–57

Thomas, D. G., Gershenzon, M. and Trumbore, F. A. (1964) *Physical Review*, **133**, A269–A279

Thomaz, M. F. and Davies, G. (1978) *Proceedings of the Royal Society*, **A362**, 405–419

Van Enckevort,W.J.P. and Lochs,H.G.M. (1988) *Journal of Applied Physics*, **64**, 434–437

Vavilov, V. S., Gippius, A. A., Dravin, V. A. *et al.* (1982) *Soviet Physics Semiconductors*, **16**, 1288–1290

Walker, J. (1979) *Reports on Progress in Physics*, **42**, 1605–1659

Wight, D. R., Dean, P. J., Lightowlers, E. C. and Mobsby, C. D. (1971) *Journal of Luminescence*, **4**, 169–193

Chapter 5

The morphology of diamond

Diamonds, natural and synthetic, exhibit a wide variety of crystal forms and shapes. A useful and extensive description of the form of natural diamonds has been given by Orlov (1977). We now describe how, despite this variety, it is possible to classify most diamonds into one of a few morphological forms according to their principal features.

5.1 Crystal structures

Crystals often exhibit regular shapes bounded by flat surfaces meeting in sharp edges. These geometric shapes arise because the crystals have grown as a regular lattice of atoms. Therefore before considering the morphology of diamond we describe some general features of crystal lattices.

5.1.a Atomic planes and Miller indices

Because the atoms in an ideal crystal form a regular geometric array we can think of them as lying on sets of equi-spaced parallel planes. We illustrate this in two dimensions in Figure 5.1 which shows atoms in a two dimensional lattice lying on four different sets of parallel lines. The same principle in three dimensions is illustrated in Figure 5.2 where the lines of Figure 5.1 are replaced by planes. It is these sets of planes which give rise to the Bragg reflections when a beam of X-rays is diffracted by a crystal (Section 2.3.b).

In order to refer to a particular set of parallel planes we index them by the following standard method. The three axes in Figure 5.2 are labelled a, b, c as shown. To find the indices we move along the a axis and count the number of planes n_1 crossing the axis for each atom on the axis, then we move along the b and c axes and determine the corresponding indices n_2 and n_3. The three integers n_1 n_2 n_3 together with the axes a b c define a particular set of planes, and are known as the Miller indices of the set. For example in Figure 5.2, along the a axis $n_1 = 1$, along the b axis $n_2 = 3$, and along the c axis $n_3 = 2$, so the planes are described as (132).

In order to describe all sets of atomic planes of whatever inclination it is necessary to extend the axes in negative directions on the other side of the origin. Then if a plane intercepts a negative axis we indicate this by writing a bar over the corresponding Miller index. For example, if we begin with say (111) planes and

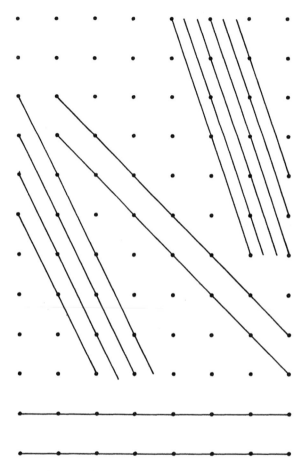

Figure 5.1 Two-dimensional lattice showing four sets of parallel lines

introduce all possible permutations of 1 and $\bar{1}$ we obtain eight different sets of indices which correspond to the eight faces of an octahedron centred on the axes. It is important to realize that the arrangement of the atoms on and around each of these eight sets of planes is similar, because diamond is a so-called cubic crystal. The axes used to specify the Miller indices are chosen so that rotation of the crystal through an angle of $\pi/2$ about any one of the three axes merely replaces one lattice point by another and does not affect the crystallographic orientation of the crystal lattice as seen by an external observer. Hence, all the eight sets of (111) planes are crystallographically similar and it is convenient to indicate any one of the eight sets by the notation {111}.

We also note that any two sets of planes in the same family, say (111) and (1$\bar{1}$1) in {111}, have the same spacings but the normals to the planes point in different directions. If we wish to refer to these directions they are completely specified by the Miller indices for the corresponding planes and are referred to as [111] and [1$\bar{1}$1], the square brackets indicating directions rather than planes. To indicate the

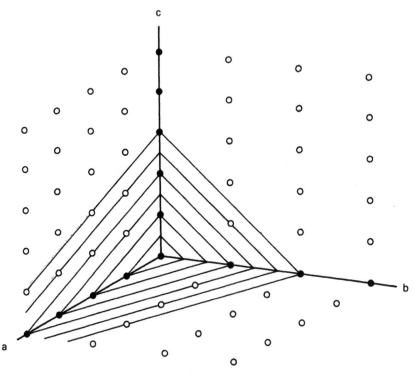

Figure 5.2 Three dimensional lattice showing a set of parallel planes with the same Miller indices defined by the axes a,b,c; see text (after Bunn, 1945)

family of directions associated with the family of planes {111} we use the standard notation <111>.

5.1.b Crystal growth

Many crystals present a precise geometric shape with smooth flat faces meeting in sharp edges. These faces give sharp reflections in an optical goniometer and the angles between the faces can be measured with considerable accuracy. Such measurements show that the angles between the faces of well formed crystals of the same species are always the same, perhaps to within minutes of arc. X-ray diffraction methods can identify the Miller indices of the different surfaces (see for example Barrett and Massalski, 1980) and it is found that the surface planes are generally characterised by low indices such as (001), (111), etc.

The shape of a particular crystal depends on how it has grown, and the factors involved are complex and not fully understood. However, it will be useful to give a simple outline of some of the principal concepts concerning the growth process. The first concept is that of layer growth. Figure 5.3 is a schematic diagram of a low index face of a crystal which is growing by deposition of material from a melt above the face. As shown in the figure the surface is plane except for an extra layer AB covering part of its area. The crystal grows as atoms from the melt attach

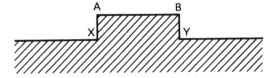

Figure 5.3 Schematic diagram of growth on a plane; see text

themselves to the solid surface, and atoms will be more strongly attracted to the surface at corner sites such as X and Y than to sites on the flat surface AB. Hence, there will be a tendency for growth to take place by the spreading of layers.

Once a layer is complete and extends to the boundary of the crystal, corner sites such as X and Y are lost. Therefore further growth can only occur if atoms from the melt nucleate a new layer on the plane surface, and this may be a very slow process in the absence of corner sites. Some further mechanism must be involved to account for the rates of growth which are actually observed. All crystals contain, to a greater or lesser extent, geometric imperfections of the crystal lattice known as dislocations. These are described in Section 6.2.a. The essential feature of the so-called screw dislocation is that its presence distorts the atomic lattice into the form of a spiral. Figure 6.9 is a schematic diagram of such a spiral emerging at a surface and shows advantageous sites for growth in the corner where the spiral cuts the surface. Therefore fairly rapid growth can proceed by a flow of atoms to the region marked A, and this growth results in the rotation of the step about the axis of the dislocation PQ. Two points about this process should be noted. First the presence of the step is maintained as new atoms are added. Second, if atoms are deposited at the same rate all along the edge A it follows that the edge will rotate faster nearer the axis of the dislocation. In this case the face grows upward more rapidly near the axis and may occasionally exhibit a characteristic spiral growth pattern such as that shown in Figure 6.11.

The growth of real crystals may be quite complex. For example growth layers may be stepped and not single, and the edges of the layers may be straight or kinked. Sometimes the steps between growth layers are large enough to be seen by the eye and these are almost certainly built up of much thinner layers which can only be detected in a high power microscope. In addition to growth on layers, and growth assisted by dislocations, there are also other possibilities. For example, if the surface is rough and irregular the surface atoms no longer form a regular array and the growth mechanisms are correspondingly modified, see for example Woodruff (1973).

Provided a crystal is not growing too rapidly, the melt and the crystal will be near to a state of approximate thermodynamic equilibrium. Therefore the preferred mode of growth, including the crystallographic orientation of the growth faces, is determined by being the one most thermodynamically favoured. It follows that the particular mode of growth and consequent form or habit of the crystal depend very much on the particular details of the system, including the chemical composition and impurity content of both the crystal and the melt, and the temperature. (Examples of these influences for a variety of crystals are given by Wanklyn (1975).) It also follows that changes in the environment during the growth of crystals may result in quite complex shapes and features.

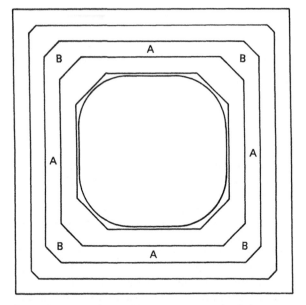

Figure 5.4 A round crystal in a supersaturated solution grows on faces with either A or B type orientation. The more rapidly growing B faces are eventually eliminated. (after Bunn, 1945)

One further aspect of the growth processes is illustrated in Figure 5.4 which shows successive stages in the growth of an initially rounded crystal placed in a supersaturated solution. Growth occurs on faces with both A and B type orientation, but the B faces grow outward faster than do the A faces. Hence the crystal assumes the shapes shown and the B faces eventually disappear. That is, the morphology of a crystal is determined by the growth planes which move outward most *slowly*. The effect of different rates of growth on crystals is further discussed by Frank (1958, 1972). For a more general review of the aspects of crystal growth mentioned above see Woodruff (1973), a useful list of references including references to diamond is given by Moore (1985).

5.1.c Possible crystal forms

As described in the previous section the growth of crystals tends to take place on low index planes such as {001}, {011}, {111} etc. The regular octahedron form of diamond is a classic example of such behaviour. As described in Section 5.1.a the indices {111} refer to a group of eight sets of planes, and by taking planes equidistant from the origin, one from each set, we define a regular octahedron as in Figure 5.5(a). Hence the common octahedral form of diamond appears to be the result of similar outward growth on all eight {111} planes.

Growth on {001} or {011} planes would give rise to rather different shapes. By considering the possible permutations of 0, 1, and $\bar{1}$ we see that the family {001} consists of six sets of mutually perpendicular planes. Hence uniform growth on each of these planes would result in a crystal with the form of a cube (Figure 5.5(b)). Similarly the {011} family consists of twelve sets of planes, and the

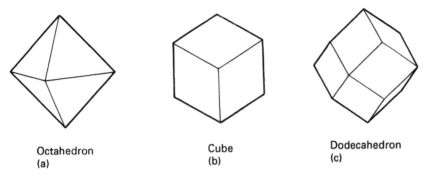

Figure 5.5 Octahedron, cube and dodecahedron crystal forms

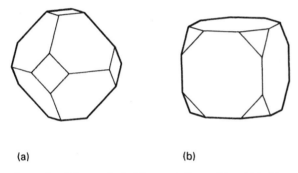

Figure 5.6 Cubo-octahedral forms of diamond in which (a), octahedron faces predominate; (b), cube faces predominate (Moore, 1985)

corresponding regular crystal has twelve sides and is known as a dodecahedron. The 12 faces are rhombic or lozenge shaped so this form is sometimes described as rhombic-dodecahedron. Note that in Figure 5.5 the crystals are drawn with faces of equal size, but this is not necessarily the case for real crystals. Variations of the rates of growth on the different faces may result in some faces being larger than others, while maintaining the same orientation.

Other crystal forms are possible in which growth has occurred on more than one family of planes. For example Figure 5.6 shows two examples of regular cubo-octahedrons in which growth has occurred on both octahedron and cube faces, in one the octahedron faces predominate and in the other the cube faces. In fact natural diamond generally grows predominantly in an octahedral habit, very occasionally in a form of cubic habit, and never in the dodecahedral form. On the other hand, synthetic diamonds are commonly either octahedrons or cubo-octahedrons occasionally with dodecahedron facets. Although virtually never observed in natural diamond, higher index faces such as (113) are occasionally observed in synthetic material.

The regular forms shown in Figures 5.5 and 5.6 are to be read only as an introduction to the morphology of real diamonds. Even on a diamond with well formed crystal faces, different faces may have developed to different extents, giving the stone a less regular shape. In addition the faces are often not plane surfaces but are modified, sometimes quite markedly, by layering arising from the growth processes and by pits and other features produced by etch processes in the earth during a previous stage of their history. We now describe these and other characteristic morphologies.

5.2 The octahedral form of diamond

5.2.a Octahedral diamond

The octahedral form of diamond is predominant in natural diamonds but only small numbers approximate to the ideal geometric form with a symmetrical shape with plane equal faces meeting in sharp edges. In general, we find many variations of the octahedron with some of the faces having developed more than others. For example, point apexes may be replaced by line edges as in Figure 5.7(a), and occasionally the octahedron assumes a flat tabular form as in Figure 5.7(b). Sometimes the octahedron faces are not planar but carry straight edged growth layers (Figure 5.8) which are of the type described in Section 5.1.b; see also Orlov (1977). In addition to these growth steps the octahedron faces show various etch features described in Section 5.3.

The angles between different faces of an ideal crystal may be calculated by relatively simple trigonometry, see for example Barrett and Massalski (1980) who

0.5mm   (a)

(b)

1.0mm

Figure 5.7 Examples of natural octahedron diamonds with (a), line apex; (b), tabular form

0.3 mm

Figure 5.8 Growth layering on the octahedron face of a natural diamond

also tabulate the angles between the various principal planes of a cubic crystal. In particular the angle between two adjacent {111} planes is given as 70° 32', the planes in ideal crystals being aligned very precisely. The faces of a diamond may also show such an exact alignment. Although the faces of many octahedral diamonds are far from perfect it is not too difficult by careful selection to find diamonds of sufficient quality to give sharp reflections of optical images, and which are free of internal distortions (Section 6.2). We then expect measurements of the angles between adjacent {111} faces to agree with the theoretical value to within about 1 minute of arc. This result is of practical importance, because if any surface is polished on the diamond, it is possible to determine its orientation with respect to the {111} planes of the crystal lattice to the same degree of precision using only the relatively simple methods of optical goniometry as described for example by Phillips (1956). (Also, precise measurements of the angles between facets can be used to identify the Miller indices of one facet if those of the others are known).

It is also possible to determine the crystallographic orientation of a face by X-ray diffraction techniques, for example by observing the reflected Laue pattern from a beam incident normally to the face, for details see Barrett and Massalski (1980). This is a useful technique but quite elaborate equipment is required to obtain measurements as precise as those given by optical goniometry. On the other hand, the X-ray method can be applied almost equally well to diamonds of any shape with or without {111} faces. (The sharpness of the reflections and therefore the precision of the measurements, using either optical goniometry or X-ray diffraction will of course depend on the quality of the diamond and the perfection of its crystal lattice.)

5.2.b The principal planes

When studying the properties of crystalline material it is often convenient to use surfaces and directions parallel to those in the crystallographic lattice. Therefore

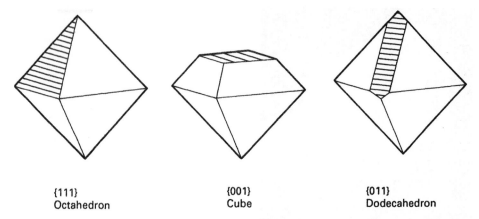

{111}   {001}   {011}
Octahedron   Cube   Dodecahedron

Figure 5.9 Schematic diagram showing the location of {111}, {001} and {011} faces on an octahedron diamond

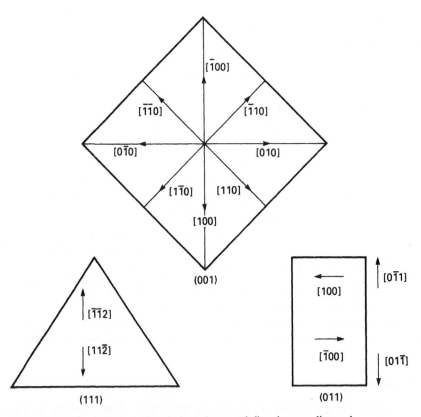

Figure 5.10 Miller indices of the principal planes and directions on diamond

most experiments on diamond surfaces have been made on surfaces approximating to {111}, {001} or {011}, and Figure 5.9 shows examples of the positions of the latter two faces polished on an octahedron. Besides being labelled {111}, {001}, or {011} the respective surfaces are often referred to as octahedron, cube or dodecahedron. In addition, the diamond trade sometimes describe these surfaces as 3 point, 4 point and 2 point respectively. This description is based on the fact that because of the symmetry of the diamond lattice a cube surface exhibits 4 directions at 90° to each other which are equivalent to each other. An octahedron surface has 3 equivalent directions at 120° to each other, while a dodecahedron surface has only two equivalent directions, at 180° to each other.

As a further point of nomenclature we note that we may describe, say, a cube surface as {001}, but to refer to a particular cube surface we use the description (001). The relative positions of all the other planes and directions are then readily specified by the appropriate Miller indices introduced in Section 5.1.a. (A helpful way of visualizing the various directions associated with Miller indices is given by the stereographic projection described for example by Barrett and Massalski (1980)). In Figure 5.10 we set out for reference the Miller indices of the principal directions on (001), (011) and (111) planes. To avoid confusion it is particularly important to note, that if we cut off the top of a regular octahedron and obtain a square cube surface as in Figure 5.9, the cube axes are *not* parallel to the sides of the square but at angles of 45° to them.

5.2.c Internal growth layering

A typical octahedron has grown outwards on {111} faces, and if the diamond were ideal its crystal structure would be quite uniform. In fact, various changes may have occurred in its environment during growth, such as variations of the temperature and of the chemical and impurity content of the parent melt. These changes can produce changes in the diamond which may be revealed by sectioning the diamond by polishing either a {001} or {011} face across the point or edge of the octahedron as in Figure 5.9. Figure 5.11 shows a CL micrograph of a polished

0.3 mm

Figure 5.11 SEM CL micrograph of a polished {001} face showing patterns due to internal growth layering.

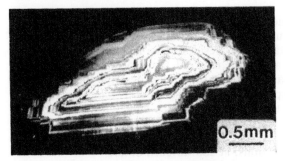

Figure 5.12 SEM CL micrograph of a polished {001} face showing complex growth layering

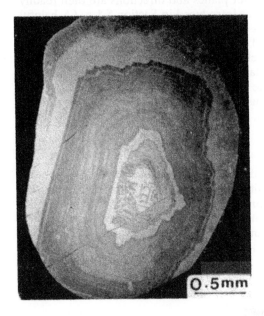

Figure 5.13 Optical micrograph of a polished {001} surface showing growth layering revealed by etching (Seal, 1963)

{001} face sectioning a natural octahedron. The intensity of the luminescence is fairly uniform over the surface except for the rectangular patterns reminiscent of picture frames. The edges of the frames are in fact traces of the {111} growth layers at the time the growth condition changed, the consequent difference in luminescence being due to a change in the impurity content. Figure 5.12 shows a more complex pattern from another diamond, the greater complexity indicating local disturbances in the melt and irregularities in the crystal. (Type II diamonds have a lower impurity content and do not show these growth bands. We also note that nearly all information on internal growth banding refers to Type Ia diamonds and that there is little information on the structure of natural Type Ib diamonds.)

The internal growth layers in a diamond may also be detected by other types of inspection. For example Figure 5.13 shows a {001} surface which has been polished and then etched in a bath of fused potassium nitrate (Seal, 1963). The rate and nature of the etch is much affected by impurities and imperfections so the

Figure 5.14 X-ray topograph of a diamond showing growth layering; the surface is close to (112). Specimen width 6 mm (Hanley, Kiflawi and Lang, 1977)

Figure 5.15 Optical micrograph of a diamond viewed through polished {011} windows between crossed polars showing growth layering revealed by birefringence

etch pattern reveals the various growth layers. The differences in impurity content will also produce small but significant changes in the density and lattice constant as mentioned in Section 3.2.a. Such changes will also produce a mismatch in the lattice at the boundary of two layers and this mismatch produces an elastic strain which may be detected by X-ray topography or birefringence techniques. For example, Figure 5.14 is a reflection topograph from an approximately (112) polished surface of a diamond which shows different growth regions bounded by the traces of the {111} growth planes.

Figure 5.15 shows an optical micrograph of a diamond between crossed polars viewed through two {011} polished windows. The different strains in the various {111} layers produce differing amounts of birefringence which are revealed by the polarizing microscope (Section 2.3.a). Four of the six sets of {111} planes cutting the polished surface do so perpendicularly and therefore the difference in birefringence shows up clearly as the growth planes are viewed edge-on. The other two sets of {111} planes make angles of only 35° with the surface and so produce much less effect in the microscope. Note also that a small difference in birefringence between layers is best seen when viewing through {011} rather than {001} windows because the relevant layers are then viewed completely edge-on.

The different growth features on a polished section of diamond have also been revealed by using a xerographic technique, that is by the electrostatic charging of the surface followed by dusting with suitable fine powders (DeVries and Tuft, 1979; Adam, Bielicki and Lang, 1981). It seems that the different growth regions attract different intensities of charge and therefore different amounts of powder. Although interesting, the method requires a careful technique and does not work equally well on all diamonds. In any case the details revealed are generally better shown by other techniques.

5.3 Effects of etch and dissolution

After the genesis of a diamond it remained deep in the earth's crust subject to chemical attack by its environment for a long period before being brought to the surface. As a consequence the faces of most natural diamonds show the effect of etching produced during this period.

5.3.a Etch features on natural diamond

Etching generally acts preferentially on corners and edges so very few diamonds are found with geometrically sharp edges, most being rounded to at least some extent. Quite often the etching process, beginning at the edges of a crystal, has proceeded far enough to produce an appreciable change in the overall shape of the diamond. In this case some authors prefer to describe the process as dissolution rather than etching.

The octahedron faces of natural diamonds show a range of etch features which are often not immediately obvious when viewed by eye, but can be seen clearly when the surface is viewed in an optical interference microscope using, say, the Nomarski technique. In this case surfaces inclined at different angles to the axis of the microscope appear with different intensities, so that the abrupt changes of slope at steps and edges are clearly delineated (Françon, 1961). Note also that before viewing any diamond surface it is essential to remove any film of dirt or grease using a suitable solvent if necessary.

Figure 5.16 shows the topographies of two {111} diamond surfaces viewed in the interference microscope; that in Figure 5.16(a) is an almost ideal surface apart from the broken corner, while that in Figure 5.16(b) is quite rough. The structure of the latter surface is obviously complicated but on other stones it is possible to resolve the effects of etch into various characteristic features. Figure 5.17 shows a

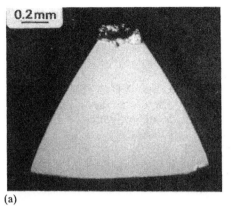

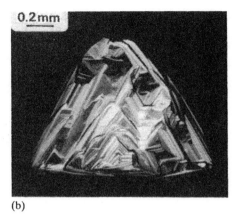

(a)   (b)

Figure 5.16 Optical interference micrographs with Nomarski technique of octahedron faces on natural diamond (a), smooth; (b), rough

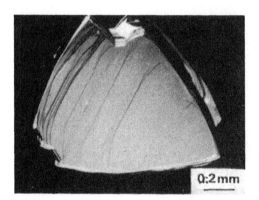

Figure 5.17 Typical octahedron face exhibiting growth layers and trigons (Nomarski technique)

diamond where straight-edged growth layers have been rounded at the ends by dissolution. Some diamonds show layers with edges completely rounded, these may be the result of the prolonged etching of growth layers, or perhaps as the result of more general dissolution as in examples given by Orlov (1977).

Etch processes also lead to effects on a smaller scale. such as the small triangular pits known as trigons visible on the octahedron face in Figure 5.17 and discussed in the following section. On other diamonds the etch patterns are on a very small scale and result in a roughening of the surface which causes the diamond to appear opaque, see Section 5.3.c.

5.3.b Trigons on natural diamond

Figure 5.18 shows a group of trigons on the octahedron face of a diamond. These trigons exhibit two main forms. The first type has the shape of an inverted shallow pyramid whose three sides are inclined at small angles (less than 2°) to the natural surface, and appear with different intensities in the interference micrograph. The

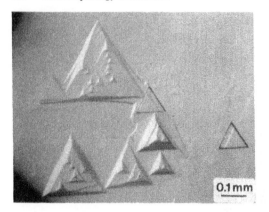

Figure 5.18 Group of trigons on a natural octahedron face (Nomarski technique)

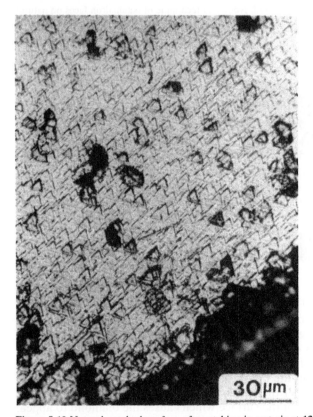

Figure 5.19 Natural octahedron face after etching in wet air at 1250°C (Evans and Sauter, 1961)

second type has a flat base which shows a uniform intensity, and steep {111} sides which appear as thick lines (Wilks, 1961).

Similar patterns can be produced on octahedral faces in the laboratory by etching with a range of etchants at various temperatures. For example Figure 5.19 shows an optical micrograph of the surface of a natural octahedron diamond after being etched by wet air at 1250°C (Evans and Sauter, 1961). The surface is covered by an array of trigons which give the diamond a matt appearance. Further examples of etch patterns on octahedron faces are given by Omar, Pandya and Tolansky (1954). Thus there is little doubt that the rather similar features seen on natural diamonds, see for example Orlov (1977) and Bruton (1981), have also been produced by etching.

The symmetry of the diamond lattice is such that a trigon may take up two possible orientations, either with its edges parallel to the edges of the octahedral faces as in Figure 5.20(a) or in the opposite orientation of Figure 5.20(b). These

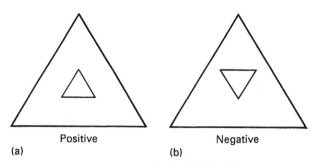

Figure 5.20 Sketch showing positive and negative orientations of trigons

two orientations are referred to as *positive* and *negative* respectively. Almost invariably the trigons on natural diamonds are of negative orientation. It so happened that the first etching experiments in the laboratory produced only trigons in the positive orientation, and it was therefore suggested that those occurring naturally were not due to etch (Tolansky, 1965). However, further experiments showed that the orientation could be altered both by using a different etchant (Frank and Puttick, 1958) and by altering the temperature of the etch (Evans and Sauter, 1961).

The factors responsible for the orientation of the trigons produced by particular conditions are discussed by Frank and Puttick (1958) and Evans and Sauter (1961) in terms of the arrangement of the carbon bonds at the surface of shallow pyramidal trigons (Frank, Puttick and Wilks, 1958). The latter authors suggested that monovalent atoms and radicals such as halogens or hydrogen might give negative orientations, but as shown by the experiments of Evans and Sauter (1961) a complete treatment must take the temperature of the etchant into account. A greater understanding of the details of these etch processes would be useful, and particularly relevant to various chemical reactions involved in the wear of diamond.

As on all crystal surfaces, etch on diamonds tends to begin at geometric imperfections. In particular a dislocation emerging at a surface offers a possible site for the attack of an etchant. Lang (1964) describes the faces of a natural octahedron diamond each of which exhibit a group of pyramidal trigons clustered mainly in the areas perpendicularly above a centre of strain birefringence in the middle of the crystal. An X-ray topograph of this diamond (Figure 6.21) shows bundles of lines spreading out from a central nucleus towards the faces of the octahedron. Lang demonstrates a correlation between the position of these dislocations and the trigons on the surface and concludes that pyramidal trigons indicate the presence of dislocations coming up to the surface.

Lang (1974a) also noted that flat-bottomed trigons were generally only observed if pyramidal trigons were also present. Even so no dislocations appeared to be directly associated with the flat bottomed trigons. Lang therefore suggested that these pits were initiated at the outcrop on the surface of dislocations issuing from points not far below the surface, and that the dislocations were eventually etched out thus leaving the trigons to spread out sideways. Subsequently Kanda *et al.* (1977) etched diamond with water at high pressure and temperature and produced pyramidal and flat-bottomed trigons similar to those on natural diamond, and in the course of the etching observed that some pyramidal trigons were replaced by somewhat larger flat-bottomed ones. A computer simulation of these etch processes has been described by Ponton, Reekie and Angus (1974). Finally we note that hexagon shaped pits have been observed on diamonds from the Argyle mine in Australia (Tombs and Sechos, 1986).

5.3.c Rounded dodecahedron diamond

Figure 5.21 shows a view of a common form of natural diamond which has a different morphology from those mentioned above. It exhibits twelve rather similar

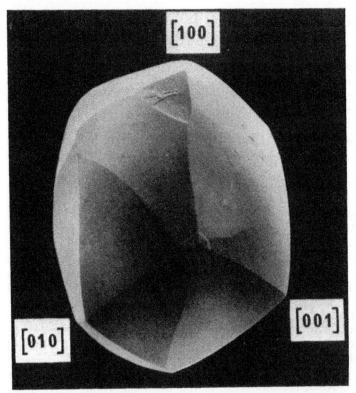

Figure 5.21 SEM micrograph of a rounded dodecahedron diamond. Specimen width 3 mm (Moore and Lang, 1974a)

surfaces which are curved and relatively smooth, and is therefore referred to as a rounded dodecahedron or sometimes just as a dodecahedron. Note, however, that this rounded dodecahedron is quite different from a regular dodecahedron with plane faces (Figure 5.5(c)), and that natural diamonds virtually never exhibit flat {011} faces.

It is now clear that rounded dodecahedra have been produced not by growth but by the dissolution of octahedral diamonds which had grown in the usual way. Thus, we find diamonds with a range of morphologies intermediate between those of the octahedron and the rounded dodecahedron, including rounded dodecahedrons with small plane octahedron faces. In addition X-ray topographs of rounded dodecahedra show the presence of growth layers aligned on {111} layers as in normal octahedra growth, see Figure 5.22 in which the orientation is such that the

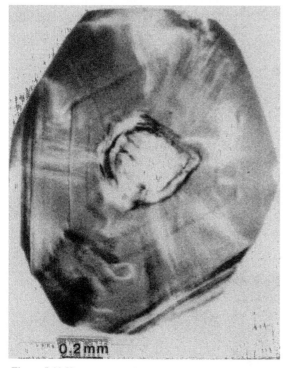

Figure 5.22 X-ray topograph of a rounded dodecahedron diamond showing traces of internal growth layers on {111} planes (Moore and Lang, 1974a)

traces of {111} planes appear as hexagons (Moore and Lang, 1974a). These authors also discuss the orientations of dislocations approximately normal to the growth layers and show that these too are consistent with the dissolution process.

The faces of dodecahedron diamonds exhibit a range of topographical features, described by Emara and Tolansky (1957), which result from the dissolution processes. These features generally produce a roughening of the surface which becomes dull and somewhat opaque, although the surfaces of some rounded dodecahedra are quite bright. In another study of 27 rounded dodecahedra the faces of some of the stones appeared brighter and more metallic than on others, and micrographs showed that the more matt appearances were due to a fine scale of surface structure (Wilks and Wilks, 1980). An examination using the light scattering technique described in Section 2.3.c showed that the stones which produced the least scattering had the brightest surfaces. Hence as the light scattering is an indication of defects or impurities it appears that these defects also gave rise to a greater number of etch sites on the surface.

5.4 Synthetic diamond

5.4.a Morphology and synthesis

Natural diamonds grow almost universally on octahedron planes so it might seem that this mode is specially favoured by the diamond lattice. In fact the octahedral mode must result from the particular conditions in the earth's crust during the genesis of the diamonds, because synthetic diamonds may be grown with a variety of morphologies. Synthetic diamonds may be produced by the high pressure process as nice crystals with the form of octahedra, or cubes, and in a variety of intermediate cubo-octahedron forms. It is also possible by careful choice of the synthesis conditions to produce crystals with true plane {011} surfaces (Yamaoko et al., 1977; Kanda et al., 1982). (Diamonds grown by vapour deposition also exhibit a variety of morphologies as in Figure 1.14.)

The cubo-octahedron form arises because growth may occur in layers on both {001} and {111} planes. The growth mode of a synthetic octahedron is similar to that of a natural octahedron, and an {001} section of such a diamond may exhibit the usual type of rectangular growth patterns shown in Figure 5.11 which are characteristic of growth on {111} planes. However, Figure 5.23 shows an optical micrograph of view through two opposite cube faces of a synthetic cubo-octahedron in which we see four sets of growth layers which make angles of 45° to the sides of

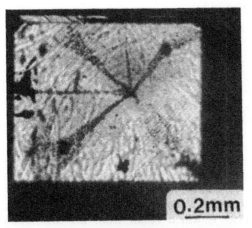

0.2mm

Figure 5.23 Optical micrograph of a synthetic cubo-octohedron diamond viewed between crossed polars through {001} faces to show traces of internal growth layers on {100} planes

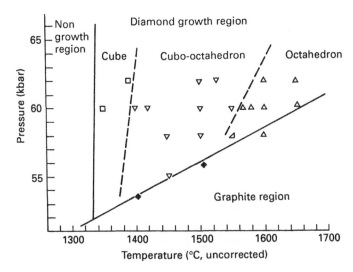

Figure 5.24 Pressure-temperature diagram showing regions producing characteristic growth forms. Habits: □, cube; ▽, cubo-octahedron; △, octohedron; ◆, graphite (after Yamaoka *et al.*, 1977)

the cube face; bearing in mind the note at the end of Section 5.2.b we see that growth has taken place on {001} faces. Further studies of these growth modes are described in the next section.

The morphology and growth modes of a synthetic diamond are determined by the particular conditions of synthesis. Yamaoko *et al.* (1977) and Muncke (1979) discuss the effect of pressure and temperature, and the results of Yamaoko *et al.* are summarized in Figure 5.24. This figure shows that large regions of the P-T diagram give rise to cube, cubo-octahedron, and octahedron forms respectively, the authors also state that dodecahedron forms may be formed near the diamond–graphite equilibrium line. The constitution of the melt and the presence of impurities also produce important effects on the morphology including the production of {011} and {113} faces (Strong and Chrenko, 1971). Kanda *et al.* (1989) have observed similar preferred habits, and modes of layer growth, in diamonds grown from different nickel and nickel based alloys. They explain the preference for a particular growth mode in terms of the chemical affinities of the atoms in the melt, and the possibility of their bonding to the structures of carbon atoms presented by the different crystallographic faces, following a treatment by Hartman and Perdok (1955).

Not all synthetic crystals are nicely formed with recognizable crystallographic faces. The wide range of possible conditions for synthesis leads to an equally wide range of morphologies, from grossly imperfect crystals, full of inclusions, with rough surfaces and no very obvious morphology to fine crystals of gem quality (Wedlake, 1979). Synthetic diamonds are particularly liable to contain inclusions of solvent metal, as for example in the diamond shown in Figure 5.23 where the prominent cross-like structure seen in plan view is formed of a large number of small metallic inclusions lying on directions spreading out from the centre of the stone; see also Section 2.4.a.

5.4.b Growth sectors

The presence of more than one growth mode in synthetic cubo-octahedra is readily seen in the SEM. Figure 5.25(a) shows a diamond with almost equally sized cube and octahedron faces viewed in the normal mode of the SEM, and Figure 5.25(b) shows the same stone viewed in the CL mode. We see that the two types of growth sectors give quite different luminous intensities. This difference is presumably due to the different arrangements of the carbon bonds in the growing $\{111\}$ and $\{001\}$ surfaces which result in different affinities for impurity atoms.

The differences in luminescence between growth sectors present a convenient method of studying the growth history of synthetic diamond. For example, Figure 5.26 shows CL micrographs of a polished section cutting a cubo-octahedron near

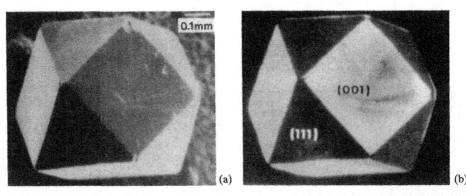

(a)   (b)

Figure 5.25 Cubo-octahedron diamond viewed in (a), normal; (b), CL mode of the SEM. The bright faces in (b) are $\{001\}$ and the dark faces are $\{111\}$ (Pipkin, 1980)

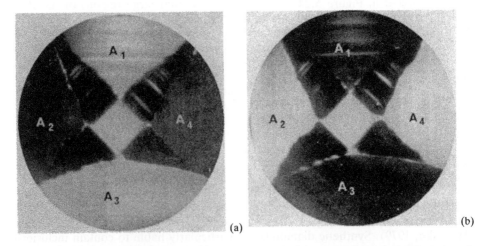

(a)   (b)

Figure 5.26 SEM CL micrograph of a polished section of a synthetic diamond showing different growth regions. (a), Polarizing filter is set to give maximum intensity in the areas A_1 and A_3; (b), polarizing filter is set to give maximum intensity in the areas A_2 and A_4. Field diameter 400 μm (Woods and Lang, 1975)

but not through its centre (Woods and Lang, 1975). For both pictures (a) and (b) the light from the crystal was passed through a polarizing filter before reaching the detector. For picture (a) the filter was set to give maximum intensity from the areas A_1 and A_3, the electric vector then being parallel to the white lines crossing these sectors, while for picture (b) the filter was set for an electric vector parallel to the line features in the areas A_2 and A_4. The four areas A_1 to A_4 were readily identified as cube zones by reference to the external morphology of the diamond. It follows that the light linear features indicate growth layering on {001} planes. This cubic growth is characterized by a bright yellow-green luminescence with the electric vector lying in the {001} layers. The central bright area in pictures (a) and (b) is presumably also a region of cubic growth, which because it is being viewed normally to the growth layers emits unpolarized light. The four smaller and darker regions between A_1 and A_4 are readily identified as octahedron zones from the external morphology, the white lines being traces of growth layering on {111} planes.

Quite commonly the growth structure of synthetic diamond is more complex. For example, Figure 5.27 shows a CL micrograph of another diamond sectioned

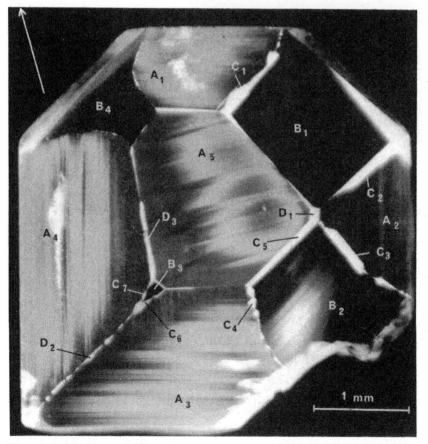

Figure 5.27 SEM CL micrograph of a polished section of a synthetic diamond showing complex patterns of growth (Woods and Lang, 1975)

by a polished surface about 10° away from the central {001} plane A_5. The various zones are now less symmetrically disposed, but we can identify four {001} zones A_1–A_4 as previously with four {111} zones B_1–B_4 between, the zone B_3 being much reduced in size. In addition there are two other sets of growth zones, C_1–C_7 which emit a whitish blue luminescence and D_1–D_3 which emit a steel-blue luminescence. Each of the four sets of zones ABCD is characterised by luminescence of a particular colour and by a relative intensity which varies to some extent with the intensity of the electron beam (Woods and Lang, 1975). All the zones exhibit growth layering under suitable excitation conditions. Therefore, as the A and B zones are identified as {001} and {111} from the external morphology, the C and D zones can be indexed from the orientation of their growth layers and are found to be {113} and {011} respectively.

A further example of a complex growth pattern is shown in Figure 5.28 which is a CL micrograph of one face of a rectangular block of diamond with

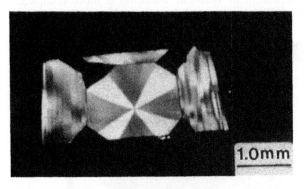

Figure 5.28 SEM CL micrograph of an unusual growth pattern in a synthetic diamond

approximately {001} faces. This pattern is discussed by Frank *et al.* (1990) who concluded that over the greater part of the face the growth was essentially on {001} planes, and accounted for the radiating pattern by postulating vicinal surfaces on the growth layer which make small angles with {001} and take up different concentrations of impurity content.

5.4.c Surface features

Because synthetic diamonds have not spent long periods deep in the earth we do not expect to see the surface features due to etch and dissolution which characterise natural stones. Even so, their faces may show very varied topographies depending on the conditions of synthesis which, for example, may produce either a single large crystal or a multitude of small ones. Therefore we make no effort to summarize all the possibilities and note only some of the more common topographical features.

Figure 5.29 shows four micrographs of faces on different cubo-octahedron diamonds, taken using the Nomarski interference technique to bring out details of the surface structure (Section 5.3.a). Viewed by eye or with a low power lens all

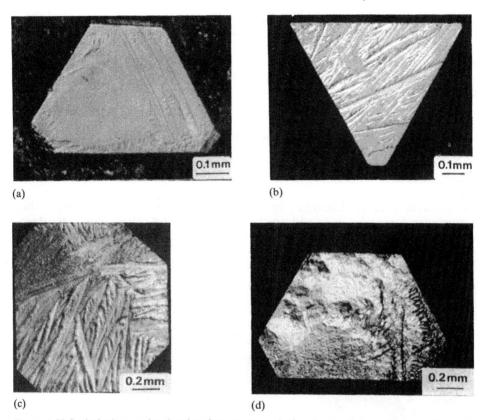

(a)   (b)

(c)   (d)

Figure 5.29 Optical micrographs of surface features on synthetic cubo-octahedron diamonds (Nomarski technique)

four surfaces appear smooth and flat, but the micrographs reveal considerable differences. A large part of the octahedral surface in Figure 5.29(a) is almost devoid of detail indicating that it is flat and smooth. On the right of the face the roughly parallel line features appear to mark the edge of growth planes and short line features may be seen near the edges. These latter features appear to be similar to those in the octahedral face of Figure 5.29(b) where they are much more prominent. Patterns of this type arise at the end of the synthesis when the temperature of the diamond and surrounding molten metal is reduced. As the metal cools it begins to condense and cover the diamond with dendritic like crystals, but at the same time carbon atoms in molten metal still in contact with the surface continue to deposit on the diamond, thus giving rise to the patterns observed (Chrenko and Strong, 1975). Figure 5.29(c) shows a cube face on which the above type of pattern appears as rows of closely spaced furrows and Figure 5.29(d) shows a more irregular type of surface quite commonly observed.

The smoothest areas on the faces of synthetic diamonds lie very close to crystallographic planes as is the case with good natural diamonds. However, the various surface features which are generally present result in an overall flatness which is usually less than that of the best natural surfaces. For example, a good natural surface will give an optical reflection in a goniometer with a spread of the order of only one minute of arc, whereas a surface similar to that of Figure 5.29(a) would show a spread of about 20 min of arc. Finally we note that other features besides dendritic details are often observed, as in Figure 5.29(c), no doubt arising from other variations in the synthesis process, other examples of surface features are given by Bovenkerk (1961) and Chrenko and Strong (1975).

5.5 Cuboid growth in natural diamond

We have described the main growth forms of natural and synthetic diamond, but growth has sometimes occurred in other modes and we now consider some of these variants.

A small number of diamonds, from a limited number of sources, have the approximate form of a cube and are known as diamond cubes. However, their surfaces are much less plane and smooth than the {001} surfaces on cubo-octahedron and cube diamonds grown synthetically. Figure 5.30 shows an X-ray topograph of a natural diamond cube and we see that the edges of the cube are

Figure 5.30 X-ray topograph of a natural diamond cube. Width of cube 1.8 mm. (Moore and Lang, 1972)

not straight but curve through angles of up to 5° or more away from the mean inclination. This departure from a planar geometry has not been caused by dissolution processes because the internal growth layers, which were protected from dissolution, show the same geometry with rounded corners (Moore and Lang, 1972). The growth morphology is obviously significantly different from that due to layer growth on crystallographic cube planes commonly seen in synthetic cubes and cubo-octahedra. Therefore the growth form seen in Figure 5.30 is described as cuboid rather than cubic. For a general discussion of the mechanisms underlying curved non-crystallographic forms of growth see Herring (1951) and Frank (1972).

Besides the cuboid type of growth seen in diamond cubes a rather similar form of growth may sometimes have occurred within diamonds of the normal octahedral shape. Figure 5.31 shows an optical micrograph of a polished {001} section of an

Figure 5.31 Optical micrograph of a polished {001} section of an octahedron diamond after etching (Harrison and Tolansky, 1964)

octahedron diamond after etching (Harrison and Tolansky, 1964) and Figure 5.32 a detail of the centre of a rather similar pattern observed by Seal (1965). In the latter diamond it is clear that growth initially spread out from a central nucleus in two modes (Frank, 1967). In the arms of the central cross, growth took place on {111} layers a few of which are visible in the figure, and between the arms of this cross the growth was cuboid with curved growth horizons. Eventually the {111}

Figure 5.32 Optical micrograph of a polished {001} section of diamond after etching. The parallel diagonal lines are polishing grooves. Width of specimen 0.9 mm. (Seal, 1965)

Figure 5.33 X-ray topograph of a central section of diamond showing cuboid growth sectors which are pinched out as the crystal grows. Height of specimen section 5 mm (Susuki and Lang, 1976a)

growth took over from the cuboid growth to give the normal octahedron form. These so-called centre-cross patterns may also be detected by X-ray topography as in Figure 5.33 which gives a section passing through the centre of a diamond. Again there are two modes of growth but here the leaf shaped {111} regions have been overgrown by the cuboid region to give an overall cuboid form (Suzuki and Lang, 1976a).

A particular feature of the cuboid growth regimes is a tendency to incorporate impurity particles, as seen in examples given by Shah and Lang (1963), Suzuki and Lang (1976b) and Van Enckevort and Seal (1987). Thus Shah and Lang describe a diamond containing precipitates of diameter 1 µm to 5 µm which give the centre cross part of the stone a cloudy appearance. For a discussion of further details of centre-cross patterns see Lang (1974a) and Suzuki and Lang (1976a, 1976b). Although very little statistical information is available, Lang (1979) suggests that perhaps one diamond in 1000 will show such a centre-cross pattern.

5.6 Miscellaneous growth modes

5.6.a Fibrous growth and coated diamond

Some diamonds appear to be completely opaque, and some dull or cloudy, because an outer layer of the diamond contains a high level of impurities. On polishing away this outer layer the crystal below is generally transparent and quite possibly of good quality. The outer layer is referred to as a coat and coated diamonds are found with a wide variety of morphologies, see Orlov (1977), also Custers (1950).

Figure 5.34 is an X-ray topograph of a slab cut from a diamond of approximately octahedral form, the plane of the slab is approximately {001}, and the cube axes

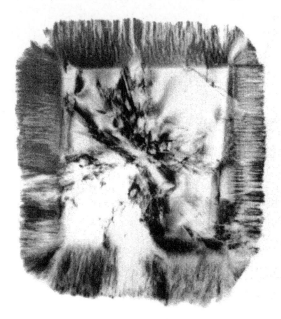

Figure 5.34 X-ray projection topograph of a plate cut from a coated diamond. Diameter of the section is 5 mm (Kamiya and Lang, 1965)

coincide with the diagonals of the square section. The structure of the outer coat is obviously quite different from that of the interior (Kamiya and Lang, 1965). The coat appears to have grown in a fibrous mode with columns of crystal growing outwards. Note in particular that the radial growth at the corners is beginning to cover the edges of the underlying octahedron and might eventually lead to a completely rounded form.

The opaqueness of the coat is believed to arise from the presence of fine particles of impurities generally of sub-micron size (Kamiya and Lang, 1965), but only a few stones have been examined to determine the constitution of the particles. Chemical analyses have been made by Seal (1966) and measurements of the infrared absorption by Angress and Smith (1965) and Chrenko, McDonald and Darrow, (1967). Seal and Chrenko, McDonald and Darrow suggest that the impurity consists of carbonates or other carbonaceous matter. Measurements have also been made of the electron spin resonance by Faulkner, Whippey and Newman (1965) and of the isotopic constitution of the carbon by Swart *et al.* (1983).

Kamiya and Lang (1965) comment on the abruptness of the transition to the fibrous mode of growth seen in Figure 5.34, and propose that the change was caused by fine particles coming into contact with the growing crystal. On the other hand some rounded dodecahedron diamonds with similar growth patterns, that is with normal octahedron growth inside and radial fibrous growth outside as in Figure 5.35, show no sign of any opaqueness in the outer layer, and in fact a beam

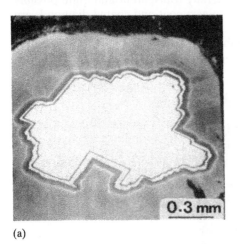
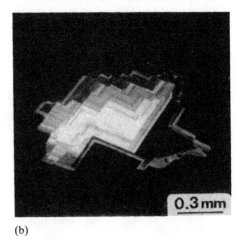

(a)   (b)

Figure 5.35 SEM CL micrographs of an apparently normal diamond with the exposure adjusted to reveal detail in (a), coat; (b), core (Wilks and Wilks, 1987)

of light was more scattered in the centre of the crystal than in the coat (Wilks and Wilks, 1987). It therefore appears that the change to fibrous growth can be caused by conditions other than just the presence of particles in the coat.

5.6.b Branching growth

Another growth mode which is certainly not at all common has been observed in a few natural diamond cubes. Figure 5.36 is a section topograph through the centre

Figure 5.36 X-ray section topograph cutting through centre of a natural cube diamond. Edge length about 5.5 mm. (Lang, 1974b)

of such a cube and shows a growth pattern quite different from any described above (Moore and Lang, 1972; Lang, 1974b). The main impression is of outward-going fibres and Lang (1979) suggests that the crystal has grown by the repeated branching of the fibres or columns in <111> directions. The topograph also shows faint traces with a roughly square outline with rounded corners and sides which are not quite linear. The obvious interpretation of these square patterns is that they are growth bands indicating the outline of the diamond at different stages of its growth. Another example of branching growth is given by Machado, Moore and Woods (1985) who describe a coated cubo-octahedron with additional facets which at first sight appear to be {011} but are in fact made up of small {111} facets produced by a form of branching growth.

5.7 Miscellaneous geometries

5.7.a Twins

Twins result from some variation of the growth process which produces a change in the crystallographic orientation of the growth planes. As a result of this change it is possible to identify a *twin plane* in the crystal such that the morphology on one side of the plane is the mirror image of the morphology on the other. Figure

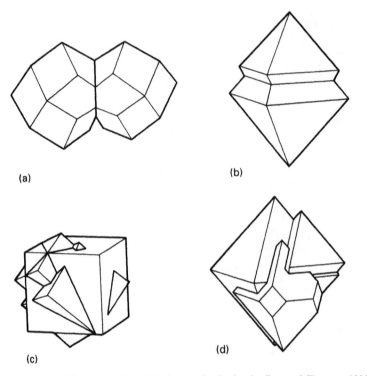

(a)

(b)

(c)

(d)

Figure 5.37 Examples of possible forms of twinning in diamond (Bruton, 1981)

5.37(a) shows a simple form of twinning about the central vertical plane and Figure 5.37(b) an example where the morphology changes at each of two horizontal twin planes. For a description of twinning in terms of the arrangement of the atoms in the diamond lattice, see Section 6.1.b.

Sometimes twinning occurs over only part of the growing surface thus producing quite complex crystals as for example in Figures 5.37(c) and (d). Photographs and drawings of a wide variety of morphologies due to twinning are given by Orlov (1977) and Bruton (1981). Generally speaking the more complex and irregular specimens are not very common except in diamonds from certain areas, particularly Australia. The two most usual forms are the macle and the naat. The macle is essentially a flattened octahedron with one opposite pair of {111} faces larger than the others, and with a twin plane parallel to the larger pair of faces as in Figure 5.38. A naat is a quite small part of a diamond which is twinned with respect to the rest of the lattice, it is generally not visible in a simple optical inspection of an unpolished diamond but can cause problems during polishing (Section 9.5.b).

Less usual forms include cyclic twinning which sometimes leads to crystals with an apparent five-fold symmetry. This form is very rare in natural diamond (Casanova, Simon and Turco, 1972) but is not uncommon in synthetic material (Wentorf, 1963; Pipkin and Davies, 1979). Finally we note that twinning may often produce re-entrant features in the surface of the diamond as in Figures 5.37 and 5.38. On the other hand re-entrant features are not always a sign of twinning. For

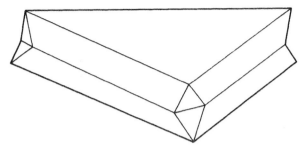

Figure 5.38 Macle shaped diamond

example, if the growth conditions encourage faster growth at the edges of a face than at the centre the face may become hollow or *hoppered*. It is even possible for the tip of a diamond to appear as an empty inverted pyramid. Examples of these so-called *negative* crystals are given by Bruton (1981).

5.7.b Irregular and Type II diamonds

Although many diamonds show a morphology which falls into one of the above classifications, others have less regular and well defined features and are classified as irregular. Such diamonds are very common. For example, Harris *et al.* (1975) described a diamond as irregular if less than 50% of its volume followed a recognizable crystal form, and then made an analysis of the output from four South African mines. They found that these stones amounted to between 8% and 30% of the total output, and for most mines the figure is probably nearer to the larger value.

Irregularity may be due to damage by fracture either in the earth at some time in the diamond's earlier history or during mining and extraction. The latter sort of damage can generally be recognized because of the characteristic cleavage surfaces produced by fracture (Section 7.4). As described in Section 5.3 dissolution processes may greatly amend the morphology, so variations in the magma surrounding the diamond will also lead to irregular forms. Another cause of irregularity is that the diamond may have grown in a region confined by the surrounding rock, with its shape determined by the contours of the rock (Bovenkerk, 1961).

Although irregular stones may make up about 30% of the output of a mine, they are usually in a minority. It is therefore surprising to find that nearly all Type II diamonds are irregular in shape and generally exhibit one or more fracture surfaces. However, the morphology of a crystal may be much influenced by the presence of impurity atoms (Section 5.4), so it appears likely that diamonds of good octahedral habit are only obtained if sufficient nitrogen is present. We note also that natural Type II stones generally exhibit fracture surfaces even though there is no evidence that Type II diamonds are more likely to fracture than Type I.

There are at least three other points of interest regarding Type II diamonds which are not well understood. Nearly all the Type II stones which have been examined contain an appreciable density of dislocations and have a brownish colouration probably associated with the dislocations (Section 3.5). This might suggest that all Type II diamonds have suffered appreciable plastic deformation in the earth perhaps because of their somewhat greater plasticity (Section 8.1). However, it must be remembered that nearly all Type II diamonds which have been studied would have been selected only from diamonds for industrial use, and therefore from stones which are predominately brown and dislocated. Hence it is quite possible that there are also Type II diamonds which are free of both colour and dislocations.

Tolansky (1973) inspected several thousand small diamonds of diameter 0.5 mm to 1 mm coming from three South African mines and characterised them as Type I or Type II according to their transparency to the 253 nm mercury line as judged by a simple photographic technique. He found that the proportions of Type II stones from two of the mines were 1.5% and 6.1% respectively but from the other mine (the Premier) the proportion was 70%, very much greater than that observed elsewhere and much greater than the proportion of Type II material in the larger Premier stones. These results have been confirmed and discussed by Moore and Lang (1974b) but are not understood. Finally we note that Milledge *et al.* (1983) have commented on the relatively high concentrations of the isotope ^{13}C in Type II diamonds.

5.8 Polycrystalline natural diamond

We now describe some relatively uncommon forms of diamond found only in a few locations and which differ from all the forms described above as they are not single crystals but an aggregate mass of fine crystals of the order of 10 μm diameter. The polycrystalline form is a very common state of matter, most metals consist of a great number of small crystals or *grains* which are joined to adjacent grains at the *grain boundaries*. Polycrystalline material is readily identified by X-ray diffraction methods. Unlike a single crystal, polycrystals give no Laue pattern (Section 2.2.b) but if placed in a beam of monochromatic X-rays give rise to a characteristic

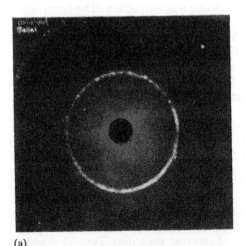

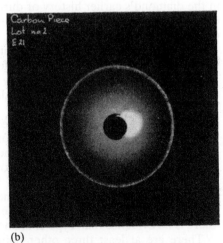

(a)   (b)

Figure 5.39 X-ray diffraction pattern obtained from (a), ballas; (b), carbonado diamond. (Reversal prints from X-ray negative film.) (Wilks, 1979)

diffraction ring or rings (see for example Barrett and Massalski, 1980). Figure 5.39 shows such diffraction rings from two specimens of polycrystalline material which can be identified as diamond by further X-ray studies to determine their lattice spacing and structure.

The state of aggregation of polycrystalline materials may vary considerably. For example in metals the whole surface of each grain is generally joined to the surface of adjacent grains leaving no empty voids. The density of the metal is then equal to that of the individual crystals. On the other hand, polycrystalline material may consist of grains much less well compacted, with voids, a lower density, and a much reduced strength. In fact natural polycrystalline diamond exists in a range of forms characterized by different degrees of compaction.

Polycrystalline diamond is commonly described as either ballas or carbonado but these terms are often used with very little precision, see for example Wilks (1979). Nevertheless, it is quite possible to distinguish two categories, ballas being a highly compacted material and carbonado appreciably less so, with other intermediate states between. Ballas usually has a round globular form up to perhaps 5 mm diameter, with a moderately smooth surface and appears translucent, with a light to dark grey colour. Carbonado looks more obviously like an aggregate of crystals, perhaps about 5 mm to 10 mm across, is usually opaque, and black or grey in colour. However, the essential differences and characteristics of ballas and carbonado are best brought out by a few simple physical measurements as described below.

Figure 5.40 shows the surface of facets polished on specimens of ballas and carbonado and viewed with the Nomarski technique to bring out the topography at the surface. The different orientations of the grains affect the rate at which they polish, so some grains are left standing higher than others. Note, in particular, the comparatively large size of the crystallites in the ballas, and that the grains are in contact with each other over their whole perimeter. The grains of the carbonado are appreciably smaller but the most notable features are holes or pores in the material which appear black in the micrograph and may be up to $200\,\mu m$ across. As this specimen was polished the holes disappeared as the level of the surface fell but at the same time new holes appeared at other sites. Thus this specimen contains voids on an extensive scale.

(a)   (b)

Figure 5.40 Optical micrographs of polished faces of (a), ballas; (b), carbonado diamond using the Nomarski technique (Wilks, 1979)

The presence of voids is also reflected in the density of the material. Measurements are generally made by weighing in a liquid or by a flotation method as described by Mykolajewycz, Kalnajs and Smakula (1964). Note the possibility that the liquid may penetrate the pores of the material, indeed bubbles of air can sometimes be seen leaving a specimen after immersion. Therefore to avoid incorrect values of the density it is advisable to coat the specimens with wax (Trueb and Butterman, 1969). Various results quoted by Trueb and Butterman, Bochko and Orlov (1970) and Wilks (1979) show that the density of ballas approaches to within 1% of that of the single crystal whereas that of carbonado may fall below the single crystal value by 5% or more. (Note that the interpretation of density measurements may be complicated by the presence of both inclusions and impurities particularly along the grain boundaries.)

The greater porosity of carbonado implies that the grains are not so strongly bonded together as in ballas. In fact if specimens of the two materials are fractured the form of the fracture is generally entirely transgranular in ballas (Trueb and Barrett, 1972) whereas in carbonado some intergranular fracture is also observed (Trueb and Butterman, 1969). This difference also becomes apparent in abrasion measurements. Using the rotating wheel abrasion tester described in Section 9.2.b the cuts on ballas appear quite similar to those on single crystal diamond, whereas on specimens of carbonado the edges of the cuts are more ragged, and the material shows a tendency to break up except at very low wheel speeds (Wilks, 1979).

Other differences are observed when thin plates of ballas and carbonado are studied in the transmission electron microscope (Moriyoshi et al. 1983). These authors observed systems of dislocations in both materials but in the carbonado the dislocations were randomly oriented whereas in the ballas they showed a polygonized structure of the type often produced by the annealing of dislocations. Hence the authors suggest that ballas may sometimes have formed from carbonado under conditions of high pressure and temperature. On the other hand, DeVries and Robertson (1985) suggest that a ballas type diamond has grown in this form whereas the carbonado has been formed by some sintering type process.

Further references to work on polycrystalline diamond are given in a review by Jeynes (1978). Detailed accounts of particular samples of ballas have been given by Trueb and Barrett (1972), of samples of carbonado by Trueb and Butterman (1969) and Trueb and De Wys (1971), and of an intermediate type of material known as framesite by DeVries (1973). Finally we note that the uneven intensity of the diffraction ring due to the ballas in Figure 5.39(a) suggests a preferred orientation of the grains (Trueb and Barrett, 1972). (On the other hand a ballas diamond described by Fischer (1961) does not show any orientation.)

References

Adam, R. C. G., Bielicki, T. A. and Lang, A. R. (1981) *Journal of Materials Science*, **16**, 2369–2380

Angress, J. F. and Smith, S. D. (1965) *Philosophical Magazine*, **12**, 415–417

Barrett, C. A. and Massalski, T. B. (1980) *Structure of Metals*. (3rd edn revised). Pergamon Press, Oxford

Bochko, V. A. and Orlov, Yu. L. (1970) *Soviet Physics- Doklady*, **15**, 204–207

Bovenkerk, H. P. (1961) In *Progress in Very High Pressure Research*. (eds F. P. Bundy, W. R. Hibbard Jr. and H. M.Strong), John Wiley, New York, pp. 58–69

Bruton, E. (1981) *Diamonds*, (2nd edn revised). N.A.G. Press, London

Bunn,C.W. (1945) *Chemical Crystallography*. Clarendon Press, Oxford

Casanova, R., Simon, B. and Turco, G. (1972) *America Mineralogist*, **57**, 1871–1873

Chrenko, R. M., McDonald, R. S. and Darrow, K.A. (1967) *Nature*, **213**, 474–476

Chrenko, R. M. and Strong, H. M. (1975) *Physical Properties of Diamond*, Report No.75CRD089 General Electric Company, Schenectady, New York)

Custers, J. F. H. (1950) *American Mineralogist*, **35**, 51–58

DeVries, R. C. (1973) *Materials Research Bulletin*, **8**, 733–742

DeVries,R.C. and Robertson,C. (1985) *Journal of Materials Science Letters*, **4**, 805–807

DeVries, R. C. and Tuft, R. E. (1979) *Journal of Materials Science*, **14**, 2650–2658

Emara, S. H. and Tolansky, S. (1957) *Proceedings of the Royal Society*, A239, 289–295

Evans, T. and Sauter, D. H. (1961) *Philosophical Magazine*, **6**, 429–440

Faulkner, E. A., Whippey, P. W. and Newman, R. C. (1965) *Philosophical Magazine*, **12**, 413–414

Fischer, R. B. (1961) *Nature*, **189**, 50

Françon, M. (1961) *Progress in Microscopy*. Pergamon Press, Oxford

Frank, F. C. (1958) In *Growth and Perfection of Crystals*, (eds R.H.Doremus, B.W.Roberts and D.Turnbull), John Wiley, New York, pp. 411–419

Frank, F. C. (1967) In *Science and Technology of Industrial Diamonds*, Volume One: Science, (ed. J. Burls), De Beers Industrial Diamond Division, London, pp. 119–135

Frank, F. C. (1972) *Zeitschrift fur Physikalische Chemie*, **77**, 84–92

Frank.F.C., Lang,A.R., Evans,D.J.F., *et al.* (1990) *Journal of Crystal Growth*, **100**, 354–376

Frank, F. C. and Puttick, K. E. (1958) *Philosophical Magazine*, **3**, 1273–1279

Frank, F. C., Puttick, K. E. and Wilks, E. M. (1958) *Philosophical Magazine*, **3**, 1262–1272

Hanley, P. L., Kiflawi, I. and Lang, A. R. (1977) *Philosophical Transactions of the Royal Society*, **284**, 329–368

Harris, J. W., Hawthorne, J. B., Oosterveld, M. M. and Wehmeyer, E. (1975) In *Physics and Chemistry of the Earth*, Vol. 9, (eds L. H. Ahrens, J. B. Dawson, A. R. Duncan and A. J. Erlank), Pergamon Press, London, Oxford, pp. 765–783

Harrison, E.R. and Tolansky,S. (1964) *Proceedings of the Royal Society*, A279, 490–496

Hartman,P. and Perdok,W.G. (1955) *Acta Crystallographica*, **8**, 49–52

Herring, C. (1951) *Physical Review*, **82**, 87–93

Jeynes, C. (1978) *Industrial Diamond Review*, **38**, 14–23

Kamiya, Y. and Lang, A. R. (1965) *Philosophical Magazine*, **11**, 347–356

Kanda,H., Ohsawa,T., Fukunaga,O. and Sunagawa,I. (1989) *Journal of Crystal Growth*, **94**, 115–124

Kanda, H., Setaka. N., Ohsawa, T. and Fukunaga, O. (1982) *Journal of Crystal Growth*, **60**, 441–444

Kanda, H., Yamaoko, S., Setaka, N. and Komatsu, H. (1977) *Journal of Crystal Growth*, **38**, 1–7

Lang, A. R. (1964) *Proceedings of the Royal Society*, A278, 234–242

Lang, A. R. (1974a) *Proceedings of the Royal Society*, A340, 233–248

Lang, A. R. (1974b) *Journal of Crystal Growth*, **23**, 151–153

Lang, A. R. (1979) In *The Properties of Diamond*, (ed. J. E.Field), Academic Press, London, pp. 425–469

Machado, W. G., Moore, M. and Woods, G. S. (1985) *Journal of Crystal Growth*, **71**, 718–727

Milledge, H. J., Mendelssohn, M. J., Seal, M. *et al.* (1983) *Nature*, **303**, 791–792

Moore, M. (1985) *Industrial Diamond Review*, **45**, 67–71

Moore, M. and Lang, A. R. (1972) *Philosophical Magazine*, **26**, 1313–1325

Moore, M. and Lang, A. R. (1974a) *Journal of Crystal Growth*, **26**, 133–139

Moore, M. and Lang, A. R. (1974b) *Diamond Research 1974*, supplement to *Industrial Diamond Review*, pp 16–25

Moriyoshi, Y., Kamo, M., Setaka, N. and Sato, Y. (1983) *Journal of Materials Science*, **18**, 217–224

Muncke, G. (1979) In *The Properties of Diamond*, (ed. J. E.Field), Academic Press, London, pp. 473–499

Mykolajewycz, R., Kalnajs, J. and Smakula, A. (1964) *Journal of Applied Physics*, **35**, 1773–1778

Omar, M., Pandya, N. S. and Tolansky, S. (1954) *Proceedings of the Royal Society*, A225, 33–40

Orlov, Yu. L. (1977) *The Mineralogy of the Diamond*. Wiley, New York

Phillips, F. C. (1956) *An Introduction to Crystallography*, (2nd edn). Longmans, London

Pipkin, N.J. (1980) *Industrial Diamond Review*, **40**, 58–62

Pipkin, N.J. (1980) *Industrial Diamond Review*, **40**, 58–62

Pipkin, N. J. and Davies, G. J. (1979) *Philosophical Magazine A*, **40**, 435–443

Ponton, J. W., Reekie, T. D. R. and Angus, J. C. (1974) *Diamond Research 1974*, supplement to *Industrial Diamond Review*, pp 33–38

Seal, M. (1963) In *Proceedings of the First International Congress of Diamonds in Industry*, (Paris 1962), (ed. P.Greene), De Beers Industrial Diamond Division, London, pp. 361–375

Seal, M. (1965) *American Mineralogist*, **50**, 105–123

Seal, M. (1966) *Philosophical Magazine*, **13**, 645–648

Shah, C. J. and Lang, A. R. (1963) *Mineralogical Magazine*, **33**, 594–599

Strong, H. M. and Chrenko, R. M. (1971) *Journal of Physical Chemistry*, **75**, 1838–1843

Suzuki, S. and Lang, A. R. (1976a) In *Diamond Research 1976*, supplement to *Industrial Diamond Review*, pp. 39–47

Suzuki, S. and Lang, A. R. (1976b) *Journal of Crystal Growth*, **34**, 29–37

Swart, P. K., Pillinger, C. T., Milledge, H. J. and Seal, M. (1983) *Nature*, **303**, 793–795

Tolansky, S. (1965) In *Physical Properties of Diamond*, (ed. R. Berman), Clarendon Press, Oxford, pp.135–173

Tolansky, S. (1973) In *Diamond Research 1973*, supplement to *Industrial Diamond Review*, pp. 28–31

Tombs, G.A. and Sechos, B. (1986) *Australian Gemmologist*, **16**, May, 41–44

Trueb, L. F. and Barrett, C. S. (1972) *American Mineralogist*, **57**, 1664–1680

Trueb, L. F. and Butterman, W. C. (1969) *American Mineralogist*, **54**, 412–425

Trueb, L. F. and De Wys, E. C. (1971) *American Mineralogist*, **56**, 1252–1268

Van Enckevort,W.J.P. and Seal,M. (1987) *Philosophical Magazine A*, **55**, 631–642

Wanklyn, B. (1975) In *Crystal Growth*, 1st edn (ed B. R. Pamplin), Pergamon Press, Oxford, pp. 217–286

Wedlake,R.J. (1979) In *The Properties of Diamond*, (ed. J.E.Field), Academic Press, London, pp.501–535

Wentorf, Jr. R.H. (1963) In *The Art and Science of Growing Crystals*, (ed. J.J.Gilman), John Wiley, New York, pp.176–193

Wilks, E. M. (1961) *Philosophical Magazine*, **6**, 1089–1092

Wilks, E. M. (1979) *Industrial Diamond Review*, **39**, 156–161

Wilks, E.M. and Wilks, J. (1980) *Industrial Diamond Review*, **40**, 8–13

Wilks, E. M. and Wilks, J. (1987) *Mineralogical Magazine*, **51**, 743–746

Woodruff, D. P. (1973) *The Solid-Liquid Interface*. Cambridge University Press, Cambridge

Woods, G. S. and Lang, A. R. (1975) *Journal of Crystal Growth*, **28**, 215–226

Yamaoka, S., Komatsu, H., Kanda, H. and Setaka, N. (1977) *Journal of Crystal Growth*, **37**, 349–352

Chapter 6

Geometric defects in the diamond lattice

The atomic lattice of a real diamond is seldom as regular as the ideal geometric lattice. Even if there are no impurity atoms, and only carbon atoms are present, the crystal may still be imperfect with some of the atoms displaced from the points of the geometric lattice. Two types of geometric defects, the vacancy or missing atom and the interstitial carbon atom, have already been mentioned in Section 3.7.a. On a completely different scale of sizes the lattice may be interrupted by cracks,either completely internal or running to the surface, some visible to the eye or with a ×10 lens. The regularity of the lattice may be interrupted by voids of linear dimensions ranging from nanometers to micrometers, and which may be filled with condensed gas (Section 2.2.b). In addition there are two more types of geometric defect, dislocation lines and twins, which are important for their effects on mechanical properties. Finally, we note that in any crystal the atomic vibrations displace the atoms from the geometric lattice points.

6.1 Stacking faults and twinning

6.1.a Stacking faults

We first consider the lattice imperfection known as a stacking fault. The schematic diagram in Figures 6.1 shows a perfect diamond lattice viewed normally to a set of {011} planes, each carbon atom being linked to its neighbours by four covalent bonds. We can regard the atoms as lying in horizontal layers of kinked chains labelled a,b,c according to their relative position with respect to the adjacent chains. In a perfect crystal the sequence of chains runs a b c a b c etc. right through the crystal. However, it is quite possible for the sequence to either include an extra row or to omit a row thus giving singularities of the form a b c a a b c or of the form a b c b c. This type of discontinuity in the sequence represents a defect known as a stacking fault.

Stacking faults are common defects in many materials and may be detected by preparing thin specimens and viewing in the transmission electron microscope. Each fault gives rise to a set of characteristic parallel fringes (Whelan and Hirsch, 1957), as in Figure 6.2 a micrograph of a synthetic diamond showing a fringe system on three parallel fault planes (Woods, 1971). (The fringes in the background arise independently of the stacking faults.) Some examples of stacking faults in natural diamonds are given by Lang (1974), and in synthetic diamond by Woods (1971);

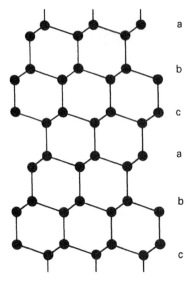

a

b

c

a

b

c   **Figure 6.1** Schematic view of a diamond lattice projected on a {011} plane, see text (after Hornstra, 1958)

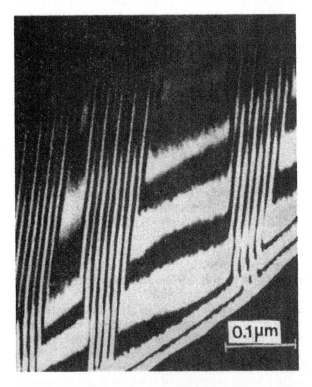

Figure 6.2 Three stacking faults (running vertically) seen in a dark field TEM micrograph (Woods, 1971)

see also a brief review by Lang (1979). Examples of high resolution X-ray topograph studies of single stacking faults have been made by Jiang and Lang (1983) and Kowalski *et al.* (1989) by working with diamonds very free of other imperfections.

6.1.b Twinning

In another geometric defect the sequence a b c a b c does not run through the whole crystal but at some point is interrupted to give a sequence of form:

a b c a b c a b | b a c b a c b a

where the right hand side is a mirror image of that on the left hand side. This effect is known as twinning and the central layer about which the two sides are reflected is known as the plane of twinning. The two regions to the left and right of the twin plane may extend to the boundaries of the crystal, but quite commonly a second twin plane results in a sequence of the type:

a b c a b c a b | b a c b a . . . c b a c b | b c a b c a b c

which corresponds to a perfect crystal modified by a thin slab of twinned material, sometimes called a *twin*

We see from the above that the junction between twinned and untwinned material constitutes a stacking fault. (For further details see Hornstra (1958).) Note that the twinned material itself is a perfect crystal with a correct sequence of layers. However, because this sequence of layers runs in the opposite direction to that in the untwinned region the twin has a different crystallographic orientation to that of the main crystal, as shown schematically in Figure 6.3. Thus even though a twin consists of perfect diamond its lattice is rotated with respect to the rest of the crystal and this results in various effects which indicate the presence of twinning.

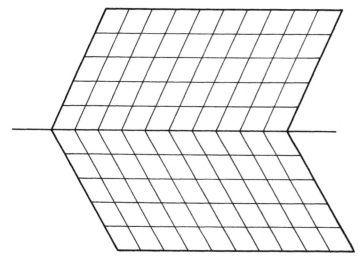

Figure 6.3 Schematic diagram to illustrate the twinning of a crystal lattice with a simple (non-diamond) structure. The centre line lies on the plane of twinning (after Cottrell, 1964)

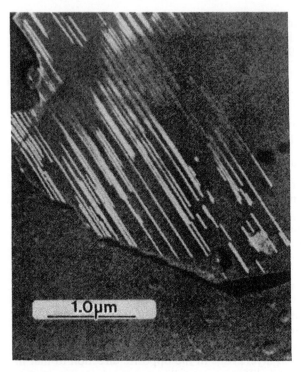

Figure 6.4 TEM micrograph showing rather dense set of microtwins (Woods,1971)

As already mentioned in Section 5.7.a one effect of twinning may be to bring about striking changes in the morphology of diamond. Perhaps the most common example is the twinned octahedron or the so-called macle shown in Figure 5.38. This macle is at first sight rather like a flattened octahedron with two opposite faces much larger than the others but it differs from the octahedral form in that the other faces although {111} make re-entrant angles with their neighbours. Apart from the gross twinning just described a diamond may contain quite small regions of twinned material, for example Figure 6.4 is an electron micrograph showing an array of twinned lamellae about 50 nm wide in a small synthetic diamond (Woods,1971). For a discussion of the mineralogy of twinning in a diamond see Slawson (1950).

6.2 Dislocations

A crystal lattice may deform either elastically or plastically. In elastic deformation the lattice remains essentially unchanged while the applied stress produces small changes in the interatomic spacings and bond lengths. Much larger changes may be produced by plastic deformations involving the motion of dislocations which we now describe.

6.2.a Geometry of a dislocation

During the plastic deformation of a single crystal one part of the crystal slides bodily over the other as in Figure 6.5, this process being known as *slip*. Because the crystal is not completely rigid slip will tend to initiate over some small region and then spread out over the whole slip plane. Figure 6.6 shows a view of a slip plane on which slip has occurred over the area A but not yet over the remaining area B. The two regions are separated by the boundary LMN. All the material inside and outside the boundary is good crystal with the ideal geometric lattice. However, it is obvious that along the line of the boundary the regularity of the lattice must be broken to accommodate the change from the slipped to the unslipped region. This line discontinuity is known as a *dislocation*.

The form of the discontinuity associated with a dislocation line is illustrated in Figure 6.7 which shows a simulation of close-packed atoms in a two dimensional solid. At first sight the array appears to be perfectly regular, but by viewing the figure along the line XY a discontinuity becomes apparent near the centre of the picture. In fact the lattice shown is not perfect because the upper half contains an additional row of atoms parallel to XY. To pursue this example further consider a simple three-dimensional lattice which we view normally from the side as in

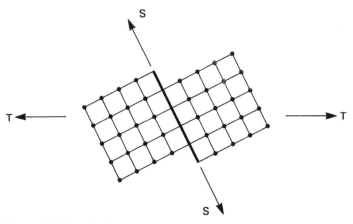

Figure 6.5 Schematic diagram to illustrate plastic deformation by one part of the lattice sliding over another. A tension T produces slip along the plane SS (after Cottrell, 1964)

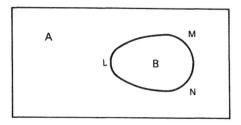

Figure 6.6 Schematic diagram to define a dislocation, see text

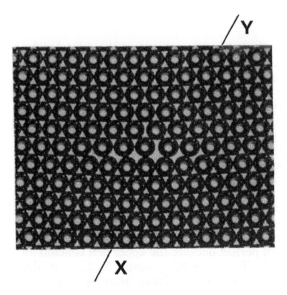

Figure 6.7 Dislocation in a two-dimensional bubble raft. The dislocation is best seen by viewing along the line XY (W.M.Lomer)

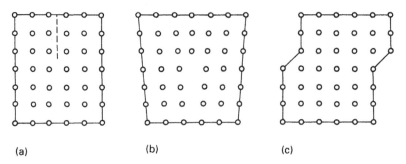

(a)   (b)   (c)

Figure 6.8 Schematic diagram of a lattice containing an edge dislocation. Moving the dislocation to the right produces a relative shear displacement of the two halves of the crystal

Figure 6.8(a). We make a plane cut along the dotted line and then insert an extra *half plane* of atoms as in Figure 6.8(b). Over most of the crystal all the atoms are in register but there is a discontinuity near the centre where the extra half plane ends abruptly. Suppose we now move the half plane to the right hand edge, we thus obtain a perfect crystal but with the upper half slipped with respect to the lower half (Figure 6.8(c)). That is, as the half plane moves across the crystal it produces slip. Thus the discontinuity around the end of the half plane is a *dislocation* as defined in Figure 6.6.

Dislocations may have a very considerable effect on the strength of crystals. If we shear a perfect crystal so that the two halves slide over each other as in Figure

6.5 then the applied stress must be sufficient to move every atom on one side of the slip plane away from the corresponding atom on the other side of the slip plane at the same time. However, we can see from Figure 6.8 that only a relatively few atoms need be rearranged at any one moment if we move a half plane step by step across the crystal. Hence, if we apply a shear stress to the crystal of Figure 6.8 the crystal may shear by the dislocation moving across at a much lower stress than would be the case in a perfect crystal.

Dislocations take a variety of forms. For example we may distort the crystal lattice by making a cut as before and then instead of inserting an extra half plane we displace one side of the cut relative to the other by one or more atomic spacings as in Figure 6.9. The effect of this displacement is to distort the lattice planes into

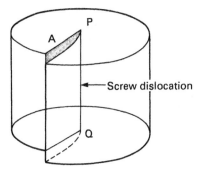

Figure 6.9 Schematic diagram of the distortion associated with a screw dislocation PQ

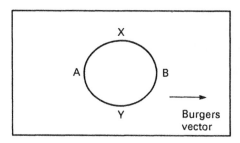

Figure 6.10 Schematic diagram of a loop of dislocation to indicate edge orientations (A,B) and screw orientations (X,Y), see text

the form of spiral ramps ascending round the line PQ. The resulting discontinuity in the crystal along the line PQ is a type of dislocation known as a *screw* dislocation in contrast to that of Figure 6.8 which is known as an *edge* dislocation.

For any dislocation the most important parameter is its Burgers vector which defines the relative displacement between undeformed crystal in the slipped and unslipped regions. This is the relative displacement produced as the dislocation, in say Figure 6.6, expands over the slip plane under the influence of an applied shear stress. Figure 6.10 is quite similar to Figure 6.6 but shows schematically the direction of the Burgers vector and also indicates four segments of the dislocation line which are either parallel or perpendicular to the Burgers vector. The form of the distortion around the dislocation line due to the transition between slipped and unslipped material is obviously different along the segments A and B than along the segments X and Y. In fact the discontinuities at A and B are edge dislocations and those at X and Y are screw dislocations, while the discontinuities at other parts on the line take up intermediate forms. Hence, edge and screw dislocations are not separate entities, and one should talk of a dislocation line as having an edge, a screw, or an intermediate *orientation*.

As we have just seen, a dislocation is the discontinuity accommodating the shift specified by the Burgers vector between slipped and unslipped material. In fact this accommodation can sometimes be made in two steps, first from the unslipped state to an intermediate condition and then to the slipped state and this two step geometry may have the lower energy and is then the preferred form. In this case the dislocation is described as two *partial dislocations* a small distance apart. For a comprehensive survey of the properties of dislocations see Hull and Bacon (1984).

Figure 6.11 An {001} face of synthetic diamond showing a growth spiral associated with a screw dislocation (Strong and Hanneman, 1967)

Finally we mention one further point concerning screw dislocations. Besides giving rise to the possibility of crystals deforming under much reduced shear stresses, they may also promote the growth of a crystal as described in Section 5.1.b. In general the growth of a crystal involves several or many dislocations, but occasionally good examples are seen of growth associated primarily with one dislocation. Thus Figure 6.11 shows an {001} face of a synthetic diamond with a good example of spiral growth as described in Section 5.1.b.

6.2.b Individual dislocations

The next two sections describe methods of viewing dislocations in diamond. We first consider diamonds with only a small number of dislocations where it is possible to identify each dislocation individually. Information is obtained using three

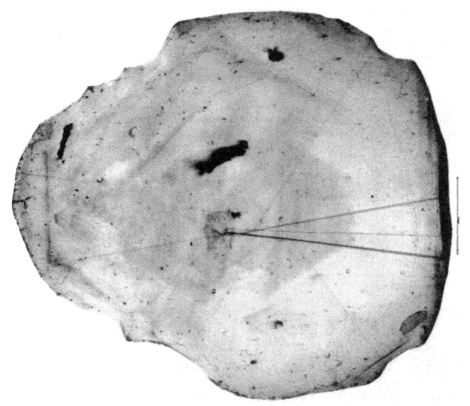

Figure 6.12 X-ray topograph showing single dislocations in a diamond with an unusually low density of dislocations. Image width 6 mm (Lang, 1974)

principal techniques, transmission electron microscopy (TEM), the cathodolumi-nesce mode (CL) of the scanning electron microscope (SEM), and X-ray topography.

Figure 6.12 shows an X-ray projection topograph of a diamond with a very low number of dislocations. The lines are visible as the result of the contrast produced by the diffraction of the X-rays from the relatively large volume of elastically strained material around the dislocation (see for example Tanner, 1976). The original plate shows some eighteen lines radiating from the central nucleus. For further details of this micrograph see Lang (1974).

The topograph technique used to obtain Figure 6.12 is quite time consuming, the crystal must be accurately positioned so that the beam is incident at the correct angle for the Bragg reflection, and the exposure may take several hours. It is therefore worth noting that topographs (Section 2.3.b) may now be taken with X-rays generated by synchrotron radiation (SR). This radiation is both very intense and well collimated to give an almost parallel beam with a diameter of about 20 mm. Hence the whole face of the crystal can be illuminated at the same time, and topographs obtained with very short exposure times. For example Figure 6.13 shows a projection topograph of a diamond very free of dislocations taken with an exposure of only 2 s (Ward, Wilks and Wilks, 1983).

1.0 mm

Figure 6.13 X-ray SR topograph of a diamond with high lattice perfection showing a few single dislocations (Ward, Wilks and Wilks, 1983)

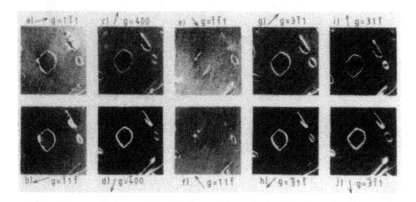

Figure 6.14 High resolution TEM micrographs of dislocation loops viewed in different orientations; see text. Diameter of central loop ~250 nm. (Hirsch, Pirouz and Barry, 1986)

A particular feature of the SR radiation is that it has a continuous range of wavelengths. Therefore to obtain a section topograph it is not necessary to align the crystal precisely for a particular Bragg reflection because, as in a Laue picture, a reflection will come up at the appropriate wavelength. Using conventional topography a distorted region such as that round a dislocation produces a darkening of the plate, but if part of the crystal is very distorted the Bragg reflection may be lost altogether and the plate then shows no darkening, as if the region were perfect. However, using SR radiation with its continuous range of wavelengths the Bragg reflection is never lost, so the topograph may be easier to interpret.

The transmission electron microscope permits the study of the structure of dislocations with much greater resolution than the X-ray methods. Figure 6.14

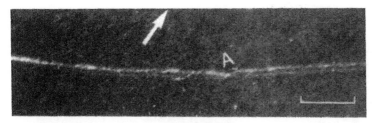

Figure 6.15 TEM micrograph of a dislocation line split into two partial dislocations. The arrow indicates the direction of the Burgers vector, and A a constriction in the dislocation. The scale marker is 50 nm (Pirouz *et al.*, 1983)

shows a series of pictures of a group of dislocation loops as the specimen is tilted at various angles (Hirsch, Pirouz and Barry, 1986). The relative intensity of the loops varies with the orientation, some loops disappearing in some positions. These changes give information on the Burgers vector and other details of the dislocations. Figure 6.15 due to Pirouz *et al.* (1983) shows a dislocation split into two partial dislocations (Section 6.2.a).

A third technique for the inspection of dislocations is to use the CL mode of the SEM as many dislocations luminesce when irradiated by the electron beam. Most observations have been made with Type II diamonds where almost all the luminescence from the diamond arises from dislocations and is not masked by other luminescence due to the impurities in Type I material like that in Figures 5.11 and 5.12. A detailed account of this luminescence is given by Hanley, Kiflawi and Lang (1977) who made observations using the normal CL technique in the SEM and also took colour photographs of the luminescence. These authors observed two types of luminescence associated with dislocations each with a characteristic spectrum of radiation. Figure 6.16 shows the luminescence from the same area of diamond photographed through different optical filters. Figure 6.16(a) was obtained with a filter passing only the region of the blue band A luminescence described in Section 4.2.b and appears to give images of individual dislocations. Figure 6.16(b) was obtained with another filter which cut out the band A luminescence but passed the wavelengths of the H3 system (Section 3.7.c), the luminescence in the image arises from sets of intersecting lines which we discuss in the next section.

Individual dislocations have been viewed by Pennycook, Brown and Craven (1980) in the scanning transmission electron microscope with a facility for viewing in the CL mode at the same time. It was found that most, but not all, of the dislocations luminesced. Figure 6.17(a) shows dislocation lines and loops taken in the normal mode, and Figure 6.17(b) the same area viewed in the CL mode using only the broad band A luminescence. Although the geometries of the dislocations are not resolved by the CL the areas luminescing are localized in regions around some of the dislocations. Other information on the luminescence is given by studies of the polarization (Hanley, Kiflawi and Lang, 1977; Yamamoto, Spence and Fathy, 1984) which show that the band A luminescence is polarized with its E vector parallel to the dislocation line. A detailed discussion of the mechanism which might be responsible for this luminescence is given by Yamamoto, Spence and Fathy. who conclude that the emission probably arises from donor–acceptor pairs (Section 4.2.b) sited on or near the dislocation lines. In this case the relatively poor definition of dislocations in the CL mode might arise from the spatial separations of the donors and acceptors in the pairs.

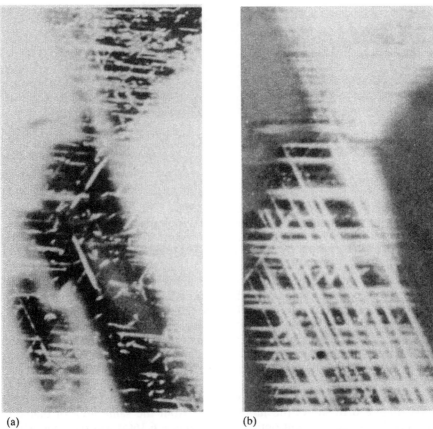

(a)   (b)

Figure 6.16 Cathodoluminescence micrographs of dislocations in a Type II diamond taken with filters to distinguish (a), blue luminescence; (b), yellow-green luminescence. Field width 200 µm (Hanley, Kiflawi and Lang, 1977), see text

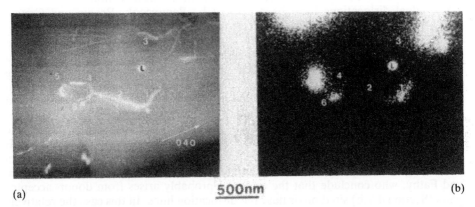

(a)   500nm   (b)

Figure 6.17 TEM micrographs of a region of a Type IIb diamond showing (a), dark field images of dislocations; (b), luminescence from the same regions (Pennycook, Brown and Craven, 1980)

6.2.c Slip bands

Many natural diamonds contain large numbers of dislocations created in a previous stage of their history. The presence of a dislocation in any crystal encourages the possibility of slip and plastic deformation as shown in Figure 6.8, but after it has passed across the crystal to produce a step in the far surface, it has in fact eliminated itself. Therefore the scale of plastic deformation is generally determined by various mechanisms whereby a moving dislocation generates further dislocations both on its own slip plane and on other slip planes. These additional dislocations can produce further slip and greater plastic deformation. Note also that the increasing number of dislocations may impede the motion of any particular dislocation through the crystal and the material becomes less plastic, a process known as work-hardening. For details of these various processes see Hull and Bacon (1984).

Plastic deformation often occurs on a number of closely adjacent parallel slip planes thus giving rise to the so-called slip bands on the surface of the crystal. Each band consists of a marked step in the surface made up of the steps due to each of the individual slip planes. These bands appear as one or more fine lines visible either to the naked eye or with a ×10 lens provided the surface is not too rough, and occasionally running right round the diamond. Examples of such slip bands, or glide lines, are given in diagrams by Orlov (1977). A survey by Harris, Hawthorne and Oosterveld (1984) of diamonds from seven South African mines showed slip bands on the order of 15% of the stones.

Slip systems may also be detected by viewing the luminescence from polished surfaces in the SEM. Figure 6.16(b) shows a micrograph obtained in this way with patterns of luminescence lying along two sets of intersecting {111} planes, the principal slip planes in diamond. Dislocations on slip planes will also give rise to elastic strain fields which may be seen by either the birefringence or X-ray topograph techniques described in Chapter 2. Figure 6.18 shows a birefringence micrograph of a Type I diamond viewed through polished {011} faces. Two sets of {111} slip planes are viewed edge on, so the strain fields associated with them are picked out by the birefringence, showing two intersecting slip systems. An

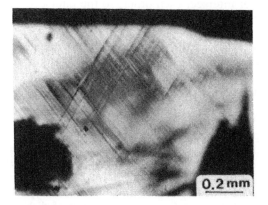

0.2 mm

Figure 6.18 Birefringence micrograph of a Type I diamond, viewed through polished {011} surfaces, showing slip bands on {111} planes (Wilks and Wilks, 1987)

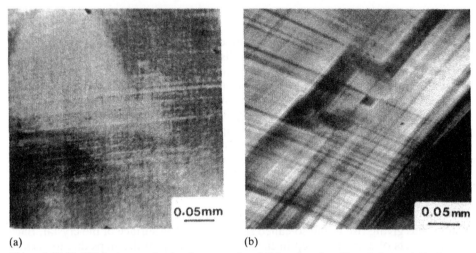

(a)   (b)

Figure 6.19 CL SEM micrographs showing cross-slip of dislocations in a Type I diamond. In the diamond of figure (a) the dislocations appear lighter and in the diamond of figure (b) darker than the background (Wilks and Wilks, 1987)

example of a similar system viewed by X-ray topography is given by Hanley, Kiflawi and Lang (1977).

The CL micrograph of slip lines in Figure 6.16(b) was obtained from a Type II diamond where almost all the luminescence is associated with the dislocations. Therefore the dislocations appear bright against a dark background. However, the position is rather different in Type I diamond which may show considerable luminescence due to impurity atoms. As might be expected, some diamonds show slip bands rather like those of Figure 6.16(b) superposed on a background of other luminescence. For example, Figure 6.19(a) is a CL micrograph of a polished {001} face showing bright slip bands on two intersecting sets of {111} planes which cut the surface. Note that the CL pattern of slip bands in diamond may sometimes appear rather similar to the patterns given by growth layers described in Section 5.2.c. However, growth bands can never intersect each other, so the intersections of the line systems in Figure 6.19(a) imply that at least one of these systems is not due to growth.

Besides patterns of the type shown in Figures 6.19(a) many Type I diamonds show patterns of intersecting slip bands which appear darker than the background luminescence, as in Figure 6.19(b). A striking example of this effect is shown in Figure 6.20 which gives two CL micrographs of the same polished face of a diamond, the contrast being adjusted for Figures 6.20(a) and 6.20(b) to bring out detail in different parts of the face. We see the same system of intersecting slip lines in both figures but these lines appear darker than the background in (a) and lighter in (b). It seems likely that during processes of plastic deformation the dislocations interacted with the optical centres responsible for the background luminescence thus modifying their behaviour to give either more or less emission. For example, Sumida and Lang (1981) have described a slip band in a Type II diamond modifying the broad band luminescence from a background of dislocations. The details of these various processes have yet to be explained.

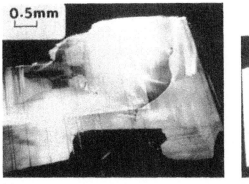

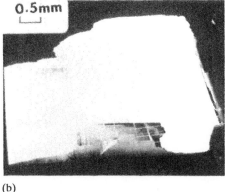

(a)   (b)

Figure 6.20 CL SEM micrographs of a Type I diamond surface showing cross-slip; (a), with the contrast adjusted for the larger area; (b), with the contrast adjusted to reveal detail in the small rectangular area. (Wilks and Wilks, 1987)

6.2.d Dislocation networks

The micrographs in the previous section showing slip bands viewed more or less edge on present a rather regular arrangement. However, if the crystal is viewed in other directions in order to obtain a three dimensional picture the dislocations form more varied patterns. Figure 6.21 is an X-ray projection topograph which shows a large number of dislocations spreading out from a central growth nucleus. Figure 6.22 is a projection topograph taken with X-ray SR and shows an irregular mass of dislocations in a brown Type I diamond, the irregularity being compounded by the fact that a projection topograph presents superimposed images of dislocations in different slip planes. Figure 6.23 is a micrograph of a third diamond taken in

Figure 6.21 X-ray projection topograph of a diamond showing dislocations spreading out from a central nucleus. Edge length 5.2 mm (Lang, 1964a)

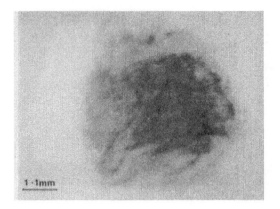

Figure 6.22 X-ray SR projection topograph of a brown Type I diamond showing a dense irregular mass of dislocations (Ward, Wilks and Wilks, 1983)

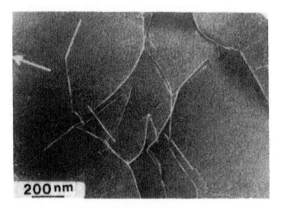

Figure 6.23 TEM micrograph of a dislocation net (Pirouz *et al.*, 1983)

the TEM at much higher magnification and shows a set of dislocations in a particular plane.

A form of dislocation network characteristic of Type II diamonds is shown in Figure 6.24, a CL SEM micrograph of a polished surface (Hanley, Kiflawi and Lang, 1977). This type of net divides the whole volume of the crystal into cells which are often of the order of 10 μm in diameter, appreciably larger than those in Figure 6.23. This mosaic-like network has the appearance of having been formed by plastic deformation followed by polygonization (see for example Honeycombe (1984)). Both Hanley, Kiflawi and Lang (1977) and Lang (1979) refer to a texture which is often observed in the birefringence patterns given by Type II diamonds with dislocation networks. This texture is also apparent in Figure 6.25, which is a topograph taken with X-ray synchrotron radiation, and should be compared with Figure 6.22 a topograph of a plastically deformed Type I diamond viewed in the same way.

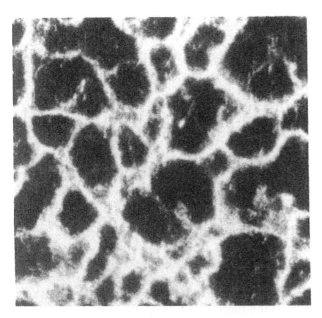

Figure 6.24 CL SEM micrograph of a net of dislocations of type commonly found in Type II diamonds. Field width is 250 μm (Hanley, Kiflawi and Lang, 1977)

1·0 mm

Figure 6.25 X-ray SR projection topograph of a Type II diamond with mass of dislocations showing some texture (Wilks and Wilks, 1987)

6.2.e Lattice distortion

As dislocations are surrounded by a quite extensive elastic strain field any appreciable density of dislocations will produce a significant distortion of the crystal lattice. For a dislocation array such as that in Figure 6.21 the distortion will be greatest near the nucleus from which the lines radiate. For the diamond of Figure 6.24 the position is rather different and the mosaic pattern divides the crystals into cells all with slightly different orientations, the geometric mismatch between cells being taken up by the dislocations at the boundaries. These differences in orientation within the crystal lattice may be quite large and produce a broadening of the spots in a Laue diffraction pattern, but if the distortion is small the misorientations are better observed by taking a so-called rocking curve. A narrow beam of X-rays is obtained by a Bragg reflection from a crystal and is then reflected from the specimen crystal to a detector which measures the change in the intensity of the beam as the specimen is turned slowly through the Bragg angle. The width of the beam measured in this way then gives a measure of the angular distortions in the lattice, see for example Tanner (1976).

Although some diamonds have suffered much plastic deformation others show only a very few dislocations as in Figure 6.12. In this case we expect the lattice to be regular and free of the disorientations described above. A striking example of the high degree of lattice perfection which may be found in diamond was given by an experiment to produce polarized photons of high energy (Jackson, 1975). Photons are produced when an electron beam is passed through a crystal, and a significant polarization can be obtained by a particular orientation of the crystal with respect to the beam. However, to achieve success the crystal must be oriented with respect to the beam to within less than 0.005°. Diamond was an obvious choice for the target crystal because the atoms are close packed giving a high number density and because the low level of the thermal energy (Section 6.3) produces a minimum of geometric irregularity when the crystal is heated by the beam. However, it was also essential that any lattice distortions in the crystal were small compared to the angular precision of the experiment.

The diamonds which finally gave the best performance were selected solely by optical criteria. A large number of high quality octahedron gem diamonds were viewed to select stones with very low levels of birefringence. These specimens were then further selected for the absence of trigons, the triangular etch pits which generally indicate the presence of dislocations emerging at the surface (Wilks and Wilks,1971; Wilks, 1976). Diamonds selected in this way appeared to have lattice distortions of less than 0.003° both in rocking curve experiments and in the production of polarized photons (Jackson, 1975).

6.2.f Scattering of light

Dislocations may also be detected by the light scattering technique as described in Section 2.3.c. The experimental arrangement shown in Figure 2.11 consists of a narrow collimated beam of light which passes in and out of the crystal through faces carefully polished to reduce scattering to a minimum. Any light scattered within the specimen is viewed on a line perpendicular to the beam using a microscope focused on the beam. As described previously, inclusions in the diamond may produce quite intense scattering depending on their size, but in addition another form of scattering is often observed.

Figure 6.26 shows micrographs of a beam of light passing through two parallel-sided slabs of diamond. The beam enters from the left and the outlines of the diamond are just discernable due to light scattered from the boundaries. The area of light to the left of the diamond is due to light scattered from the exit window of the collimating system and is of no consequence but serves to indicate the position of the beam. In diamond F2 the beam is clearly visible throughout the diamond whereas in the micrograph of diamond A46 taken under the same conditions the

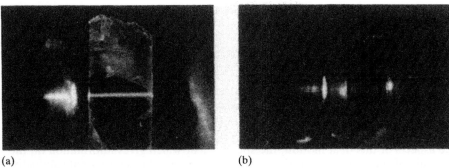

(a)   (b)

Figure 6.26 Micrograph of beam of light passing through (a), diamond F2 with many dislocations; (b), diamond A46 with very few dislocations (Wilks and Wilks, 1980)

beam is almost invisible except for some scattering where it enters and leaves the diamond. The two micrographs correspond to a difference in beam intensity of about a factor 30 and differences of this order appear to be quite common (Wilks and Wilks, 1980).

In a survey of 39 selected diamonds, Ward, Wilks and Wilks (1983) measured the intensity of the beam in the diamond, and took synchrotron radiation topographs to obtain estimates of the density of dislocations. Thirteen of the diamonds gave topographs similar to that of Figure 6.13 showing a very low density of dislocations and all thirteen produced only weak scattering of light. Another fifteen diamonds gave topographs similar to that of Figure 6.22 with a high density of dislocations and all fifteen gave strong scattering. The eleven other diamonds gave topographs and scattering intermediate between these two limits. We conclude that the scattering of the light was caused primarily by the dislocations, in a similar way to the scattering of light in quartz reported by Moriya and Ogawa (1980).

During the above experiments the diamonds were traversed across the beam in order to scan different parts of the stone. Generally the intensity of the scattered light did not vary greatly over the diamond but occasionally regions of appreciably different scattering power were observed. The intensity of the scattered light may also depend on the orientation of the beam in the diamond if the dislocations have some preferred orientation. For example, if the dislocations lie mainly on one set of parallel slip planes they may act rather like a mirror so far as the scattered light is concerned. Hence, slip planes inclined to the vertical will tend to scatter light upward for one direction of the beam and to scatter light downward if the direction is reversed (Van Enckevort and Seal, 1987; Wilks and Wilks, 1987).

As we have already mentioned in Section 5.3.a, dislocations in diamond often indicate their existence by the presence of trigons or triangular etch pits where they intersect the surface of the crystal. Another effect of this type was observed during as study of some 27 rounded dodecahedron diamonds with rounded natural faces typical of dissolution processes (Section 5.3.b). The faces on some of these stones appeared brighter and more metallic than those on others, and it was quite possible to group all 27 stones into sets of different brightness. It was then seen, to a good degree of correlation, that the greater light scattering in a diamond was accompanied by less bright surfaces. Micrographs of the surfaces showed that the more matt appearance of some of the surfaces was due to the fine scale of the surface structure. Hence it seems that the dislocations associated with the greater light scattering give rise to a greater density of etch sites on the surface and so to a less bright appearance (Wilks and Wilks, 1980).

6.3 Thermal vibrations

Quite apart from the distortions due to dislocations, point defects and foreign atoms, the atoms in a real crystal are never positioned exactly at geometric lattice points because of the thermal vibrations of the atoms about their mean positions, the amplitude of the vibrations increasing with temperature. We now briefly review three aspects of this heat motion; the specific heat which gives information on the magnitude of the heat motion, the thermal expansion, and the thermal conductivity.

6.3.a The specific thermal capacity

Measurements of the specific heat, or specific thermal capacity, have been reviewed in detail by Berman (1965, 1979), and the general form of the results is shown in Figure 6.27. The specific heat of diamond, like that of other crystalline

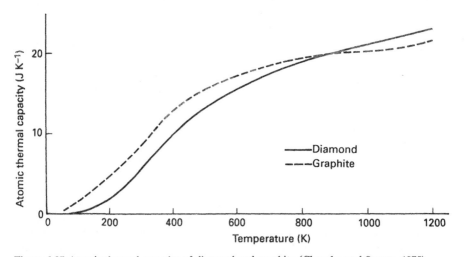

Figure 6.27 Atomic thermal capacity of diamond and graphite (Chrenko and Strong, 1975)

materials, has a value of approximately 25 J mol^{-1} at relatively high temperatures and falls off towards zero at lower temperatures. However, in nearly all other materials the specific heat does not fall below 25 J mol^{-1} until well below room temperature (300 K) whereas the specific heat of diamond begins to fall below about 1200 K. This difference arises because the very strong binding forces between the carbon atoms produce much higher frequencies of vibration and the heat motion is correspondingly more difficult to excite.

The total thermal energy of any crystal at an absolute temperature T' is equal to the area under the specific heat curve between temperature zero and temperature T'. For diamond at room temperature this area is much less than for other materials and so is the heat motion. Hence diamond was chosen for the experiment described in Section 6.2.e which required a crystal as free as possible from irregularities produced by a thermal motion. We also note that the thermal properties of insulating crystalline materials are governed to a large extent by one parameter, the Debye constant θ, which determines the general form of the specific heat curve, and the magnitude of the thermal expansivity and the thermal conductivity. The value of θ for diamond is approximately 1900 K, much greater than the values of 200 K and 300 K typical of many common solids, and higher than that of any other solid.

6.3.b Thermal expansion

Because of the low thermal energy the coefficient of thermal expansion of diamond is an order of magnitude smaller than that of most materials. Values of this coefficient have been obtained by making X-ray determinations of the lattice constant which can be measured to a high degree of precision. Various sets of measurements are reviewed by Berman (1979) and give a value of $0.8 \ 10^{-6} \ K^{-1}$ at room temperature. A theoretical discussion of the correlation between the expansion coefficient and the atomic vibrations is given by Parsons (1977). The low value of the expansion coefficient has practical implications when diamonds are bonded to metals in the manufacture of tools, because it results in a differential expansion between the diamond and the metal which may cause difficulties during brazing or if the tool runs hot during use. This differential expansion may also produce deleterious effects in PCD polycrystalline diamond at high temperatures, as the metal solvent/catalyst will expand more than the diamond (Section 12.2.d)

6.3.c Thermal conductivity

The thermal conductivity of diamond at room temperature is remarkably high especially for an insulating material. The conductivity of a typical Type I diamond is comparable to that of copper, while a Type II has a considerably greater conductivity. The magnitude of this conductivity depends on the temperature and impurity content. Figure 6.28 shows the thermal conductivity of some diamonds as a function of temperature (Berman, Hudson and Martinez, 1975). The maximum in the curve is characteristic of most insulating crystalline solids (see for example Berman, 1976), but because of the extreme value of the Debye parameter θ for diamond the maximum occurs at a relatively high temperature where the conductivity has an extremely high value. Roughly speaking the maximum arises because below 100 K any increase of temperature produces more thermal motion and so increases the conductivity. Above about 100 K, however, the increasing thermal motion also produces more irregularities in the crystal which impede the flow of heat, and this latter effect predominates, so the conductivity falls. Values at temperatures between 300 K and 400 K are given by Burgemeister (1978).

Various attempts have been made to correlate the thermal conductivities of different diamonds with the concentrations of the various nitrogen defects. None of these are entirely satisfactory because the best procedures to resolve an infrared spectrum into its various components (Section 3.3) were not then available, so the concentrations of the A, B, single nitrogen, and platelets are often rather uncertain. However, there is certainly a strong correlation in Type Ia diamonds between the thermal conductivity and the value of the maximum infrared absorption between 8.30 μm and 8.55 μm (Figure 6.29). Hence in these diamonds

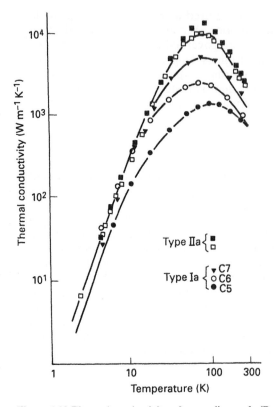

Figure 6.28 Thermal conductivity of some diamonds (Berman, Hudson and Martinez, 1975)

the thermal resistance appears to arise primarily from the A and B centres but there is also some evidence that the presence of platelets reduces the conductivity (Burgemeister, 1978). This result is in line with the experimental observations that Type II diamonds, both Type IIa and Type IIb, have the highest conductivity. Note, however, that Type Ia diamonds with infrared absorptions of about $1 \, cm^{-1}$ or less are almost as good conductors as Type II stones. For further details see Burgemeister (1978) and Berman (1979).

There is at present not much information on the conductivity of Type Ib diamonds partly perhaps because good quality natural Ib diamonds are not too common. Burgemeister (1978) made measurements on three Type Ib diamonds with infrared absorptions at $7.8 \, \mu m$ and $8.85 \, \mu m$ of about $1.6 \, cm^{-1}$ and observed conductivities which were almost as large as those of Type II diamonds. As there is now an increasing supply of good quality synthetic crystals which are generally Type Ib, it would be of interest to study in some detail how the conductivity depends on the concentration of the single nitrogen atoms as indicated by the infrared absorption. (Claims that Type Ib diamonds have as high a conductivity as Type II may be somewhat misleading because this is probably only true if their nitrogen content is relatively low.)

The maximum conductivity in diamond is determined by the fact that about 1% of the carbon atoms are not the normal isotope ^{12}C but the heavier ^{13}C. These

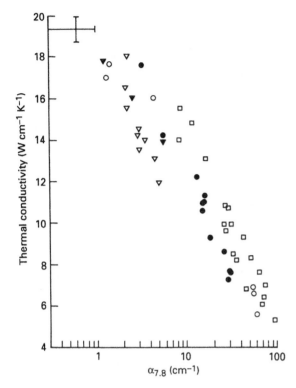

Figure 6.29 Thermal conductivity of Type I diamonds at 320 K plotted against the maximum value of the absorption coefficient at 7.8 μm. The bars in the top left corner indicate values obtained for Type II diamonds. (Burgemeister, 1978)

atoms are distributed randomly in the diamond and therefore produce irregularities in the vibrations of the lattice. Moreover it is known that in other crystals these irregularities lead to a small thermal resistance which have a significant effect on high thermal conductivities (see for example Berman, 1976). Hence diamonds grown recently with a nominal concentration of ^{13}C of only 0.07%, see Section 1.4.c, exhibit thermal conductivities about 50% higher than observed previously (Anthony *et al.* 1990).

The high conductivity is important in several industrial applications. For example, when a diamond is used as a cutting tool the conductivity removes the frictional heat generated at the cutting edge and spreads it over a larger volume, thus permitting the tool to be run under more severe conditions. Another important application is the use of small diamond slabs to act as thermal bases or heat sinks for various solid-state semiconducting devices as described in Section 17.6.d.

6.4 Annealing of diamond

The properties of most materials may be modified by annealing processes, that is by holding the specimen at an elevated temperature for some period, and diamond is no exception. In particular it is found that both geometric defects and impurity atoms become mobile at a high enough temperature and the mobility leads to important changes in the various absorption centres associated with nitrogen. We have already mentioned in Section 3.7 that any interstitial carbon atoms in a diamond are very probably mobile at room temperature. Therefore they will have either diffused to the surface or combined with other defects which anneal out in the regions of 300°C and 500°C. It was also noted that any vacancies in the diamond lattice are probably mobile at temperatures above about 600°C.

Further changes beyond the simple annealing of vacancies and interstitials are produced at still higher temperatures. However, various precautions are necessary in order to study these changes. At temperatures over 600°C the diamond must be held in a high vacuum in order to prevent oxidation and graphitization of the surface (Section 13.3.a). Moreover if diamond is heated to temperatures above 1500°C it will convert to the more stable graphite form at an appreciable rate (Section 1.2.a). Therefore to anneal the diamond at these temperatures it must be held under a sufficient pressure, of the order of 6 GPa at 1500°C, to ensure that the diamond is more stable than graphite. Therefore any annealing must be carried out in presses similar to those used for the synthesis of diamond. Examples of internal cracking and degeneration when diamonds are heated under too low a pressure are given by Evans and Rainey (1975).

6.4.a Aggregation of nitrogen

The first experiments on the mobility of the nitrogen defects were made on synthetic Type Ib diamonds which were deep yellow in colour. These were annealed at temperatures in the region of 1900°C under pressures of 5.5 GPa to 6.5 GPa, above or close to the diamond–graphite equilibrium line (Chrenko, Tuft and Strong, 1977; Strong, Chrenko and Tuft, 1979). It was observed that after the treatment the deep yellow colour associated with single nitrogen defects had faded significantly. Measurements of the optical absorption spectra indicated that the treatment had resulted in single nitrogen atoms combining to produce both A nitrogen and N3 nitrogen centres. By studying the rate of production of the A centres as a function of the annealing temperature the authors deduced an activation energy for the process of about $250 \, \text{kJ mol}^{-1}$.

Shortly afterwards Brozel, Evans and Stephenson (1978) annealed some typical Type IaA diamonds at temperatures in the range 1900°C to 2350°C under pressures of 8.5 GPa. Initially the nitrogen in these diamonds was mainly in the A form, with a very low level of any single nitrogen defects as shown by electron paramagnetic resonance measurements. However, after annealing there was an appreciable concentration of single nitrogen atoms typical of Type Ib diamonds. No significant change was observed in the strength of the absorption spectrum due to the A centres, nor would be expected as the nitrogen in the new single form amounted to only 0.5% of the total estimated nitrogen in the A form.

In another experiment to study the aggregation of the nitrogen Allen and Evans (1981) increased the mobility of the single nitrogen atoms in synthetic Type Ib diamonds by a prior irradiation with 2 MeV electrons to produce vacancies and interstitials (Collins, 1978). Subsequent annealing at temperatures up to 2200°C under 8.5 GPa pressure then produced A centres, N3 centres, platelets, and it appeared B centres. The A and N3 centres and the platelets were detected by their optical absorption spectra, the platelets by electron microscopy. The B centres were much less numerous than the A centres and their contribution to the optical absorption could not be resolved from that due to the A centres, but their presence was inferred from the appearance of H4 centres known to be associated with B centres (Section 3.7.c).

Finally, Evans and Qi (1982) by increasing the pressure on the diamonds to 9.5 GPa and working at temperatures up to 2700°C were able to dispense with the prior electron irradiation and also to produce a greater yield of B centres. For example, Figure 6.30 shows the result of annealing a typical Type IaA diamond at

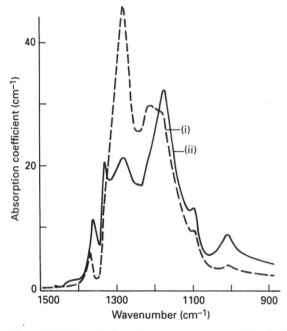

Figure 6.30 Infra-red absorption spectra of a natural Type IaA diamond (i) initially and (ii) after heating under pressure for 1 h at 2600, 30 min at 2700°C and finally 3 h at 2500°C (Evans and Qi, 1982)

temperatures above 2500°C; the spectrum after the treatment is essentially that of a Type IaB diamond. (Although according to Collins and Stanley (1985) diamonds produced in this way can show some differences in their absorption and luminescence from natural Type IaB diamonds.) Hence the various experiments described above show that the progressive annealing of diamonds with single nitrogen centres first produces A centres which then anneal to give B centres, N3 centres, and platelets.

6.4.b The platelets

We now consider the nature of platelets and how they are formed. As described in Section 2.4.b platelets appear to be an extra one or two layers of atoms which force the diamond lattice apart and create a strain field. The work of Allen and Evans (1981) described in the previous section showed that platelets are produced when diamonds containing A nitrogen centres are annealed at temperatures high enough to make the centres mobile. The same authors also showed that a similar heat treatment of Type II diamond containing a minimum of nitrogen did not produce any platelets. These results suggest that the platelets consist at least partly of nitrogen but this is still a matter of discussion, as described below. We also note that Brozel, Evans and Stephenson (1978) observed that annealing at very high temperatures (2350°C) caused a large reduction in the 1370 cm^{-1} infrared absorption peak due to the platelets.

In view of the above results it seems likely that the different concentrations of A,B, and N3 centres and of platelets found in natural diamonds are at least partly the result of different heat treatments received by the diamonds while in the earth. To study this point Brozel, Evans and Stephenson (1978) measured the strength of the optical absorption due to the A centres, B centres, and platelets in a selection of diamonds showing a wide range of absorptions. Figure 6.31 shows the strength of the platelet absorption in these diamonds normalized to the total nitrogen content and plotted against the relative amounts of nitrogen in the A or B form. The diagram shows that for these stones there are few or no platelets if the nitrogen is all in the A form, and the authors assume that this is because these diamonds had never been in conditions which caused the A centres to aggregate. The diagram also shows that diamonds with an increasing fraction of B centres tend to show an approximately proportional increase in the strength of the platelet absorption,

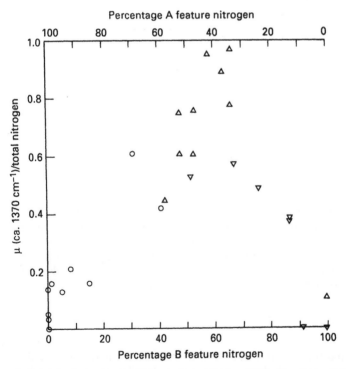

Figure 6.31 Plot of the absorption due to platelets divided by the total nitrogen concentration in arbitrary units against the relative amounts of A and B feature nitrogen (Brozel, Evans and Stephenson, 1978); see text

suggesting that both B centres and platelets have been produced by the aggregation of A centres. However, in diamonds where the B centres predominate the platelet absorption is much reduced, suggesting that a further stage of annealing had resulted in the decomposition of any platelets which had formed previously.

It has also been known for some time that diamonds containing platelets may also contain loops of dislocation line which are of comparable size to the platelets and also lie on {100} planes. In fact electron microscope studies have now shown that the loops appear to be formed during the decomposition of platelets (Hirsch, Pirouz and Barry, 1986). These authors studied the same diamond which exhibited the dislocation loops shown in Figure 6.14. (Note that the loops are not necessarily in the plane of the figure, and that some are incomplete as they have been cut off by the top of the specimen during the thinning process prior to inspection.) Figure 6.32 is a micrograph of another feature in the same diamond lying on a {100} plane

500 nm

Figure 6.32 TEM micrograph of a platelet in course of disintegration, see text (Hirsch, Pirouz and Barry, 1986)

inclined to the plane of the figure, and which the authors interpret as a platelet in the course of disintegration. The two light areas are the images of still unchanged platelet, the curving white line joining their upper edges is a dislocation line which will eventually form a complete loop. The white spots in the central area are the small defects known as voidites already described in Section 2.4.c, which also arise from the decomposition of the platelets. Further details of the mechanisms involved are given by Hirsch, Pirouz and Barry (1986).

In a further analysis, somewhat similar to that of Brozel, Evans and Stephenson (1978), Woods (1986) studied 50 diamonds of various nitrogen content using the method of Clark and Davey (1984) to resolve the infrared absorption at $1282\,cm^{-1}$ into components associated with the A and B features. He then found that his diamonds divided into two groups, the 'regular' diamonds which would fall on the left hand side of Figure 6.31 where the platelet concentration is roughly proportional to the concentration of the B defects, and the 'irregular' diamonds which would fall on the right hand side of Figure 6.31 where any platelets originally present had suffered some 'catastrophic degradation'. Woods noted in particular that in all the regular diamonds the strength of the platelet peak was approximately proportional to the absorption at $1282\,\mathrm{cm}^{-1}$ due to the B centres.

Thus far we have a fairly simple picture of single nitrogen atoms aggregating to form A centres which then aggregate to give B centres, N3 centres and platelets, with the possibility of the platelets eventually degenerating into smaller units. An obvious inference is that the platelets consist of one or two layers of nitrogen as for example in a model proposed by Lang (1964b), but the position as summarized by Woods (1986) and Sumida and Lang (1988) is still uncertain. Woods (1986) stresses the fact that in the regular diamonds the number of atoms in the platelets is of the same order of magnitude as the number in the B centres. This result is somewhat surprising because if all these atoms are nitrogen atoms from the A centres why should they form about equal numbers of such very different structures as the B centres and the platelets. Woods therefore proposed that a B centre is formed by the aggregation of two A centres in a process suggested by Loubser and van Wyk (1981 personal communication) which gives rise to a carbon interstitial, and that the platelets are formed by the aggregation of these carbon interstitials.

The above picture receives some confirmation from the experiments of Bruley and Brown (1989) who used electron energy loss spectroscopy (EELS) to study over 80 platelets in two diamonds viewed in the scanning transmission electron microscope. No nitrogen could be detected in any of the platelets (although Bruley (1989) comments that the measurements may have been affected by radiation damage). On the other hand, the study of Hirsch, Pirouz and Barry (1986) appeared to show the disintegration of platelets accompanied by the production of voidites, while Bruley and Brown (1989) state that the voidites contain only nitrogen. To resolve this apparent contradiction Bruley and Brown suggest that nitrogen atoms may condense on dislocations surrounding the platelets and thus provide the nitrogen in the voidites. (This picture might also account for the observation of Evans and Rainey (1975) that annealing between 2200°C and 2250°C reduced the optical absorption at $1370\,\mathrm{cm}^{-1}$ by 40% or more, even though the X-ray spike intensity produced by the platelets was unaffected, the change in absorption perhaps being caused by a migration of the nitrogen atoms.) Quite clearly further work is necessary to obtain a full understanding of the structure of both the B centre and the platelets. For a review of the present situation and earlier work see articles by Sumida and Lang (1988), by Bursill and Glaisher (1985), and by Humble, Lynch and Olsen (1985).

We also mention two other points noted in the detailed paper of Woods (1986). First, the width and precise position of the platelet absorption peak varied linearly with the total absorption at $1282\,\mathrm{cm}^{-1}$. Secondly, the absorption due to the N3 centres in the regular diamonds also increased in proportion to the strength of the B features. However, the number of atoms in the N3 form is very small compared with the number in the A and B forms, so Woods regards their formation as a secondary effect in which two A centres yield an N3 centre and a carbon and nitrogen interstitial

References

Allen, B. P. and Evans, T. (1981) *Proceedings of the Royal Society*, A**375**, 93–104

Anthony,T.R., Banholzer,W.F., Fleischer,J.F., *et al.* (1990) *Physical Review B*, **42**, 1104–1111

Berman, R. (1965) In *Physical Properties of Diamond*, (ed. R.Berman), Clarendon Press, Oxford, pp. 371–393

Berman, R. (1976) *Thermal Conduction in Solids*. Clarendon Press, Oxford

Berman, R. (1979) In *The Properties of Diamond*, (ed. J. E.Field), Academic Press, London, pp. 3–22

Berman, R., Hudson, P. R. W. and Martinez, M. (1975) *Journal of Physics C*, **8**, L430–L434

Brozel, M. R., Evans, T. and Stephenson, R. F. (1978) *Proceedings of the Royal Society*, **A361**, 109–127

Bruley,J. (1989) *Ph.D Thesis*, Cambridge University

Bruley,J. and Brown,L.M. (1989) *Philosophical Magazine*, **A59**, 247–261

Burgemeister, E. A. (1978) *Physica*, **93B**, 165–179, **94B**, 366

Bursill, L. A. and Glaisher, R. W. (1985) *American Mineralogist*, **70**, 608–618

Chrenko,R.M. and Stroug,H.M. (1975) *Physical Properties of Diamond*. Report No.75CRD089, General Electric Company, Schenectady, New York

Chrenko, R. M., Tuft, R. E. and Strong, H. M. (1977) *Nature*, **270**, 141–144

Clark, C. D. and Davey, S. T. (1984) *Journal of Physics C*, **17**, 1127–1140

Collins, A. T. (1978) *Journal of Physics C*, **11**, L417–L422

Collins, A. T. and Stanley, M. (1985) *Journal of Physics D*. **18**, 2537–2545

Cottrell,A.H. (1964) *The Mechanical Properties of Matter*. John Wiley, New York

Evans, T. and Qi, Z. (1982) *Proceedings of the Royal Society*, **A381**, 159–178

Evans, T. and Rainey, P. (1975) *Proceedings of the Royal Society*, **A344**, 111–130

Hanley, P. L., Kiflawi, I. and Lang, A. R. (1977) *Philosophical Transactions of the Royal Society*, **284**, 329–368

Harris, J. W., Hawthorne, J. B. and Oosterveld, M. M. (1984) *Annales Science, Université de Clermont-Ferand, II*, **74**, 1–13

Hirsch, P. B., Pirouz, P. and Barry, J. C. (1986)*Proceedings of the Royal Society*, **A407**, 239–258

Honeycombe, R. W. K. (1984) *The Plastic Deformation of Metals*, (2nd edn), Edward Arnold, London

Hornstra, J. (1958) *Journal of the Physics and Chemistry of Solids*, **5**, 129–141

Hull, D. and Bacon D.J. (1984) *Introduction to Dislocations*, (3rd edn) Pergamon Press, Oxford

Humble, P., Lynch, D.F. and Olsen, A. (1985) *Philosophical Magazine*, **A52**, 623–641

Jackson, A. (1975) *Nuclear Instrumentation and Methods*, **129**, 73–83

Jiang,S-S, and Lang,A.R. (1983) *Proceedings of the Royal Society*, **A388**, 249–271

Kowalski,G., Lang,A.R., Makepeace,A.P.W. and Moore,M. (1989) *Journal of Applied Crystallography*, **22**, 410–430

Lang, A. R. (1964a) *Proceedings of the Royal Society*, **A278**, 234–242

Lang,A.R. (1964b) *Proceedings of the Physical Society*, **84**, 871–876

Lang, A. R. (1974) *Journal of Crystal Growth*, **24/25**, 108–115

Lang, A. R. (1979) In *The Properties of Diamond*, (ed. J. E. Field), Academic Press, London, pp.425–469

Moriya,K. and Ogawa,T. (1980) *Philosophical Magazine A*, **41**, 191–200

Orlov, Yu. L. (1977) *The Mineralogy of the Diamond*. John Wiley, New York

Parsons, B. J. (1977) *Proceedings of the Royal Society*, **A352**, 397–417

Pennycook, S. J., Brown, L. M. and Craven, A. J. (1980) *Philosophical Magazine A*, **41**, 589–600

Pirouz, P., Cockayne, D. J. H., Sumida, N. *et al*. (1983) *Proceedings of the Royal Society*, **A386**, 241–249

Slawson, C. B. (1950) *American Mineralogist*, **35**, 193–206

Strong, H. M., Chrenko, R. M. and Tuft, R. E. (1979) *United States Patent No.4,174,380*, 13 November 1979

Strong, H. M. and Hanneman, R. E. (1967) *Journal of Chemical Physics*, **46**, 3668–3676

Sumida, N. and Lang, A. R. (1981) Philosophical Magazine A, **43**, 1277–1287

Sumida,N. and Lang,A.R. (1988) *Proceedings of the Royal Society*, **A 419**, 235–257

Tanner, B. K. (1976) *X-ray Diffraction Topography*. Pergamon Press, Oxford

Van Enckevort,W.J.P. and Seal,M. (1987) *Philosophical Magazine A*, **55**, 631–642

Ward, R. C. C., Wilks, E. M. and Wilks, J. (1983) *Industrial Diamond Review*, **43**, 137–141

Whelan, M. J. and Hirsch, P. B. (1957) *Philosophical Magazine 2*, 1303–1324

Wilks, E. M. (1976) *Nature*, **262**, 570–571

Wilks, E. M. and Wilks, J. (1971) *Industrial Diamond Review*, **31**, 238–242

Wilks,E.M. and Wilks,J. (1980) *Industrial Diamond Review*, **40**, 8–13
Wilks, E. M. and Wilks, J. (1987) *Wear*, **118**, 161–184
Woods, G. S. (1971) *Philosophical Magazine*, **23**, 473–484
Woods, G. S. (1986) *Proceedings of the Royal Society*, **A407**, 219–238
Yamamoto, N., Spence, J. C. H. and Fathy, D. (1984) *Philosophical Magazine B*, **49**, 609–629

Mechanical properties

We now consider the principal mechanical properties of diamond both natural and man-made. As in previous chapters the majority of the experimental studies have been made with natural diamond.

Mechanical properties

Chapter 7

Strength and fracture

7.1 Brittle and plastic materials

Most materials fail at stresses below and sometimes far below the theoretical strength calculated for a perfect crystal. This is because a specimen generally contains two types of imperfections: cracks and dislocations. The effect of a crack is to enhance the magnitude of any stress in the material at the tip of the crack, perhaps by an order of magnitude depending on the size and geometry of the crack. Hence the fracture of a specimen can be initiated at the tip of a crack by an overall applied stress much less than the ultimate tensile strength. For example, suppose

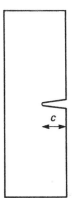

7.1 Schematic diagram showing a small crack of length c in a uniform bar; see text. (Size of crack greatly exaggerated.)

that a uniform bar containing a small crack of length c (Figure 7.1) is loaded to produce a uniform tensile stress σ_0. By applying standard elastic theory it is readily shown that the actual tensile stress at the tip is

$$\sigma \sim \sigma_0 \, (c/2\rho)^{1/2} \tag{7.1}$$

where ρ is the radius of the crack, see for example Cottrell (1964) and Kelly and Macmillan (1986). The radius of the tip of the crack is probably of the order of the atomic spacing so Equation (7.1) implies that the tensile stress at the tip of a microscopic crack only 1 μm long may be ×100 greater than the mean stress in the bar. Hence any such flaws may cause a material to fracture at loads much lower than expected from their theoretical strength. Thus, Griffith (1921) in a classic experiment greatly increased the strength of glass fibres by etching the surface to remove cracks. More detailed accounts of these topics are given by Cottrell (1964) and Kelly and Macmillan (1986).

The strength of a material may also be reduced by the presence of dislocations. We described in Section 6.2 how the application of a stress may result in the motion and generation of dislocations, thus producing a permanent deformation which is not recovered on removing the load. This so-called plastic deformation tends to relax the stress, but if the load is maintained the material continues to flow or deform, and may eventually fail by necking. This type of deformation and failure is of course very common as dislocations often move quite readily under the influence of applied stresses.

Materials are described as brittle or plastic respectively if they fail by fracture or by plastic flow. A typical example of a plastic material is soft annealed copper, while glass is well known as a brittle material. Various authors have discussed the relative probability of crystals deforming by fracture or plastic deformation. Kelly, Tyson and Cottrell (1967) considered a range of crystals and showed that at one end of the spectrum plastic flow is greatly favoured in metals such as copper and aluminium, while in diamond at the other end of the spectrum fracture is the preferred mode though not by a large margin. The deformation of brittle materials is discussed by Sinclair and Lawn (1972) and Sinclair (1972, 1975). These authors consider the mechanics of the atomic lattice at the tip of a crack and show that in some crystals it should be possible for a crack to move forward without producing any plastic deformation. Their analysis is confirmed by studies of cracks in silicon, germanium, silicon carbide and alumina (Lawn, Hockey and Wiederhorn, 1980).

It is necessary to remember that the distinction between plastic and brittle materials is not entirely straightforward because brittle fracture is often accompanied by some plastic deformation. For example, if a Vickers indenter is forced slowly into a plate of glass some plastic flow generally occurs before the plate fractures (Marsh, 1964). It appears that the compressive stresses produced under the indenter inhibit cracking and thus encourage plastic flow, and this behaviour is observed in such nominally brittle materials as alumina, silicon and silicon nitride (Hockey, 1971; Lawn, Hockey and Richter, 1983). When such plastic flow occurs it must, of course, be taken into account when calculating the forces required to produce fracture because it relaxes the internal stresses in the material, see for example Tabor (1970).

The distinction between brittle and plastic materials is further blurred because the motion of dislocations can be assisted by the thermal vibrations. Hence materials tend to behave more plastically as the temperature is raised, for example glass bends easily after heating to a quite moderate temperature. Therefore the amount of plastic deformation produced during the fracture of a nominally brittle material depends very much on the temperature of the specimen. The transition from an almost completely brittle fracture to completely plastic flow is often spread over a range of temperature, and the details of these transitions are not well understood. Even so, the behaviour of a range of materials suggests that the transition usually occurs at an absolute temperature of about half the absolute melting temperature, see for example Atkins and Tabor (1966).

At room temperature diamond is far below its melting point so we expect it to show brittle behaviour. Moreover, the very directional nature of the strong carbon–carbon bonds greatly restricts the rearrangements of the atoms which are necessary to permit the motion of dislocations. Therefore diamond at room temperature behaves almost entirely as a brittle material. For example a diamond is readily split by cleavage, and sharp edges on polished gems and tool stones may be chipped by contact with other hard materials. Gane and Cox (1970) pressed two diamond wedges against each other cross-wise to give a point contact, and viewed the resulting damage in the SEM. On applying an increasing load to force the wedges together no damage at all was observed until very obvious fracture occurred as shown in Figure 7.2(a); note the flaking of the diamond, shown more clearly in Figure 7.2(b) and that there is no sign of any plastic deformation.

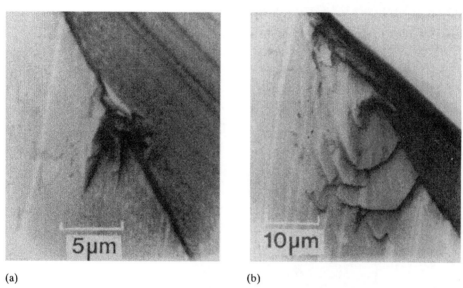

(a)   (b)

Figure 7.2 SEM micrographs of damage produced by pressing together crossed diamond wedges: (a), under a load of 8 g wt; (b), under a higher load (Gane and Cox, 1970)

The tendency of diamond to fracture rather than to flow plastically is also shown in an experiment by Moriyoshi *et al.* (1983). These authors produced fine cracks in various ceramic foils by pricking with a sharp needle and then viewed them in the transmission electron microscope. For most ceramics the crack was accompanied by an array of nearby dislocations as in the foil of magnesium oxide shown in Figure 7.3. However, the appearance of a pricked foil prepared by thinning down a specimen of diamond was quite different. Figure 7.4. shows a ribbed pattern suggestive of fracture but no signs of any dislocations. We conclude from a wide range of evidence that the usual mode of failure of diamond at room temperature is by fracture, although there is one example of apparent plastic deformation under very high pressures (Section 7.3.a). We discuss this and other possible examples of plastic flow at room temperaure in Chapter 8.

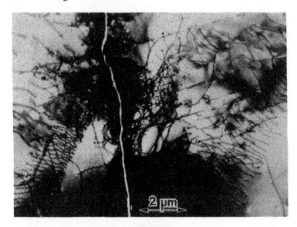

Figure 7.3 TEM micrograph of a thin foil of MgO single crystal showing dislocations around a microcrack (Moriyoshi *et al.*, 1983)

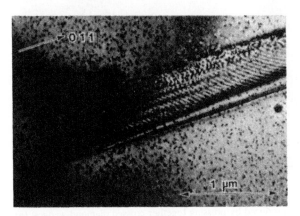

Figure 7.4 TEM micrograph of thin foil of single crystal diamond showing microcracks but no dislocations (Moriyoshi *et al.*, 1983)

7.2 Elastic constants and ultimate strength

7.2.a The elastic moduli

We now consider the elastic moduli of diamond which specify the ratio of the stress to the strain in reversible elastic deformations. If a material is isotropic its response to an applied stress is completely determined by three moduli of elasticity, the Young modulus E, the shear modulus σ and the bulk modulus K, relating respectively to tensile, shear and compressive stresses. We will also be concerned with a derived constant ν the Poisson ratio, which is completely specified by the values of E, σ and K, and which is the ratio of the lateral contraction to the longitudinal extension in a uniaxial test. However, diamond like other crystals is not isotropic and the relations between stress and strain are therefore more complicated.

In order to define the strain in a small volume element we need in general nine *strain components* and to define the stresses nine *stress components*. (Although in each case only six of the components have independent values.) The relationships between the various components of stress and strain are now quite complicated and in the completely general case must be represented by a fourth-order tensor. However, because of the symmetry of real crystals this relationship simplifies greatly, and the elastic behaviour of a cubic crystal may be completely specified by only three moduli, or stiffness constants. These are c_{11} associated with tensile stresses along a cube axis, c_{12} associated with shear in a cube plane, and c_{44} associated with purely compressive stresses. For a fuller discussion of elastic stress and strain see for example Cottrell (1964) and Auld (1973).

The moduli of diamond are not readily measured directly because only small strains are produced by a given stress and the dimensions of a diamond are usually small. In addition, because diamond is hard and brittle and often of irregular shape it is generally difficult to apply forces so as to produce a uniform stress. However, the three stiffness constants c_{11}, c_{12}, and c_{44} are closely related to the velocity of compressional and shear waves, see for example Auld (1973). These relationships are particularly simple for waves propagating along the <100> or<110> axes as shown in Table 7.1. Hence by measuring the velocity along a <110> axis for a

Table 7.1 Wave velocities of elastic waves in terms of the stiffness constants c_{11}, c_{12} and c_{44}

Direction of propagation	Type	Velocity
<001>	Compressional	$(c_{11}/\rho)^{1/2}$
	Shear <100>	$(c_{44}/\rho)^{1/2}$
<110>	Compressional	$[(c_{11}+c_{12}+2c_{44})/2\rho]^{1/2}$
	Shear <$\bar{1}$10>	$[(c_{11}-c_{12})/2\rho]^{1/2}$
	Shear <001>	$[c_{44}/\rho]^{1/2}$

Table 7.2 Values and temperatures dependence of the stiffness constants of diamond (McSkimin and Andreatch, 1972)

	Value in GPa	Temp. coeff. $\times 10^5$
c_{11}	1079 ± 5	-1.37 ± 0.2
c_{12}	124 ± 5	-5.70 ± 1.5
c_{44}	578 ± 2	-1.25 ± 0.1

compressional and two differently polarized acoustic waves it is possible to obtain values for the three constants. The measurements are not entirely straightforward as the velocity is high and the time of flight is short, for details of the method using a slab of diamond about 12 mm thick see McSkimin, Andreatch and Glynn (1972). The results of these authors which differ somewhat from earlier results in the literature are given in Table 7.2. These values are in good agreement with values deduced from the so-called Brillouin scattering of light where the frequency of some of the light is shifted by the elastic thermal vibrations (Grimsditch and Ramdas, 1975). Results for the variation of c_{11}, c_{12} and c_{44} with temperature between $-196°C$ to $+50°C$ are given by McSkimin and Andreatch (1972) and shown in Table 7.2.

The Young modulus of diamond measured along a cube axis E_{11} and the bulk modulus K are given by:

$$E_{11} = \frac{(c_{11} - c_{12})(c_{11} + 2c_{12})}{c_{11} + c_{12}}$$

$$K = (c_{11} + 2\,c_{12})/3$$

Because of the anisotropy of diamond the values of both the Young's modulus and the shear moduli vary by about 10% with the crystallographic directions of the measurements (Ruoff, 1979). Therefore it is generally simpler and advantageous to work with stresses directed along principal directions of the crystal, but values of the moduli for any orientation may be obtained in terms of c_{11}, c_{12} and c_{44} by mathematical transformations described by Auld (1973).

Although the Young and other moduli do not vary greatly with direction this is not so for the Poisson ratio which is often an important parameter in the mechanics of fracture. The Poisson relation is determined by the three stiffness constants (see Auld (1973) but the position is complicated because its value depends both on the direction of the longitudinal extension and on the direction in which the lateral contraction is measured. Particular values calculated for diamond using the methods described by Auld (1973) vary considerably with direction ranging at least from 0.01 to 0.20.

Finally we note that the elastic constants of a material are generally taken to be independent of the amplitude of the strain but this assumption is valid only if the strain is small; it does not extend to strains of the order of 10% which generally precede the onset of fracture. Therefore a complete study of the conditions of fracture requires information on the relationships between stress and strain for stresses comparable with the fracture stresses which are of the order of 30 GPa or more. The available information is quite limited: McSkimin and Andreatch (1972) give values for the pressure coefficients of c_{11}, c_{12} and c_{44} obtained in the pressure range up to 0.14 GPa, and Grimsditch, Anastassakis and Cardona (1978) give values for the six third-order coefficients c_{111}, c_{112}, c_{123}, c_{144}, c_{166} and c_{456} determined from observations of Raman scattering (Section 3.1.c) at pressures up to 1 GPa. (If the stress–strain curve is linear, the elastic energy is determined by the coefficients c_{11}, c_{12} and c_{44} together with quadratic terms in the stresses. In the third-order approximation the energy includes additional contributions determined by the six third-order coefficients together with terms cubic in the stresses, see for example Kelly and Macmillan (1986).)

7.2.b Theoretical strength

In this section we consider the maximum strength to be expected from a diamond lattice free of dislocations, cracks, and other defects. Consider a rod loaded in uniform tension which eventually splits between two adjacent planes of atoms AA and BB (Figure 7.5). In order to produce fracture it is necessary to supply a minimum amount of energy, sufficient to break all the bonds joining the atoms in

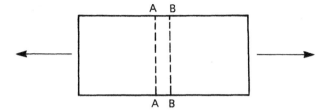

Figure 7.5 Schematic diagram to illustrate cleavage; see text

Table 7.3 Theoretical cleavage energies for diamond (Field, 1979) (To obtain a fracture surface energy, γ, divide by 2.)

Plane	Angle between plane and (111) plane	Cleavage energy Jm^{-2}
111	0° and 70°32′	10.6
332	10° 0′	11.7
221	15°48′	12.2
331	22° 0′	12.6
110	35°16′ and 90°	13.0
322	11°24′	13.4
321	22°12′	14.3
211	19°28′	15.0
320	36°48′	15.3
210	39°14′	16.4
311	29°30′	16.6
100	54°44′	18.4

the plane AA to those in the plane BB. This energy will be equal to the number of bonds between particular planes multiplied by the energy of the carbon–carbon bond which is known to be 5.8×10^{-19} J (Pauling, 1960). Values of this *cleavage energy* calculated for different sets of crystallographic planes are shown in Table 7.3. (The cleavage energy U_c is sometimes written as 2γ where γ is descibed as the *fracture energy* for one plane of atoms. Note, however, that this fracture energy is not the surface energy of the free surface as after fracture the broken bonds will interact to produce at least some reduction of energy, so the energy of fracture is generally greater than the surface energy.)

To obtain an estimate of the stress needed to produce cleavage Orowan (1948) pointed out that the energy to break the bonds must be provided by the stored elastic energy in the region adjacent to the plane of fracture. He thus obtained the relationship:

$$(\tfrac{1}{2}\,\sigma_c^2/E)2b \sim U_c \tag{7.2}$$

where σ_c is the critical stress, E the Young modulus, U_c the cleavage energy, and b the initial spacing between the fracture planes. Equation (7.2) was put forward by Orowan as an order of magnitude estimate because the choice of the value $2b$ to specify the effective volume of the crystal which provides the energy to break the bonds, though not unreasonable, is arbitrary. Nevertheless this simple approach gives a tolerable estimate for many materials, and for diamond gives a tensile strength of~ 200 GPa. (Even so Equation (7.2) should not be written as an equality as in some accounts!)

Subsequently Tyson (1966) and Kelly, Tyson and Cottrell (1967) made estimates of the critical tensile and shear stresses by considering the relative displacements of the atoms based on a postulated law of force for a pair of atoms. This method gave comparable results to those of Orowan, but like his does not allow for the non-linearity of the material at the higher stresses. A more recent approach developed by Ruoff (1978) uses the third-order elastic coefficients (Section 7.2.a) to evaluate the elastic energy as a function of the strain, and shows that the stress–strain curve has the general form shown in Figure 7.6. Once the stress reaches the

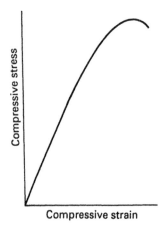

Compressive strain

Figure 7.6 Schematic diagram of estimated form of stress–strain curve of diamond (after Nelson and Ruoff, 1979)

Table 7.4 Estimated strength of perfect diamond (Whitlock and Ruoff, 1981)

Direction of stress	Compressive strength (GPa)	Tensile strength (GPa)
<100>	220	98
<110>	560	54
<111>	280	53

maximum value, the system is unstable against any further increase in stress, and the material fractures.

The most complete calculations of the strength of diamond are those of Whitlock and Ruoff (1981) which use the third-order elastic coefficients given by Grimsditch, Anastassakis and Cardona (1978), and supersede earlier calculations by Nelson and Ruoff (1979). Their results, given in Table 7.4, appear to be the best estimates available, but the measurements of the non-linearity on which they are based extend only up to 1 GPa so it is possible that further higher order elastic coefficients should be used in the calculation. As described in the next section, compressive strengths of the order shown in Table 7.4 are observed in diamond anvils used to obtain ultra-high pressures, but in other applications diamonds fail at much lower loads, generally because of tensile stresses acting on small cracks or flaws.

7.3 Compression of single crystals

The maximum compressive stress which a diamond will withstand without failure is known as its compressive strength. Because strength tests generally result in the fracture of the diamond into one or more pieces not many measurements are made on sizeable diamonds of good quality. Most of the available information comes from experiments using large good quality stones as diamond anvils, and from the study of small industrial diamonds less than 1 mm across.

7.3.a Diamond anvils

Figure 7.7 shows an arrangement used by Mao and Bell (1978) to subject small crystals of ruby to very high pressures by applying a load across anvils which concentrate the force over a small area. The steel gasket confines the specimen between the faces of the anvils, and permits the diamonds to bed down without producing local stresses at high spots on the surface. Considerable attention must be paid to mounting and supporting the anvils and maintaining their alignment as the load is increased. Details of the techniques employed are given by Caveney (1979); Seal (1984, 1987); Bundy (1986); and Onodera, Furono and Yazu (1986).

The shape of the anvils must be carefully chosen so that high pressures can be produced over the working face without creating excessive stresses elsewhere which

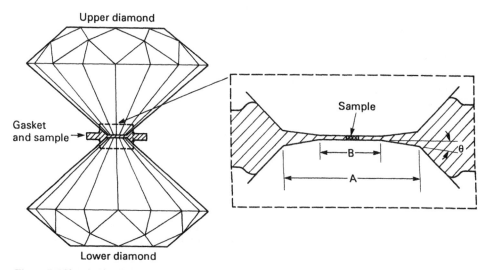

Figure 7.7 Sketch of a diamond anvil pressure cell. Height from top surface of upper diamond to bottom surface of lower diamond approximately 5 mm. Inset shows enlarged view of gasket-sample assembly. A and B indicate the outer and inner diameters of the bevelled region, the sample width is 250 μm (Mao and Bell, 1978)

cause failure. Adams and Shaw (1982) solve linear elastic equations for various different geometrical forms of the anvil and its supports. They show that the largest compressive and shear stresses occur near the working face, that the largest tensile stresses are produced by unsupported windows, and give examples of how these stresses depend on the shape of the anvils. Calculations by Bruno and Dunn (1984) and Moss and Goettel (1987) show that the performance of the anvils may be improved by an appropriate design of the bevels around the working face (Figure 7.7). The stresses in anvils have also been studied experimentally by observing the birefringence in a two-dimensional model using a pair of anvils cut from a thin diamond plate (Seal, 1984).

Pressures up to the order of 30 GPa are now readily obtained and experiments on many materials have been made at these pressures, see for example reviews by Jayaraman (1983) and Ferraro (1984). The pressure is generally obtained by observing the fluorescence of a ruby crystal, the wavelength of which is shifted by the pressure, see for example Mao and Bell (1978). Much higher pressures have been obtained by paying particular attention to the alignment of the anvils. Mao and Bell (1978) and Mao *et al.* (1979) describe four runs in which they estimate that the pressure reached values between 137 GPa and 172 GPa. In three of these runs one or both anvils failed completely but in the run to 172 GPa both anvils survived. However, an inspection in the SEM showed a depression on one of the faces (Figure 7.8) which a Talysurf profilometer showed to be about 0.5 μm deep.

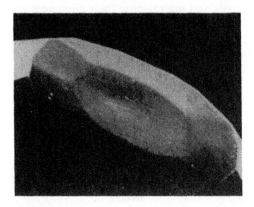

Figure 7.8 SEM micrograph of face of a diamond showing a depression produced after loading under a pressure of 172 GPa (Mao and Bell, 1978)

This result appears to give an example of plastic deformation at room temperature. The authors also observed an increase in the maximum value of the birefringence in the central area of the diamond from 6.10^{-5} to 6.10^{-3}, a change consistent with the onset of plastic deformation. (At the maximum pressure the centre of the face showed 'a light shade of brown colouration' which disappeared on releasing the pressure below 80 GPa; this result is not understood.)

Subsequently other authors have reported higher pressures without any accompanying deformation of the anvils. Goettel, Mao and Bell (1985) in an experiment up to 275 GPa observed no deformation of the anvils, no increase in the birefringence, nor any brown colouration. Subsequently Xu, Mao and Bell (1986) reported pressures up to 550 GPa. These various experiments using diamond anvils show that it is possible, under carefully controlled conditions, to obtain compressives stresses approaching the theoretical strength of the diamond.

Finally we note that at present there may be some uncertainty in the measured values of the higher pressures, as these depend on an extrapolation of the behaviour of the fluorescence of ruby from its behaviour at lower pressures. Also, calculations of the pressure distributions within the anvils assume that the elastic behaviour is linear which is not the case at high pressures. Hence it is not yet possible to make exact comparisons between experiment and theory particularly as some experimental accounts do not specify the crystallographic orientation of the anvils.

7.3.b Strength of grits

We now describe measurements on small diamonds, or *grits*, less than 1 mm across, of the type widely used in industrial applications. These grits are generally sorted for size by passing through a range of sieves and are often described by the mesh number of the finest sieve through which they pass, the US mesh number being the number of divisions per inch in the sieve. The strength of a grit is measured by placing it between steel or diamond anvils and applying an increasing load until fracture occurs, see for example Field *et al.* (1974). Measurements of this kind are reviewed by Field (1979) and by Field and Freeman (1981) who also give references to other experiments by Russian workers.

Most measurements have been made on irregularly shaped grits but two sets of results have been obtained on synthetic diamonds with well formed faces. By placing opposite faces carefully between the anvils and measuring their area it was possible to obtain values for the critical compressive stress as well as for the critical load. Field *et al.* (1974) studied two batches of grit of size 35/40 US mesh, containing respectively 'low' and 'high' concentrations of magnetic impurities. The results given in Table 7.5 show that the observed critical loads are greater using steel anvils, presumably because the grits bed into the steel and spread the load over a greater area of contact, thus avoiding premature failure at local high spots. However, even the loads observed with the steel anvils are far below the theoretical values. The table also shows that the strength of the grits depends on the level of impurity, being less in the more impure specimens. This is as expected because the magnetic impurities are almost certainly inclusions of solvent metal from the synthesis which give rise to internal cracks and strain.

It is more difficult to measure the strength of many industrial grits because they have a less uniform morphology and cannot be nicely positioned between anvils.

Table 7.5 Measured fracture loads of synthetic diamonds of size 35/40 US mesh with well defined cube faces (Field *et al.*, 1974)

Material	Steel anvils		Diamond anvils	
	Load, kg	Stress, GPa	Load, kg	Stress, GPa
High magnetic impurity	23 ± 3	4.5 ± 0.6	19 ± 5	3.8 ± 0.9
Low magnetic impurity	66 ± 7	13.0 ± 1.5	49 ± 10	9.7 ± 1.9

Table 7.6 Fracture load measurements on Debdust of varying grit sizes (Field *et al.*, 1974)

Size (US mesh)	Steel anvils (kg)	Diamond anvils (kg)
16/20	52 ± 5	29 ± 12
20/30	41 ± 3	24 ± 9
30/40	26 ± 4	11 ± 2.5
40/50	19.5 ± 3	8 ± 2.5
50/60	16 ± 3	6.5 ± 1.5

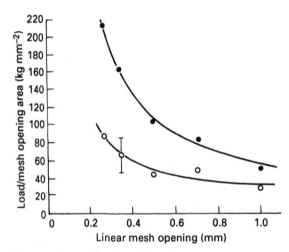

Figure 7.9 Strength of diamond grit measured by compressing single grits between anvils. ●, steel anvils; ○, diamond anvils. (Field *et al.*, 1974)

Even so, Field *et al.* (1974) were able to obtain quite consistent results in experiments to measure the average fracture loads for a range of natural grits of different sizes, see Table 7.6. Field *et al.* make estimates of the critical pressures by dividing the loads by the area of the appropriate mesh opening, and these values are plotted in Figure 7.9 against the linear size of the mesh opening. As in Table 7.5 the pressures are considerably below the theoretical strength, but there is also a marked size effect in that the critical pressure is larger, the smaller the grit. The size effect has also been observed in similar measurements on synthetic diamond by Mukhin, Yarmak and Popov (1974), Ziminov, Nikulin and Yarmak (1974), Novikov, Mal'nev and Voronin (1985) and Kolchemanov, Aparnikov and Bezrukov (1987), and in dynamic measurements by Stupkina (1971) and Feng and Field (1989).

Size effects somewhat similar to those described above have been observed in a variety of materials, see for example Kendall (1978). Kendall argues that smaller particles are more resistant to cracking because there is less elastic energy available in the particle to drive a crack forward. Alternatively, larger grits may contain larger flaws, as in the case of samples quoted by Novikov, Mal'nev and Voronin

(1985). There is not much information on this point save that similar size effects have been observed in both natural and synthetic grits coming from a range of sources. (Another size effect concerning the Hertz indentation of diamond is discussed in Section 7.5 in terms of the mechanics of the fracture process.)

Although the size effect is not well understood measurements of the compressive strength between anvils provide a good method of *comparing* the strengths of different grits particularly if the grits are of the same size. Two sets of such measurements are reported by Field (1979) for the compression of natural 20/30 mesh grits between steel anvils. The critical loads for grits with a smooth surface were about 40% greater than for rough grits, and the loads for blocky shaped grits were about 40% greater than for needle shaped grits. Other experiments by Field and Freeman (1981) suggest that the strength of grits may be increased by up to 30% by etching to remove surface roughness.

7.4 Cleavage

The diamond shown in Figure 7.2 exhibits a typical irregular fracture surface but diamond like many other materials may also fracture by a regular *cleavage* along a crystallographic plane with low Miller indices. In fact, diamond is readily cleaved along {111} planes in a process which divides the stone into two parts each with an approximate plane cleavage surface.

7.4.a Technique of cleaving

In the traditional method of cleaving a diamond, the stone is cemented to the end of a small wooden stick, and the cleaver then makes a nick or *kerf* to position his blade or knife (Figure 7.10). Until recently this kerf was generally produced by rubbing a sharp edge of a piece of diamond against one of the edges of the diamond to be cleaved in order to make a nick a fraction of a millimeter deep. Today, the use of a laser beam permits a more precise positioning of a better shaped kerf. The cleaver then places the edge of a steel blade in the kerf, with the plane of the blade coincident with the desired {111} cleavage plane. The blade is given a light tap with a small wooden mallet and if all goes well the diamond splits into two approximately along a {111} plane. Further practical details are given by Watermeyer (1982) and Vleeschdrager (1986).

Field (1979) has used high speed photography to observe the propagation of fracture in a diamond following different forms of impact. Figure 7.11 shows the

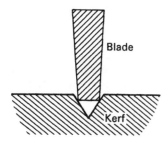

Figure 7.10 Sketch of a cleaver's knife located in a 'kerf' on a diamond prior to cleavage (after Vleeschdrager, 1986)

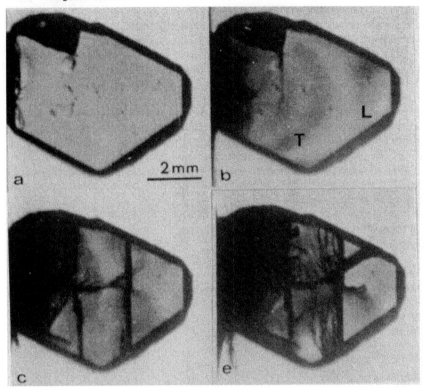

Figure 7.11 Fracture of a diamond slab after impulse loading at the mid–point of the left edge, viewed at successive instants. L indicates the region of the longitudinal wave and T the region of the transverse wave (Field, 1979)

fracture of a diamond slab about 1 mm thick by the impact of a small explosive charge mounted on a light plate attached to the left hand edge of the slab. The first frame in Figure 7.11 shows the slab just before the charge was set off. The next frame shows that the stress wave moving from the left has produced a crack running down from the top surface. The modulation of intensity in the picture also shows the longitudinal and transverse stress waves arriving at the far edge of the slab in the regions marked L and T. On reaching the boundary of the slab these waves are reflected, giving rise to enhanced tensile stresses, and then travel back through the slab to produce more damage. The third frame shows a vertical black band on the right which indicates a fracture running across the whole slab, and we also see a good example of a crack which has branched into two components. Finally the fourth frame shows extensive damage in almost the whole slab. In contrast Figure 7.12 shows photographs of a more controlled cleavage where the diamond was placed on a steel base plate and a small explosive charge drove a steel chisel into a kerf cut on the diamond, producing a clean split into only two parts.

It is well known that the quality or flatness of a cleaved diamond surface is very dependent on the skill and technique of the cleaver. However, it is only recently

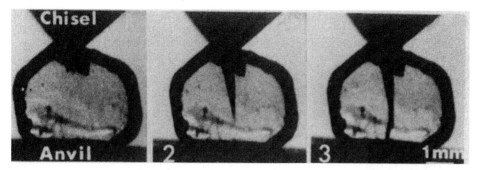

Figure 7.12 Cleavage of a diamond mounted on a steel anvil (Field, 1979)

that some account has been given of the physical principles involved. It is obviously important to apply the blade in the correct position so that the maximum stress is oriented to split the crystal along a {111} plane but other considerations are involved. When fracture begins the length of the initial crack increases and therefore the stress at its tip also increases, see Equation (7.1) in Section 7.1. Thus the crack tends to accelerate and measurements from photographs such as those in Figures 7.11 and 7.12 show that it may reach speeds of the order of half the velocity of the dilatational stress waves (Field, 1979). Not surprisingly, cracks of higher energy tend to produce rougher surfaces, so it is important for the cleaver to use as little force as possible

When the cleaver strikes the steel blade he generates a stress pulse which passes into the diamond and the quality of the cleavage depends on the length of this pulse. It must be long enough to maintain the stress until the diamond is completely cleaved, so as the crack velocity is typically of the order of $500\,\text{ms}^{-1}$ the pulse length must be of the order of $20\,\mu\text{s}$. In fact an estimate of the length of pulse produced by the traditional cleaver's mallet hitting the blade gives a time of this order (Field, 1979). It is also important that the length and width of the steel blade are such that no further stress pulses reflected from the sides of the blade arrive at the diamond before cleavage is complete. Finally we note that even when diamonds are most carefully cleaved the quality of the cleaved surface may vary considerably because of differences between the diamonds themselves (Section 10.3).

7.4.b Preferred cleavage planes

When diamonds are divided by cleavage the cleaver always splits them across {111} planes but it is also possible by suitably orienting the diamond to produce cleavage on some other planes particularly {011}. Ramaseshan (1946) studied 15 crystal fragments and observed cleavage faces corresponding to several low index crystallographic planes. However, the preferred cleavage is predominantly on {111} planes, and we now briefly discuss why this cleavage is preferred.

Until recently explanations of the preference for {111} cleavage followed Ramaseshan (1946) who considered cleavage to be most likely across planes with the lowest energy of cleavage. The values in Table 7.3 indicate that the {111} has the lowest energy, but the table also shows that other cleavage systems have energies not greatly different, so do not explain the very marked preference seen in practice for the {111} cleavage. In fact the most preferred cleavage will be that which requires the least applied *stress* rather than that which involves the least energy, and estimates of the stress required to produce tensile fracture were given in Table 7.4. These show that the critical stress for cleavage along {111} planes is only about half that for cleavage along {001} or {011} planes, and thus accounts for the preferential cleavage.

7.5 Hertzian fracture

7.5.a The Hertz test

A common method of assessing the strength of brittle materials, the so-called Hertz test, consists of forcing a hard sphere or spherically tipped indentor against a flat surface on the specimen to produce a crack running round the area of contact. The crack runs downwards and outwards to form a cone extending into the material as shown in Figure 7.13. The load required to produce the crack gives a measure of

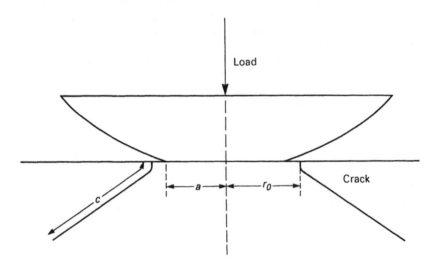

Figure 7.13 Schematic diagram of a fully formed Hertzian cone crack

the strength of the material but the relationship between strength and load is quite involved. (It is worth noting that a spherical indenter avoids the problems of positioning a flat indenter so as to obtain a uniform load on the specimen.)

When the load is applied in a Hertz test both specimen and indenter deform elastically and a stress field is set up in each. The stresses in isotropic materials were computed by Hertz (1881), see for example Johnson (1985). All the stresses are compressive except the radial stresses outside the area of contact which are tensions with their maximum value in the surface at the perimeter of the area of contact. The maximum compressive stress occurs on the surface under the axis of the indenter and has the value:

$$p_{max} = 1.5 \, p_{mean} \tag{7.3}$$

where

$$p_{mean} = P/\pi a^2 \tag{7.4}$$

P being the load on the indenter and a the radius of the area of contact. When the indenter and specimen are of the same material the value of this radius is given by

$$a^3 = 1.5 \, (1 - v^2) \, PR/E \tag{7.5}$$

where v is the Poisson ratio, R the radius of the indenter, and E the Young modulus. The maximum tensile stress is

$$\sigma_m = \tfrac{1}{2} \, (1 - 2v) \, p_{mean} \tag{7.6}$$

As the load on the indenter is increased a cone crack suddenly appears, extending downwards for a distance comparable to the diameter of the area of contact. On the simplest interpretation, fracture occurs when the maximum tensile stress equals the tensile strength of the material but this approach is unsatisfactory in at least three ways. First, the critical loads are generally substantially lower than the estimated values. Second, Equations (7.4), (7.5) and (7.6) imply that with a given material the critical load should be proportional to the square of the radius of the indenter, that is $P_c \propto R^2$, whereas the observed values often approximate to the so-called Auerbach relationship $P_c \propto R$. Finally the cracks generally initiate not on the perimeter of the contact area where the stress is greatest but at a radial distance from the centre about 10% greater as in Figure 7.13, see for example Wilshaw (1971) and Nadeau (1973).

There is no doubt that the observed strengths are well below the thoretical estimates because of the presence of small cracks which act as stress concentrators (Section 7.1). However, two different approaches have been used to account for the Auerbach relationship. Several authors have given explanations based on a model in which surface cracks or flaws exhibit a range of lengths, the smaller flaws being more common (Oh and Finnie, 1967; Tsai and Kolsky, 1967; Fisher, 1967; Hamilton and Rawson, 1970). The larger the indenter used in a Hertz test the larger the stressed area, so a larger indenter is more likely to encounter a longer crack, and an analysis of this situation leads to an Auerbach type relationship. However, although the distribution of crack lengths is an important factor, these treatments are not always satisfactory because the statistical nature of the argument leading to an Auerbach relationship also implies the existence of more scatter in the values of the critical load than is often observed (Harrison and Wilks, 1978). The treatments also assume that the cracks begin on the perimeter of the area of contact, and do not consider the form of the stress field. For further criticisms see Mouginot and Maugis (1985).

The other approach begins with an analysis of the stress field acting at the tip of a microscopic crack already present in the surface (Frank and Lawn, 1967). The application of the load causes the crack to grow, although at this stage it is still too small to be seen. As the crack penetrates deeper below the surface the stress field due to the applied load decreases rapidly, and this variation must be taken into account and leads to a two stage process as observed by Mikosza and Lawn (1980). Frank and Lawn's treatment accounts for the Auerbach relationship but assumes that the crack originates on the perimeter of the area of contact, and its derivation of the Auerbach effect depends too critically on the value of the Poisson ratio Wilshaw (1971).

7.5.b Theory of Mouginot and Maugis

The most complete treatment of Hertzian fracture is that of Mouginot and Maugis (1985). These authors consider an isotropic material with a distribution of flaw cracks of length c and consider the formation of a cone crack with circular symmetry about the axis of the indenter. That is, they assume that once a flaw crack begins to grow it immediately spreads round to form a ring, and that the subsequent growth of this ring crack can be discussed without reference to its initial formation. The authors emphasize that the rate at which the stress field falls off below the surface of the flat specimen is much more rapid near the edge of the area of contact than further out. Hence the stress at the tip of a crack of given length c may in fact be greater if the crack lies somewhat outside the circle of contact. Mouginot and Maugis therefore consider the general case of a circular crack, as in Figure 7.13, extending for a length c, and whose radius at the surface is r_0 where $r_0 \geqslant a$ the radius of the area of contact (Figure 7.13). The motion of this crack is determined by the strain energy release rate G defined as the energy

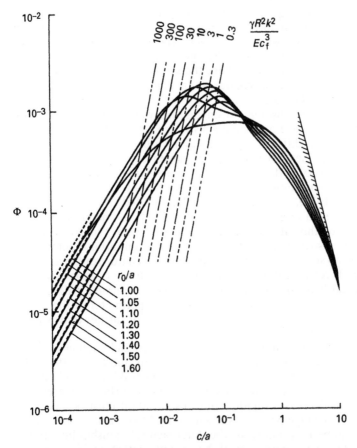

Figure 7.14 Function $\Phi(c/a)$ for several values of the parameter r_0/a and $\nu = 0.22$ as calculated by Mouginot and Maugis (1985); see text

provided by the stress field per unit extension of the length of the crack, and the authors show that

$$G = \frac{4}{\pi^3} \frac{1 - \nu^2}{E} \frac{P^2}{a^3} \Phi \, (c/a) \tag{7.7}$$

where ν is the Poisson ratio, r_0 the initial radial coordinate of the crack, $\Phi \, (c/a)$ is a function of the elastic field, and Figure 7.14 shows curves of $\Phi \, (c/a)$ for several values of r_0/a, calculated for $\nu = 0.22$.

The authors then assume that a crack initially of length c_f will extend if the energy, provided by the stress field is sufficient to provide the additional fracture energy, that is $G \geq 2\gamma$ where γ is the fracture energy of each new surface. It then follows that

$$\Phi_{\text{crit}} = \frac{9\pi^3 (1 - \nu^2)}{8} \frac{\gamma R^2 k^2}{Ec_f^3} \left[\frac{c_f}{a} \right]^3 \tag{7.8}$$

where $k=1$ for a completely rigid indenter and $k=2$ when the indenter and specimen are of the same material. Hence a plot of $\ln(\Phi)_{\text{crit}}$ against $\ln(c_f/a)$ for an arbitrary value of c_f gives a straight line of slope 3 and a set of these lines are shown in Figure 7.14 for different values of the parameter $(\gamma R^2 k^2/Ec_f^3)$. The cone crack initiates when G rises to the value 2γ, and the corresponding value of Φ must lie on both sets of curves in Figure 7.14. For example, for a flaw of size c_f such that $(\gamma R^2 k^2/Ec_f^3) = 10$ the critical value of Φ is given where the line labelled 10 cuts the curve in the other set which gives the highest value of Φ at the intersection, in this case the curve for $r_0 = 1.10a$. Then using Equation (7.5) to substitute for a^3 it follows that

$$P_c = (3\pi^3/4 \, \Phi_{\text{crit}}) \, \gamma R \tag{7.9}$$

hence by taking a range of values of c_f it is possible to construct a curve of $P_c/\gamma R$ versus $(R/c_f^{3/2})$ as in Figure 7.15. Under the critical load the crack begins to extend downwards, the value of (c/a) increases and both Φ and G increase so that the crack expands irreversibly until the condition $G = 2\gamma$ is again met at a greater value of (c/a). This second value of (c/a) gives the length of the crack formed under the critical load and is shown in Figure 7.15. We see that the calculated values of the critical loads fall into at least two distinct regions. First the region near minimum in the curve of $P_c/\gamma R$ versus $R/c_f^{3/2}$ corresponds to the Auerbach regime where $P_c \propto R$. Second, the region with high values of $R/c_f^{3/2}$ corresponds to large indenters and small flaws where the stress field remains approximately constant over the length of the crack. In this case the tensile stress

$$\sigma \propto P/a^2 \propto P^{1/3} \propto R^{2/3}$$

so that $P_c \propto R^2$ as shown by the line marked undiminished stress field. The authors describe various experiments on non-diamond materials by themselves and other workers which appear to be well described by their treatment.

7.5.c Cone cracks in diamond

Hertz indentations are usually made with indenters harder than the specimen but as diamond is the hardest of all materials tests on diamond must be made with a diamond indenter. This gives the possibility that the indenter may fail before the

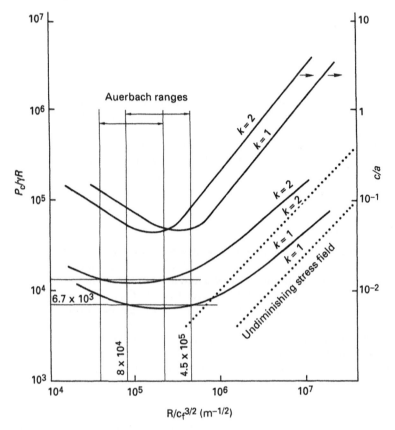

Figure 7.15 Critical loads P_c and equilibrium lengths of crack c in reduced co–ordinates as a function of the radius of the indenter (after Mouginot and Maugis, 1985)

specimen. On the other hand the use of a diamond indenter reduces the complications of frictional effects between specimen and indenter which must be taken into account when they are of different materials (Johnson, O'Connor and Woodward, 1973).

Measurements of the critical load as a function of the radius of the indenter have been made by Howes (1962), Ikawa, Shimada and Ono (1976), Ikawa, Shimada and Tsuwa (1985) and Ikawa and Shimada (1981, 1983). Ikawa and co-workers indented both natural and roughly and finely polished faces, and measured the critical loads required to produce fracture. They present their results as the critical tensile stresses at the edge of the circle of contact calculated from the critical loads by the Hertz relationships. These stresses show considerable scatter, of the order of a factor 2 indicating a much greater scatter on the measured loads which are proportional to the cubes of the stresses. Even so plots of the mean loads against the radii suggest that the critical load is at least approximately proportional to the radius for indenters of radius 5 μm to 500 μm; see also Seal (1958).

In an attempt to obtain results with less scatter Acton and Wilks (1989) made measurements on {011} surfaces polished in <100> and <110> directions to give

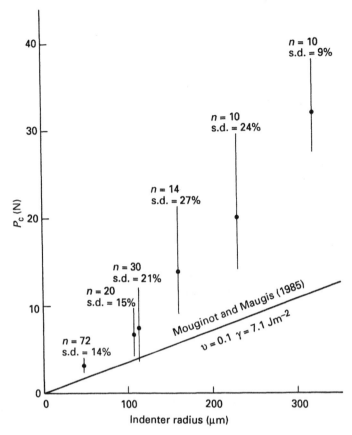

Figure 7.16 Range, mean value and standard deviation of loads to produce cone cracks on an {011} diamond surface with indenters of different radii. The surface had been polished with 0–1 μm powder in an <0$\bar{1}$1> direction to give a relatively rough finish; n is number of indents (Acton and Wilks, 1989).

smooth and rough finishes respectively. Indentations on the smooth surfaces caused damage to the indenter, but on the rougher surfaces the critical loads were less by a factor of about 4, and a series of experiments could be made without causing any observable damage to the indenter. Values of the loads obtained on the rougher surface are shown in Figure 7.16. The scatter is considerably less than in earlier measurements but still appreciable, and probably arises in several ways including variations in the surface polish of the specimen, unobserved damage to the indenter, and of course local variations in the diamond. These results show that the critical load varies linearly with the radius for the smaller radii and then rises somewhat faster than linearly for the larger radii. That is, the loads follow an approximate Auerbach relationship. The fact that the loads rise more quickly for the higher radii suggests that the experimental points correspond to that part of the curve for $P_c/\gamma R$ in Figure 7.15 which lies near to and just to the right of the minimum.

It must be remembered that Mouginot and Maugis assume that the specimen is isotropic, whereas diamond is appreciably anisotropic. Figure 7.17 shows the

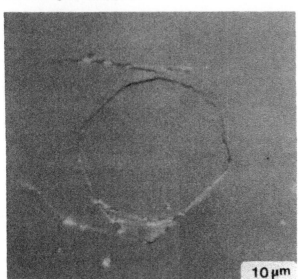

Figure 7.17 Ring crack on a finely polished {011} surface of diamond viewed in interference contrast optical microscope (Acton and Wilks, 1989)

outline of a cone crack on a finely polished {011} surface, the crack is not circular but is outlined by the traces of {111} cleavage planes. We also see a pair of subsidiary cracks which run out from the two sides of the ring lying in approximately <011> directions. These additional cracks appear to arise because the crack tends to propagate along the two cleavage planes with <011> traces, besides following the stress field.

To obtain an estimate of the critical loads in diamonds from the treatment of Mouginot and Maugis we must take account of the anisotropy, particularly the fact that the value of the Poisson ratio varies considerably with direction. An exact calculation is hardly possible but an estimate for the Auerbach regime extrapolated from the calculations of Mouginot and Maugis, using the value $\nu = 0.1$ suggested by Field (1986), is shown as a straight line in Figure 7.16. The agreement between this line and the experiments is probably as good as can be expected because of the uncertainties arising from the anisotropy of the diamond. It is possible to estimate the size of the initial flaws c_f from the position of the minimum in Figure 7.15, and this leads to a value of about 1 µm which is of the order suggested by previous and more simplistic estimates (Section 7.1). For further details of these various points see Acton and Wilks (1989).

The above discussion shows that the value of the critical load in a Hertz test depends in a complex way on the size of the flaws and the radius of the indenter. Therefore indentation experiments do not lead to a simple value of 'the strength of diamond'. The behaviour in the Auerbach regime is particularly interesting. The strength of diamond is generally much below its theoretical tensile strength and is determined by the presence of small cracks or flaws. Both Ikawa and Shimada (1983) and Acton and Wilks (1989) give examples of rougher surfaces exhibiting lower critical loads, as might be expected. Yet the above treatment implies that in the Auerbach regime the value of the critical load does *not* depend appreciably on the value of the initial flaw length c_f.

Finally we note that in the Auerbach regime the pressure under the indenter at the critical load increases as the indenter radius decreases. This suggests that indenters of very small radius may be able to produce stresses approaching the ultimate strength of the diamond. For example, a finely polished {001} surface under an indenter of radius 103 μm can withstand a mean pressure p_{mean} of at least 40 GPa (Acton, 1989). If the Auerbach relationship were to hold down to a radius of 5 μm as in the experiments of Ikawa *et al.* then it follows from the Hertz equations that the mean pressure at fracture would be ~110 GPa, corresponding to a maximum pressure of ~165 GPa and a maximum tensile stress of ~55 GPa. Critical stresses of this order have been reported by Ruoff and Wanagel (1977) using indenters of radius 2 μm to 20 μm. There is also the further consideration that with very small indenters of micron-size radii the areas of contact become so small that the indenter may be testing material free of flaws.

7.5.d Energy of cleavage

Other experiments on cone cracks have been made to evaluate the energy of cleavage. If the load on a fully formed cone crack is increased the crack extends steadily with increasing load, and this situation is more easily analysed than the irreversible formation of the crack. The experiment generally uses a truncated cone as an indenter rather than a spherical tip so that the area of contact remains constant, thus avoiding the production of further cone cracks. Note that the stress field is no longer Hertzian because of the presence of the crack.

We first consider an isotropic material in which a cone crack extends along the line of the principal stress trajectory, the angle of the cone depending on the value of the Poisson ratio (Finnie and Vaidyanathan, 1974). The cleavage energy has been calculated by Roesler (1956) by determining the elastic energy release rate for a cone, and thus finds that

$$2\gamma = \frac{C P^2 (1 + \nu) \sin \alpha}{2 \pi E R_B^3} \tag{7.10}$$

where γ is the fracture energy, P the load, α the semi-angle of the cone, R_B the radius of the base of the cone, and C a constant which is a function of α and the Poisson ratio ν. The value of the constant C must be determined by solving the elastic equations for the stresses. This calculation was made by Roesler for a crack of semi-angle $\alpha = 68° 12'$ in isotropic material with $\nu = 0.25$, but this result was amended by a factor 0.81 in a further calculation by Finnie and Vaidyanathan (1974).

The above result was applied to diamond by Field and Freeman (1981) who observed the extension of cone cracks on {001} surfaces and give a plot of P^2 versus R_B^3 (Figure 7.18) which shows the proportionality predicted by Equation (7.10). The semi-angles of the cones were about 48° which appear to be in reasonable agreement with the values predicted by Finnie and Vaidyanathan if as before we take $\nu = 0.1$. However, the only values available for the constant C were those calculated by Roesler and by Finnie and Vaidyanathan for cones of semi-angles 68° or 72° in isotropic material with $\nu = 0.25$. For lack of a better figure Field and Freeman took Finnie and Vaidyanathan's value for a 68° cone and

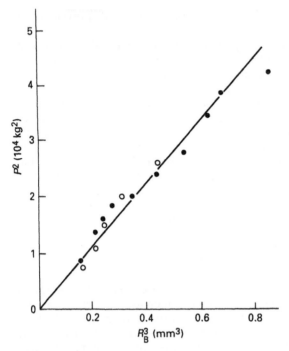

Figure 7.18 Plots of P^2 versus R_B^3 for Type Ia diamond indented on a {001} face; see text (Field, 1979)

hence obtained a mean value for the cleavage energy of $11.0\,\mathrm{J\,m^{-2}}$. This figure is quite close to the value of $10.6\,\mathrm{J\,m^{-2}}$ in Table 7.3 for {111} cleavages, but the agreement is somewhat fortuitous. The semi-angle of the cone was about 48° whereas the {111} cleavage planes make angles of only 35° with the vertical. Also the value of the constant C is taken from calculations not exactly applicable to diamond.

Field and Freeman (1981) also made a few indentations on polished surfaces lying within a few degrees of {011}. These cracks had semi-angles of about 60° and like those mentioned in the previous section were not as regular as the cracks on the {001} surfaces. The authors quote a mean cleavage energy of $14.2\,\mathrm{J\,m^{-2}}$.

7.6 Dynamic tests

So far we have described strength tests which are essentially static or quasi-static, but in many situations strength against impact very important. However, the measurement of the impact strength of brittle materials presents difficulties and only a limited amount of information is available on diamond.

7.6.a Sliding indenters

Experiments with spherical indenters sliding on glass (Gilroy and Hirst, 1969) and on titanium carbide (Powell and Tabor, 1970) show that the load required to produce ring cracks may be considerably reduced if the indenter is sliding across the surface. This comes about because the sliding results in additional tensile stresses induced by the friction forces, and the maximum tensile stress in the surface is increased by a factor:

$$F = 1 + \left(\frac{3\pi}{8}\right)\left(\frac{4 + \nu}{1 - 2\nu}\right)\mu \qquad (7.11)$$

where ν is the Poisson ratio and μ the coefficient of friction (Hamilton and Goodman, 1966). In a Hertz type indentation the stresses are proportional to the cube root of the load, hence with a moving indenter we expect the critical load to be reduced by a factor F^3, and the above experiments gave results in reasonable agreement with this expression.

The effect of sliding on the indentation of diamond has been observed in experiments which measured both the coefficient of friction and the critical load required to produce a cone crack (Acton, 1989). Measurements were made on an {001} surface polished with 0–1 μm diamond powder in a <100> direction to give a fine finish, with an indenter of radius 161 μm sliding either in <110> or <100> directions with a velocity of about 0.07 mm s^{-1}. At the start of the experiment the indenter was loaded statically and made six indentations at loads of 50 N without producing cracks. After the sliding experiments, a further static indentation damaged the indenter, but experiments on other indenters suggest that the critical load was probably no more than 70 N. Values of the critical loads and coefficients of friction are given in Table 7.7 together with values of the factor F^3 calculated

Table 7.7 Critical load P_c required to produce ring cracks for an indenter of radius 161 μm on a polished {001} surface for static and sliding indentations. The table also shows measured values of the sliding friction and values of the factor F^3 calculated from Equation (7.11), see text

	P(N)	μ	F^3
Static	>50		
Sliding <011>	19	0.034	1.7
Sliding <100>	3.4	0.10	4.2

by taking $\nu = 0.1$. We see that the reduction of the critical load when sliding in a <110> direction is of the same order as predicted by Equation (7.11), but that the reduction in the <100> direction is much greater than that predicted. A more complete treatment of the critical load under a sliding indenter is given by Mouginot (1987) based on the static treatment of Mouginot and Maugis (1985), and for small values of the coefficient of friction the analysis leads to results similar to those given by the simpler treatments. Hence it appears that the critical loads observed in a <100> direction are anomalously low, probably because of the unusual nature of polished diamond surfaces (Section 9.4) which is reflected in measurements of the friction of diamond on diamond (Section 11.4).

7.6.b High strain rates

Material may often withstand a greater load before fracturing if the load is applied for only a short time because the fracture process takes time to develop, see for example Cottrell (1964). This effect has been observed in diamond by Levitt and Nabarro (1966) who observed the mutual impact of diamond spheres of about 2 mm diameter obtained by tumbling 700 selected dodecahedron diamonds of good quality in an air mill. Six different methods were used to bring the two diamonds together ranging from a slow compression in a tensometer to firing one from an air gun, thus giving contact times ranging from 10^3s to 10^{-6}s.

About 25 runs were made by each method to determine the critical stress when either or both diamonds broke into two or more pieces. A summary of the results is shown in Table 7.8 which gives the estimated time of contact and maximum

Table 7.8 Average nominal breaking stress of diamond measured at different rates of strain (Levitt and Nabarro, 1966)

Contact time (s)	Breaking stress (GPa)
4.2 10^3	63
2.4 10^1	65
1.1 10^{-2}	115
1.6 10^{-4}	129
6.1 10^{-6}	108
1.1 10^{-6}	70

stress between the balls calculated from the velocities and mechanics of the systems together with the assumption that the elastic field is given by the Hertz equations.

Table 7.8 shows that the stress required to produce fracture first increased at shorter times of impact but then decreased again. This unexpected decrease appeared to be accompanied by a change in the mode of fracture. The diamonds which fractured at the four lower rates of loading showed a ring crack, usually an incomplete cone crack, and cleavage on one or more {111} planes. Hence the final fracture of these diamonds appeared to have developed from a growing cone crack at a pressure approximately twice that needed to initiate a cone. However, for the two shortest times of contact the broken diamonds showed no cone cracks and often no ring cracks even though unbroken stones showed normal ring cracks. Also on several occasions the cleavage followed {110} planes.

The authors suggest that the development of a cone crack is a relatively slow process taking a time T_c. Hence as the time of contact decreases and becomes less than T_c a cone crack is only formed if the stress is raised sufficiently to develop the crack in the time available. The authors also note that the production of a cone will reduce the stress at the top of the crack below the value given by the Hertz equations. Therefore if the time of impact is reduced and the cone does not fully develop, the stress at the tip of the crack may be correspondingly greater and total fracture may occur at a smaller load. More recently Feng and Field (1989) measured the compressive strength of synthetic cubo-octahedral diamond grits in the size range 350 μm to 1000 μm, both statically and by an impact method giving a time of contact 'usually less than 5 μs', and observed not much difference between the static and dynamic strengths. This result is not necessarily inconsistent with those of Levitt and Nabarro in which the values for the slowest and fastest rates of impact were also quite similar. Further experiments would be of interest.

7.6.c Testing of grits

Much of the production of industrial diamond is in the form of grits of size ranging from 0.05 mm to 0.5 mm which are used in the manufacture of saws and drills. The performance of these tools depends on the performance of the diamonds under conditions of impact so some assessment of the impact strength of grit is most desirable. Yet the measurement of the impact strength of extremely large numbers of very small diamonds presents serious problems.

A widely used method of assessing impact strength, the Friatester, developed by Belling and Dyer (1964) and Belling and Bialy (1974), simulates some of the working conditions of the grits and is relatively simple and reliable. The diamond grit is first sieved through two screens to obtain a sample of fairly uniform size, which is placed in a cylindrical steel capsule of about 12 mm internal diameter and 22 mm long, together with a steel ball about 6 mm diameter. The capsule is then oscillated to and fro along its axis for a fixed number of cycles at a frequency of about 40 Hz. At the end of the test the grit is again sieved to determine what fraction of the material has not been reduced in size, and the value of this fraction is taken as a measure of the impact strength.

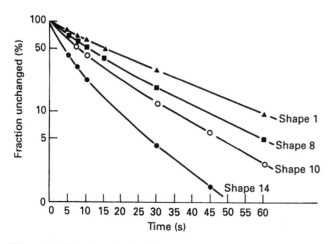

Figure 7.19 Fraction of grits of the same size but different shape remaining unchanged in size after different periods of time in a Friatester; see text (Belling and Dyer, 1964)

Figure 7.19 shows an example of results obtained with a Friatester in an experiment to investigate the relative performance of grits of the same size but different shape. These grits were initially sorted for shape on a vibrating table developed by De Beers which separates the diamonds into a range of 14 different shapes (see Section 15.2.a). Shape 1 consists of particles of almost symmetrical blocky shape; shapes of higher index number contain particles of progressively greater departure from blockiness, until finally shape 14 contains very thin flat and needle shaped stones. Figure 7.19 shows the fraction of grits remaining unchanged in size after various periods of time in the Friatester. The more blocky material suffered less breakdown. These results are not surprising but give a good example how the Friatester can be used to distinguish between grits of different strength.

The Friatester does not provide a precise measurement of the strength of diamond but like many industrial tests makes useful comparisons between different samples. However, for comparisons to be meaningful it is essential to ensure that all the various conditions of the test remain constant. These points of experimental detail, including the efficiency of the sieving of the grits, the velocity of the motion, and the mass of grit employed are discussed in detail by Belling and Bialy (1974). These authors also outline methods of analysing the shape of curves such as those in Figure 7.19 in order to obtain more precise values from the tests. Thus the Friatester has now been developed to detect quite small differences in performance and is a valuable means of quality control. (References to other experiments to measure the strength of grits, including several Russian papers, are given by Field and Freeman (1981).)

7.7 Fatigue

The strength of a material is determined by applying an increasing stress until it fails. If the test is repeated with a somewhat lower stress, say 75% of the critical stress, the material does not fail. However, if this lower stress is applied repeatedly for a large number of cycles the material may eventually fail. Repeated stressing tends to weaken the material and this phenomenon is known as fatigue.

The best known examples of fatigue concern metals particularly in aircraft. Repeated stresses too small to cause failure are often sufficient to move dislocations, thus leading to the production of point defects including vacancies. These defects eventually coalesce to form holes and cracks which concentrate and increase the local stresses in the material, eventually to the critical value. Fatigue in brittle materials arises by other mechanisms and has not been studied so extensively. We now discuss fatigue in diamond at room temperature where the diamond is virtually completely brittle.

7.7.a Fatigue at room temperature

Bowden and Tabor (1965) briefly describe two investigations in which diamond surfaces were stressed repeatedly. In the first N.L.Hancox (unpublished) observed the production of ring cracks after several thousand impacts by tungsten balls. Subsequently Cooper (1961) produced ring cracks by 2000 repeated static loadings using a stress only 25% of that required to produce a crack by one loading. Levitt and Nabarro (1966) give several examples of fracture after repeated loadings at stresses below the critical level, but do not report any fracture at loads as low as those used by Cooper. In experiments described below Bell et al. (1977) observed ring cracks produced by repeated loadings up to 75% of the critical stress after either 100 or 1000 cycles. Adams and Shaw (1982) refer to diamond anvils failing under pressure after several applications.

Evidence of fatigue is also provided by two experiments in which diamonds were rubbed against other materials. Tolansky (1965) turned a long spiral groove on a glass cylinder with a glazier's knife and observed that the edge of the diamond remained virtually unchanged until it suddenly deteriorated after ruling a length of about 2 km. Later, in a set of detailed experiments, Crompton, Hirst and Howse (1973) measured the wear of diamond rubbing on a range of metal surfaces of different hardnesses. In all cases the diamond was observed to wear steadily but only after an induction period during which the wear was virtually zero, the induction period being shorter the harder the material of the disc, as described in more detail in Section 13.2.

7.7.b Subsurface damage

Valuable information on the mechanism of fatigue may be obtained with the scanning electron microscope used in the cathodoluminescent (CL) mode (Section 4.1). As an introduction to this technique we describe some experiments by Casey *et al.* (1977). Figure 7.20 is an optical micrograph of the ring of a cone crack on a natural {111} surface of diamond, the crack being a hexagon because of the six-fold sets of cleavage planes coming up to the face. The micrograph shows only the surface of the diamond but of course a cone crack extends downwards as is seen in Figure 7.21 which shows the central core of a cone crack on a polished {001} face exposed by subsequent damage which removed the surrounding material. Figure 7.22 shows the surface of the diamond of Figure 7.20 viewed in the CL mode of the SEM. As discussed by Casey *et al.* the dark shadows arise because the luminescent light excited by the electron beam may be totally reflected by the crack and not reach the collector.

The above technique is a valuable method of detecting subsurface cracking but it is necessary to pay attention to several points of detail. Because the light may be reflected internally one or more times within the diamond before emerging to the detector, care is needed in relating the form of the shadowing to the geometry

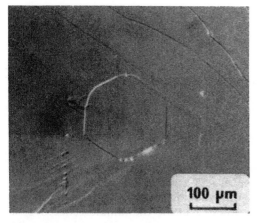

Figure 7.20 Ring crack on a natural {111} diamond surface silvered and viewed in an optical interference microscope (Casey *et al.*, 1977)

Figure 7.21 SEM micrograph of a polished {001} diamond surface showing the central core of a cone crack exposed by subsequent damage. The almost flat sides of the core are approximately {111} planes (Casey *et al.*, 1977)

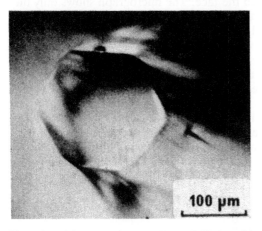

Figure 7.22 Ring crack shown in Figure 7.20 viewed in the CL mode of the SEM (Casey *et al.*, 1977)

of the cracks. It is often instructive to observe how the shadow pattern changes if the diamond is rotated, see Casey et al. (1977). To obtain the best results it is obviously necessary to select diamonds which give an appreciable and fairly uniform luminescence.

The appearance of the damage in the micrograph is particularly sensitive to the operating conditions of the microscope. Image contrast and resolution are affected by (i), accelerating voltage; (ii), beam current; (iii), focusing; (iv), tilt of the specimen with respect to the incident beam and the light detector; and (v), the details of the electronic signal processing. In general the best images are obtained by reducing the accelerating voltage in order to excite the luminescence over a relatively small depth, and increasing the beam current to maintain the intensity of the luminescence. In practice the lower useful limit to the beam voltage is set by the increasing amount of noise as the voltage is reduced. For further details see Casey *et al.* (1977). The technique is also described by Enomoto and Tabor (1981) and by Enomoto, Yamanaka and Saito (1986).

7.7.c Cumulative damage

Evidence of the cumulative damage responsible for the fatigue of diamond was given by an experiment by Bell *et al.* (1977). Small spheres of diamond of radius about 0.5 mm were used to make repeated indentations on polished {001} faces of diamond selected to be as free as possible from both birefringence and any polishing flaws. Ideally, it would have been preferable to study the progressive change at one site with an increasing number of indentations, but it would have been difficult to replace the diamond specimen in exactly the same position after each inspection. Therefore the authors adopted the procedure of making approximately 1, 10, 100 and 1000 indentations at four adjacent sites. Particular care was taken to mount the indenter on a high precision bearing to prevent any lateral motion or rotation.

Figure 7.23(a) shows a surface indented for 1, 11, 104 and 985 cycles viewed in an optical microscope with the Nomarski technique. The dashed circles indicate the location of the test sites and we see that almost no damage is visible after 1 and 11 indentations, very little after 104, but quite appreciable damage after 985 cycles. Figure 7.23(b) shows the same area as Figure 7.23(a) viewed in the CL

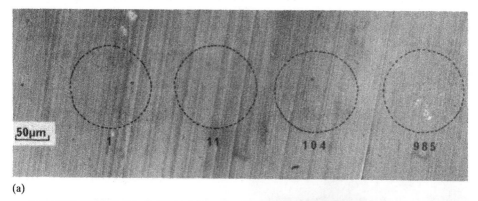

(a)

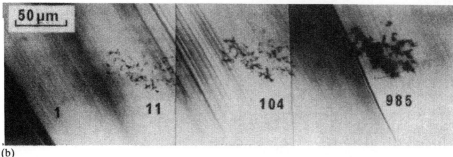

(b)

Figure 7.23 Damage on a polished diamond surface produced by 1,11,104 and 985 repeated indentations, viewed (a), in optical interference microscope; (b), in CL mode of SEM (Bell *et al.*, 1977)

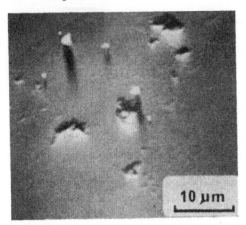

Figure 7.24 SEM micrograph of shallow pits produced by repeated indentations. Small white particles of dirt indicate the direction of shadowing given by the beam (Bell *et al.*, 1977)

mode of the SEM; the damage is now more obvious and is seen to increase steadily with the number of indentations.

Figure 7.24 shows a detail at higher magnification of some typical damage, together with small pieces of dirt on an adjacent part of the surface. Both the dirt and the damage are shadowed as a consequence of the relative positions of the electron beam, the surface, and the electron collector. By comparing the shadowing on the two features we see that the damage has the form of small depressions. Figure 7.23(b) also shows that although the damage is more pronounced after a greater number of indentations, the pattern of damage is very similar at each test site. It therefore seems probable that this pattern arises from the particular rugosities on the surface of the indenter used in all four tests.

Most of the damage appears to have the same general character but there is also a second form of damage which is only observed after many indentations. This damage appears as thin black lines in Figure 7.23(b) particular after 985 cycles and is shown at higher magnification in Figure 7.25. The faint banding visible over the

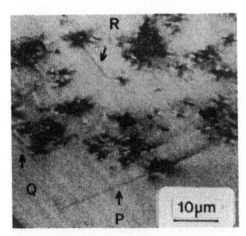

Figure 7.25 Details of damage after 985 indentations; see text (Bell *et al.*, 1977)

whole of Figure 7.25 is due to {111} growth layers which intersect the polished surface of the diamond. Hence the crack R lies in a {111} plane and, after allowing for the tilt of the specimen in the microscope, it follows that the line P also lies approximately in another {111} plane.

7.7.d Mechanism of fatigue

The above results show that localized damage builds up and eventually produces line cracks which will grow to form a complete ring crack. However, we still have to consider why the initial damage builds up in the first place. There are three obvious possibilities. First, the damage extends because of minute differences in the location of successive indentations. Second, the presence of debris from previous indentations will result in additional concentrations of stress under the indenter. Third, as the depressions are formed by the cracking out of material, they will in places be deeper than the protuberances on the indenter, so when the load is reapplied there is a possibility of further motion. Whatever the case, there is a tendency for these depressions to grow in size, and eventually coalesce to form a continuous crack.

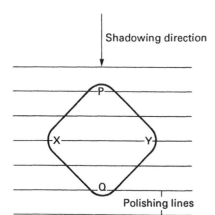

Figure 7.26 Sketch to indicate positions on a ring crack relative to the polishing lines; see text (Bell *et al.*, 1977)

Further information on the mechanism of fatigue is given by a study of fully developed ring cracks using shadowed replicas examined in the TEM (Bell *et al.*, 1977). Figure 7.26 shows a schematic diagram of a ring crack on an {001} surface showing the direction of the polishing lines and Figure 7.27 a TEM micrograph of a part of such a crack running in a <110> direction and cutting across the polishing lines running parallel to <100>. We see that even at this high magnification, the crack runs in a straight line, presumably because this direction lies in a {111} plane which is the preferred cleavage plane. Figure 7.28 shows micrographs of the opposite sides of the same crack at the positions P and Q in Figure 7.26. The appearance of these cracks is quite different, their general direction is parallel to the polishing lines, that is to <100>, but they are made up of short lengths parallel to <110> directions. The fracture has been produced by the diamond cleaving on the preferred {111} planes, while still proceeding on average in a <100> direction.

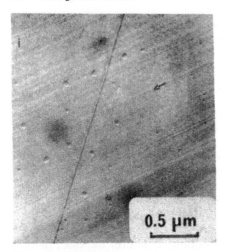

Figure 7.27 Detail of a ring crack on a {001} polished face of diamond running in an <110> direction. Viewed with replica technique and TEM (Bell *et al.*, 1977)

Figure 7.28 Details of the ring crack in Figure 7.27 to illustrate incomplete closure. The two lengths of crack shown are located at opposite ends of a diameter; see text. The arrow indicates the direction of the polishing lines, 'i' the inner side of the crack (Bell *et al.*, 1977)

The upper crack in Figure 7.28, corresponding to position P in Figure 7.26, appears light and the lower crack corresponding to position Q appears dark. The replica was preshadowed by evaporating tungsten at an angle of 45° to the surface from the direction indicated by the arrow in Figure 7.26, so the change of intensity of the micrograph at the crack indicates a step which either collects additional tungsten, or shields an area from receiving any. It is clear from Figures 7.26 and 7.28 that for both cracks the height of the surface is less on the inner side of the crack than the outer. Hence, the diamond does not recover its initial state on removing the load.

Further information is obtained by observing the polishing lines as they cross the zig-zag cracks in Figure 7.28. The polishing was carried out before producing the crack, and the lines were then straight and continuous. However, there are now obvious displacements of the lines as they cross the cracks, particularly at the lower end of the upper crack, and at both P and Q the direction of the displacement is such that the two sides of the crack have moved away from each other. That is, even after the load has been removed the crack remains jammed open, with the consequent height discontinuity discussed above. Note also that this displacement of the polishing lines is most marked at positions P and Q, and is imperceptible at positions X and Y in Figure 7.26, because here the crack opens in a direction parallel to the polishing lines, and so produces no discontinuities in the lines. These results confirm those of Lawn and Komatsu (1966), who studied the surface topography near a ring crack with multiple beam interferometry, and the state of strain in the adjacent material by X-ray topographs, and concluded that the strain field arose because the cone crack had not closed completely when the load was removed; see also Section 8.2.a.

It is not surprising that there should be incomplete closing of cracks in diamond on removal of the load, as this is a comparatively common phenomena in brittle crystals, see for example Williams, Lawn and Swain (1970). As discussed by these authors, the most likely mechanism preventing closure is the presence of debris. Tolansky (1965) has given examples of the cleavage of diamond where the two cleaved surfaces do not match exactly, so minute fragments of diamond must have been removed during the cleavage. Under the experimental conditions of the Hertz test such fragments would almost certainly prevent closure of a cone crack. In addition in many materials the intense stress at the tip of an expanding crack may cause sufficient plastic strain to prevent the crack closing but the possibility of plastic deformation in diamond at room temperature can generally be ignored as discussed in the next chapter.

References

Acton, M. R. (1989) *D.Phil Thesis*, Oxford University

Acton, M. R. and Wilks, J. (1989) *Journal of Materials Science*, **24**, 4229–4238

Adams, D. M. and Shaw, A. C. (1982) *Journal of Physics D*. **15**, 1609–1635

Atkins, A. G. and Tabor, D. (1966) *Proceedings of the Royal Society*, **A292**, 441–459

Auld, B. A. (1973) *Acoustic Fields and Waves in Solids, Vol. 1*, Wiley, New York. Chapter 3

Bell, J. G., Stuivinga, M. E. C., Thornton, A. G. and Wilks, J. (1977) *Journal of Physics D*, **10**, 1379–1387

Belling, N. G. and Bialy, L. (1974) *Industrial Diamond Review*, **34**, 285–291

Belling, N. G. and Dyer, H. B. (1964) *Impact Strength Determination of Diamond Abrasive Grit*, (Booklet) De Beers Industrial Diamond Division, London

Bowden, F. P. and Tabor, D. (1965) In *Physical Properties of Diamond,* (ed. R. Berman), Clarendon Press, Oxford, pp. 184–220

Bruno, M. S. and Dunn, K. J. (1984) *Review of Scientific Instruments,* **55**, 940–943

Bundy, F.P. (1986) *Physica,* **139** and **140B**, 42–51

Casey, M., Lewis, A. G., Thornton, A. G. and Wilks, J. (1977) *Journal of Physics D,* **10**, 1877–1881

Caveney, R. J. (1979) In *The Properties of Diamond,* (ed. J. E. Field), Academic Press, London, pp. 619–639

Cooper, R. (1961) *Ph.D. Thesis,* Cambridge University

Cottrell, A. H. (1964) *The Mechanical Properties of Matter,* John Wiley, London

Crompton, D., Hirst, W. and Howse, M. G. W. (1973) *Proceedings of the Royal Society,* **A333**, 435–454

Enomoto, Y. and Tabor, D. (1981) *Proceedings of the Royal Society,* **A373**, 405–417

Enomoto, Y., Yamanaka, K. and Saito, K. (1986) *Wear,* **110**, 239–254

Feng, Z. and Field, J.E. (1989) *Industrial Diamond Review,* **49**, 104–108

Ferraro, J. R. (1984) *Vibrational Spectroscopy at High External Pressures, The Diamond Anvil Cell.* Academic Press, Orlando Fl.

Field, J. E. (1979) In *The Properties of Diamond,* (ed. J.E. Field), Academic Press, London, pp. 281–324

Field, J.E. (1986) In *Proceedings of the Second International Conference on the Science of Hard Materials,* (Rhodes 1984), Institute of Physics Conference Series No 75. (ed. E.A. Almond, C.A. Brookes and R. Warren), Adam Hilger, Bristol, pp. 181–205

Field, J. E. and Freeeman C. J. (1981) *Philosophical Magazine,* **43**, 595–618

Field, J. E., Hauser, H. M., Hutchings, I. M. and Woodward, A. C. (1974) *Industrial Diamond Review,* **34**, 255–259

Finnie, I. and Vaidyanathan, S. (1974) In *Fracture Mechanics of Ceramics,* Vol. 1, (ed. R. C. Bradt, D. P. H. Hasselman and, F. F. Lange), Plenum, New York, pp. 231–244.

Fisher, G.M.C. (1967) *Journal of Applied Physics,* **38**, 1781–1786

Frank, F. C. and Lawn, B. R. (1967) *Proceedings of the Royal Society.* **A299**, 291–306

Gane, N. and Cox, J. M. (1970) *Journal of Physics D,* **3**, 121–124

Gilroy, D.R. and Hirst, W. (1969) *Journal of Physics D,* **2**, 1784–1787

Goettel, K. A., Mao, H. K. and Bell, P. M. (1985) *Review of Scientific Instruments,* **56**, 1420–1427

Griffith, A. A. (1921) *Philosophical Transactions of the Royal Society,* **221**, 163–198

Grimsditch, M. H., Anastassakis, E. and Cardona, M. (1978) *Physical Review B,* **18**, 901–904

Grimsditch, M. H. and Ramdas, A. K. (1975) *Physical Review B,* **11**, 3139–3148

Hamilton, G. M. and Goodman, L. E. (1966) *Transactions of the ASME Journal of Applied Mechanics,* **33**, 371–376

Hamilton, B. and Rawson, H. (1970) *Journal of the Mechanics and Physics of Solids,* **18**, 127–147

Harrison, J. and Wilks, J. (1978) *Journal of Physics D,* **11**, 73–81

Hertz, H. (1881) *J. reine angew. Math.* **92** 156. An English translation appears in H. Hertz Miscellaneous Papers, (1896) Macmillan, London, Chapters 5 and 6

Hockey, B. J. (1971) *Journal of the American Ceramic Society,* **54**, 223–231

Howes, V. R. (1962) *Proceedings of the Physical Society,* **80**, 78–80

Ikawa, N., Shimada, S. and Ono, T. (1976) *Technology Reports of the Osaka University,* **26**, 245–254

Ikawa, N. and Shimada, S. (1981) *Technology Reports of the Osaka University,* **31**, 315–323

Ikawa, N. and Shimada, S. (1983) *Technology Reports of the Osaka University,* **33**, 343–348

Ikawa, N., Shimada, S. and Tsuwa, H. (1985) *Annals of the CIRP,* **34**, 117–120

Jayaraman, A. (1983) *Reviews of Modern Physics,* **55**, 65–108

Johnson, K. L. (1985) *Contact Mechanics,* Cambridge University Press, Cambridge, pp. 90–104

Johnson, K. L., O'Connor, J. J. and Woodward, A. C. (1973) *Proceedings of the Royal Society,* **A334**, 95–117

Kelly, A. and Macmillan, N. H. (1986) *Strong Solids,* (3rd edn), Clarendon Press, Oxford

Kelly, A., Tyson, W. R. and Cottrell, A. H. (1967) *Philosophical Magazine,* **15**, 567–586

Kendall, K. (1978) *Proceedings of the Royal Society,* **A361**, 245–263

Kolchemanov, N. A., Aparnikov, G. L. and Bezrukov, G. N. (1987) *Stanki i Instrument, No.* **2**, 19–20

Lawn, B. R., Hockey, B. J. and Wiederhorn, S. M. (1980) *Journal of Materials Science,* **15**, 1207–1223

Lawn, B. R., Hockey, B. J. and Richter, H. (1983) *Journal of Microscopy,* **130**, 295–308

Lawn, B.R. and Komatsu, H. (1966) *Philosophical Magazine*, **14**, 689–699

Levitt, C. M. and Nabarro, F. R. N. (1966) *Proceedings of the Royal Society*, **A293**, 259–274

Mao, H. K. and Bell, P. M. (1978) *Science*, **200**, 1145–1147

Mao, H. K., Bell, P. M., Dunn, K. J., *et al.* (1979) *Review of Scientific Instruments*, **50**. 1002–1009

Marsh, D. M. (1964) *Proceedings of the Royal Society*, **A279**, 420–435

McSkimin, H. J. and Andreatch, P. (1972) *Journal of Applied Physics*, **43**, 2944–2948

McSkimin, H. J., Andreatch, P. and Glynn, P. (1972) *Journal of Applied Physics*, **43**, 985–987

Mikosza, A.G. and Lawn, B.R. (1971) *Journal of Applied Physics*, **42**, 5540–5545

Moriyoshi, Y., Kamo, M., Setaka, N. and Sato, Y. (1983) *Journal of Materials Science*, **18**, 217–224

Moss, W. C. and Goettel, K. A. (1987) *Applied Physics Letters*, **50**, 25–27

Mouginot, R. (1987) *Journal of Materials Science*, **22**, 989–1000

Mouginot, R. and Maugis, D. (1985) *Journal of Materials Science*, **20**, 4354–4376

Mukhin, M. E., Yarmak, M. F. and Popov, V. V. (1974) *Almazy i Sverkh. Materialy*, No. **8**, 4–6

Nadeau, J. S. (1973) *Journal of the American Ceramic Society*, **56**, 467–472

Nelson, D. A. and Ruoff, A. L. (1979) *Journal of Applied Physics*, **50**, 2763–2764

Novikov, N. V., Mal'nev, V. I. and Voronin, G. A. (1985) *Industrial Diamond Review*, **45**, 17–18

Oh, H.L. and Finnie, I. (1967) *Journal of the Mechanics and Physics of Solids*, **15**, 401–411

Onodera, A., Furono, K. and Yazu, S. (1986) *Science*, **232**, 1419–1420

Orowan, E. (1948) *Reports on Progress in Physics*, **12**, 192–194

Pauling, L. (1960) *The Nature of the Chemical Bond.* Cornell University Press, Ithaca, NY

Powell, B.D. and Tabor, D. (1970) *Journal of Physics D*, **3**, 783–788

Ramaseshan, S. (1946) *Proceedings of the Indian Academy of Science*, **A24**, 114–121

Roesler, F. C. (1956) *Proceedings of the Physical Society*, **B69**, 55–60

Ruoff, A. L. (1978) *Journal of Applied Physics*, **49**, 197–200

Ruoff, A. L. (1979) In *High Pressure Science and Technology*, Vol. 2, (ed. K. D. Timmerhaus and M. S. Barber), Plenum, New York, pp. 525–548

Ruoff, A.L. and Wanagel, J. (1977) *Science*, **198**, 1037–1038

Seal, M. (1958) *Proceedings of the Royal Society*, **A248**, 379–393

Seal, M. (1984) *High Temperatures - High Pressures*, **16**, 573–579

Seal, M. (1987) In *High-Pressure Research in Mineral Physics*, (ed. M. H. Manghnani and Y. Syono), American Geophysical Union, Washington D.C., pp. 35–40

Sinclair, J. E. (1972) *Journal of Physics C*, **5**, L271–L274

Sinclair, J.E. (1975) *Philosophical Magazine*, **31**, 647–671

Sinclair, J. E. and Lawn, B. R. (1972) *Proceedings of the Royal Society*, **A329**, 83–103

Stupkina, L. M. (1971) *Soviet Physics - Crystallography*, **15**, 728–730

Tabor, D. (1970) *Review of Physics in Technology*, **1**, 145–179

Tolansky, S. (1965) In *Physical Properties of Diamond*, (ed. R. Berman), Clarendon Press, Oxford, pp. 135–173

Tsai, Y.M. and Kolsky, H. (1967) *Journal of the Mechanics and Physics of Solids*, **15**, 29–46

Tyson, W. R. (1966) *Philosophical Magaine*, **14**, 925–936

Vleeschdrager, E. (1986) *Hardness 10: Diamond.* Gaston Lachurié, Paris

Watermeyer, B. (1982) *Diamond Cutting.* (2nd edn) Centaur, Johannesburg

Whitlock, J. and Ruoff, A. L. (1981) *Scripta Metallurgica*, **15**, 525–529

Williams, J.S., Lawn, B.R., and Swain, M.V. (1970) *physica status solidi (a)*, **2**, 7–29

Wilshaw, T. R. (1971) *Journal of Physics D*, **4**, 1567–1581

Xu, J. A, Mao, H. K. and Bell, P. M. (1986) *Science*, **232**, 1404–1406

Ziminov, N.V., Nikulin, K.K. and Yarmak, M.F. (1974) *Almazy i Sverkh Materialy*, No. **9**, 5–8

Chapter 8

Plastic deformation of diamond

Although diamond behaves as a brittle material at room temperature it is plastic at high temperatures when the thermal energy enables dislocations to become mobile. As mentioned in Section 7.1 we expect brittle materials to become plastic at a temperature equal to about half the melting temperature (in degrees Kelvin). However, the melting of diamond is not easily observed. If heated in air it begins to oxidize and graphitize at about 600°C, and if heated in an inert atmosphere or a high vacuum it begins to convert to graphite above about 1500°C, see Section 13.3.a. Therefore we cannot observe melting under zero or atmospheric pressure, but an estimate of an effective melting temperature may be obtained by extrapolating the melting curve into the graphite-stable region of the phase diagram (Muncke, 1974, 1979). This author gives a value for an effective melting temperature of about 3300 K, in reasonable agreement with an estimate by Korsunskaya, Kamentskaya and Aptekar (1972) using a different method. These estimates suggest that diamond should become plastic at about 1650 K or 1350°C. We first describe experiments in which diamond exhibits fully plastic behaviour at high temperatures, and then discuss its behaviour at ambient and intermediate temperatures.

8.1 Plasticity of diamond

The first experiment to demonstrate plastic flow was probably that of Phaal (1964) who pressed a Vickers pyramid indenter against a diamond flat at temperatures of about 1800°C and observed the production of slip lines (Figure 8.1). Subsequently Evans and Wild (1965, 1966) mounted small polished plates of diamond approximately 5 mm × 3 mm × 0.5 mm on two tungsten wedges as shown in Figure 8.2 and then loaded the upper side of the plate by a third wedge. These experiments were also made at 1800°C this temperature being chosen to be as high as possible without producing an excessive amount of graphitization. A plate of Type I diamond was bent through 9° without fracturing and a plate of Type II diamond through 21°. Experiments on Type II diamonds showed that the transition from brittle fracture to plastic bending occurred at a lower temperature of about 1600°C suggesting that Type I diamonds are more resistant to plastic deformation.

In other experiments by Evans and Sykes (1974) the edge of one octahedron diamond was pressed against the edge of another mounted at right angles so as to give a point contact as in Figure 8.3. The diamonds were forced against each other

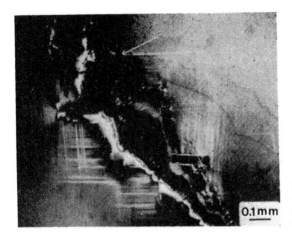

Figure 8.1 Optical micrograph of a diamond surface after a Vickers indentation at 1800°C showing numerous slip lines (Phaal,1964)

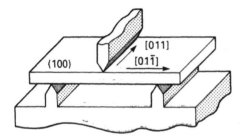

Figure 8.2 Orientation of a diamond plate in a bending test using 3 tungsten wedges (Evans and Wild,1965)

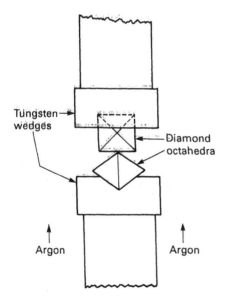

Figure 8.3 Schematic diagram of two diamond octahedra for mutual indentation experiments. The two edges in contact are at right angles to each other (Evans and Sykes, 1974)

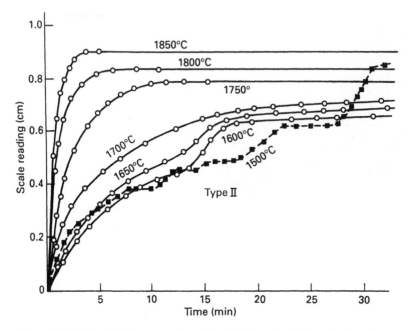

Figure 8.4 Relative displacement of two Type II diamonds during an indentation experiment, in arbitrary units (Evans and Sykes, 1974)

by loading the upper diamond through a tungsten rod and their relative displacement monitored by observing the position of the upper end of the rod. The temperatures of the diamonds and the lower end of the rod were maintained by a furnace purged with pure argon to avoid oxidation. Some results with Type II diamonds are shown in Figure 8.4 where the abscissae give the time after applying a load of 16.4 kg, and the ordinates the relative displacement of the diamonds measured in arbitrary units by an optical lever connected to the top of the rod.

At the highest temperatures employed the full deformation is produced in about 5 min, but this time increases when measurements are made at lower temperatures. The form of the curves are similar in the range 1850°C to 1700°C but at 1650°C and below the curves are qualitatively different and an examination of the diamonds in the optical microscope showed that some cracking had occurred. These lower curves (and particularly that for 1500°C) show some level sections which indicate that the plastic deformation was only taking place intermittently, so the authors concluded that the transition from brittle to plastic behaviour occurred at a temperature of about 1650°C, comparable to that for the bending of the slabs. The increase of the plastic deformation with time is a typical example of the phenomenon of creep well known in metals, and may be analysed in terms of a standard creep equation (Mott, 1953; Tabor, 1970).

$$p^{-m/3} - p_0^{-m/3} = A \exp\left[\frac{-Q}{3RT}\right](t^{1/3} - t_0^{1/3}) \tag{8.1}$$

Where p is the mean pressure under the indenter at the time t, p_0 the mean pressure at the time t_0 when the full load is applied, $m = 5$, A is a constant, and Q an

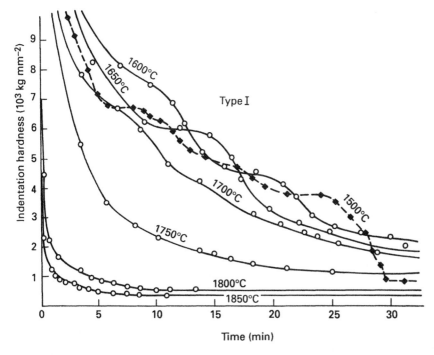

Figure 8.5 Indentation hardness of Type I diamonds at various times after the imposition of a full load (Evans and Sykes, 1974)

activation energy close to the energy for self diffusion of the atoms. The results at the higher temperatures in Figure 8.4 give straight lines when plotted in this way and lead to a value at Q of 10.7 eV.

Experiments were also made with Type I diamonds. Some results are shown in Figure 8.5 plotted as the indentation hardness as a function of time, the indentation hardness being the load divided by the projected area of the indentation in a plane parallel to the slab. We see in Figure 8.5 the same features of behaviour as seen in Figure 8.4 but the rate of deformation and creep was less at all temperatures; the results could again be represented by Equation (8.1) but with an activation energy of about 14 eV. Cracking was now observed at temperatures up 1700°C, and the behaviour was only fully plastic at 1750°C. Hence it appears that, as in the beam experiments, Type I diamonds are somewhat more resistant to plastic deformation than Type II. This difference must arise both because of the larger number of dislocations generally found in Type II stones and because the impurity atoms or groups of atoms in Type I stones will tend to pin dislocations and restrict their movement (Wild, Evans and Lang, 1967).

8.2 Deformation at room temperature

The behaviour of diamond at room temperature described in Chapter 7 is that of an almost completely brittle material but there is also some evidence, based on a few experiments, that dislocation motion may be produced under certain severe conditions. In particular Mao and Bell (1978) reported plastic deformation when two diamond anvils were forced together under extremely high pressure (Section 7.3.a). Other authors have claimed that plastic flow due to dislocation motion may be observed under the tips of indenters used to make hardness measurements. To discuss these observations we first give a brief outline of some hardness measurements on non-diamond materials.

In a typical hardness test of a metal a pyramidal-shaped indenter much harder than the specimen (and usually of diamond) is pressed down on a polished surface under a given load. At first the area of contact is small and the pressure under the indenter is high so the metal flows plastically, thus increasing the area of contact. Eventually this increase in area reduces the pressure under the indenter until it is insufficient to produce further plastic flow. Hence by determining the final pressure from the load and the area of the indentation one obtains a measure of the material's resistance to plastic flow, that is its hardness. For details of various geometries see Tabor (1970) and Westbrook and Conrad (1973).

Hardness measurements are often made on brittle materials by using Knoop indenters which have the shape of an elongated pyramid with one base diagonal approximately five times the length of the other. This geometry has the advantages that the material is less likely to crack, and also permits an examination of any variation of the hardness with the azimuthal orientation of the indenter. Experiments using Knoop indenters have been made on various hard crystals such as alumina, magnesium oxide and quartz, and even at room temperature it appears that some indentations can be interpreted entirely in terms of plastic deformation (Brookes, O'Niell and Redfern, 1971). These authors also suggested that Knoop indentations of diamond can be interpreted in the same way, and authors such as Ruoff (1979 a,b) and Doi et al. (1986) have since assumed that Knoop indentations of diamond can be discussed entirely in terms of plastic flow. As the mechanical properties of a diamond surface are of practical importance we now discuss two types of indentation experiments where the authors claim to have observed plastic flow.

8.2.a Hertz indentations

The first report of plastic flow in an indentation experiment on diamond appears to be that of Howes and Tolansky (1955). These authors pressed a diamond ball of radius 0.39 mm onto an octahedron face of a diamond and measured the load required to produce a cone crack of the type described in Section 7.5.a. Besides measuring the load the authors also investigated the profile of the cracked surface using multiple beam interferometry, and two of their profiles are given in Figure 8.6. They claim that the material heaped up on either side of the indentation has been pushed there by plastic deformation and provides 'strong evidence for the existence of plastic flow'.

It is difficult to accept this interpretation. The base of the indentation in Figure 8.6(a) lies at the same level as the main surface outside, so the profile implies that the volume of material has increased. The jagged profiles in Figure 8.6(b) are certainly not characteristic of plastic flow. If plastic flow had occurred one would expect the base of the indent to reflect the curvature of the ball but it is shown as flat within a few nanometers. (Note that the difference in the vertical and horizontal scales greatly exaggerate the vertical heights and gives a misleading impression of the surface topography.) In fact if the ball was pressed into soft

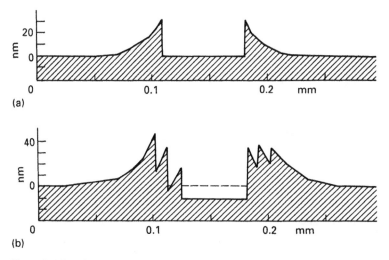

Figure 8.6 Depth profiles of two hexagonal ring cracks on diamond obtained by multiple beam interferometry (Howes and Tolansky, 1955)

material to give the same diameter of contact ($\sim 50\,\mu m$) the centre of the impression would be about 700 nm below the level of the surface!

It appears that the deformation in Figure 8.6 arises in the way described by Lawn and Komatsu (1966) which does not involve any plastic flow. The essential concept is that under load a cone crack is formed as shown in Figure 8.7(a) and that when the load is removed the crack does not close completely because of debris from the cleavage which remains in the crack. (Evidence of debris preventing the closure of cone cracks is also given in Section 7.7.d.) The authors measured the profile of cracks on octahedron and cube faces using multiple beam interferometry and calculated the expected slope of the surface using the crack model of Figure 8.7, and found good agreement as shown in Figure 8.8. The authors also obtained X-ray reflection topographs which show contrast arising from the cracks and the associated displacements of the lattice, but give no sign of any contrast characteristic of dislocations.

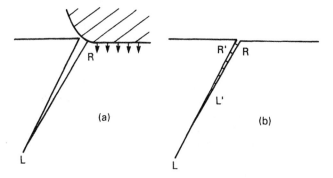

Figure 8.7 Formation of a pressure crack (a), the crack RLR' is formed under an applied load; (b), on removing the indenter the debris prevents the crack closing completely (Lawn and Komatsu, 1966)

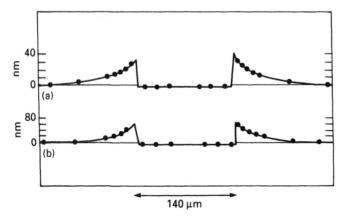

Figure 8.8 Full lines show the measured profiles of sections of two ring cracks, and the black circles the profiles computed by Lawn and Komatsu (1966)

8.2.b Knoop indentations

Evidence to demonstrate plastic flow during indentations with the Knoop type indenter mentioned above has been put forward Brookes (1970), who presents a photomicrograph of indentations on a polished {001} face with the long axis of the indenter in <001> and <110> directions and presents two arguments to show that these were formed by plastic flow. He first notes that the indentations in the <110> direction are much larger, as would be expected from the lie of the slip planes if the flow was plastic. However, this argument is insufficient because the <110> directions will also favour fracture because they coincide with the traces of the octahedron cleavage planes.

The second argument for plastics flow is based on the appearance of the indentations. Brookes states that indentations made with loads at 500 g were too small to measure accurately and that loads of 2 kg produced significant cracking, so most of the measurements were made with a load of 1 kg. A micrograph of an indentation made with this load shows no sign of cracking but the resolution is low and does not even reveal the polishing lines on the polished surface. It may be that under this particular load some plastic deformation has occurred but the evidence in the micrograph leaves the question open.

Other evidence for plastic flow under a Knoop indenter is presented by Humble and Hannick (1978, 1979) but again the evidence is not conclusive. They used ion milling to thin down a diamond and thus obtained a TEM micrograph of the region immediately below the indenter which shows the presence of dislocation lines. However, these may have been present before the indentation. Other features described by the authors as dislocations or cracks look similar to the crack system produced by a sliding indenter. The authors also give two micrographs showing details of these cracks before and after heating to about 1000°C. This annealing produced various small changes to the crack system, one being the production of what appears to be two small loops of dislocation line adjacent to other damage. If this is the case, as seems likely, then the dislocations must have been moved to this position by the stress fields surrounding the indentation with the assistance of the thermal energy associated with the higher temperature. However, it does not

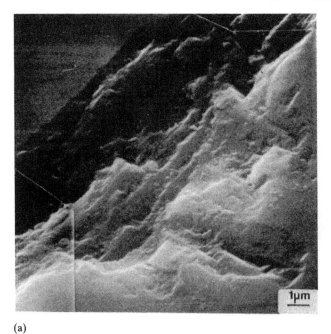

(a)

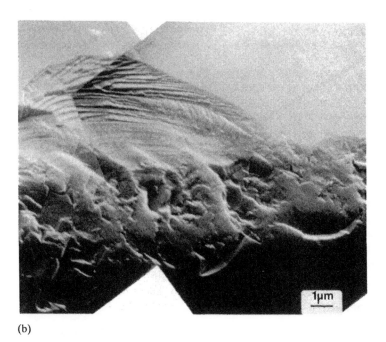

(b)

Figure 8.9 Damage produced by the same indentation (a), on a polished diamond surface; (b), on the Knoop indenter

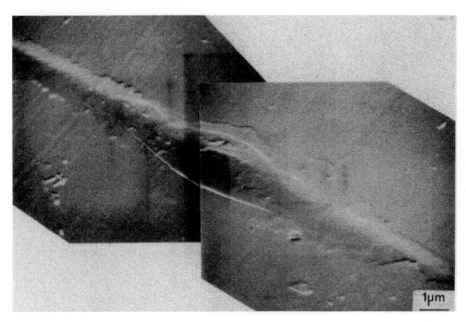

Figure 8.10 Damage on a diamond surface after Knoop indentation showing traces of {111} cleavage planes along the edges of the indentation

follow that the dislocations had been generated by the indentation, as they might have been present previously. We also note that Gane (1971) observed certain features produced by pressing one diamond edge against another but this evidence is also ambiguous.

None of the above experiments give an unequivocal demonstration of plastic flow under an indenter. On the other hand there is no evidence for the complete absence of plastic flow. The question remains open but there is no ground for assuming that the room temperature indentation of diamond by a Knoop indenter can be discussed solely in terms of plastic deformation. A common result of pressing an indenter on to a flat is to produce cracking as shown for example in Figure 8.9. Even indentations which at first sight appear smooth generally reveal cracks when inspected more closely. For example, Figure 8.10 shows an SEM micrograph of an indentation where a crack is only clearly recognizable because it follows a zig-zag path determined by the positions of the {111} cleavage planes. Hence there is considerable difficulty in deciding how best to interpret the results of indentation experiments of this type and how to compare the results obtained on diamond with those obtained on other more plastic materials, see for example a discussion by Bakul, Loshak and Mal'nev (1973) and also the following section.

8.3 Effect of temperature

The experiments of Evans and Sykes (1974) described in Section 7.2 show that at a temperature of 1800°C diamond behaves quite plastically and that below this figure and particularly below 1500°C the plasticity decreases significantly. However, it appears that plasticity may be observed at rather lower temperatures

Figure 8.11 Optical interference micrograph of the surface of a diamond crystal subjected to high pressures and temperatures (DeVries, 1975)

if the diamond is subjected to an overall compressive stress. For example, DeVries (1975) made experiments in which small diamond crystals embedded in diamond powder were subject to high pressures and temperatures, and at the higher temperatures this treatment produced series of visible lines on the surface of the crystals. These lines are brought into sharper relief when the diamond is polished (Section 10.2.a) and Figure 8.11 shows an optical interference micrograph of such a polished surface. The lines in the micrograph lie in {111} planes and appear to be traces of either slip planes or deformation lamellae, as discussed by DeVries and also by Lee, DeVries and Koch (1986). Figure 8.12 based on a series of runs by DeVries at various temperatures and pressures indicates the conditions which produced visible slip or deformation, and it is seen that resistance to plastic flow increases greatly below about 1200°C.

The variation of plasticity with temperature down to room temperature has been studied by several authors by making measurements of the indentation hardness. Figure 8.13 from Bakul, Loshak and Mal'nev (1978) summarizes some of these experiments and shows that the hardness steadily increases as the temperature falls. Rather similar values obtained by Harrison (1973) are quoted by Brookes (1979). All the various results give hardness values which depend on temperature in the same general way but the agreement in detail is only moderate. This is at least partly because the results are influenced by the different geometries of the various indenters and perhaps more importantly because the size of an indentation may depend on the time of application of the load as well as on the temperature, see Section 8.1. (The results in Figure 8.13 appear to have been obtained in commercial hardness testers using standard conditions without regard to time effects. The authors also state that the indenters suffered damage but give no details of this damage.)

Despite the above uncertainties, all the curves in Figure 8.13 show that some change occurs in the form of the temperature dependence in the region of 1100 K. As discussed above, the plasticity at the higher temperatures is associated with the increased mobility of the dislocations made possible by thermal activation. This

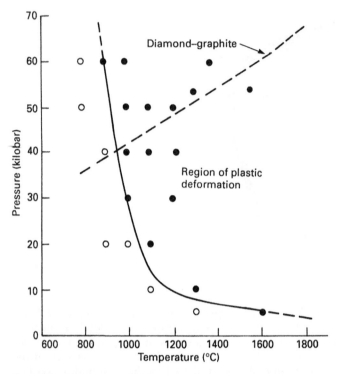

Figure 8.12 Diagram indicating pressure and temperature conditions producing visible slip in diamond. (●), visible slip lines; (○), no slip visible. The dashed line shows the diamond-graphite equilibrium line (DeVries, 1975)

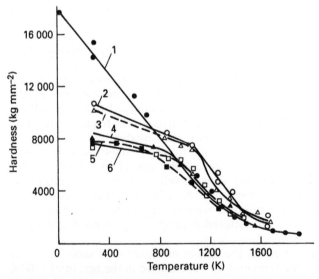

Figure 8.13 Summary of various measurements of indentation hardness as a function of temperature (Bakul, Loshak and Mal'nev, 1978)

suggests that the temperature dependence of the hardness will include a Boltzmann factor $e^{-q/RT}$ where q is an activation energy and T the absolute temperature, so in Figure 8.14 we show the hardness values given by curve 2 of Figure 8.13 plotted against $1/T$. We see that the transition in the region of 1100°C is now even more marked.

The variations in the strength of diamond indicated by the above indentation experiments are not fully understood, even though somewhat similar patterns of behaviour are observed in other crystals. For example Atkins and Tabor (1966) have described measurements on the hardness of magnesium oxide which when plotted against $1/T$ show the same general type of temperature dependence as in Figure 8.14. Atkins and Tabor interpret the upper and lower temperature regions

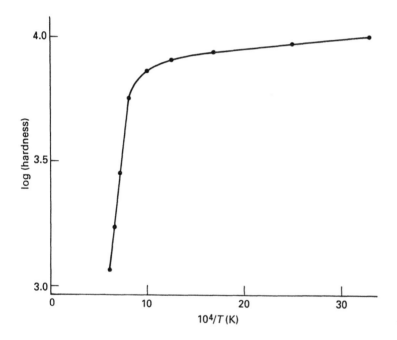

Figure 8.14 Indentation hardness values given by curve 2 of Figure 8.13 plotted against $1/T$

as characteristic of plastic and brittle fracture, and suggest that the still appreciable temperature dependence in the brittle region arises from small changes in the elastic constants. However, in the case of diamond at room temperature the fractional changes in the elastic constants are of the order of $2.10^{-5}\,\text{K}^{-1}$ (McSkimin and Andreatch, 1972), and such changes are too small to account for the slopes of the curves in Figure 8.13. We also note that measurements of the compressive strength of small diamond crystals by Novikov, Mal'nev and Voronin (1985) show virtually no temperature dependence below about 800°C (Figure 8.15), so the rises shown in Figure 8.13 may lie within the uncertainty of the experiments.

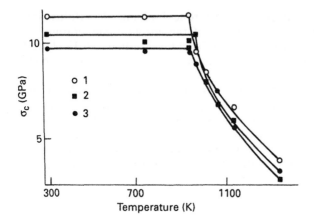

Figure 8.15 Strength of synthetic diamond crystals measured in compression as a function of temperature; σ_c is the average contact pressure at fracture. Curves (1), (2) and (3) refer to crystals grown with Fe/Ni, Fe/Co and Ni/Mn solvents respectively (Novikov, Mal'nev and Voronin, 1985)

References

Atkins, A. G. and Tabor, D. (1966) *Proceedings of the Royal Society*, **A292**, 441–459

Bakul, V. N., Loshak, M. G. and Mal'nev, V. I. (1973) *Sinteticheskiye Almazy*, **6**, 5–11 (In Russian)

Bakul,. V, N., Loshak, M. G. and Mal'nev, V. I. (1978) *Sinteticheskiye Almazy*, **1**, 7–11 (In Russian)

Brookes, C. A. (1970) *Nature*, **228**, 660–661

Brookes, C. A. (1979) In *The Properties of Diamond*, (ed. J.E. Field) Academic Press, London, pp. 383–402

Brookes, C. A., O'Niell, J. B. and Redfern, B. A. W (1971) *Proceedings of the Royal Society*, **A322**, 73–88

DeVries, R. C. (1975) *Materials Research Bulletin*, **10** , 1193–1200

Doi, Y., Sato, S., Sumiya, H. and Yazu, S. (1986) In *Proceedings of the Second International Conference on the Science of Hard Materials*, (Rhodes, 1984) Institute of Physics Conference Series No 75. (eds E.A.Almond, C.A.Brookes and R.Warren), Adam Hilger, Bristol, pp. 235–237

Evans, T. and Sykes, J. (1974) *Philosophical Magazine*, **29**, 135–147

Evans, T. and Wild, R. K. (1965) *Philosophical Magazine*, **12**, 479–489

Evans, T. and Wild, R. K. (1966) *Philosophical Magazine*, **13**, 209–210

Gane, N. (1971) *Diamond Research 1971*, supplement to *Industrial Diamond Review*, pp. 16–19

Harrison, P. (1973) *PhD Thesis*, Exeter University

Howes, V. R. and Tolansky, S. (1955) *Proceedings of the Royal Society*, **A230**, 287–293

Humble, P. and Hannink, R. H. J. (1978) Nature, **273**, 37–39

Humble, P. and Hannink, R. H. J. (1979) In *Physics of Materials*, (eds D. W. Borland., L. M. Clarebrough and A. J. W. Moore), CSIRO and University of Melbourne, Melbourne, pp. 145–153

Korsunskaya, I. A., Kamentskaya, C. and Aptekar I. L. (1972) *Doklady Academy Nauk SSSR*, **204**, 909

Lawn, B. R. and Komatsu, H. (1966) *Philosophical Magazine*, **14**, 689–699

Lee, M., DeVries, R. C. and Koch, E. F. (1986) In *Proceedings of the Second International Conference on the Science of Hard Materials*, Institute of Physics Conference Series No.75, (eds E.A.Almond, C.A.Brookes and R.Warren), Adam Hilger, Bristol, pp. 221–232

Mao, H. K. and Bell, P. M. (1978) *Science*, **200**, 1145–1147

McSkimin, H. J. and Andreatch, P. (1972) *Journal of Applied Physics*, **43**, 2944–2948

Mott, N. F. (1953) *Philosophical Magazine*, **44**, 742–765

Muncke, G. (1974) *Diamond Research 1974*, supplement to *Industrial Diamond Review*, pp. 7–10

Muncke, G. (1979) In *Properties of Diamond*, (ed. J. E.Field) Academic Press, London, pp. 473–499

Novikov, N. V., Mal'nev, V. I. and Voronin, G. A. (1985) *Industrial Diamond Review*, **45**, 17–18

Phaal, C. (1964) *Philosophical Magazine*, **10**, 887–891

Ruoff, A. L. (1979a) *Journal of Applied Physics*, **50**, 3354–3356

Ruoff, A. L. (1979b) In *High Pressure Science and Technology Vol. 2*, (eds K. D. Timmerhaus and M. S. Barber), Plenum, New York, pp. 525–548

Tabor, D. (1970) *Review of Physics in Technology*, **1**, 145–179

Westbrook, J. H. and Conrad, H. (1973) (eds). *The Science of Hardness Testing and its Research Applications*. American Society for Metals, Metals Park, Ohio

Wild, R. K., Evans, T. and Lang, A. R. (1967) *Philosophical Magazine*, **15**, 267–279

Chapter 9

Polishing and shaping diamond

When diamonds are used as jewellery or in technology they must often be shaped to a particular geometric form and this is usually done by a form of grinding generally described in the diamond trade as polishing. This grinding or polishing process shows some unusual features.

We first note that during the grinding of metals, material is removed by the grinding grits gouging out the metal, and that the finer finishing polishes generally involve the plastic deformation and smearing of the surface. However, ceramic materials are generally more resistant to plastic deformation, and brittle fracture produced by the polishing grains may play a large or even dominant role in the grinding process. Surveys of the grinding and polishing of ceramics are given in books by Samuels (1971) and Hockey and Rice (1979).

The normal procedure in grinding is to use grit or powder of a harder material than the workpiece but there is nothing harder than diamond so diamond must be polished with diamond. There is also another significant way in which the polishing of diamond differs from polishing other materials. As described in Chapters 7 and 8 diamond at ambient temperature lies at the far end of the spectrum of brittle materials and exhibits an almost totally brittle behaviour, and this is reflected in its behaviour during polishing.

9.1 Technique of polishing

The traditional method of polishing diamond is by grinding on a flat wheel or *scaife* typically about 300 mm in diameter and made from cast-iron of carefully selected porosity. This scaife is charged with diamond powder ranging in sizes from less than 1 μm to about 40 μm, the larger sizes giving faster removal rates but a rougher finish. Therefore it is common practice to begin polishing with coarse powder and finish off with say 0–1 μm powder to give a smoother surface. The diamond powder is mixed with olive oil, or some other base, to form a paste or suspension which is rubbed over the metal scaife and then left for some time for the suspension to be absorbed by the pores. The diamond to be polished is usually mounted in a metal holder known as a dop where it is held in place by two or more metal claws, or sometimes by a low melting point metal. The surface to be polished is placed against the scaife rotating at perhaps 2500 rpm under a load of the order of 1 kg. Accounts of the various details of the polishing process are given in books by Watermeyer (1982) and Vleeschdrager (1986).

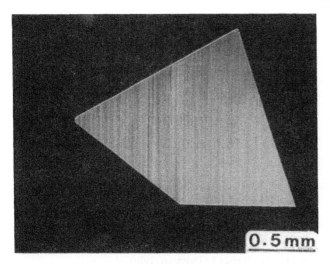

Figure 9.1 Optical interference micrograph of the surface of a polished gem stone

The success of the polishing operation depends quite critically on the orientation of the diamond but if all goes well a suitable polish is readily obtained which will appear smooth when inspected with a jeweller's ×10 lens. In fact a more detailed examination shows that the surface is covered with fine polishing lines made by the abrading particles. For example Figure 9.1 shows a polished surface viewed in an optical microscope using the Nomarski interference technique. This technique, which involves the use of a special prism and polarized light is such that if a surface is perfectly plane its image is of uniform intensity. However, the intensity of the image varies as the plane is tilted with respect to the axis of the microscope. Hence the modulation of tone in Figure 9.1 indicates that the surface is not truly flat but is covered with polishing grooves with sloping sides. The microscope is readily calibrated by observing the changes when a flat surface is tilted on a goniometer stage. In fact the polishing grooves are generally quite shallow with gentle sloping sides inclined to the surface at angles of the order of 1°.

The use of coarser diamond powder produces a more marked system of polish lines but both fine and coarse powder produce similar types of patterns. There will generally be systems of finer polish lines beyond the resolution of Figure 9.1. For example Figure 9.2 is an electron micrograph of polishing lines at a magnification of ×10,000 obtained by a replica technique described in Section 9.4.b. Even at this much greater magnification the form of the pattern is quite similar to that of Figure 9.1.

Success in the polishing of diamond depends on attention to detail. The scaife must run smoothly, free of vibration, and the diamond must be held rigidly. Particular attention must be given to the surface of the scaife which is initially machined to give a plane but rough surface with an open porous finish which accepts the diamond powder, see Watermeyer (1982). As polishing proceeds, the surface of the wheel becomes shiny and apparently quite smooth but continues to work. Eventually, however, the pores of the iron become closed up and it is necessary to remachine or 'skim' the scaife, see Figure 9.3. (When using a new

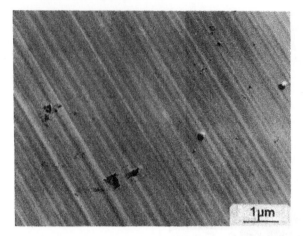

Figure 9.2 High magnification electron micrograph of a polished diamond surface obtained by a replica technique. (The black spots are debris from the replica.) (Thornton, 1976)

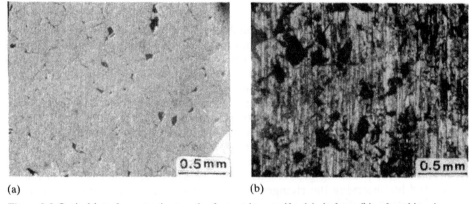

(a)   (b)

Figure 9.3 Optical interference micrograph of a cast-iron scaife: (a), before; (b), after skimming

scaife it is usual to run it in by polishing some diamond for an hour or so to achieve a steady condition.)

Diamond is very resistant to polish if presented to the scaife in certain orientations. It is therefore necessary to avoid these orientations as the diamond will tend to remove the powder and score the wheel, and particular care must be taken not to force a sharp corner of the diamond on and into the scaife. It is also necessary to avoid overheating the diamond by applying too great a load, as it is quite possible under poor conditions for a diamond to become red hot. In this case an outer layer of the diamond will be burnt and damaged, and must be polished away. In addition the decomposition products generally lead to an inferior performance of the scaife and must be cleaned off.

Today the cast-iron scaife is often replaced by a wheel of similar geometry in which the diamond powder is bonded in a metal alloy, or sometimes an organic resin, on the surface of the wheel. This type of scaife is in fact a particular form of the grinding wheels described in Chapter 15, but there is not much published information on the relative performance of the bonded wheels used for polishing diamond. A diamond bonded scaife costs much more than a cast-iron scaife and it can be easily damaged, but its use avoids the interruption and labour involved in keeping a cast-iron scaife fully charged with powder. This is an important consideration on production lines particularly when using automated techniques (Section 9.6.a). The bonded wheel will also polish diamonds in the more difficult directions and is particularly useful for polishing polycrystalline PCD products (Section 12.3).

As with cast-iron scaifes, good polishing with bonded scaifes is only achieved by paying attention to detail. The scaife must be kept free of debris coming from the metal bond. The diamond tends to run hotter than on a cast-iron scaife and care must be taken to avoid damage to the wheel. Even after careful use the scaife eventually becomes glazed and the diamond powder not sufficiently proud of the surface. It is then necessary to open up the surface by some form of grinding process, perhaps by rubbing with abrasive sticks, perhaps using an alumina wheel.

9.2 Measurements of polishing rates

As already mentioned, a diamond may be polished much more readily in some directions than others. Consider, for example a diamond which is being polished on a cast-iron scaife so that the powder moves across an {001} face in a <100> direction. Material is removed steadily and the surface eventually shows a good smooth finish as on a gemstone. Suppose now that, without lifting the diamond from the scaife, it is rotated through an angle of 45° about an axis passing through its centre and normal to the scaife. One might think this rotation would make no difference because the polished {001} surface appears so smooth. In fact it results in the rate of removal being reduced by a factor of the order of 100. That is, the scaife now produces almost no effect on the diamond ! To discuss the behaviour of diamond during polishing we begin by describing two techniques for measuring rates of polish.

9.2.a Scaife tests

The first systematic measurements of polishing rates were made by Tolkowsky (1920) who polished diamonds on a cast-iron scaife and measured the loss in weight as a function of time and other parameters. By observing losses of up to about 1 ct (0.2 g) of diamond he was able to obtain reliable results for the dependence of the removal rate on the speed of the wheel, on the load on the diamond, and on the orientation of the diamond. Other authors have observed the increase in an area of a face during polishing, and hence the volume removed (Whittaker and Slawson, 1946; Denning, 1953; Hukao, 1955).

An alternative method of measuring the amount of material removed is described by Hitchiner, Wilks and Wilks (1984). Three abrasion cuts about 10 μm deep were made on the surface to be polished (Figure 9.4) and the depths measured with a profilometer. The depth of each cut was then observed after each of three or four periods of polishing each sufficient to reduce the depths by 1 μm or 2 μm. Ideally the depths of all three cuts should decrease at the same rate, except possibly for the first period of polish when a difference will arise if initially the surface is

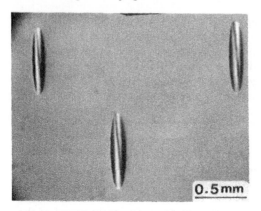

Figure 9.4 Optical interference micrograph of three abrasion cuts on a polished diamond surface

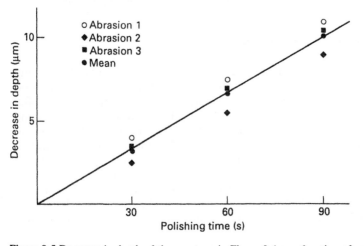

Figure 9.5 Decrease in depth of three cuts as in Figure 9.4 as a function of polishing time

not exactly parallel to the wheel. Some typical measurements are shown in Figure 9.5. The points for the three cuts do not quite coincide probably because the effective radius of the scaife was not quite the same for each cut, but we see that the mean of the three depths decreases proportionately to the time of polish. Hence, the volume removed per second is obtained by measuring the slope of this line and the area of the polished surface.

Various precautions are necessary when measuring polishing rates on a scaife. The experiments of Hitchiner, Wilks and Wilks (1984) mentioned above also underline the importance of keeping a scaife or wheel adequately charged with diamond powder. This matter is rather complicated and is further discussed in Section 9.3.b. As the rates are often very sensitive to the crystallographic orientation of the diamond this must be accurately determined, with the diamond mounted in a suitable goniometer head sufficiently massive to hold it rigidly. It is also necessary to make periodic calibration checks to guard against changes in the condition of the scaife, and one must remember that small defects in the diamond such as naats (Section 5.7.a) may be very abrasion resistant and considerably effect the measured removal rate.

9.2.b Micro-abrasion tests

In this test described by Wilks and Wilks (1965, 1972, 1982) the diamond is abraded by a wheel, of radius about 10 mm, with a V-shaped edge with an included angle of about 100°. The wheel may be of either cast-iron or diamond-bonded material. The cast-iron wheels were made from the type of cast-iron used to make scaifes, and charged with diamond powder of 0–1 μm diameter mixed with olive oil. The bonded wheels had a matrix of cobalt, and a volume concentration of about 30% diamond, of size 4 μm to 28 μm. Both types of wheels were generally run at speeds between 1000 rpm and 10,000 rpm. The diamond under test is mounted on a balance arm and held against the abrading wheel under a known load of the order of 100 g wt for the order of 20 s, a typical abrasion cut being 0.5 mm long, 0.1 mm wide and about 1 μm deep.

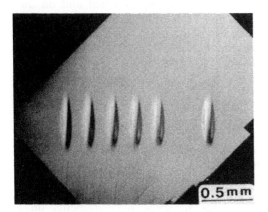

Figure 9.6 Six abrasion cuts made on a polished diamond surface. The depths are the same to ± 5%

The depths of the cuts are measured either by optical interference techniques, or by a profilometer. By paying close attention to the smooth running of the wheel it is possible to produce a series of cuts whose depths are reproducible to at least ± 5% (Figure 9.6). Interferometric studies also show that when the abrasions are made in the normal direction of polish the surface is generally smooth and of comparable quality to that of a similarly polished flat surface. Because the diamond surface is much harder than the cast-iron wheel the edge of the wheel is soon blunted, so the cuts are quite shallow, with sides generally making angles of no more than 5° with the flat surface. It follows that the abrading conditions during the micro-abrasion test are quite similar to those during a scaife test. As described in detail by Wilks and Wilks (1972) the volume of material removed by not too deep cuts increases linearly with time as in flat grinding, and this volume is roughly proportional to the square of the depth. Hence a factor x between the depths of two cuts corresponds to a factor of approximately x^2 in the rates of removal of material.

Micro-abrasion tests do not give as exact values for removal rates as can be obtained in flat grinding tests but they offer several advantages. They are appreciably less time consuming than scaife tests. Several cuts may be made on the same face to check for reproducibility and to detect local imperfections, such as naats. It is also easier to position the diamond with the required orientation and to make changes in that orientation than when grinding on a scaife.

9.3 Directional effects

We now summarize the unusual dependence of the abrasion resistance of diamond on its crystallographic orientation before discussing the reasons for this behaviour in Section 9.4.

9.3.a Grain of diamond

The resistance of a diamond to polish, or its *abrasion resistance*, depends greatly both on the crystallographic orientation of the face being polished and on the direction of polish on the face. This dependence on orientation has long been known to diamond polishers who explained the effect in terms of 'the grain' of a diamond. This 'grain' was a rather diffuse concept but was generally taken as the direction of easy polish on a particular face. However, some of the statements concerning grain given in the literature, see for example Grodzinski (1953), are quite inconsistent with the physical structure of diamond.

Consider a polished cube {001} face ABCD as shown schematically in Figure 9.7. This face polishes relatively easily in directions parallel to any of the four <100> axes lying in the surface, but is very difficult to polish in directions at 45° to these axes. It follows from the basic symmetry of the crystal lattice of diamond that all four <100> directions lying in the surface should be equally easy to abrade. Yet experienced diamond polishers used to believe that one of these four directions polished more rapidly, and this direction was known as 'the grain'. To resolve this conflict between theory and reality we refer to some measurements on a

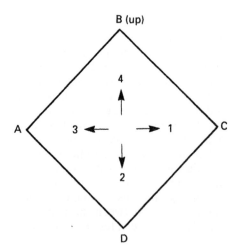

Figure 9.7 Schematic diagram of a polished {001} face to describe the effect of tilting a facet, see text. The arrows indicate the <100> directions

rectangular block of diamond which had been polished with the aim of producing six {001} faces (Wilks and Wilks, 1954). Micro-abrasion measurements were made with cast-iron wheels in the four <100> directions on all six faces. The expected symmetry was observed on only two of the faces, and on the other four faces the depths of the cuts in different directions varied by factors of up to 3. However, goniometer measurements showed that some of the faces were tilted away from {001} through angles of up to 2°.

To analyse the above results we note that provided the angles involved are small any tilt of the surface away from {001} can be described as the result of two independent tilts about the axes AC and DB in Figure 9.7. It turns out that if the abrasion is made in say the direction AC then a small tilt about AC produces little effect, but that a tilt about BD produces significant changes. Thus a tilt such that the direction of abrasion is *downhill* with respect to the {001} planes as shown in Figure 9.8 results in a deeper cut, while a tilt in the opposite *uphill* direction leads

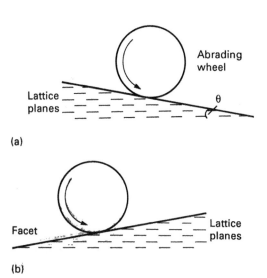

(a)

(b)

Figure 9.8 Diagram to illustrate possible orientations of polished facets with respect to the abrading wheel and to {001} planes, (a), 'downhill'; (b), 'uphill', see text (Wilks and Wilks, 1954)

to a smaller cut. That is cuts in similar but opposite directions have different depths. Figure 9.9 shows the results for all six faces expressed as the ratios of the depths of corresponding downhill and uphill cuts plotted against the relevant angle of tilt θ. We see that as the polished surface tends to coincide with {001} the cuts tend to the same depths to within the accuracy at the experiment. However, tilts of only 2° can produce large differences in the depths of parallel and anti-parallel cuts, and this effect accounts for the apparently anomalous behaviour associated with 'the grain'.

Figure 9.9 shows the extreme importance of the accurate orientation of the polished surfaces when making measurements of polishing rates. As mentioned in Section 5.2.a the orientation is best determined by optical goniometry, and this requires that the specimen diamonds be of good quality and give sharp reflections from their natural {111} faces. It is also necessary to use high quality goniometer

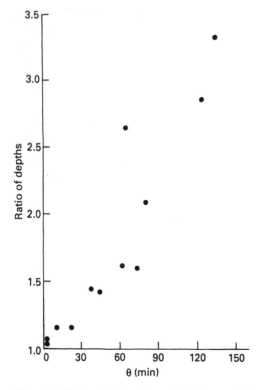

Figure 9.9 Ratio of depths of parallel but opposite abrasions on a {001} face plotted against θ the angular misorientation of the face (Wilks and Wilks, 1954)

heads to maintain the exact orientation of the stone during polishing. However, by careful adjustments, it is quite possible to orient faces to within 5 minutes of arc or better.

9.3.b Azimuthal variations on principal planes

The abrasion resistance of each of the three principal planes of diamond (cube, dodecahedron, octahedron) depends on the direction of abrasion. Figure 9.10 shows the rates of polish as given by Tolkowsky (1920) for the three faces as a function of the azimuthal angle, measurements being made every 12° or so, and the results averaged to reduce the effects of any misorientations. The values for the cube, dodecahedron and octahedron faces show a 4-fold, 2-fold and 3-fold symmetry as is to be expected from the symmetry of the diamond. Note that the variation in the values of the polishing rates are very considerable. Tolkowsky polished for a fixed time and observed losses of mass varying from about 300 mg down to 4 mg and less, that is variations of the order of 100:1.

Results similar to the above have been reported by Slawson and Kohn (1950), Denning (1953), Hukao (1955) and Hitchiner, Wilks and Wilks (1984). The latter authors also observed that the lower polishing rates were greatly increased if the scaife was frequently recharged with diamond powder. Table 9.1 shows the rates

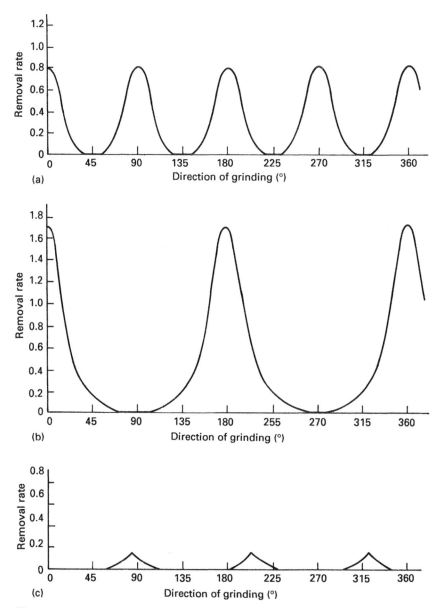

Figure 9.10 Rate of removal (in the same arbitrary units) during polish on a cast-iron scaife as a function of azimuthal angle for (a), {001}; (b), {011}; and (c), {111} faces (after Tolkowsky, 1920)

of polish of an {011} face of a diamond in the directions of easy and hard abrasion, (a), without recharging and (b), with recharging after every 240 revolutions of the scaife. When polishing in an easy <100> direction recharging made no significant difference, but in the hard direction frequent recharging greatly increased the polish rate up to about the value quoted. Therefore we take the figure of 0.66 in Table 9.1 as a measure of the diamond's intrinsic abrasion resistance in a <100>

Table 9.1 Effect on removal rate of recharging scaife when polishing an {011} face of diamond. (Wheel speed, 2932 rev min^{-1}; load, 2015 g wt) (Hitchiner, Wilks and Wilks, 1984)

Direction of polish	Removal rates ($\times 10^{-4}$mm^3 s^{-1})	
	Without recharging	With recharging
<100>	5.7	5.9
<0$\bar{1}$1>	0.02	0.66

Table 9.2 Effect of powder concentration on the removal rate when polishing diamond in an easy direction on an {011} face. The normal concentration was 0.2 g of 5 μm–10 μm powder mixed with 0.6 g olive oil

Concentration	Rate (10^{-4}mm^3 s^{-1})
Normal	6.5
×5	9.2
÷5	6.8
÷25	4.4

Table 9.3 Summary of removal rates when polishing an {001} face of diamond on a cast-iron scaife and a bonded wheel. (Wheel speed, 2932 rev min^{-1}, load 2015 g wt) (Hitchiner, Wilks and Wilks, 1984)

Direction	Removal rates ($\times 10^{-4}$mm^3 s^{-1})	
	Cast-iron scaife	Bonded wheel
<100>	5.9	3.3
<110>	0.66	0.14

direction whereas the figure of 0.02 obtained without recharging gives only some measure of how much powder has been lost. Because this point was often ignored in earlier measurements the values obtained there for the removal rates in the hard directions of abrasion should be treated as useful indications rather than exact values.

Although the removal rates in hard directions depend critically on the concentration of powder on the scaife the position is quite different when polishing in an easy direction. Table 9.2 shows the polishing rates in an easy <100> direction on an {011} face for different concentrations of diamond powder and we see that the removal rate changes relatively slowly with concentration. These results suggest that once sufficient powder is provided to fill the available pores in the cast-iron further powder produces only a limited effect. Of course some powder will be lost from the scaife during polishing and eventually some recharging of the scaife may become necessary. However, it seems that with the relatively high removal rates obtained in the easy directions of polish there is sufficient debris on the diamond surface to provide enough powder to maintain the charge on the wheel.

Some indication of the performance of a bonded wheel is given by Table 9.3 which shows polishing rates on a {011} face produced by both a bonded wheel and a fully recharged cast-iron scaife. Of course the performance of a bonded wheel depends considerably on the size and concentration of the powder and the nature of the binder, so the absolute magnitudes of the removal rates given in the table are of no great significance. However, we see that the rates obtained with the bonded wheel although smaller than those given by the scaife are still of the same order of magnitude. The importance of the bonded wheel is that the rate of polish is maintained without any replenishment whereas if the cast-iron scaife is not

Table 9.4 Depths of cuts made by bonded wheels, normalized to 10 units in the easiest direction, corresponding to a depth of 2.73 μm (Wilks, 1961)

	Face					
	Dodecahedron {011}		Cube {001}		Octahedron {111}	
Direction	<100>	<0$\bar{1}$1>	<100>	<110>	<11$\bar{2}$>	<$\bar{1}\bar{1}$2>
Depth	10.0	1.8	8.5	1.1	3.9	2.1

continually recharged the rate falls to 0.02 or lower (Table 9.1). Hence the bonded wheel offers considerable advantage for the removal of material in hard directions.

The variations of polishing rates with azimuthal variations of direction have also been observed using the micro-abrasion tester described in Section 9.2.b, and Table 9.4 gives a summary of some depths of cuts made with bonded wheels (Wilks, 1961). The depths of the cuts are normalized to the <100> direction on an {011} face in order to take account of differences between the wheels. As some of these cuts were rather deep, up to 5 μm, the values relate only approximately to the polishing rates obtained on a flat bonded wheel, but give some indication of the relative polishing rates on different faces and in different directions.

9.3.c Effect of tilting a face

We saw in Section 9.3.a that the polishing rates on a cube face may be much altered by small tilts of the face away from {001}. We now describe the effects of larger tilts, and also similar effects near {011} and {111} faces.

Figure 9.7 shows a cube face with the four <100> directions labelled 1 to 4. Micro-abrasion tests using a cast-iron wheel were made in these four directions and as expected the cuts were all of equal depth. The measurements were then repeated on facets obtained by swinging the face about an axis parallel to AC through an angle θ so that B was lifted above the cube plane. Figure 9.11 shows the depths of the cuts now obtained in the four directions as a function of θ (Wilks and Wilks, 1959). As mentioned above, small angles of tilt do not cause much change in the depths of the cuts in the directions 1 and 3 parallel to the axis of tilt. However, large changes are observed in directions 2 and 4 which correspond to the downhill and uphill directions in Figure 9.8. We see from Figure 9.11 that a tilt of only 2° is sufficient to result in one of the four directions being markedly less resistant to abrasion than the other three. This result reconciles the diamond polisher's perception of 'the grain' described in Section 9.3.a with the known crystal symmetry of the diamond, because in the past diamond faces were generally aligned by eye with an accuracy seldom better than 1° or 2° and often much worse.

Figure 9.12 shows the effect of tilting a surface away from an {111} plane. The measurements were similar to those just described except that the abrasions were made by bonded rather than cast-iron wheels because of the greater abrasion resistance of the {111} face (Wilks and Wilks, 1972). Abrasions were made in the two opposite directions A_x and A_0 on an octahedron face as shown in Figure 9.12(a), these directions being respectively the most and the least abrasion resistant. Figure 9.12(b) shows that as the face is tilted about the side MN extremely rapid variations of the abrasion resistance are observed. On a correctly aligned {111} plane the cut in direction A_0 is approximately twice as deep as that

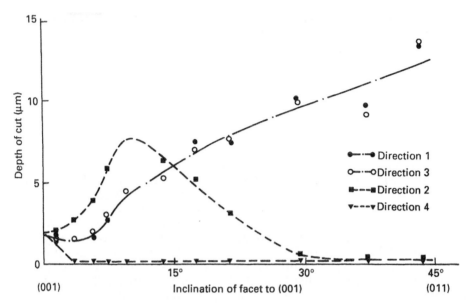

Figure 9.11 Depths of abrasion cuts on facets inclined to {001}. The directions 1–4 refer to Figure 9.7 (Wilks and Wilks, 1959)

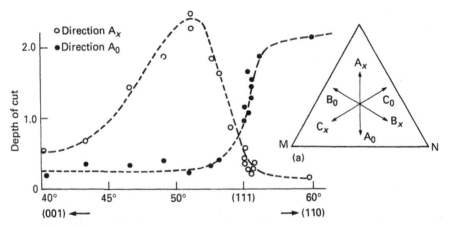

Figure 9.12 Depths of abrasion cuts on facets near {111} (in arbitary units). Inset (a) shows the directions A_x and A_0 and the <110> axis MN about which the octahedron plane was tilted (Wilks and Wilks, 1972)

in direction A_x but a tilt of only 1° towards the cube face reverses the position, and the A_x cut is twice as deep as the A_0 cut. (This behaviour makes it very difficult to polish an exactly aligned {111} face because the diamond prefers to polish on a less resistant facet. Therefore the results in Figure 9.12 for planes close to {111} were obtained on surfaces selected from a range of cleavage planes (Wilks and Wilks, 1967).)

Further measurements on the effects of tilting cube and dodecahedron faces through angles of about 5° and 10° are given by Wilks and Wilks (1972). Table 9.5

Table 9.5 Summary of main features of the abrasion resistance (Wilks and Wilks, 1972)

Face	Hard directions	Easy directions
Cube {001}	<110>	<100>
Dodecahedron {011}	<0$\bar{1}$1>	<100>
Octahedron {111}	<$\bar{1}\bar{1}$2>	<11$\bar{2}$>

Face	Direction of abrasion	Direction of easier abrasion on tilted surface
Cube	<100>	Downhill
Cube	<110>	Uphill
Dodecahedron	<100>	Uphill
Dodecahedron	<0$\bar{1}$1>	Uphill
Octahedron	See Figure 9.12	

summarizes some principal features of the various measurements. It shows the least and most abrasion resistant directions on each face and whether the less resistant direction on a tilted face is 'uphill' or 'downhill' as described above. We discuss these results in Section 9.4.

9.3.d Quality of polish

Besides determining the rate of removal of material during polishing, the orientation of a face and the direction of abrasion also affects the quality of the polished surface. Thus polishing in an abrasion resistant direction generally produces a rougher polish than is obtained in an easier direction. For example Figure 9.13 shows the {011} surfaces obtained by polishing by the same cast-iron

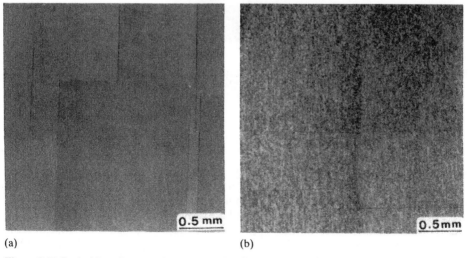

(a)   (b)

Figure 9.13 Optical interference micrograph of {011} surfaces polished by the same cast-iron scaife with 0–1 μm powder in (a), an easy direction; (b), a hard direction (Acton and Wilks, 1989)

(a)

(b)

Figure 9.14 Optical interference micrograph of the same diamond surface polished in (a), an easy direction of abrasion; (b), in the reverse direction (Wilks, 1973)

scaife with 0–1 μm powder in an easy <100> direction and a hard <110> direction. Figure 9.14 shows an even more marked difference. Figure 9.14(a) is a micrograph of a facet which has been polished on a bonded scaife and which is tilted 2° away from (001), and Figure 9.14(b) shows the same facet after polishing in the same manner but in the opposite direction. We see that considerable cracking was produced during the latter polish. No such effect was observed when polishing on an adjacent true (001) plane. Hence tilts of only 2° can result in severe imperfections on a polished surface.

9.4 Mechanism of polish

We now consider how the polishing process removes material from the diamond surface and gives rise to the remarkable dependence of the abrasion resistance on the crystallographic orientation of the surface.

9.4.a Polishing as micro-chipping

After an extensive study of the polishing process Tolkowsky (1920) concluded that material is removed during polishing by processes of micro-cleavage and that the abrasion resistance of the diamond was dependent on the position of the cleavage planes relative to the surface being polished. It is well known that cleavage can be easily initiated by a light blow in the appropriate direction but that impacts in other directions are much less effective. Hence it follows that the force required to break away a chip depends greatly on the direction in which the force is applied relative to the cleavage planes.

To predict the directions of easy abrasion or polish Tolkowsky made use of three dimensional models constructed from octahedral and tetrahedral blocks to display the positions of all possible sets of {111} cleavage planes (Figure 9.15). Note particularly that these models are solely a means of visualizing the positions of the cleavage planes. They must not be taken to imply that a real diamond, which is generally a single homogeneous crystal is in any way built up from a set of blocks as in the model. With this introduction we now describe Tolkowsky's discussion of the differences in abrasion resistance commencing with the variation on a {001} face as a function of the azimuthal angle of abrasion.

Suppose we prepare a cube face by grinding down the top of the regular octahedron shown in Figure 9.15(a) in a <110> direction parallel to one of the square edges. After removing octahedra and tetrahedra we eventually arrive at the type of surface shown in Figure 9.15(b) where the elementary blocks tend to lie in ridges parallel to the direction of abrasion. Each block presents a minimum surface area normal to the abrading direction and is buttressed against displacement by the adjacent blocks in the ridge. On the other hand if we abrade at 45° to this direction, along <100>, then the ridges will be attacked from the side and the surface eventually takes the form shown in Figure 9.15(c). The octahedra on this surface present an appreciable area to the abrading particles, are less buttressed, and so more easily removed. That is, material will be removed more rapidly by polishing in <100> directions than in <110> directions.

Rather similar arguments account for the hard and easy directions of abrasion on {011} faces where polishing in a hard direction again produces a system of ridges rather like those in Figure 9.15(b). On an octahedron {111} face one set of cleavage planes lies parallel to the surface, so the abraded surface will tend to present only flat cleavage surfaces to the abrading particles so all directions will be difficult to abrade. For further discussion of the details of the model see Wilks and Wilks (1972).

Of course, in practice, the size and shape of the fragments chipped off from the surface will not be entirely regular in form nor of uniform size. Nevertheless, we expect to see general trends in both the geometry of the debris and of the resulting polished surface of the form indicated by the model. We can thus account for the hard directions on cube and dodecahedron planes by the tendency of the abraded surface to take up the ridge-like structures shown in Figure 9.15(b). The

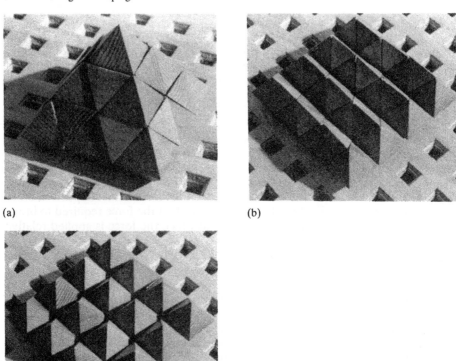

(a)   (b)

(c)

Figure 9.15 Block models of diamond showing (a), an octahedron tip; (b), an {001} face polished in a hard direction, (c), an {001} face polished in an easy direction, see text (Wilks and Wilks, 1972)

topography of a real surface is obviously very complex, so it is impossible to make any numerical estimates of the abrasion resistance except to say that the variations with orientation could be large.

The essential features of the above discussion are confirmed by a variety of observations both of the abrasion process and of the nature of the polished surfaces which are produced, as is discussed below. We remark also that Tolkowsky's account of the dependence of polishing rates on direction was mainly restricted to azimuthal variations on the principal {001}, {011} and {111} planes, whereas the rates may also be much changed if the polished surface is tilted away from a principal plane by even a small angle (Section 9.3.c). Although it is difficult to make quantitative estimates, Tolkowsky's treatment can be extended to predict whether a particular tilt of a surface increases or decreases the resistance, and predictions made in this way are in line with the experimental results summarized in Table 9.5 (Wilks and Wilks, 1972).

The polishing process has also been discussed by Wentorf (1959) in terms of observations of the cracks produced on a diamond surface by a sliding diamond stylus. These cracks lie on {111} planes transverse to the direction of sliding, and their formation leads to the removal of material. The author thus predicts similar variations in abrasion resistance to those given by Tolkowsky. The polishing process has also been discussed by Jeynes (1983) in terms of micro-chipping processes modified by plastic deformation but such deformation appears improbable as discussed in Chapters 7 and 8.

9.4.b The polished surface

According to the above discussions a polished surface of a diamond is formed by processes of brittle fracture on a microscopic scale. Hence we expect the surface left by the removal of material to be rough and jagged on a microscopic scale with rugosities delineated at least to some extent by {111} cleavage planes. Moreover because diamond is resistant to plastic deformation except at high temperatures it appears that this surface roughness is not smoothed over in the polishing process as might be the case with other materials. We now describe experiments on the friction of diamond and studies with the electron microscope which show the very unusual nature of the polished surface.

The friction of diamond sliding on diamond is described in Chapter 11, but we note here an experiment which measured the coefficient of friction for the tip of an octahedral diamond sliding over a polished {001} face (Casey and Wilks, 1973). In Figure 11.2(a) the coefficient varies by a factor of about 2 with changes in the azimuthal angle of sliding, being greatest in <100> and least in <110> directions. The experiments were made in the open laboratory so both stylus and polished surface would be covered with films of absorbed air, but even so the values of the friction reflect the symmetry of the atomic structure of the underlying diamond.

In a further experiment the diamond surface was repolished with the abrading particles moving in a <110> direction, at an angle of 45° to the direction normally used, that is the repolishing was in the direction most resistant to abrasion. The friction was now appreciably different as shown in Figure 11.2(b). In particular the symmetry of the friction was not four-fold but two-fold, although the underlying crystal structure can hardly have charged. This result is difficult to understand on conventional theories of friction but follows naturally from the picture of the polished surface implied by the abrasion measurements. As will be discussed in Chapter 11 the friction of diamond on diamond is the consequence of dissipative processes as the stylus moves irreversibly over the jagged surface. Further, according to the discussion of the previous section the microstructure of the surface is determined by the direction of polish, so that faces polished in <110> and <100> directions have different topographies. In fact, the difference in symmetries shown by the friction measurements are entirely in accord with a Tolkowsky type treatment. It is presumably such differences in surface topography which are responsible for the change in the sound of the polishing process often observed as the polisher moves onto an easy direction of polish.

The topography of polished surfaces has also been examined by electron microscopy. High resolution optical techniques show that the scale of any roughness is very small, less than 10 nm, so high resolution shadowed replicas were developed for study in the transmission electron microscope (Thornton and Wilks, 1976). A difficulty inherent in the preparation of these replicas arises, because in order to replicate the surface structure, the film must closely contour the jagged surface, on the other hand the replica must strip away from the surface without breaking. Particular care is also necessary because the structure turns out to be

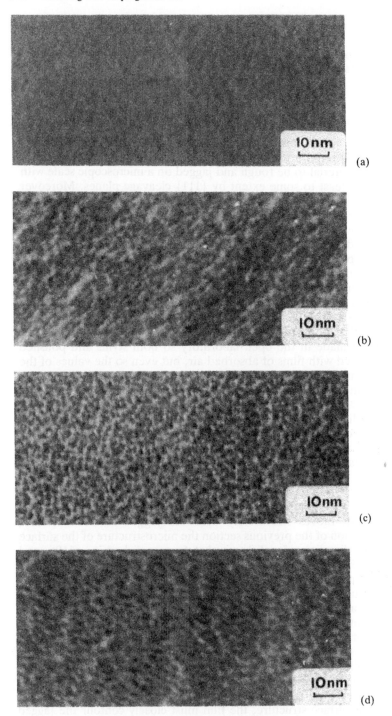

Figure 9.16 Electron micrographs of replicas of (a), cleaved mica; (b), polished {011}; (c), polished {111}; and (d), polished {001} diamond faces (Thornton and Wilks, 1976)

irregular, so that it is very important that the micrographs be in focus and free of astigmatism.

All four micrographs in Figure 9.16 were obtained from replicas prepared under carefully controlled and similar conditions, save that (a) was taken from a freshly cleaved surface of mica, and the others from polished surfaces of diamond. The contrast in the micrograph corresponds to height variations on the surface, so the polished surfaces are much rougher than the mica. The horizontal scale of the irregularities is seen to be of the order of 5 nm, while the effects of changing the shadowing conditions implied that the height variations in the surface are of the same order. Thus, the micrographs confirm the rough and jagged microscopic nature of polished diamond surfaces, which is implied by the abrasion and friction experiments.

A complete interpretation of the micrographs involves a consideration of the resolving power of the system. That is, we must first ask if the detail in the micrographs corresponds to the structure of the diamond surface, or merely reflects the structure of a particular replica. Figure 9.17 shows a feature on a diamond surface replicated by two different replicas; the outline of the feature appears to

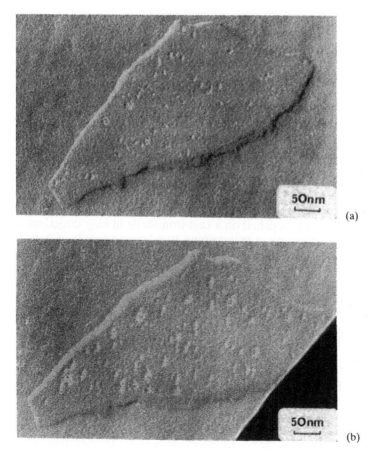

(a)

(b)

Figure 9.17 Same area of a diamond reproduced on different replicas (Thornton and Wilks, 1976)

be the same in each micrograph to within about 3 nm. Further experiments suggested that this resolution is set primarily by the migration of the tungsten shadowing material before it comes to rest on the surface.

We now return to the micrographs in Figures 9.16, bearing in mind the above limit on the resolution. The micrographs were prepared using the same replica technique from cleaved mica and from {011}, {111} and {001} faces polished on the same diamond using a metal-bonded wheel containing diamond powder about 30 μm in size. Besides showing much more contrast than the cleaved mica all three micrographs from the diamond appear characteristically different from each other. Similar differences were again observed on taking replicas from similarly polished {011}, {111} and {001} faces on a second diamond (Thornton and Wilks, 1976). Thus the micrographs confirm the very unusual structure of polished diamond surfaces already implied by the various studies of abrasion and friction described above. Finally we note that it would be of considerable interest to investigate the structure in more detail with techniques of greater resolving power such as the tunnelling electron microscope (Binnig and Rohrer, 1982; Golovchenko, 1986) and the force microscope (Binnig, Quate and Gerber, 1986). (Note however, a comment in Section 17.6.b.)

9.4.c Thermal effects

Diamonds may become quite hot while being polished because the work done by the horizontal friction force on the diamond is dissipated as heat. This force increases with both load and speed so that under severe conditions of heavy loading and high speeds the diamond may become hot enough to burn or graphitize. In this case the surface finish greatly deteriorates so these conditions are generally avoided, and in fact all the results discussed above refer to conditions within the normal range. Even so there have been suggestions in the literature, see for example Seal (1958) and Bowden and Tabor(1965), that all diamond polishing is largely brought about by thermal wear. We now discuss this point in some detail, beginning with some experiments on how the rate of abrasion depends on the speed of the wheel.

The most complete experiments on the effect of wheel speed are those of Tolkowsky (1920) for diamonds polished on a cast-iron scaife in easy directions of abrasion where there is no difficulty in maintaining an adequate supply of powder on the wheel. These experiments showed that the removal rate is proportional to the linear velocity of the wheel over the surface of the diamond up to speeds of at least $100 \, \text{ms}^{-1}$ (Figure 9.18). This result is confirmed by experiments using a wheel tester running at somewhat lower linear velocities where the rate was proportional to the speed from about $0.4 \, \text{ms}^{-1}$ to $4 \, \text{ms}^{-1}$ (Wilks and Wilks, 1959). Hence it follows that each revolution of the wheel removes the same amount of material irrespective of its speed.

We now return to the suggestions mentioned above that the abrasion of diamond is essentially a thermal process. These proposals were based on the well known fact that the abrasion of many materials is assisted by the temperature of local hot spots generated at the tips of asperities on abrading surfaces, see Section 13.3.b. In order to account for the great dependence of the abrasion rates on the orientation of the diamond Seal (1958) pointed out that as the coefficient of friction varied with direction by about a factor 2 one might expect the temperature of the hot spots to vary by a factor of the same order. Hence as thermal wear generally

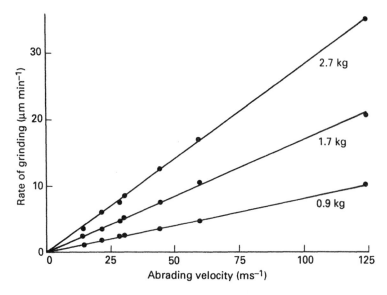

Figure 9.18 Rate of removal of material during polishing as a function of linear velocity at three different loads (after Tolkowsky, 1920)

increases exponentially with the absolute temperature one might thus account for the very large differences in abrasion rates. However, this picture implies that the rate of abrasion should rise very rapidly with the speed of the wheel. Therefore the fact that over a wide range of wheel speeds the volume of material removed varies only directly as the speed of the wheel, shows that the hot spots in the surface play no significant role in the abrasion process.

Despite the above discussion one hears reports from time to time that polishers have obtained faster removal rates by heating the scaife to perhaps 200°C or 300°C. However,it is only recently that detailed information has become available. Yarnitsky et al. (1988) descibe measurements on the removal rates observed when using a conventional scaife with facilities for heating either the diamond or the scaife. Heating the diamond did not affect the removal rate but heating the scaife did. Figure 9.19 shows the removal rate as a function of the temperature of diamond as measured by a thermocouple on the diamond close to the polished surface and the scaife. A temperature rise of about 100°C was accompanied by over a threefold increase in removal rate but heating to higher temperatures caused the rate to fall back, eventually to a much lower level than at room temperature. Hence it is most unlikely that the initial rise in removal rate is due to either straightforward thermal wear or to a greater plasticity of the diamond. The authors suggest that, as the temperature first increases, the scaife becomes less hard and is therefore better able to retain the abrading powder so that the removal rate increases. However, as the temperature rises further the scaife becomes steadily softer so the diamond powder will be pressed further into the iron and becomes progressively less effective.

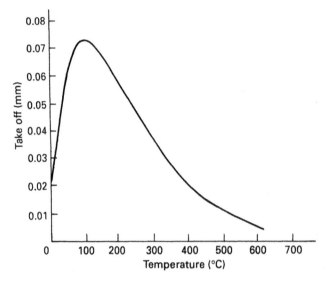

Figure 9.19 Removal of diamond by a cast-iron scaife and powder heated to different temperatures, plotted against the temperature of the diamond near to the scaife (Yarnitsky *et al.*, 1988)

9.4.d Effect of load

The removal rates during polishing were studied as a function of the load on the diamond by Tolkowsky (1920) who found that for loads up to about 2.5 kg the rate was proportional to the load, and similar results were obtained in micro-abrasion tests by Wilks and Wilks (1972). These results are probably best explained by recalling that as the surfaces of both diamond and scaife are rough, actual contact only occurs between the tips of asperities. Hence, because the pressures used in polishing are relatively low, the area of true contact is much less than the nominal area of contact. An analysis of this situation (Greenwood, 1967; Johnson, 1985) shows that as the pressure is first increased the number of contacts increases, as might be expected. In addition, however, it appears that the *mean* area of the individual contacts remains approximately constant. (The mean area of each contact increases with increasing load but at the same time new contacts with smaller areas are formed.) Therefore the principal effect of increasing the load on the diamond is to increase the number of contacts more or less in proportion to the load without much changing their area. In this case we expect the removal rate to be proportional to the number of contacts and therefore to the load, as is observed. Tolkowsky also observed that, at a given speed and load, the removal rate was independent of the area of diamond being polished. This result also follows from the above discussion. The same load on a smaller area produces a greater number of contacts per unit area, but the total number of contacts and therefore the removal rate remains unchanged.

9.5 The polished surface

The micro-structure of the polished diamond surface described in Section 9.4.b is on so small a scale that it can only be viewed with the electron microscope.

However, a polished surface also shows other structure which can be seen in the optical microscope.

9.5.a Polish lines

Although the surface of a diamond polished by normal techniques may appear smooth to the eye it is always marked by an array of parallel polish lines produced by the abrasive particles moving over the surface. As already mentioned these lines are shallow grooves with a wide range of widths (Figures 9.1 and 9.2). Naturally the use of larger size powder on the scaife produces larger and deeper lines, but even with a given size powder the prominence of the lines may depend considerably on the method of polish. For example Figure 9.20(a) shows quite pronounced polish lines on a diamond polished on a bonded wheel, while Figure 9.20(b) shows the same surface after polishing in exactly the same way except that the diamond was traversed to and fro across the wheel during polishing. We see that the system of polishing lines has been much reduced by the traversing.

It is quite possible to polish a face so that the polish lines are hardly visible as in Figure 9.13(a), or even not apparent at all, but to obtain good results requires both care and experience. It is absolutely essential for the scaife or wheel to run in high precision bearings with a complete absence of vibration. The scaife or wheel must be kept clean, clear of any accumulation of debris, and kept charged with the appropriate amount of powder. The diamond must be brought up to the wheel in the correct orientation because even small changes in orientation may greatly affect the quality of polish.

(a)   (b)

Figure 9.20 Optical interference micrograph of a diamond surface polished on a bonded wheel (a), diamond held steady; (b), diamond traversed with a reciprocating motion across the wheel

As when polishing other materials, it is general expedient to begin with larger powder to remove material more quickly and then use one or two grades of smaller powder to produce a smoother final finish, perhaps using lighter loads. It is important to note that the pattern of polish lines seen on a face has been formed by only the last few revolutions of the scaife just before the removal of the diamond. Therefore a skilled polisher can do much to achieve a smooth finish by the way he lifts the diamond away from the wheel.

9.5.b Defects of polish

Polishing lines, more or less marked, are always produced by the normal polishing process. In addition a polished surface may exhibit other structural features produced either by poor polishing or by defects in the diamond. For example on a well polished stone the polish lines run continuously across the surface but on the diamond shown in Figure 9.21 we see lines parallel to the direction of polish

0.5mm

Figure 9.21 Optical interference micrograph of a polished {001} surface showing the onset of score lines. The arrow indicates the direction of polish

beginning abruptly at points within the surface. It seems that these coarse polish lines have been produced by larger than average particles of diamond which have broken away during the polishing process. The outbreak of the coarser lines at the top of the micrograph occurs along the edge of a growth layer (Section 5.2.c) which appears to have caused some structural weakness.

Figure 9.22 shows other features on the polished surface of a diamond containing many small inclusions. Besides the continuous polish lines running in the direction of polish (from right to left) three sets of coarser lines are also visible. The defect responsible for the middle set of lines cannot be identified from the micrograph but the upper set is seen to initiate where a small inclusion intersects the surface. The lower set of lines initiates at a larger feature containing a small inclusion. Apart from the presence of the inclusion, this feature appears to be a typical example of

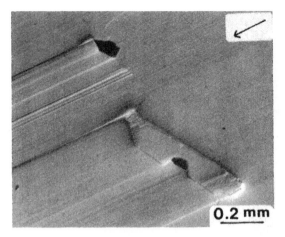

Figure 9.22 Optical interference micrograph of a polished surface with naat and inclusions, see text. The arrow indicates the direction of polish

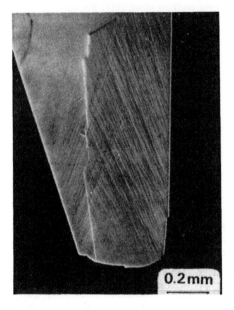

Figure 9.23 Optical interference micrograph of a polished macle diamond showing a line of twinning

a so-called naat, that is a small volume of diamond which is appreciably misorientated with respect to the bulk of the diamond crystal. It is therefore more abrasion resistant and stands slightly above the surrounding surface. The misorientation often takes the form of twinning (Section 5.7.a) but may also occur with less specific forms. Occasionally twinning occurs on a larger scale, with perhaps the whole crystal being twinned, as in the diamond shown in Figure 9.23. Finally we note that polished surfaces often exhibit slightly raised patterns associated with growth layers in the diamond, these are described in Section 10.2.a.

Figure 9.24 Optical interference micrograph of an {001} surface polished with 4–8 μm powder showing cracking along {111} planes

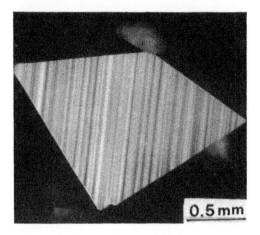

Figure 9.25 Optical interference micrograph of a facet on a gemstone which has been poorly polished and exhibits zuiting

In addition to the above imperfections polishing sometimes produces quite appreciable cracking as in Figures 9.14 and 9.24. This seems most likely to occur when polishing surfaces which do not coincide with the principal crystallographic planes, as in the example shown in Figure 9.14. Such cracking may often be reduced by changing the direction of polish.

Figure 9.25 gives an example of a polished face showing polishing lines and a particularly marked banding effect known to polishers as zuiting. This appears to result from poor polishing procedures, including dirty conditions or insufficient powder. Very similar patterns can be produced by rubbing a polished diamond surface on a clean cast-iron scaife free of powder, see for example Thornton and Wilks (1974).

9.5.c Subsurface damage

The character of a polished surface is not completely described by its surface topography. Polishing proceeds by micro-chipping, so the polishing process must inevitably cause cracking below the surface which will affect the strength of the diamond. For example, as described in Section 7.5, polishing with larger size powder produces a surface less resistant to Herz indentations presumably because of the presence of deeper subsurface cracks.

Figure 9.26 shows an X-ray topograph of a nicely polished (001) surface of diamond on which three cuts had been made with a micro-abrasion tester, the surface of these cuts being somewhat rougher than the polished surface (Frank *et al.*, 1967). The darkening on the topograph indicates the presence of strains associated with the abrasions. The form of the strain is discussed in detail by the authors who conclude that the surface of the cuts contains cracks opened during the abrasion and which have not fully closed. Therefore the surface layer has expanded somewhat and is in a state of compression, constrained by the surrounding material. The topograph also shows that with careful polishing the level of residual strain may be quite low, as over most of the polished surface. The low level of damage in other well polished surfaces has also been shown by

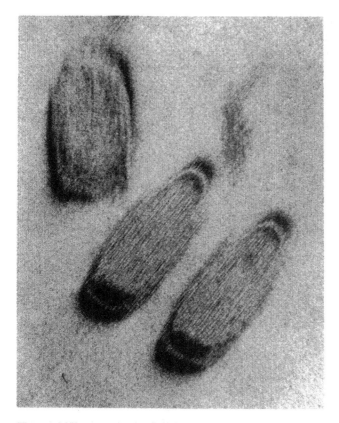

Figure 9.26 X-ray topograph of three abrasion cuts on a polished {001} face. The single cut runs parallel to <110> and the pair of cuts to <100> (Frank *et al.*, 1967)

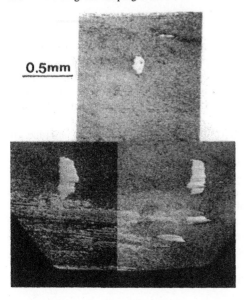

0.5mm

Figure 9.27 Optical micrograph of an {011} face of a diamond initially carrying abrasion cuts, after polishing in a hard direction; for details see text (Hitchiner, Wilks and Wilks, 1984)

experiments on the ion-channelling of beams of protons and α particles (Derry, Fearick and Sellschop, 1981).

Subsurface damage plays an important role in determining the strength and fatigue of diamond as described in previous sections. Therefore the response of a surface to a mechanical test or polishing procedure depends not only on its orientation and on its topography but also on any subsurface damage produced by previous events. An example of the effect of subsurface damage is given by Hitchiner, Wilks and Wilks (1984). In the course of other experiments these authors polished an {011} surface of a diamond in a hard <011> direction of abrasion and then made three abrasion cuts similar to those of Figure 9.4 in order to measure polishing rates. After these measurements had been completed the polishing was continued in the same <0$\bar{1}$1> direction to remove the abrasion cuts completely before beginning a new experiment. (The polishing was made at quite a low speed to study other effects but this is of no consequence to this discussion.)

Figure 9.27 is an optical micrograph of the above surface taken during the final polish to remove the three cuts. The position of the cuts is now marked by the three largest white areas which appear smoother than the rest of the surface. Quite surprisingly, profilometer measurements showed that these areas stood slightly *above* the rest of the surface. Similar results were observed on other diamonds, as in Figure 9.28. The authors account for this effect by noting that the cuts were

I 5 μm

200 μm

Figure 9.28 Profilometer trace taken along the long axis of an abrasion cut, similar to those of Figure 9.27, in the course of being polished away (Hitchiner, Wilks and Wilks, 1984)

made with a rotating wheel abrading in an easy <100> direction, whereas the main body of the surface was polished in a harder <0$\bar{1}$1> direction. Therefore we would expect the surface of the cuts to be smoother with less subsurface damage than the rest of the surface. Hence, when the diamond was polished down after the cuts had been made, the material under the cuts was more abrasion resistant and therefore at one stage of the polishing stood above the general level. (Of course, if polishing is continued until all the more resistant material is removed, the whole surface eventually becomes uniform.)

The above experiments show that the polishing process exhibits a form of hysteresis effect in that a surface does not take up its final equilibrium polish until all the previous structure and damage has been removed. We have already seen that the difference in polishing rates in <100> and <110> directions on {001} faces is due to differences in the surface structures *generated by the polishing processes themselves* (Figure 9.15). Hence, if we change the direction of polish there must be a period of adjustment during which the surface structure changes and the polishing rate moves to a new value. An example of this hysteresis effect is seen in an experiment to investigate how under certain conditions the rate of polish is reduced by reducing the pressure of the surrounding air (Hitchiner, Wilks and Wilks, 1984). These authors found that the full change in polishing rate after changing the pressure was only obtained after a further few micrometers of material had been completely removed, see Section 13.4.a and Figure 13.8.

9.5.d Extra-fine finish

The skilful use of the conventional polishing techniques described above can produce surfaces on which the polish lines are hardly visible in the optical microscope. We now mention other techniques which have been used to obtain even finer polishes for the fabrication of microtome knives and other high precision cutting tools.

One common approach is to use diamond powder of smaller size for the final polishes. There are no very detailed accounts of this technique but it has been developed over a number of years to produce smooth finishes and thus very sharp edges on diamond knives, see for example Fernandez-Moran (1956, 1985) who used an ultracentrifuge to prepare powder sufficiently fine and uniform. To obtain the desired results the scaife must be machined flat to a high degree of accuracy and run in massive precision bearings which permit a high speed with great stability despite the relatively high forces generated by the abrasion of diamond. It is also important to avoid vibration and shock effects by locating the polishing machine on a suitable site and by the use of shock free mountings. Finally, various attempts have been made to polish diamond with non-diamond powders, see for example Grodzinski (1953). With such softer powders it may be possible to achieve a smoother finish but the rate of removal of material will be so much reduced that extremely long periods of polish will be necessary.

A quite different way of obtaining a fine finish is to complete the polishing process by some form of chemical etch or dissolution process which smooths any surface roughness. For example Thornton and Wilks (1974) polished a diamond surface by rubbing the diamond on a cast-iron scaife free of any diamond powder but charged with a solution of either potassium nitrate or potassium chlorate. The bulk temperature of the diamond rose by up to 180° on adding potassium nitrate solution (Figure 9.29) and the polish lines began to disappear as shown in Figure

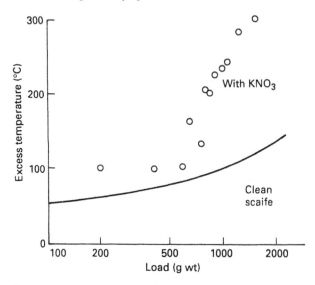

Figure 9.29 Bulk excess temperature of a diamond rubbed under different loads in the presence of potassium nitrate solution (Thornton and Wilks, 1974)

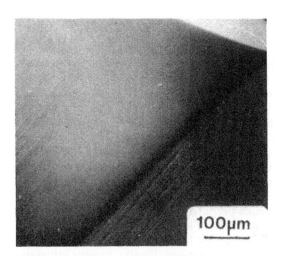

Figure 9.30 Optical interference micrograph of {001} surface of diamond polished in the usual way, and then rubbed on a cast-iron scaife in the presence of potassium nitrate (Thornton and Wilks, 1974)

9.30. This method appears promising but the reaction products tend to accumulate on the scaife, so it will be necessary to devise some way of removing them in order to maintain a continuous polishing process. (We note in Section 11.1.b that the friction of a diamond surface was reduced by this type of polish, which suggests that the nanometer scale structure has been affected as well as the polishing lines.)

A simpler method of chemical polish might be to immerse the surface in a bath of the chemical reagent. However, the agent must be chosen to act preferentially on the tips of the asperities, and there is not much information available on this

Figure 9.31 Interference micrograph of facets produced on a spherical diamond surface by a chemical polish (Acton, 1989)

point. Chemical dissolution would be particularly convenient for polishing curved surfaces were it not that dissolution processes tend to produce facets. This tendency can result first in changes of the curvature and then in shallow junction lines of the type seen in Figure 9.31.

Another possibility for a chemical type of polish depends on the fact that diamond in air begins to oxidize at temperatures above about 600°C. If a diamond becomes too hot the surface begins to burn and blacken, and polishers take care to keep away from these conditions. However, if the temperature could be maintained just below the point where serious burning begins conditions might then favour the production of a smoother finish. We ourselves have obtained an unusually smooth polished face on a diamond which had run very hot and which showed clear signs of overheating on other faces. A patent application due to Hall, Hall and Lauridsen (1987) gives details of a practical process on these lines which, it is claimed, can greatly reduce the time required to polish scratch free surfaces.

9.6 Shaping diamond

9.6.a Traditional methods

The shaping of diamond to produce gem stones and tools involves sawing, bruting and faceting. We describe these processes briefly, fuller accounts are given in books by Bruton (1981), Watermeyer (1982) and Vleeschdrager (1986).

A diamond saw consists of a circular blade of phosphor-bronze about 75 mm in diameter and about 0.1 mm thick. This rotates at high speed on a spindle to which it is clamped by two metal flanges. The narrow edge of the blade is charged with diamond powder and brought against the diamond to be sawn. The charged edge removes diamond in much the same way as a polishing scaife, and the blade cuts through the diamond under the action of a gravity feed. Sawing is a slow process taking perhaps several hours according to the size of the diamond. Hence, it is important to choose the best orientation for the cut, often along an {001} or {011} plane, and to give the blade a good start by making a nick or *kerf* on the diamond, using either a metal blade, a sharp edge of diamond, or a laser beam. To obtain better performance, sawing machines are now being developed to measure the load on the saw with a suitable sensor and thus control its rate of descent. It should be noted that sawn surfaces are usually quite rough and generally show parallel lines of saw marks.

Bruting is a chipping process used to rough out a diamond into the circular form of many gem stones. Typically, a regular octahedron is sawn across the central {001} plane to produce two half stones each with an approximately square face to serve as the *table* or main face of the gem. Each half is then mounted on a lathe and turned against the sharp edge of another diamond so as to convert the square outline into an approximately circular one. A bruted stone is readily recognized by the rough conchoidal form of the chipping around the rim or *girdle*.

To produce, for example, the well known brilliant cut, the diamond must first be sawn and bruted and then polished to give a total of 58 different facets (Section 17.1). In the past the successful polishing of gemstones depended on the polisher's skill and experience to obtain the correct orientations of the facets and to select the easiest direction of polish on each of them. Today, polishing is increasingly being performed by machines which are programmed to give the required orientations of the facets, and which are able to select an easy direction of polish on each facet. This choice of direction is usually made by polishing for short intervals in set directions and at the same time measuring the rate of polish, usually by observing the movement of the diamond down towards the scaife with a sensitive displacement indicator.

Cleaving provides a ready method of dividing a diamond along {111} planes as described in Section 7.4. Although this type of division is not generally used to shape the brilliants described above, cleavage presents a useful and rapid alternative to sawing on several occasions. In particular it is widely used to deal with diamonds of difficult and irregular shape. For example, it permits the division of a macle diamond (Section 5.7.a) along the plane of twinning thus producing two stones each much easier to polish. Cleaving is also useful for removing small parts of a diamond containing unwanted cracks or inclusions.

The preparation of cutting tools involves rather different procedures. The typical shape of many tools is shown in Figure 14.2. The essential features are a flat rake face or *table*, side *flanks* which come together at a rounded nose, and a sharp cutting edge at the nose of the tool between the flanks and the table. To obtain the curved flanks at the nose the diamond is polished with a radiusing machine which swings the diamond about an axis parallel to the scaife, the radius of the nose being determined by the final distance of the axis from the scaife. When polishing the curved parts of a tool it is necessary to remember that the abrasion resistance of the surface changes with its crystallographic orientation,so different parts of the surface will polish at different rates and this may affect both the geometry and quality of the surface. Therefore the radiusing machine must be strongly constructed with the diamond mounted rigidly on an arm swinging on massive bearings. It is also usual for the rotating scaife to move on a planetary path, see for example Teather (1969), so that the direction of abrasion is continually varying through a range of angles to ensure that the polishing is not held up by one particularly hard direction.

9.6.b. Electric discharge methods

Various attempts have been made to increase polishing rates by using some form of electric discharge or current. Peters, Nefflen and Harris (1945) used a normal polishing machine with a cast-iron scaife and diamond powder with the diamond held in a copper dop by a high melting point solder. When a 5000 V output from 60 Hz transformer was connected between the dop and the scaife a bluish arc was observed at 'the contact' of the diamond and the scaife. Under these conditions the authors observed increases in the polishing rates on {001} and {011} faces of the order of 2:1. They also state that rates of sawing could be increased in a rather similar way.

Peters, Nefflen and Harris comment that more information is required in order to understand the effect of the discharge. Presumably the discharge strikes the asperities on the diamond surface and the forces thus generated assist the polishing process to remove material. However, no detailed study of this effect has been reported. Nor is there any information regarding the path of the discharge; diamonds are generally highly insulating and it is not clear how the discharge passes through the diamond. Peters *et al.* (1947) described the use of a high voltage applied to the pointed tip of a platinum/iridium electrode for the drilling of the initial hole in a wire drawing die. Another description of spark erosion methods has been given by Levitt (1968). However, the high voltages involved in all these methods present a considerable hazard, so the methods are seldom employed.

Peters *et al.* (1947) also drilled dies by immersing the die and a pointed electrode in a bath of an electrolyte, either potassium nitrate or sodium chloride. In this case electric potentials of the order of 100 V were sufficient to produce drilling rates much faster than could be obtained by purely mechanical methods, for further details, see the original paper. Note that any electrolytic method is limited to some extent by the usually very high resistance of a diamond to an electric current. It is, however, now being used quite extensively to drill dies of PCD diamond materials which are electrical conductors because of the metallic phase between the crystallites, see for example Eder (1983, 1986) and Finnigan (1987).

9.6.c Ion beam milling

The ion beam technique is used to shape and thin a variety of materials. A beam of ions, often argon ions, is generated in a gaseous plasma and accelerated by an anode held at some voltage between 1 keV and 20 keV. The accelerated beam then strikes the surface of the specimen and the high energy ions displace the atoms of the specimen causing many of them to be ejected or *sputtered* from the surface. Provided that we are dealing with a uniform material such as a single crystal the removal of material occurs evenly over the surface on a fine scale.

The method was first applied to diamond by Spenser and Schmidt (1972) who were able to drill holes about 1 mm in diameter through a diamond 1.6 mm thick, but the rates of penetration were no more than $2 \mu m \, min^{-1}$, much less than can be attained by electric discharge and laser methods of drilling. The ion technique was subsequently developed to machine sharp edges of diamond and diamond styli of very small radius, see Miyamoto and Taniguchi (1983), and Miyamoto (1987). These authors first polished the diamond to a carefully specified shape and then subjected it to a uniform beam of ions. The action of the beam is best shown by two examples.

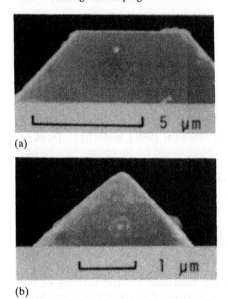

(a)

(b)

Figure 9.32 SEM micrographs to illustrate formation of a knife edged stylus by ion milling: (a), initial shape polished mechanically; (b), after milling for 4.9 h (Miyamoto, 1987)

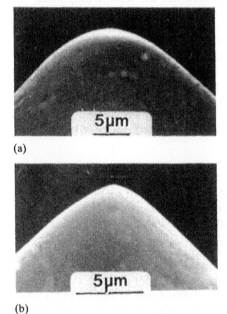

(a)

(b)

Figure 9.33 SEM micrographs of a polished diamond stylus: (a), polished mechanically to a radius of about 10 μm; (b), after ion milling (Miyamoto and Taniguchi, 1983)

Figure 9.32 illustrates the production of a sharp edged diamond knife. The diamond is prepared by polishing two sloping sides and a truncated top as shown in the cross section of the knife in Figure 9.32(a). The diamond is then placed symmetrically in the beam so that material is removed evenly from both the sloping sides. Material is of course also removed from the flat truncated area but more slowly because the removal rate depends on the angle of incidence of the beam on the surface. Hence, as machining proceeds, the two sides move in towards each other, and eventually meet to give a sharp edge as in Figure 9.32(b). The edges may be so fine that special techniques are needed to assess their effective radii; see Suganuma (1985) and Asai *et al.* (1988) as described in Section 14.6.c. In fact Asai *at al.* report measurements of the edge of polished diamond knives with radii of no more than 20 nm.

Figure 9.33(a) shows the outline of an SEM micrograph of a diamond stylus mechanically polished to a radius of about 10 μm. As just mentioned, the rate of sputtering depends on the angle of incidence of the beam, and for angles up to 45° the rate increases with the angle. Therefore an ion beam removes material more rapidly from those parts of the surface furthest from the tip, with the result that the stylus sharpens at the tip (Figure 9.33(b)). Further details including the effect of rotating the specimen and the use of computer simulation to predict the effect of the beam are given by Miyamoto, Davis and Kawata, (1989), Davies (1989) and Taniguchi (1989).

Ion milling is also a useful technique for polishing away surface cracks on diamonds with little risk of causing them to run deeper into the stone. Even so it must be noted that in all forms of ion milling the ions penetrate the surface and inevitably produce a layer showing some subsurface damage, the thickness of the layer being greater with higher energy electrons. Therefore the electron voltage must be selected to afford the best compromise between speed of removal, which increases with voltage, and quality of surface finish.

9.6.d Laser machining

Lasers are used for drilling and cutting a wide variety of materials. A laser beam can concentrate a pulse of energy on to a very small area of a specimen in a very short interval of time. Hence, before the heat generated has time to spread to the rest of the specimen the material under the beam may receive sufficient energy to cause it to vaporize. For example, the energy in a pulse of 10 J is sufficient to vaporize a cylinder of aluminium about 0.3 mm in diameter and 3 mm long (Ready, 1978). It is of course necessary that the laser beam be absorbed by the specimen as only the absorbed light is converted into heat but for these and other details see Ready (1978).

Lasers are commonly used to drill through diamonds in the first stage of producing wire drawing dies, see for example Eder (1983, 1986), Finnigan (1987) and Niederhäuser (1986). They can pierce the diamond much more quickly than other processes, and without the hazard of any broken drills. On the other hand laser drilling does not produce a very exact geometry, and inevitably leaves some damage below the machined surface to a depth of perhaps 20 μm (Finnigan, 1987). For both these reasons the holes pierced by lasers must be fine polished by more conventional methods to produce the finished dies. (Lasers can also drill through the polycrystalline diamond material PCD which is used to make the larger sizes of wire drawing dies. However, this material is electrically conducting and may be drilled by electric discharge methods (Section 9.6.b) which give a more exact geometry and a better surface finish.)

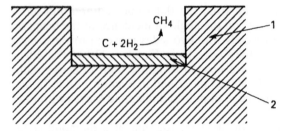

Figure 9.34 Schematic diagram to indicate a method of working diamond with metal, see text. 1 indicates the diamond, 2 the metal foil (Grigoriev and Kovalsky, 1984)

9.6.e Working with metal

As described in Chapter 13 diamond wears at quite an appreciable rate when rubbed on ferrous metals. This behaviour suggests that it should be possible to polish a diamond by rubbing at high speed with a steel or cast-iron wheel. In fact the process is very slow compared to normal rates of polish, but may be of use for particular applications. For example, natural polycrystalline diamond such as ballas is extremely resistant to polish by the normal methods but we have observed that it may be polished on a steel wheel at rates similar to those produced by a diamond bonded wheel.

A different way of working diamond with metal has been described by Grigoriev and Kovalsky (1984). The principle of their method is shown in Figure 9.34. A piece of iron or nickel foil is cut to shape and placed on the surface of the diamond which is then heated to about 1000°C in an atmosphere of hydrogen. At this temperature the carbon atoms in contact with the lower surface of the foil diffuse through the foil to the top surface where they react with the hydrogen and are carried away as methane. Hence the foil sinks into the diamond as shown, engraving its shape on the surface. The authors describe variants of this method which they have developed for various purposes including sawing and drilling.

References

Acton,M.R. (1989) *D.Phil Thesis*, Oxford University

Acton,M.R. and Wilks,J. (1989) *Journal of Materials Science*, **24**, 4229–4238

Asai,S., Taguchi,Y.,Kasai,T. and Kobayashi,A. (1988) In Programme and Abstracts *First International Conference on the New Diamond Science and Technology* Japan New Diamond Forum, Tokyo, pp. 152–153

Binnig,G., Quate,C.F. and Gerber,Ch. (1986) *Physical Review Letters*, **56**, 930–933

Binnig,G. and Rohrer,H. (1982) *Helvetica Physica Acta*, **55**, 726–735

Bowden, F. P. and Tabor, D. (1965) In *Physical Properties of Diamond*, (ed R. Berman), Clarendon Press, Oxford, pp. 184–220

Bruton, E. (1981) *Diamonds* (2nd edn), revised, N.A.G. Press, London

Casey, M. and Wilks, J. (1973) *Journal of Physics D*, **6**, 1772–1781

Davies,S.T. (1989) *Industrial Diamond Review*, **49**, 201–203

Denning, R. M. (1953) *American Mineralogist*, **38**, 108–117

Derry, T. E., Fearick, R. W. and Sellschop, J. P. F. (1981) *Physical Review B*, **24**, 3675–3680

Eder, K. G. (1983) *Industrial Diamond Review*, **43**, 200–204

Eder, K. G. (1986) *Wire Industry*, **53**, 696–700

Fernandez-Moran, H. (1956) *Industrial Diamond Review*, **16**, 128–133; *Journal of Biophysical and Biochemical Cytology*, **2**, 29–30

Fernandez-Moran, H. (1985) In *The Beginnings of Electron Microscopy*, (ed. P. W. Hawkes), Academic Press, Orlando, pp.178–223

Finnigan, G. (1987) *Industrial Diamond Review*, **47**, 31–33

Frank, F. C., Lawn, B. R., Lang, A. R. and Wilks, E. M. (1967) *Proceedings of the Royal Society*, **A301**, 239–252

Golovchenko,J.A. (1986) *Science*, **232**, 48–53

Greenwood, J. A. (1967) *Transactions of the ASME, Journal of Lubrication Technology*, **89**, 81–91

Grigoriev, A. P. and Kovalsky, V. V. (1984) *Indiaqua*, **39**, No 3 pp. 47–54

Grodzinski, P. (1953) *Diamond Technology*, (2nd edn) N.A.G.Press, London

Hall, D. R., Hall, H. T. and Lauridsen, C. L. (1987) *United States Patent*, Number 4, 662, 348

Hitchiner, M. P., Wilks, E. M. and Wilks, J. (1984) *Wear*, **94**, 103–120

Hockey, B. J. and Rice, R. W. (1979) *The Science of Ceramic Machining and Surface Finishing II*, (eds B.J.Hockey and R.W.Rice), National Bureau of Standards, Special Publication 562. U.S.Government Printing Office: Washington D.C.

Hukao, Y. (1955) *Industrial Diamond Review*, **15**, 107–109

Jeynes, C. (1983) *Philosophical Magazine A*, **48**, 169–197

Johnson, K. L. (1985) *Contact Mechanics*, Cambridge University Press, Cambridge, pp. 411–416

Levitt, C. M. (1968) *Review of Scientific Instruments*, **39**, 752–754

Miyamoto, I. (1987) *Precision Engineering*, **9**, 71–78

Miyamoto,I., Davies,S.T. and Kawata,K. (1989) *Nuclear Instruments and Methods in Physics Research*, **B39**, 696–699

Miyamoto, I. and Taniguchi, N. (1983) *Precision Engineering*, **5**, 61–64

Niederhäuser, H. R. (1986) *Wire Industry*, **53**, 709–711

Peters, C. G., Emerson, W. B., Nefflen, K. F., *et al.* (1947) *Journal of Research of the National Bureau of Standards*, **38**, 449–464

Peters, C. G., Nefflen, K. F. and Harris, F. K. (1945) *Journal of Research of National Bureau of Standards*, **34**, 587–593

Ready, J. F. (1978) *Industrial Applications of Lasers*, Academic Press, New York, pp. 398–427

Samuels, L. E. (1971) *Metallographic Polishing by Mechanical Methods* (2nd edn) Pitman, London

Seal, M. (1958) *Proceedings of the Royal Society*, **A248**, 379–393

Slawson, C. B. and Kohn, J. A. (1950) *Industrial Diamond Review*, **10**, 168–172

Spencer, E. G. and Schmidt, P. H. (1972) *Journal of Applied Physics*, **43**, 2956–2958

Suganuma, T. (1985) *Journal of Electron Microscopy*, **34**, 328–337

Taniguchi,N. (1989) (ed.), *Energy-Beam Processing of Materials*. Clarendon Press, Oxford

Teather, H. G. (1969) *Industrial Diamond Review*, **29**, 56–59

Thornton,A.G. (1976) *D.Phil Thesis*, Oxford University

Thornton,A.G. and Wilks, J. (1974) *Diamond Research 1974*, supplement to *Industrial Diamond Review*, pp. 39–42

Thornton, A. G. and Wilks, J. (1976) *Journal of Physics D*, **9**, 27–35

Tolkowsky, M. (1920) *DSc Thesis*. University of London

Vleeschdrager, E. (1986) *Hardness 10: Diamond*, Gaston Lachuriè, Paris

Watermeyer, B. (1982) *Diamond Cutting*, (2nd edn) Centaur, Johannesburg

Wentorf,Jr.R. H. (1959) *Journal of Applied Physics*, **30**, 1765–1768

Whittaker, H. and Slawson, C. B. (1946) *American Mineralogist*, **31**, 143–149

Wilks, E. M. (1961) *Philosophical Magazine*, **6**, 701–705

Wilks, E. M. and Wilks, J. (1954) *Philosophical Magazine*, **45**, 844–850

Wilks, E. M. and Wilks, J. (1959) *Philosophical Magazine*, **4**, 158–170

Wilks, E. M. and Wilks, J. (1965) In *Physical Properties of Diamond*, (ed R. Berman), Clarendon Press, Oxford, pp. 221–250

Wilks, E. M. and Wilks, J. (1967) In *Science and Technology of Industrial Diamonds*, Vol. 1: Science, (ed. J. Burls), De Beers Industrial Diamond Division, London, pp. 93–103

Wilks, E. M. and Wilks, J. (1972) *Journal of Physics D*, **5**, 1902–1919
Wilks, E. M. and Wilks, J. (1982) *Wear*, **81**, 329–346
Yarnitsky,Y., Sellschop,J.P.F., Rebak,M. and Luyckx,S.B. (1988) *Materials Science and Engineering*, **A105/106**, 565–569

Chapter 10

Mechanical differences between diamonds

This chapter reviews differences between diamonds which affect their mechanical strength. We confine our discussions to temperatures well below 1000°C where the diamond behaves as a completely brittle crystal.

10.1   Inclusions and cracks

Diamonds often contain inclusions which show a wide range of sizes and constitution as described in Chapter 2. Top quality gem stones are apparently free of all inclusions while other diamonds may contain large numbers, like that in Figure 10.1. The inclusions will generally be surrounded by regions of strain set up by the misfits between the inclusions and the diamond. A typical example of strain round an inclusion is shown in Figure 2.7 where the strained area is revealed by the birefrigence it produces. The internal stresses associated with these strains may add to an applied stress and cause a diamond to fail more readily. In addition, cracks and voids round an inclusion may concentrate an applied stress field, which also makes the diamond more liable to fracture. We have already noted that the

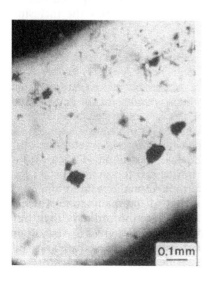

0.1mm

Figure 10.1 Diamond containing many inclusions viewed through polished {011} faces. The cracks are aligned on {111} planes

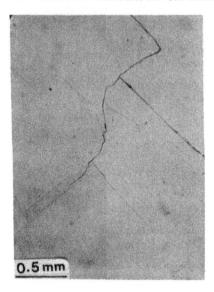

Figure 10.2 Cracks on the surface of a gem stone which appeared during polish

Figure 10.3 Optical micrograph of the diamond in Figure 10.2 viewed through the polished face and focussed on the inclusion responsible for the crack

strength of synthetic diamond grits may be reduced by the presence of inclusions of solvent metal (Section 7.3.b).

Internal stress in a diamond is often revealed during polishing. As a face is polished down towards an inclusion the thickness of strained diamond above the inclusion is reduced and may not to be able to contain the internal stresses. Thus Figure 10.2 shows a crack on the surface of a gem stone which opened during polishing, and Figure 10.3 a view through the the surface to show the inclusion responsible, with cracks running out from it.

It is not always easy to distinguish between internal cracks and inclusions. Light may be totally reflected at the surface of a crack so that it appears completely dark in the field of view. However, the two types of defect can generally be resolved by viewing the diamond and varying its orientation, preferably using a stereoscopic microscope with a long working distance. Under these conditions the appearance of the diamond changes and in some positions the cracks can be clearly identified, often with various fracture marks visible on their surfaces. For example Figure 10.4 shows the cracks in Figure 10.3 viewed in a somewhat different orientation with the focus adjusted to show details of one of the cracks. (The depth of focus of the microscope prevents the whole of a crack being in focus at the same time.)

Occasionally an inclusion in a diamond is found to be a small crystal of diamond with a crystallographic orientation different from that of the main stone. In this case the inclusion may be completely invisible when viewed in ordinary light but can be revealed by other techniques. For example, Figure 10.5 gives a view of an apparently high quality octahedron diamond taken between crossed polars; the image produced by strain birefringence indicates an inclusion near the centre of

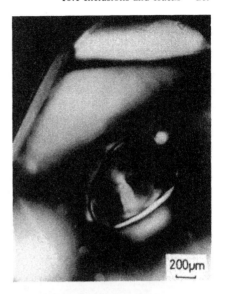

Figure 10.4 Cracks and inclusion in Figure 10.3 viewed at a different angle

Figure 10.5 View through crossed polars of an apparently good quality octahedron diamond (Wilks, 1980)

Figure 10.6 SR X-ray topograph of the diamond shown in Figure 10.5 (Ward, Wilks and Wilks, 1983)

the stone. The same inclusion is also shown in an SR X-ray topograph (Figure 10.6) where the darkening of the inclusion arises from its different crystallographic orientation. Such inclusions of diamond may or may not give rise to serious cracking on polished surfaces but some indication of their presence can generally be seen when the surface is carefully inspected. The so-called naat described in Section 9.5.b is a particular example of this type of included material.

10.2 Abrasion resistance

10.2.a Polish patterns

Different diamonds sometimes exhibit appreciable differences in their abrasion resistance, and differences are sometimes observed between different parts of the same diamond. These latter differences may give rise to differences in surface levels on polished surfaces. Figure 10.7(a) is a CL SEM micrograph of a polished {001} face with a central area exhibiting well marked boundaries and a low level of luminescence. Figure 10.7(b) shows the same area viewed with an interference optical microscope which delineates changes in surface level; a profilometer inspection of the surface showed that the non-luminescing material stands about 50 nm above the general level. That is the non-luminescing material is more abrasion resistant, and subsequent measurements with a micro-abrasion tester showed that cuts on it were about 20% less deep than on the surrounding material (Wilks and Wilks, 1982).

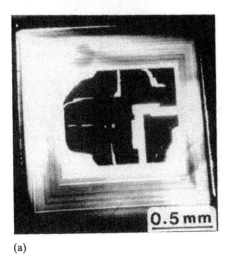

(a)

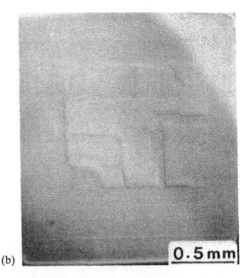

(b)

Figure 10.7 Polished {001} surface of a diamond viewed (a), in CL mode of SEM; (b), in optical interference microscope (Wilks and Wilks, 1982)

An inspection of carefully polished {001} faces of natural diamonds often reveals a pattern of height differences associated with the 'picture frame' type of patterns of luminescence, described in Section 5.2.c, which arises from changes in impurity content during the growth of the diamond. Examples of these and other patterns are given by Wilks (1969), Hanley, Kiflawi and Lang (1977) and Wilks and Wilks (1978,1982). The surface pattern shown in Figure 10.7(b) is sufficiently large to permit measurements with a micro-abrasion tester, but the bands of differing height are often quite narrow and it is difficult to make detailed measurements. We discuss these differences in abrasion resistance further in Section 10.5.

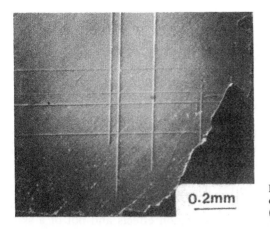

Figure 10.8 Optical interference micrograph of a polished surface showing ridges; see text (Wilks and Wilks, 1978)

Besides the above patterns another type of prominent line feature is sometimes observed on polished surfaces. Figure 10.8 is an optical interference micrograph of part of a polished surface which shows well marked narrow ridges about 100 nm high, much higher than the steps of 10 nm or so associated with the picture frame patterns due to growth layering (Wilks and Wilks, 1978). The lines in Figure 10.8 coincide with the traces of {111} planes in the surface, but they also intersect each other and end within the body of the crystal; therefore they cannot indicate the boundaries of growth layers, so it seems likely that they are associated with slip bands produced in the diamond at some earlier stage of its history. In addition the quality of the polish on and near these lines was much rougher than on the surrounding diamond, in contrast to the polish on raised growth layers which is generally similar to that on the surrounding material. These factors suggest that the lines were produced by deformation twinning which would produce a layer of diamond in a less favourable orientation for polishing, see also DeVries (1975).

10.2.b Methods of testing

We now describe the comparison of the abrasion resistance of different diamonds using the micro-abrasion tester described in Section 9.2.b. These comparisons would be relatively straightforward were it not for the fact that the abrasion resistance is extremely sensitive to the crystallographic orientation of the diamond surface (Section 9.3). Therefore it is essential when comparing two diamonds to polish faces coinciding with principal planes to within a few minutes of arc, using for example optical goniometry as described in Section 5.2.a. The effect of any remaining small deviations from the true plane can then be minimised by abrading both surfaces in the downhill directions defined in Section 9.3.c where the resistance changes more slowly with the angle of tilt (Figure 9.11).

When making comparisons the abrasions must be made under similar standard conditions with no undue wear of the wheel. Therefore a measurement to compare diamonds A and B usually consisted of six cuts in the same direction in the order shown in Figure 10.9, and was only accepted as satisfactory if there was no significant wear between the first and last pairs of cuts. As a further precaution the measurements were then repeated interchanging A and B. The variations generally observed on the depths of cut are of the order of a few parts per cent so that a difference of 10% between the two stones may be detected. For further

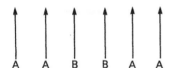

Figure 10.9 Schematic diagram showing the order of testing when comparing the abrasion resistance of diamonds A and B

details including the use of cleavage surfaces to orient diamonds not having any good plane natural surfaces see Wilks and Wilks (1982, 1984).

The importance of accurate orientation is underlined by an opinion sometimes held by diamond polishers that a diamond has a 'skin'. That is, if a natural {111} face of even a good quality diamond is polished away then the material inside polishes more easily. However, we have already seen that a diamond polishes faster at small angles away from {111} (Figure 9.12). Therefore any polishing of a natural face generally creates a facet which is not exactly {111} and therefore not so abrasion resistant, but this is the result of orientation effects and not of a skin (Wilks and Wilks, 1967).

10.2.c Differences between diamonds

As an example of differences in the abrasion resistance of different diamonds we now describe experiments in which brown and Type II diamonds were compared with a typical good quality Type I diamond (Wilks and Wilks, 1987a). This Type I diamond was a colourless stone and had previously been compared with several other similar Type I diamonds and showed a similar abrasion resistance (Wilks and Wilks, 1982). The brown stones were selected from 35 Type I diamonds which all showed a marked brown colouration together with an optical absorption spectra in the visible region similar to that in Figure 3.9. From this group a selection of nine stones was made to obtain the best {111} faces for optical goniometry. The Type II diamonds were selected on the basis of their infrared absorption spectra and as is often the case with Type II stones all showed some brownish colouration; they were all of irregular shape and it was therefore necessary to create {111} cleavage surfaces to permit their correct orientation. The result of the comparison is shown in Table 10.1 which gives the ratio of the depths of cuts on the brown and Type II stones to those on the colourless stone.

Table 10.1 Abrasion resistances of brown and Type II diamonds compared with that of a typical good quality colourless Type I diamond. The table gives the ratio between the depths of cuts made on each stone to those on the standard Type I

Brown diamonds		Type II diamonds	
Diamond	Ratio	Diamond	Ratio
B2	0.85	F2	0.81
B4	0.92	F5	0.81
B5	0.87	F12	0.79
B15	0.79	F22	0.81
B19	0.80	T2	0.80
B54	0.94	T4	0.81
B68	0.92		
B69	0.82		
B70	0.86		

Table 10.1 shows that the brown and Type II diamonds are all more abrasion resistant than the typical colourless diamond. Subsequently, a selection of these and other brown, Type II, and selected high quality colourless Type I diamonds were surveyed using synchrotron radiation X-ray topography (Section 6.2.f). The selected Type I diamonds showed particularly low densities of dislocations but all the brown and Type II diamonds showed high densities, as in the topographs shown in Figure 6.22 and 6.25. It therefore appears that dislocations are responsible for an increased abrasion resistance, (as well as for the scattering of light described in Section 6.2.f).

The removal of material by abrasion depends on the propagation of large numbers of small cracks, and the propagation of a crack depends on the maintenance of a sufficient stress concentration at its tip to ensure the breaking of atomic bonds. There are many references in the literature to the blunting and arresting of cracks by processes which reduce the stress field, see for example Latanision and Pickens (1981). It seems probable that the enhanced abrasion resistance of dislocated diamond is due to regions of lattice disorder, associated with immobile dislocations, which reduce the stress at the tip of a crack, and hence restrict its progress.

10.3 Cleavage

The cleavage of diamond was discussed in Section 7.4 in terms of cleavage along {111} planes but a cleavage plane seldom coincides exactly with a lattice plane, and the fracture surface produced by the cleavage of any material usually exhibits a quite complex topography. We now discuss correlations between the topography of cleaved diamond surfaces and other characteristics of the diamond. We start with a brief outline of some features of fracture patterns on diamond which are common to the fracture of most brittle materials.

10.3.a Cleavage patterns

Figure 10.10 gives a black and white reproduction of a Nomarski interference micrograph of a cleaved surface of diamond. On the original micrograph differences in the slope of the surface appear as differences in colour. Thus an

0.5 mm

Figure 10.10 Optical interference micrograph of the cleaved surface of a diamond; see text (Wilks and Wilks, 1987b)

absolutely smooth surface appears with a quite uniform tint, with a particular colour depending on the value of any small angle between the normal to the surface and the optical axis of the microscope. Hence roughness on a surface is delineated either by a range of colours or a range of intensities as in Figure 10.10. On this diamond the cleavage began in the top left hand corner and spread across the crystal to the bottom right. The central area of the figure lying near the diagonal is of fairly uniform intensity, and near the start of cleavage there is a triangular region which shows very little structure. Then about half way down the diagonal we see the commencement of so-called river lines running roughly normal to the crack front as it passes through the crystal. Above and below the diagonal we see more pronounced river lines mostly running out from near the start of the cleavage.

All the river lines indicate a step in the cleavage surface. That is, the cleavage crack on one side of the line ran through the crystal at a lower level than on the other side of the line. Despite these differences in levels, the uniform intensity of much of the right hand half of Figure 10.10 indicates that the surfaces between the steps are all quite closely parallel to each other. Even a very marked step at the very top right of the picture makes little difference to the orientation of the surface. To the left, below the diagonal, there are changes of intensity which indicate various tilts of the cleavage plane away from its position in the upper half of the figure. These changes are readily related to the differences in slope by observing a plane surface mounted at different angles on a goniometer head. For example, the range of intensity in Figure 10.10 corresponds to a maximum difference in angle of about 2°.

The form of cleavage patterns like those in Figure 10.10 are discussed in some detail by various authors, see for example Zapffe and Worden (1949), Field (1971), Swain, Lawn and Burns (1974) and Lawn and Wilshaw (1975). The steps marked by the river lines arise where the crack front is deflected to another plane either by imperfections in the crystal structure or by reflections of the stress waves produced by the initial blow. The latter are responsible for the so-called Wallner lines which appear approximately as arcs of circles centred on the point of initiation. They are clearly visible in the central region of the micrograph but can also be seen on the left hand side of the picture. These Wallner lines mark the position of the crack front at an instant when it encountered a reflection of the initial stress wave tending to divert the cleavage to other planes. In the centre of Figure 10.10 we see that the lines mark the start of a river system indicating differences in levels across the surface. Thus, even in a perfect crystal, we may expect to observe Wallner lines and river lines because of stress waves reflected from the boundaries. However, the features which are of primary interest at present are the intensity changes and the larger step heights which are usually due to imperfections in the lattice.

10.3.b Differences between diamonds

When discussing the cleavage of different diamonds one must remember that the topography of a cleavage surface may depend to a significant extent on the skill of the cleaver. Therefore before discussing differences between diamonds we first describe an experiment to determine whether two cleavages of the same stone are likely to produce similar types of surface.

Five gem quality stones of various types, a Type II, a brown Type I, a good quality colourless Type I, and two Type I selected for a high degree of lattice perfection as described in Section 6.2.e were studied. Each stone was sufficiently large that it could be accurately cleaved into either three or four slices, thus making it possible to compare either two or three pairs of cleavages on the same diamond. An inspection of these surfaces showed that the cleavage surfaces coming from each diamond were all quite similar, any differences being much less pronounced than the marked differences observed between diamonds of different types (Wilks and Wilks, 1987b). We conclude that it is generally meaningful to associate a particular form of cleavage surface with a particular type of diamond.

Further cleavages were made on another Type II diamond and on two of the brown Type I diamonds mentioned in Section 10.2.c. All the Type II cleavages were quite similar in appearance to that of Figure 10.11(a). Over most of the surface the variations were due principally to the river lines, although some small areas were very irregular indicating some loss of material. The cleavage surfaces on the brown stones were quite different as seen in Figure 10.11(b) which shows a typical example. The variations in intensity are more pronounced and the structure much more complicated than the simple river lines in Figure 10.11(a). As previously, the black areas indicate steps of considerable size. This difference between brown and Type II diamonds was further confirmed by observing

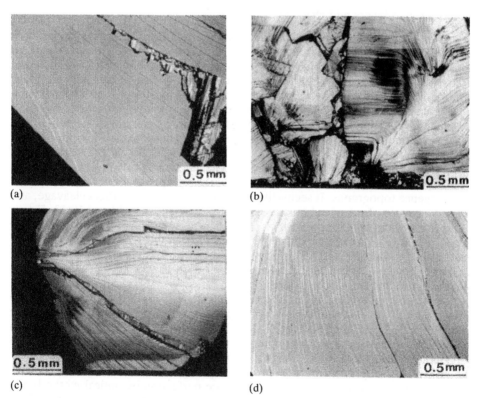

(a)   (b)

(c)   (d)

Figure 10.11 Optical interference micrographs of cleaved diamond surfaces: (a), Type II diamond; (b), brown Type I diamond; (c), good quality Type I diamond; (d), Type I diamond selected for lattice perfection (Wilks and Wilks, 1987a,b)

cleavages on another 13 Type II stones and a further 9 brown stones, all of which followed the same pattern.

Figure 10.11(c) shows a cleavage face from an apparently typical, good quality Type I colourless diamond. This surface is clearly intermediate in form between those observed on the Type II and brown diamonds (Figures 10.11(a) and (b)). That is, the Type II stones appear to give particularly smooth cleavages and the brown stones particularly rough ones. Figure 10.11(d) shows a typical cleavage from one of the Type I colourless diamonds specially selected for a high degree of lattice perfection. This is clearly much smoother than the surfaces on the unselected colourless stones and only somewhat inferior to the Type II surfaces.

It is also possible to assess the smoothness of a cleavage surface by observing the spread of the signal in an optical goniometer. Several of the brown and Type II stones were examined in this way and the difference between the cleavage of the two types was again observed. The spread of the signals from the brown stones ranged from 18' to 136' with a mean of 74', while from the Type II the spread ranged from 2' to 29' with a mean of 18'. The appearance of the cleavage surface of Type II is generally so characteristic that on at least three occasions we have identified nominally Type II diamonds to be Type I solely on the evidence of a cleavage surface.

10.3.c Growth layers and cleavage

We now ask why different diamonds exhibit the range of cleavage patterns described above, in particular why brown Type I diamonds give particularly rough surfaces and Type II particularly smooth surfaces. As these two types of diamond are both characterized by high densities of dislocations it seems unlikely that their different cleavage behaviour is associated with the dislocations. More probably the difference is due to the relatively large scale variations in the crystal structure associated with growth layers containing different concentrations of impurities, the smoother cleavage of Type II diamonds being due to the absence of impurities and growth layering.

Any discontinuities between different growth layers produce mechanical strains in a crystal, these strains reveal the presence of growth layers in X-ray and birefringence topographs. It seems likely that these strains also affect cleavage, and a good example of a particular instance is given in Figures 10.12 and 10.13. Figure 10.12 is a Nomarski micrograph showing a pronounced change of direction on the cleavage plane of a Type I diamond, while Figure 10.13 shows the same area viewed in the normal and CL modes of the SEM. It is clear that the plane of cleavage has been deflected at a boundary which marks an abrupt change in luminescence and impurity content.

Appreciable layering is often observed in Type I diamonds and particularly in brown ones. Twenty seven Type I diamonds selected solely for their brown colour were inspected in the CL mode of the SEM and all of them exhibited growth patterns more marked and more complex than those generally observed on Type I diamonds (Wilks and Wilks, 1987a). On the other hand Type II diamonds show no growth layering because of their low impurity content. In addition the two colourless Type I diamonds selected for lattice perfection by optical methods and which gave particularly smooth cleavages (Section 10.3.b) showed only minimal growth layering and luminescence in the SEM. We therefore conclude that variations in impurity content are often reflected in the smoothness of a cleavage.

Figure 10.12 Cleavage surface of a diamond viewed in an optical interference microscope (Wilks and Wilks, 1987a)

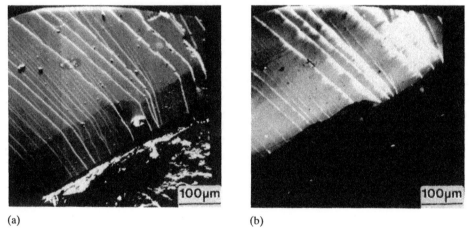

(a)   (b)

Figure 10.13 Cleavage surface in Figure 10.12 viewed in the SEM in: (a), normal mode; (b), CL mode (Wilks and Wilks, 1987a)

10.4 Synthetic diamonds

Until recently it has been difficult to compare the mechanical properties of natural and synthetic diamonds because of the small size of the synthetic specimens. However, there is now no doubt that the mechanical strength of good quality synthetic crystal is similar to that of good quality natural material, as is to be expected. For example, measurements with an abrasion tester similar to those described in Section 10.2.b showed no significant difference between good quality synthetic and natural material. For details of these results and of the various precautions required for a satisfactory comparison see Wilks and Wilks (1982).

The behaviour of synthetic crystals of lesser quality is more complex. Inclusions, particularly of solvent metal, are often present and may take a variety of forms, see for example Bezrukov *et al.* (1972). These inclusions, like those in natural material may greatly weaken the diamond (Section 7.3.b). Also, because the expansion coefficient of metals is greater than that of diamond thermal stresses may cause damage when a stone is heated to high temperatures, see for example Gargin (1982). However, it is important to note that the methods of synthesis have been and are being steadily improved so some information in the literature will apply to products which have now been superseded.

Various authors have looked for some correlation between the strength of synthetic diamonds and their nitrogen content. Nachal'naya *et al.* (1978) reported that the compressive strength of small crystals decreased with increasing content of single atom nitrogen. On the other hand, Bokii and Kirova (1975) state that doping with nitrogen results in the production of very strong crystals, see also Bokii, Kirova and Nepsha (1979). Chrenko and Strong (1975) abraded {001} faces of synthetic diamonds on a rotating aluminium oxide grinding wheel in a resistant <110> direction. They found that the abrasion resistance decreased by a factor of about 10 as the nitrogen content increased from 0.1 ppm to 200 ppm. However, these results present some difficulties of interpretation. For example, the resistance was reduced by a factor 2 by the presence of only 2 ppm of nitrogen. This result is hard to understand, as are the very low values of the abrasion resistance quoted for two natural diamonds. It must be noted, however, that the resistance of diamond to abrasion in a <110> direction is very sensitive to the exact alignment of the diamond which is not easy to achieve with small specimens. Nor was any information given on whether the diamonds were running hot.

The above experiments on the effect of nitrogen present a confused picture. However, it must be remembered that the measurements refer to particular diamonds synthesized under particular growth conditions, including temperature and pressure, which are not stated. It may be that the nitrogen atoms are only related to the strength indirectly. That is, the nitrogen may either be responsible for, or act as an indicator of, different growth conditions which give rise to crystals with different imperfections. (Both growth rates and crystal shapes are affected by the concentration of nitrogen (Strong and Chrenko, 1971), and the different growth sectors in synthetic diamonds often have different luminescence and impurity content (Section 5.4.b).)

10.5 Atomic impurities

We have seen in previous sections that the mechanical behaviour of diamond may be affected by the presence of cracks, inclusions, and immobile dislocations. We now consider the limited amount of information on how the strength may be affected by impurities on an atomic scale.

10.5.a Strength and colour

Diamond polishers have long debated whether the mechanical properties of diamonds are affected by their colour. There are many anecdotal accounts of the relative merits and behaviour of, for example, yellow and brown diamonds, but there is not much reliable information. Fracture tests generally show considerable scatter which may obscure relatively small differences, and abrasion tests may be misleading unless care is taken in the orientation of the diamonds. The only well established relationships appear to be those associating brown colouration with a greater abrasion resistance and a rougher cleavage as described above.

Several Russian authors have sought to correlate the strength of natural diamonds with their fluorescence when irradiated with ultraviolet light. Artsimovich, Vovchanovskii and Ivanov (1969) observed the wear of diamonds in drill crowns and reported that non-luminescent stones showed less wear. However, Kalinin, Kornilov and Spitsyn (1972) measured the 'compressive and impact strengths' and the wear of a range of diamonds and found that those luminescing green blue or yellow had the best wear properties but did not give much detail of the measurements. Further references to Russian work are given by Kalinin *et al.* (1976). The latter authors also give curves to show, somewhat surprisingly, a linear relationship between compressive strength and the *wavelength* of the luminescence, but apparently they did not measure the intensity of the luminescence which would have given some indication of the concentration of the impurities.

10.5.b Growth layers and sectors

We have already noted that atomic impurities are responsible for the growth layering described in Section 5.2.c. Differences in the impurity content of adjacent layers are often indicated by differences in cathodoluminesence and sometimes give rise to differences in abrasion resistance which can be identified in the type of polish patterns described in Section 10.2.a. However, these layers are not easy to study in detail because they are often quite narrow and may be inclined at an angle to the surface.

Hanley, Kiflawi and Lang (1977) observed the polished surface of a thin slab of natural diamond with a rather complex growth structure. They identified areas which stood higher than the general level of the surface, and exhibited very little cathodoluminescence and very little absorption in the ultraviolet. Hence the authors concluded that the higher areas consisted of Type II material, a result in line with other experiments which show that Type II material is more abrasion resistant (Section 10.2.c). However, as discussed in Section 10.2.c, the enhanced resistance of Type II material appears to be associated with a high density of dislocations, and it is not clear how growth layers in a diamond could have been plastically deformed without the surrounding material suffering similar treatment.

There are also other indications that a complete explanation of polishing patterns may be quite complex. In the studies just mentioned the more resistant parts of the surface appeared darker when viewed in the CL mode of the SEM but the position is different for the diamond surface shown in Figure 5.35. The roughly circular boundary in Figure 5.35(a) separates two different modes of growth (Section 5.6.a), and the abrasion resistance of the central area is appreciably greater than that of the surrounding material. However, the central area also shows much more luminescence, in contrast to the more abrasion resistant regions observed by Hanley, Kiflawi and Lang (1977).

Rather similar types of patterns are observed on the polished faces of synthetic diamond of cubo-octahedral and other morphologies (Kirk, 1973). As described in Section 5.4 these diamonds have generally grown in two or more growth modes and the polishing patterns indicate the boundaries between different growth sectors. From the height differences on the surface Kirk (1973) and Woods and Lang (1975) deduced that the relative abrasion resistance of the different growth sectors ran in the order $\{011\} > \{111\} > \{001\}$. Woods and Lang (1975) also observed {113} sectors which appeared to have a resistance intermediate between {011} and {111}. Patterns on the rather complex diamond of Figure 5.28 are described and discussed by Frank *et al.* (1990) who observed a 70 nm step between {001} and {011}. It would be interesting to correlate these differences in abrasion resistance with the impurity content of the sectors but it is difficult to make quantitative correlations between luminesence and impurity content (Section 4.3.b).

Kirk (1973) also notes that the fine metallic inclusions often found in synthetic material tend to be located preferentially in certain growth sectors, see also Strong and Chrenko (1971). A rather similar preferential location of small inclusions has also been observed in natural diamonds showing the centre cross type of growth pattern described in Section 5.3.b (Lang, 1974; Suzuki and Lang, 1976; Van Enckevort and Seal, 1987). However, it is not yet clear to what extent these small inclusions lead to variations in the mechanical strength.

10.5.c Platelets

Several authors have looked for correlations between the strength of diamond and the strength of the infrared absorption line at 1370 cm^{-1} characteristic of platelets (Section 2.4.b.). Wilks and Wilks (1978) noted that of four similar tools, two gave a consistently poor performance and showed a large 1370 cm^{-1} absorption whereas the other two tools gave good performances and showed little or no 1370 cm^{-1} absorption. Wong (1981) claimed to correlate poor tool performance with absorption at 1370 cm^{-1} which was large compared with the absorption at 1282 cm^{-1}. However, he presents few details of the measurements, and one would expect that the significant parameter would be the absolute value of the absorption at 1370 cm^{-1}. Ikawa and Shimada (1983) claim a correlation between the strength measured in a Hertz test and the optical absorption due to platelets at 1370 cm^{-1} but the differences in strength lie within the experimental scatter. Measurements of the abrasion resistance of two diamonds with considerably different absorptions at 1370 cm^{-1} (16 cm^{-1} and 2 cm^{-1} gave virtually the same values (Wilks E.M. unpublished). Ruoff and Vohra (1989) remarked that platelets are not essential in diamonds for use as high strength anvils. Hence the effect, if any, of platelets on mechanical strength is still somewhat obscure. We should, however, remember that platelets vary greatly in size from nanometers to, exceptionally, hundreds of micrometers. The larger ones, at least, might well affect mechanical strength, but the strength of the 1370 cm^{-1} absorption line gives little or no indication of the size of the platelets.

10.6 Effects of ion implantation

10.6.a Ion implantation

Ion implantation is a technique of introducing foreign atoms into the surface layer of a material. The atoms are converted to ions in some form of electrical discharge and accelerated by voltages of the order of 25 KeV to 100 KeV towards the surface of the specimen. The ions pass into the surface and collide with the atoms of the material until they eventually come to rest, with a distribution in depth having a maximum concentration at about 0.1 μm below the surface. For an account of the details of the techniques of implantation and various applications see, for example, Ziegler (1988).

Ion implantation is now widely used in two fields of application. Ions injected into semiconductors may behave as donor or acceptor atoms, so implantation provides a precise and convenient method of doping silicon and other materials. Implantation can also increase the strength of materials especially metals. Implanted ions may act as pinning points for dislocations and thus produce a hardening effect to improve the performance of cutting tools, etc. A strengthening can also be produced in brittle materials, such as sapphire at room temperature, where dislocation motion is negligibly small, see for example Burnett and Page (1985). In this case the strengthening is produced by other mechanisms, including the generation of internal compressive stresses which tend to prevent the propagation of cracks.

It is important to realize that the layer of implanted material may be much modified beyond the simple addition of foreign atoms. The collisions between the high energy incoming ions and the atoms of the specimen crystal result in a large amount of damage. Before coming to rest each incoming ion may eject about 100 atoms away from their positions in the crystal lattice, thus creating a corresponding number of vacancies and interstitial atoms. Indeed, if the diamond is irradiated at a temperature above about 800°C the atoms are sufficiently mobile that the irradiated part of the surface rises above the level of the surrounding material (Nelson et al., 1983). Note that the final state of the material will depend both on the strength of the irradiation and on the mobility of the defects created, and any reorganization they undergo during or after the irradiation. Hence, the effect of a given implant will depend on the temperature of the implant and on any subsequent annealing processes.

10.6.b Implantation and damage

The damage which accompanies the implantation of ions in diamond has been observed in several ways. Studies using X-rays (Vavilov et al., 1974 a) and electron diffraction (Zhang, Yu and Su, 1986) imply the presence of a disordered region where the diamond has been partially amorphized. This disorder in the crystal lattice will affect the thermal vibrations and therefore both the infrared absorption (Ananthanarayanan et al., 1972) and the Raman spectra (Ananthanarayanan et al., 1972; Crowder et al., 1972). In particular, the peak in the Raman spectrum broadens and decreases in amplitude while the background intensity rises. The distribution in depth of the implanted ions has been measured by Derry et al. (1988) using the technique described in Section 2.1.e.

Any disordering of the lattice also results in the breaking of carbon to carbon bonds and the production of optical centres. These may give increased absorption in the visible spectrum (Morhange et al., 1975) and make new contributions to the luminescence (Zaitsev, Gippius and Vavilov, 1982; Varichenko, Zaitsev and Stelmakh, 1986). There are also additional contributions to the electron spin resonance spectrum (Brosious et al., 1974; Flint and Lomer, 1983). Discussions of how these centres are effected by annealing treatments have been given by Bourgoin (1975), Gippius, Zaitsev and Vavilov (1982) and Varichenko et al. (1986).

Some of the most detailed experiments have been made by Prins, Derry and Sellschop (1986) who studied the rise of the implanted material above the general

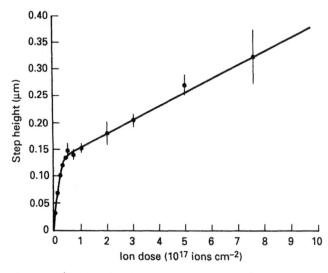

Figure 10.14 Dose dependence of the step height on a diamond surface caused by the volume expansion during implantation of 70 keV fluorine ions (Prins, Derry and Sellschop, 1986)

level of the surface. The curve in Figure 10.14 shows this increase in surface height as a function of ion dose for implantations of fluorine ions carried out nominally at room temperature but where the energy in the beam caused the temperature of the specimen 'to exceed 100°C'. The height of the step increased in two stages. Initially the increase was quite rapid but after a dose of about 6.10^{16} ions cm^{-2} the rate of increase slowed off and the step height then increased linearly with the dose.

It is fairly well established that vacancies in diamond are immobile at temperatures below 500°C. Although the position on the migration of the interstitial atoms is less clear they are very probably mobile at 100°C (Section 3.7.b). Therefore Prins, Derry and Sellschop proposed that at the lowest doses the implantations created vacancies which were immobile and interstitial atoms which moved to the surface to form the step. Higher doses produced more interstitials and vacancies but there was now an increasing possibility that the interstitials diffusing to the surface would be captured by vacancies. Eventually an equilibrium is reached in which all the additional interstitials are trapped by vacancies before reaching the surface. Hence the increase of step height at higher doses is due only to the presence of the additional fluorine ions and varies linearly with the dose. Prins, Derry and Sellschop (1987) also carried out experiments in which fluorine ions were implanted at 77K where the interstitials are probably not mobile. As would be expected the observed behaviour was different; the details of a rather complicated situation are discussed by the authors.

10.6.c Electrical conductivity

The presence of boron in some natural diamonds causes them to behave as p type semiconductors where the current is carried by holes (Section 3.4). The electrical properties of these natural diamonds are varied and unpredictable but suggest that if boron could be introduced in controlled concentrations into other diamonds they might then be used as material for the fabrication of semiconducting devices. (Although considerable difficulties remain to be overcome, see Section 17.7). The most obvious way of introducing boron is by doping during the synthesis of diamond (Section 1.2.b) but this is a difficult process. Hence various attempts have been made to produce electrical conductivity in diamond by the implantation of ions.

In one of the first implantation experiments on diamond Wentorf and Darrow (1965) bombarded diamond surfaces with ions of helium, nitrogen, argon and oxygen and produced electrical conductivities with both n and p type carriers. It is now clear that in these and similar experiments the electrical conductivity arose as a result of the damage to the crystal lattice produced by the bombarding ions (Davidson *et al.*, 1971). However, a useful semiconducting material must be free of damage so that the carriers can move easily through the lattice. This suggests that so far as diamond is concerned it will be necessary to implant donor atoms such as boron and then anneal out the damage but this annealing is difficult because of the tendency of diamond to graphitize (Kalish *et al.*, 1979, 1980). Accounts of the implantation of diamond using a variety of ions have been given by Vavilov (1974), Vavilov *et al.* (1974b), Blanchard, Combasson and Bourgoin (1976), Vavilov *et al.* (1979) and Braunstein and Kalish (1983). The latter implanted specimens with boron or carbon under similar conditions followed by similar anneals at 1400°C but obtained conductivity only in the samples implanted with boron, which they took as evidence that the conductivity was due to donor boron atoms rather than damage. Further work on these lines is described by Prins (1988).

Other experiments have bombarded diamond surfaces with carbon ions so causing the top layer of the diamond to graphitize or become amorphous (Hauser and Patel, 1976; Hauser, Patel and Rodgers, 1977; Prins, 1985). Such layers have a finite conductivity which increases with temperature as in a normal semiconductor, but the current is thought to be carried by electrons which move from atom to atom by a *hopping* process made possible by the broken bonds created by the implantation. Devices such as p–n junctions require the production of both p and n type material and little progress has been made in producing diamond material which is n type. Therefore Prins (1983a) suggested that hopping electrons might play a similar role to n type carriers. In any case the development of successful devices on these lines will require further studies of the implantation and subsequent annealing processes on the lines of experiments described, for example, by Derry and Sellschop (1981), Nelson *et al.* (1983) and Prins (1983b, 1985).

10.6.d Mechanical differences

Various attempts have been made to increase the strength of diamond by implantation techniques. For example, a patent application by Hartley and Poole (1981) refers to implantations using fluxes of 9×10^{15} ions cm^{-2} of nitrogen and 3×10^{15} ions cm^{-2} of carbon which reduced the wear of diamond on samples of titanium carbide and of poly methyl methacrylate; this work is also described by Hartley (1982). In another patent application Hudson, Mazey and Nelson (1981) describe implantations by protons of energy 60 keV to 10 keV at fluxes ranging from 10^{16} to 5×10^{17} ions cm^{-2}, with the temperature held in the range 600°C to 800°C in order to encourage annealing effects. As a result of these treatments the authors claim an increase in the scratch hardness and a decrease in the wear of the diamond against a platinum wheel. However, in spite of these apparently encouraging results, Hartley (1982) comments that the observed effects are unlikely to justify commercial exploitations and that more information is required on the basic processes involved.

Samuels (1987) has described experiments to determine the effect of implantation on the critical load required to produce ring cracks when a spherically tipped diamond stylus slides over a polish diamond surface (Section 7.6.a). Before the measurements the central part of an {001} polished surface was masked by a metal plate and the surface implanted with a flux of protons. Then the loaded stylus was moved in a <110> direction on a series of tracks cutting the boundary between the implanted region and that protected by the mask. Particular care was taken to obtain well controlled conditions including the polish and orientation of both specimen and stylus.

Figure 10.15 shows the surface of a diamond where the upper and lower sections had been implanted with a nominal flux of 7×10^{16} protons cm^{-2} at 100 keV and a nominal temperature of 250°C. The two thin horizontal lines across the micrograph mark the small difference in level between the implanted and unimplanted regions. The upper half of the figure shows ten lines of ring cracks

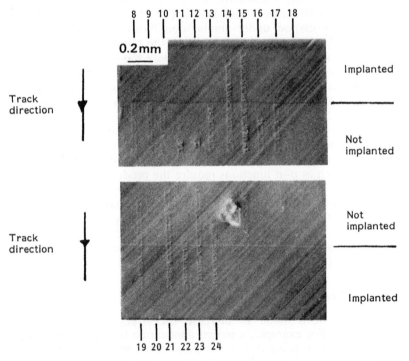

Figure 10.15 Optical interference micrograph of tracks produced by a loaded diamond stylus sliding over a diamond surface part of which had been implanted with 100 KeV protons (Samuels, 1987)

Track no	8	9	10	11	12	13	14	15	16
Load g wt	1220	1317	1414	1512	1606	1704	1801	1801	1606
No passes	3	2	4	3	5	2	1	2	1

Track no	17	18		19	20	21	22	23	24
Load g wt	1414	1220		1414	1220	1606	1220	1317	1414
No. passes	2	5		2	5	1	4	1	1

produced by the stylus as it moved across the surface under different loads, each of the tracks commencing in the upper implanted region and ending in the untreated region. The loads ranged from 1220 g wt to 1801 g wt and all produced ring cracks on the untreated section of the surface, yet only the two passes at 1801 g wt produced any cracks on the implanted surface. After another similar implantation with a flux of 14×10^{16} protons cm^{-2} the critical load on the treated surface was double that on the untreated. The lower half of the figure shows tracks commencing in the untreated region. These and other experiments show that once cracking is initiated on a track the cracks themselves encourage further cracking. In particular, if the stylus moves from an untreated to a treated region, then the presence of cracks on the untreated side of the boundary tends to lower the critical load on the implanted side.

The above experiments show that appreciable strengthening of diamond may be obtained by implantation, but the temperature of the implantation may be quite critical. Besides the implantations at 250°C Samuels (1987) also observed that two implantations at 600°C produced little effect. Also, another implantation at 250°C under conditions nominally similar to an earlier one resulted in a marked softening of the diamond, the stylus making a shallow groove on the surface. Hence it is clearly important during an implantation to monitor and maintain both the temperature of the specimen and the magnitude of the flux, and the latter measurement is not trivial. It is easy to measure the net current flow to the specimen but this is not usually equal to that of the beam flux because of secondary electron and ion emission from the specimen during the implantation. This effect is considerable and not easy to quantify. On the other hand the height of the step at the boundary of the implanted region can provide a useful check that standard conditions have been maintained.

Very little work has been done on diamond to identify the mechanisms responsible for the strengthening effect, but studies have been made of somewhat similar effects in sapphire and other materials. The implantation of sapphire with relatively low ion doses reduces radial and lateral cracking in indentation tests, while higher doses result in softening effects (Burnett and Page, 1984). In other experiments the implantation of one side of thin plates of sapphire produced a permanent curvature of the plate showing that the implant had introduced internal compressive stresses in the surface layer, the curvature passing through a maximum value for doses of the order of 10^{17} ions cm^{-2} (Burnett and Page, 1985; Hioki et al., 1986).

The above experiments on sapphire suggest that the ions introduced at the lower doses give rise to compressive stresses which produce a strengthening effect. Then at high doses the lattice begins to amorphize and the strength falls. It seems probable that similar processes occur in diamond, although Prins (private communication) suggests that the initial increase in strength may arise primarily by crack blunting due to disorder in the lattice, in a way rather similar to that suggested in Section 10.2.c to account for the enhanced abrasion resistance observed in plastically deformed diamond.

10.7 Effects of neutron irradiation

Various experiments have been made on the effects of irradiating diamond with fluxes of high energy, or fast, neutrons obtained from nuclear reactors. Because neutrons are electrically neutral they are much less likely to collide with the carbon nuclei in the diamond, so a beam irradiates the whole crystal and not just the surface. As with ion implantation, if a neutron does collide with a nucleus it will eject it from its position in the lattice with sufficient energy to produce further damage in the form of interstitials and vacancies. We now summarize some of the changes that this type of neutron irradiation may make on the properties of diamond but do not go into much detail because they generally only reduce the mechanical performance.

X-ray measurements of the lattice spacing showed that the damage produced by neutron irradiation results in an expansion of the lattice by the order of a few parts per cent for fluxes of the order of 10^{20} neutrons cm^{-2} (Primak, Fuchs and Day, 1956; Damask, 1958). The X-ray patterns also show that the damage of this order tends eventually to reduce the diamond to an amorphous form (Levy and Kammerer, 1955). The annealing of this damage has been discussed by Lonsdale and Milledge (1965) and by Vance (1971). The damage is accompanied by a considerable increase in the optical absorption both in the visible spectrum (Clark, Ditchburn and Dyer, 1956) and in the infrared (Woods, 1984). The appearance of the diamond is changed quite rapidly, first to a dark green and then to an unattractive black colouration. Electron spin resonances associated with the damage are discussed by Owen (1965) and Whippey (1972).

The first observations on mechanical effects showed that neutron irradiation increased the wear rates of diamonds abraded by silicon carbide (Damask, 1958). Fluxes of fast neutrons of the order of 10^{19} neutrons cm^{-2} reduced the density by about 2% and increased the wear rate by about a factor 2. Figure 10.16 shows the effect of irradiation on the abrasion of diamond by diamond powder in a micro-abrasion test (Wilks, 1967). All the measurements were made in an easy direction

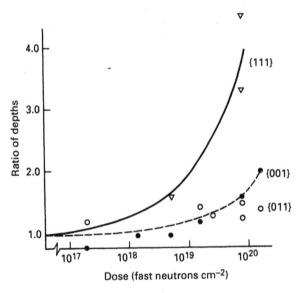

Figure 10.16 Effect of fast neutron irradiation on the abrasion resistance of {001}, {011} and {111} surfaces of diamond. The diagram gives the ratio of the depths of abrasion cuts after and before irradiation (Wilks, 1967)

of abrasion and the figure shows the ratio of the depths of the cuts before and after irradiation as a function of the neutron dose. The abrasion resistance of {001}, {011} and {111} faces decreased steadily with increasing dose. It was also noted that the more strongly irradiated diamonds gave appreciably rougher cleavage surfaces. Measurements on the strength of diamond by indentation tests also show a reduction of strength with increasing neutron dose (Brookes and Vance, 1972).

Neutron irradiation generally tends to reduce the mechanical performance of diamonds at room temperature and gives no prospect of improving the wear characteristics. However, the position is quite different at high temperatures when the strength of diamond is limited by its plasticity. Thus Evans and Wild (1967) observed an increase of over a factor 2 in the stresses required to bend diamond plates at 1800°C following irradiation with a flux of about 10^{19} neutrons cm^{-2} and then annealing at temperatures between 1000°C and 1800°C.

References

Ananthanarayanan, K. P., Borer, W. J., Plendl, H.S. and Gielisse, P. J. (1972) *Radiation Effects,* **14**, 245–248

Artsimovich, G. V., Vovchanovskii, I. F. and Ivanov, V. V. (1969) *Razvedka i Okhrana Nedr,* **(9)**, 29–35 (In Russian)

Bezrukov, G. N., Butuzov, V. P., Khatelishvili, G. V. and Chernov, D. B. (1972) *Soviet Physics Doklady,* **17**, 421–424

Blanchard, B., Combasson, J. L. and Bourgoin, J. C. (1976) *Applied Physics Letters,* **28**, 7–8

Bokii, G. B. and Kirova, N. F. (1975) *Soviet Physics Crystallography,* **20**, 386-388

Bokii, G. B., Kirova, N. F. and Nepsha, V. I. (1979) *Soviet Physics Doklady,* **24**, 83–84

Bourgoin, J. C. (1975) *Diamond Research 1975,* supplement to *Industrial Diamond Review,* pp. 24–28

Braunstein, G. and Kalish, R. (1983) *Journal of Applied Physics,* **54**, 2106–2108

Brookes, C. A. and Vance, E. R. (1972) see *The Physical Properties of Diamond* (1979) (ed. J.E. Field), Academic Press, London, pp. 394–395

Brosious, P. R., Corbett, T. W. and Bourgoin, J. C. (1974) *physica status solidii, (a)* **21**, 677–683

Burnett, P. J. and Page, T. F. (1984) *Journal of Materials Science,* **19**, 845–860, 3524–3545

Burnett, P. J. and Page, T. F. (1985) *Journal of Materials Science,* **20**, 4624–4646

Chrenko, R. M. and Strong, H. M. (1975) *Report No. 75CRD089* General Electric company, Schenectady, New York

Clark, C. D., Ditchburn, R. W. and Dyer, H. B. (1956) *Proceedings of the Royal Society,* **A234**, 363–381; **A237**, 75–89

Crowder, B. L., Smith, J. E., Brodsky, M. H. and Nathan, M. I. (1972) In *Proceedings of the 2nd International Conference on Ion Implantation in Semiconductors,* (Garmisch-Partenkirchen, 1971), (eds I. Ruge and J. Graul) Springer-Verlag, Berlin, pp 255–261

Damask, A. C. (1958) *Journal of Applied Physics,* **29**, 1590–1593

Davidson, L.A., Chou, S., Gibbons, J.F. and Johnson, W.S. (1971) *Radiation Effects,* **7**, 35–44

Derry, T.E., Prins, J.F., Madiba, C.C.P. *et al.* (1988) *Nuclear Instruments and Methods in Physics Research,* **B35**, 431–434

Derry, T. E. and Sellschop, J. P. E. (1981) *Nuclear Instruments and Methods in Physics Research,* **191**, 23–26

DeVries, R.C. (1975) *Materials Research Bulletin,* **10**, 1193–1200

Evans, T and Wild, R. K. (1967) *Philosophical Magazine,* **15**, 447–451

Field, J. E. (1971) *Contemporary Physics,* **12**, 1–31

Flint, I. J. and Lomer, J. N. (1983) *Physica,* **116B**, 183–186

Frank, F. C., Lang, A. R., Evans, D. J. F. *et al.* (1990) *Journal of Crystal Growth,* **100**, 354–376

Gargin, V. G. (1982) *Sverkh. Materialy,* **3**, 22–25 (In Russian)

Gippius, A.A., Zaitsev, A. M. and Vavilov, V. S. (1982) *Soviet Physics Semiconductors*, **16**, 256–261

Hanley, P. L., Kiflawi, I. and Lang, A. R. (1977) *Philosophical Transactions of the Royal Society*, **A284**, 329–368

Hartley, N. E. W. (1982) In *Metastable Materials Formation by Ion Implantation* (ed by S. T. Picraux and W. J. Choyke), Elsevier Science, Amsterdam, pp 295–302

Hartley, N. E. W. and Poole, M. J. (1981) *British Patent Specification*, No. 1,588,418

Hauser, J. J. and Patel, J. R. (1976) *Solid State Communications*, **18**, 789–790

Hauser, J. J., Patel, J. R. and Rodgers, J. W. (1977) *Applied Physics Letters*, **30**, 129–130

Hioki, T., Itoh, A., Ohkubo, M. *et al.* (1986) *Journal of Materials Science*, **21**, 1321–1328

Hudson, J. A., Mazey, D. J. and Nelson, R. S. (1981) *British Patent Specification*, No. 1,588,445

Ikawa, N. and Shimada, S. (1983) *Technology Reports of the Osaka University*, **33**, 343–348

Kalinin, V. D., Kornilov, N. I. and Spitsyn, A. N. (1972) *Razvedka i Okhrana Nedr*, **9**, 27–30 (In Russian)

Kalinin, V. D., Kornilov, N. I., Uvarov, V. A. *et al.* (1976) *Sovietskaya Geologiya*, **12**, 135–138 (In Russian)

Kalish, R., Bernstein, T., Shapiro, B. and Talmi, A. (1980) *Radiation Effects*, **52**, 153–168

Kalish, R., Deicher, M., Recknagel, E. and Wichert, T. (1979) *Journal of Applied Physics*, **50**, 6870–6872

Kirk, R. S. (1973) *Journal of Materials Science*, **8**, 88–92

Lang, A.R. (1974) *Proceedings of the Royal Society*, **A340**, 233–248

Latanision, R.M. and Pickens, J.R. (1981) *Atomistics of Fracture*. Plenum, New York

Lawn, B. R. and Wilshaw, T. R. (1975) *Fracture of Brittle Solids*. Cambridge University Press, Cambridge

Levy, P. W. and Kammerer, O. F. (1955) *Physical Review*, **100**, 1787–1788

Lonsdale, K. and Milledge, H. J. (1965) In *Physical Properties of Diamond*, (ed R. Berman), Clarendon Press, Oxford, pp. 12–68

Morhange, J. F., Beserman, R., Bourgoin, J. C. *et al.* (1975) In *Proceedings of 4th International Conference on Ion Implantation in Semiconductors*, (Osaka, 1974), (ed S. Namba), Plenum, New York, pp. 457–461

Nachal'naya, T. A. *et al.* (1978) *Sint Almazy*, **3**, 10–14 (In Russian)

Nelson, R. S., Hudson, J. A., Mazey, D. J. and Piller, R. C. (1983) *Proceedings of the Royal Society*, **A386**, 211–222

Owen, J. (1965) In *Physical Properties of Diamond*, (ed R. Berman), Clarendon Press, Oxford, pp. 274–294

Primak, V., Fuchs, L. H. and Day, P. P. (1956) *Physical Review*, **103**, 1184–1192

Prins, J. F. (1983a) In *Ultrahard Materials Application Technology*, Vol. 2, (ed P. Daniel), De Beers Industrial Diamond Division, London, pp. 15–25

Prins, J. F. (1983b) *Radiation Effects Letters*, **76**, 79–82

Prins, J. F. (1985) *Physical Review B*, **31**, 2472–2478

Prins, J. F. (1988) *Nuclear Instruments and Methods in Physics Research*, **B35**, 484–487

Prins, J. F., Derry, T. E. and Sellschop, J. P. F. (1986) *Physical Review B*, **34**, 8870–8874

Prins, J. R., Derry, T. E. and Sellschop, J. P. F. (1987) *Nuclear Instruments and Methods in Physics Research*, **B18**, 261–263

Ruoff, A.L. and Vohra, Y.K. (1989) *Applied Physics Letters*, **55**, 232–234

Samuels, B. (1987) *D Phil. Thesis*, Oxford University

Strong, H. M. and Chrenko, R. M. (1971) *Journal of Physical Chemistry*, **75**, 1838–1843

Suzuki, S. and Lang, A.R. (1976) *Journal of Crystal Growth*, **34**, 29–37

Swain, M. V., Lawn, B. R. and Burns, S. J. (1974) *Journal of Materials Science*, **9**, 175–183

Vance, E. R. (1971) *Journal of Physics C*, **4**, 257–262

Van Enckevort, W. J. P. and Seal, M. (1987) *Philosophical Magazine A*, **55**, 631–642

Varichenko, V. S., Zaitsev, A. M. and Stelmakh, V. F. (1986) *physica status solidi, (a)* **95**, K123–K126

Vavilov, V. S. (1974) In *Proceedings of 12th International Conference on the Physics of Semiconductors*, (Stuttgart,1974), (ed M.H. Pilkuhn), B.G. Teubner, Stuttgart, pp 277–285

Vavilov, V. S., Gukasyan, M. A., Guseva, M. I. *et al.* (1974b) *Soviet Physics Semiconductors*, **8**, 471–473

Vavilov, V. S., Konorova, E. A., Stepanova, E. B. and Trukhan, E.M. (1979) *Soviet Physics Semiconductors,* **13**, 635–638

Vavilov, V. S., Krasnopevtsev, V. V., Miljutin, Yu.V. *et al.* (1974) *Radiation Effects,* **22**, 141–143

Ward, R. C. C., Wilks, E. M. and Wilks, J. (1983) *Industrial Diamond Review,* **43**, 137–141

Wentorf, R. H. and Darrow, K. A. (1965) *Physical Review,* **137**, A1614–A1616

Whippey, P. W. (1972) *Canadian Journal of Physics,* **50**, 803–812

Wilks, E. M. (1967) *Industrial Diamond Review,* **27**, 110–115, 154–159

Wilks, E. M. (1969) *Diamond Research 1969,* supplement to *Industrial Diamond Review,* pp.7–12

Wilks, J. (1980). *Precision Engineering,* **2**, 57–72

Wilks, E. M. and Wilks, J. (1967) In *Science and Technology of Industrial Diamonds,* Volume 1. Science, (ed J. Burls), De Beers Industrial Diamond Division, London, pp. 93–103

Wilks, E. M. and Wilks, J. (1978) *Diamond Research 1978,* supplement to *Industrial Diamond Review,* pp. 2–10

Wilks, E. M. and Wilks, J. (1982) *Wear,* **81**, 329–346

Wilks, E.M. and Wilks, J. (1984) *Industrial Diamond Review,* **44**, 82–85

Wilks, E. M. and Wilks, J. (1987a) *Wear,* **118**, 161–184

Wilks, E.M. and Wilks, J. (1987b) *Industrial Diamond Review,* **47**, 17–20

Wong, C. J. (1981) *Journal of Engineering Materials and Technology,* **103**, 341–345

Woods, G. S. (1984) *Philosophical Magazine B,* **50**, 673–688

Woods, G.S. and Lang, A.R. (1975) *Journal of Crystal Growth,* **28**, 215–226

Zaitsev, A. M., Gippius, A. A. and Vavilov, V. S. (1982) *Soviet Physics Semiconductors,* **16**, 252–256

Zapffe, C. A. and Worden, C. O. (1949) *Acta Crystallographica,* **2**, 377–388

Zhang, G-L., Yu, H. and Su, N-N. (1986) *Radiation Effects,* **97**, 273–282

Ziegler, J. F. (1988) *Ion Implantation Science and Technology,* (2nd edn) Academic Press, Boston

Chapter 11

Friction

Friction is a complex phenomenon which can arise by a variety of mechanisms (see for example Bowden and Tabor, 1950, 1964; Rabinowicz, 1965; Shaw, 1984). The value of the friction when diamond slides on other materials depends both on the nature of the material and on the conditions of the motion, in particular high pressures, high sliding speeds and high temperatures all affect the friction processes. The variety of possible conditions appears to have discouraged measurements and much of the experimental work has concerned the friction of single crystal diamond sliding on single crystal diamond. These results turn out to . be of particular interest because of the nature of polished diamond surfaces.

The coefficient of friction μ associated with a stylus sliding over a horizontal surface is defined as F/W where F is the horizontal friction force and W the load on the stylus. This coefficient is usually determined by direct measurements of F and W as described in the various papers mentioned below. Experiments on the friction of diamond on diamond have usually been made in air with a diamond stylus sliding over a polished {001} face, and give quite low values of μ of the order of 0.05 to 0.1. Detailed measurements have been made by several authors but many of the earlier results did not appear consistent with each other. For example, Enomoto and Tabor (1980, 1981) reported that the coefficient increased with increasing load, Casey and Wilks (1973) that it was independent of load, and Hillebrecht (1981) that it decreased with increasing load.

We discuss the differences between the various results in the following sections. Nearly all the measurements have been made at relatively slow speeds of sliding and under these conditions it is generally agreed that the coefficient μ is independent of the speed. For example, Casey and Wilks (1973) observed that an increase in speed from $0.93\,\mu\text{m s}^{-1}$ to $90.0\,\mu\text{m s}^{-1}$ made no significant change.

11.1 Diamond sliding on diamond

11.1.a Surface films

The coefficient of friction μ usually takes values of the order 0.05 to 0.1 but if measurements are made in a high vacuum the friction is greatly increased. Figure 11.1 shows the friction measured in an experiment by Bowden and Hanwell (1966) where the spherical tip of a stylus was traversed repeatedly to and fro across a polished diamond surface under a moderately high vacuum of 5.10^{-9} torr. Initially

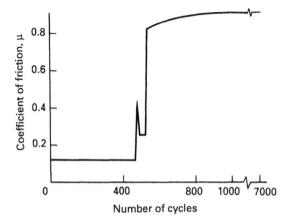

Figure 11.1 Variation of the coefficient of friction for diamond sliding on diamond with the number of traversals. (One cycle represents one traversal in each direction.) Pressure 5 x 10^{-9} torr (Bowden and Hanwell, 1966)

the friction had the same magnitude as in air but after about 500 cycles the value of μ rose by a factor of about 7. On readmitting air the friction was restored to its original value and careful measurements showed that this restoration was complete even after admitting only enough air to form one or two monolayers on the surfaces.

The above experiments suggest that adsorbed films of air on diamond act as a form of lubricant. Moreover, these films were only removed after some 500 passes in high vacuum. The efficiency of the air film as a lubricating agent is also underlined by the fact that the friction in air is not reduced or modified by the presence of a light oil in the form of either a thin film or a drop (Seal, 1958; Casey and Wilks, 1973).

The friction of diamond on diamond in conditions of high vacuum is usually of not much practical interest, so most measurements of the coefficient of friction have been made in air. Therefore in discussing these experiments we shall assume that the surfaces were generally covered by an adsorbed film of air with some lubricating action.

11.1.b Orientation and surface polish

A characteristic feature of the friction of diamond sliding on diamond is that the value of μ depends on the crystallographic orientations of the polished face and the stylus and on the crystallographic orientation of the direction of sliding. Measurements on polished {001} faces by Seal (1958) using a gramophone needle as a stylus and by Casey and Wilks (1973) using the tip of a natural octahedron diamond as a stylus showed a marked variation of μ with the azimuthal angle of sliding. For example Figure 11.2(a) shows that the friction is greater by a factor of about 2 for sliding in <100> directions than for sliding in the intermediate <110> directions.

The friction in Figure 11.2(a) exhibits a four-fold symmetry as is to be expected on an {001} face, but if the same cube face is polished, not in the usual way by

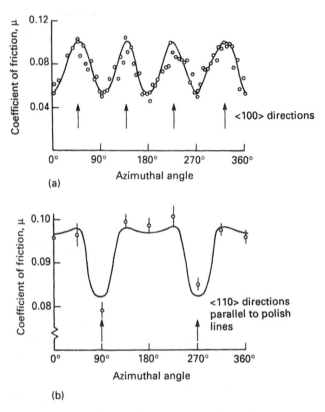

(a)

(b)

Figure 11.2 Coefficient of friction for a diamond stylus sliding over a polished {001} face of diamond as a function of direction of sliding: (a), the face polished in a normal <100> direction; (b), the face polished in a <110> direction (Casey and Wilks, 1973)

abrading in a <100> direction, but in a more resistant <110> direction the friction varies as shown in Figure 11.2(b). The values of the friction are now quite different and show only a two-fold symmetry. At first sight it may seem surprising that the symmetry should be changed in this way but polishing in a <110> direction gives rise to microstructure on a nanometer scale which has a two-fold symmetry (Section 9.4), and this symmetry reveals itself in the measurements of the friction.

In other experiments Bowden and Brookes (1966) measured the friction of {001} faces using cone-shaped styli with included angles of 60°, 120° and 170°. They found that the anisotropy of the friction was much more marked with the sharper styli (Figure 11.3) and concluded that the wider ones were unable to follow the finer details of the surface topography. Subsequently Hillebrecht (1981) made measurements using spherically tipped styli with the relatively large radius of 450 μm and observed no variation of the friction with the direction of sliding.

Several other experiments also show that the coefficient of friction is affected by the topography of the surfaces. A particularly smooth polished surface showing no signs of the usual polish lines may be obtained by the use of oxidizing agents on a cast-iron scaife (Section 9.5.d) and this treatment reduces the coefficient μ by about a factor 2 in all directions (A. G. Thornton, 1975, personal communication).

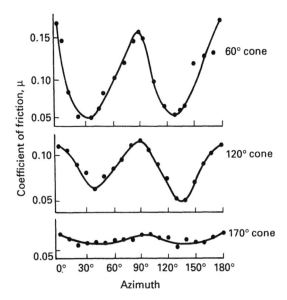

Figure 11.3 Coefficient of friction for diamond cones of different apex angle sliding on an {001} surface as a function of sliding direction (Bowden and Brookes, 1966)

Another surface polished in the normal way and then further polished by an ion-beam technique (Section 9.6.c) exhibited coefficients of friction with values of about 0.15 which were essentially independent of the direction of sliding. Enomoto and Tabor (1980, 1981) have described how the presence of polishing lines may reduce the four-fold symmetry of the friction on a polished {001} face.

11.1.c Effect of load

Further apparent discrepancies between the various experiments arise when we consider how the coefficient of friction varies as a function of the load on the stylus.

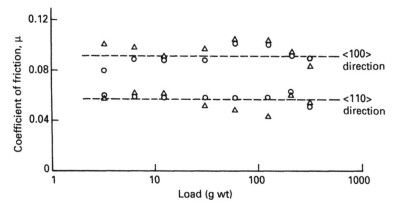

Figure 11.4 Coefficient of friction for the tip of a natural octahedron diamond sliding on a polished {001} surface as a function of load (Casey and Wilks, 1973)

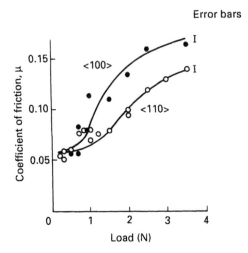

Figure 11.5 Coefficient of friction for a spherical stylus of diamond of radius 80 μm sliding on a polished {001} surface in <100> and <110> directions as a function of load (Enomoto and Tabor, 1981)

Thus Casey and Wilks (1973) using the tip of an octahedron diamond as a stylus found that μ was *independent* of the load over a wide range of loads (Figure 11.4). On the other hand Seal (1958) and Enomoto and Tabor (1981) using rounded styli observed values of μ which were considerably *greater* at higher loads (Figure 11.5). The latter authors also observed that the difference between the friction in <100> and <110> directions was only appreciable at the higher loads. Finally, Hillebrecht

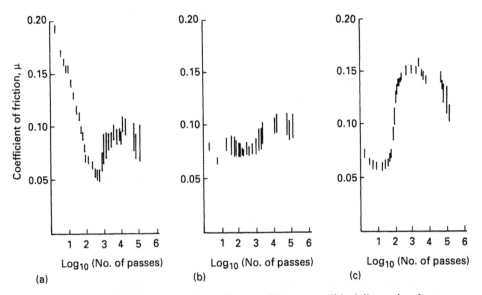

Figure 11.6 Coefficient of friction of a diamond stylus sliding on a polished diamond surface as a function of the number of passes on the same track. The tangent plane was (001) for both diamonds and the stylus was sliding in its own [110] direction; for the lower diamond the sliding directions were: (a), [100]; (b), [210]; (c), [110] (Seal, 1967)

(1981) using a rounded stylus obtained values which *decreased* somewhat at higher loads.

A further complication arises because the value of μ may be affected by the measurement. Seal (1967) made up to 100000 passes over the same track and observed considerable modifications to the friction. Figure 11.6 shows that these changes produced by repeated measurements are very dependent on the crystallographic orientations of the diamond, and that with some orientations appreciable changes occur after only a few passes; see also Casey (1972). These results imply that measurements of the friction may produce damage to the surface of the diamonds and this possibility must be kept in mind. However, we assume initially that any such effects will be relatively small provided only a limited number of passes are made on a given track.

11.2 Dependence on pressure

The friction of diamond on diamond is a complex phenomenon influenced by several parameters. Before discussing the mechanisms involved we first describe a series of experiments to determine the coefficient of friction as a function of the load on the stylus while all other parameters were kept as constant as possible (Samuels and Wilks, 1988).

11.2.a Measurements on {001} faces

Many of the styli used in earlier experiments had a rather ill defined geometry and were of uncertain crystallographic orientation. The present experiments were made with carefully polished spherically tipped styli sliding over polished flat surfaces. Spherical styli were preferred both because the form of a spherical tip can be specified more precisely than the tip of a conical slider, and because this shape reduces the possibility of the stylus gouging into the flat. It turns out that the pressure under the stylus is a more relevant parameter than the load W, and the mean pressure p over the area of contact varies only as $W^{1/3}$ (Section 7.5.1). Hence, in order to obtain a wide range of pressures it was necessary to use styli of radii ranging from 60 μm to 490 μm.

Both styli and flats were polished on gem quality colourless diamonds. Particular care was taken to establish their crystallographic orientation by using only diamonds with well formed natural faces permitting optical goniometry. The diamonds for the flat surfaces were also selected to give an appreciable and uniform cathodoluminescence to facilitate the detection of surface damage in the SEM, see Section 7.7.b. Further details of the styli and flat surfaces and of the apparatus used to measure the friction are given by Samuels and Wilks (1988).

Measurements were first made on an {001} face. The tangent plane of the stylus was an {011} face and moved so as to present an abrasion resistant direction. Three sets of friction measurements were made, two in <100> directions parallel and perpendicular to the polish lines ($\mu_\parallel$ and $\mu_\perp$), and a third set in one of the <110> directions (which all make angles of 45° with the polish lines). The value of μ was determined in all three directions as a function of pressure by 13 sets of measurements using a range of styli and loads. As discussed by the authors, it appears that the value of the friction in a given direction is largely determined by the value of the mean contact pressure rather than the radius of the stylus or the value of the load. They therefore summarize their results in a schematic diagram giving the friction as a function of pressure as shown in Figure 11.7. (The diagram labels the <100> directions as easy and the <110> direction as hard with reference to the ease of abrasion in these directions.)

11.2.b Dependence on pressure

The summary of the measurements given by Figure 11.7 suggests that at least two processes contribute to the friction, one giving rise to values of μ which decrease with pressure and the other to values which increase with pressure. The figure also

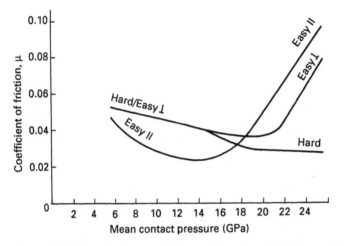

Figure 11.7 Schematic summary of the experimental results for the coefficient of friction for a diamond stylus sliding on an {001} face as a function of mean contact pressure. Easy and hard, ‖ and ⊥, refer to the directions of sliding, see text (Samuels and Wilks, 1988)

shows that the friction in a given direction may be greater than, equal to, or less than the friction in another direction depending on the pressure between the surfaces, a result which accounts for the apparently differing results obtained by previous authors. For example, Hillebrecht (1981) used a load of 0.5 N on a 450 μm stylus, corresponding to a mean pressure of 3.5 GPa, and observed little or no anisotropy in the friction. On increasing the load to produce pressures up to 11 GPa, the value of μ fell by about 10%. Both these results follow the general pattern shown in Figure 11.7. On the other hand Enomoto and Tabor (1981) used styli and loads which produced mean pressures of 14 GPa, 17 GPa and 22 GPa, and hence observed that the anisotropy of the friction increased considerably as the load was increased.

The experimental situation when using a pointed stylus is more complex. Consider first the friction given by the 170° cone used by Bowden and Brookes (Figure 11.3). The mean pressure due to elastic deformations under an ideal cone is given by:

$$p = \frac{E \cot \theta/2}{4(1 - v^2)} \tag{11.1}$$

E being the Young modulus of diamond, θ the full angle of the cone, and v Poisson the ratio (see for example Johnson (1985) p.114). Substituting in values and taking $v = 0.1$ we obtain a pressure of 23 GPa. However, this is certainly an upper limit as the tip can hardly have the form of a geometrically ideal cone. Hence, as the observed friction was almost isotropic, it seems likely that the actual mean pressure corresponded to the central region of Figure 11.7.

The assumption that the tips have an ideal conical form is even less realistic for the 120° and 60° cones. For a 120° cone Equation (11.1) gives a mean pressure of 150 GPa, and the tip of the stylus would fracture before this pressure was reached. In fact, neither Bowden and Brookes nor Casey and Wilks reported any gross cracking of the flats in their experiments, which suggests that the pressures did not exceed the order of 25 GPa (see Section 11.2.c). Thus, the pressures under 120° and 60° cones were probably in the range shown in the right hand half of Figure 11.7, in which case the higher pressures under the 60° cone would give rise to the higher anisotropy, as observed.

The exact shapes of the octahedron tips used by Casey and Wilks (1973) were not well defined but as the included angle between opposite octahedron faces is 110° we might expect the pressure to be similar to that in the sharper cones used by Bowden and Brookes. That is, the pressures would be on the right hand side of Figure 11.6 and we would expect the friction to show appreciable anisotropy, as was observed (Figure 11.2(a)).

Casey and Wilks also observed that the coefficient of friction in both <100> and <110> directions showed little or no variation with loads ranging from 3 g wt to 300 g wt (Figure 11.4). This range of load would change the pressure under a spherical tip by a factor 4.6, but under a conical tip the pressure is independent of the load (Equation 11.1). A direct application of this result would explain the constancy of the friction but the position must remain open because Equation (11.1) applies to an ideal rather than an actual stylus. Even so it is clear that the contact pressure under the stylus plays a dominant role in determining the magnitude of the friction.

11.2.c Surface damage

Particular care was taken during the measurements of Figure 11.7 to observe any changes produced on the surface of the flat. The curved styli were more difficult to inspect but were always checked for any gross damage. The most obvious damage produced on the flats took the form of cone cracks, either complete or partial, which appear on the surface as single ring cracks or series of ring cracks at pressures above about 25 GPa. These cone cracks modify the topography of the surface considerably (Sections 7.5.c and 8.2.a) so all the main observations were taken with loads below this limit.

At the lower loads no damage to the flat was visible in the optical microscope, but damage could be observed by a careful examination in the CL mode of the scanning electron microscope, if necessary by reducing the accelerating voltage in order to concentrate the induced luminescence near the surface. For example, Figure 11.8 shows a micrograph of several friction tracks produced by one, five and ten passes of a stylus over the same paths, and we see that the measurements have produced significant darkening of the tracks. As might be expected, the darkening becomes more marked with repeated passes and increasing load. It was found that some damage was always produced by a friction measurement even

Number of Passes

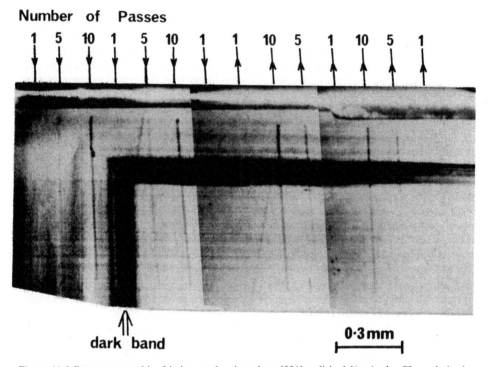

Figure 11.8 Damage caused by friction tracks viewed on {001} polished face in the CL mode in the SEM. The friction tracks are vertical. The pronounced right-angled band marks a growth layer in the diamond (Samuels and Wilks, 1988)

under the lowest pressures. The damage appears similar to that observed in fatigue measurements (Section 7.7) and probably consists of an assembly of fine cracks. No change in the form of the darkening with increasing load could be detected until the abrupt appearance of ring cracks.

The damage produced by the stylus could also be revealed by another technique. Thus, the surface of a flat showing damage in CL (Figure 11.9(a)) was gently smeared with a very small amount of grease, tracks were then clearly visible in the optical microscope (Figure 11.9(b)). It appears that surface cracking produces a roughening of the surface which results in the preferential retention of the grease. (The top ends of the tracks in Figure 11.9 are sometimes darker and somewhat displaced. Later experiments suggested that these effects were due to transitory starting up processes arising from small amounts of play in the bearings carrying the specimens.)

A further indication of the presence of damage was sometimes given by the build-up of some material in front of the stylus. This material, presumably debris, could be detected in the optical microscope (Figure 11.10) and recalls the perhaps similar 'waxy deposit' seen by Seal (1958). A build-up of debris in front of the stylus during a pass increases the force required to move the stylus and this effect is most marked when the stylus pushes the debris along the grooves of the polishing lines. A discussion of this effect and of other variations of the friction during single and repeated passes is given by the authors.

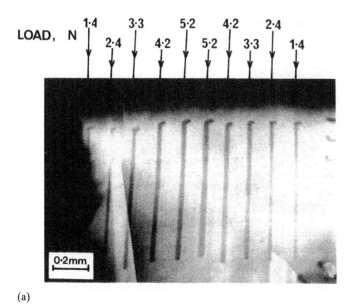

(a)

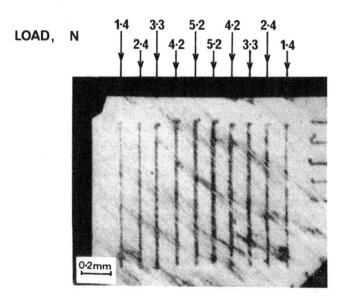

(b)

Figure 11.9 Friction tracks in <110> direction on a polished {001} face produced by a stylus of radius 340 μm viewed: (a), in the CL mode of the SEM; (b), using the grease technique described in the text (Samuels and Wilks, 1988)

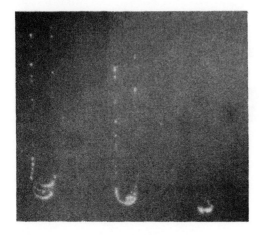

Figure 11.10 Optical interference micrograph showing debris at the end of tracks produced by a stylus of radius 181 μm sliding on an {001} surface in a <100> direction perpendicular to the polish lines (Samuels and Wilks, 1988)

11.3 Mechanisms of friction

Any account of the mechanisms responsible for the friction force must begin with the nature of the polished surfaces (Section 9.4). Even well polished surfaces are rough on a scale of the order of 5 nm with asperities outlined mainly by cleavage planes. It follows that if two such surfaces are placed together in close contact, the two sets of asperities will interlock. Any attempt to slide one surface over the other must be impeded by collisions between these asperities which thus give rise to a friction force. Continuous relative motion can only occur if the interlocking asperities either ride over each other, push past each other, or fracture. We consider each of these possibilities in turn.

11.3.a Fracture

We noted in the previous section that a measurement of the friction always produces some damage presumably by small scale fracture. If this fracture is the only mechanism involved in the friction process, then the work done against the friction force must be equal to the work done in fracturing the diamond. However, estimates of the energy of fracture show that coefficients of friction derived in this way are two orders of magnitude too small, see for example Enomoto and Tabor (1980) and Seal (1981).

A particular example of the influence of fracture on the friction was observed by Samuels and Wilks (1988) using loads near to the critical value for the production of ring cracks. Sometimes ring cracking began some way along the length of a track and the friction then rose by perhaps 30%. This result might suggest that the work of fracture makes a substantial contribution to the friction, but further measurements on the same track showed otherwise. These subsequent measurements were made with a much lower load insufficient to produce ring cracks. Nevertheless the values of μ on both the initial and damaged parts of the track were similar to those observed with the heavier load. Thus it appears that the production of ring cracks gives rise to a higher friction because of modifications to the topography of the surface, and that the fracture of the surface does not in itself make a major contribution to the friction.

11.3.b The ratchet mechanism

We now consider the position when the relative motion of two diamond surfaces is achieved by the asperities riding over each other. A detailed analysis is quite complicated because each asperity is surrounded by other asperities on the same surface which may or may not be bearing part of the load. However, suppose that as a result of the sliding motion, two asperities, one on the slider and one on the flat, meet and ride over each other. The two surfaces in the region of the asperities are thus forced apart against the load by the friction force. As the sliding continues, contact between the two asperities will be lost and the load taken by other sets of asperities. Because the asperities are rough and irregular, the transference of load to different sets of asperities will be abrupt and *irreversible*. Hence, according to the second law of thermodynamics, some of the work of separation degenerates into heat, and work must be done to maintain the sliding. Variants of this type of friction in other materials have been described by Bowden and Tabor (1950) and Rabinowicz (1965) and described as ratchet or roughness mechanisms.

The coefficient of friction for a ratchet mechanism will depend on the detailed topography of the surfaces including the steepness of the asperities. Taking the asperities to be bounded approximately by cleavage planes, Casey and Wilks (1973) show that an irreversible energy loss of about 10% of the stored energy at each encounter is sufficient to account for the magnitude of the friction. To estimate the dependence of the friction on the direction of sliding they considered the motion of a stylus over a polished {001} surface with the topography shown schematically in Figure 9.15(c). Hence the friction should be greater in the <100> direction than in the <110> direction by a factor of 1.4 which is of the correct order of magnitude.

The above approach accounts naturally for the dependence of the friction on the method of polishing the diamond surface (Figure 11.2), because different methods of polishing produce different topographies with asperities of different shape and overall symmetry. Tabor (1979) and Seal (1981) have extended this approach by introducing an adhesive force between the interacting asperities. However, it is well established that the friction of diamond on diamond is unchanged by the presence of a light oil, so the assumption of an adhesive force appears somewhat unrealistic. Also, none of the above treatments of the ratchet mechanism account for the observed dependence of the friction on pressure, nor do they take account of the effect of the polish lines.

11.3.c Elastic losses

The third way in which the asperities on the sliding surfaces can move past each other is by pushing each other aside elastically (Samuels and Wilks, 1988). The contact between two such asperities will generally be lost abruptly, leaving them with stored elastic energy which will cause them to vibrate. To make an order of magnitude estimate of this effect we calculate the deflection δ of a diamond-like pyramidal asperity bounded by {111} planes when a force f parallel to one side of the base is applied on the centre line of a face:

$$\delta = \frac{f}{Ed} \frac{2^{1/2}c(c^2 + 1.2)}{1 - c} = \frac{f}{Ed} \beta \qquad (11.2)$$

where E is the Young modulus, d the length of the side of the base, and c the fraction of the asperity height at which the force is applied. If the force is suddenly released the asperity will vibrate at a frequency at least of the order of $V/2d$ where V is the velocity of longitudinal sound waves, that is a frequency of at least 10^{12} Hz. The speed of sliding in the present experiments was about 66 μm s^{-1}, so taking the distance between collisions at the lower limit of 5 nm, the time between collisions is $\sim 10^{-4}$ s. Hence, the asperities would be free to vibrate for $\sim 10^8$ cycles between collisions, sufficient to dissipate most of the elastic energy even though the internal damping in diamond may be very low.

To estimate the effect of the above loss of elastic energy we introduce the basic equation of the friction obtained by considering a unit displacement of the stylus and equating the work done against the friction force with the energy loss in the diamond. Consider the situation in which the stylus is at rest with virtually all its asperities in contact with asperities on the flat. If we now move the slider forward a distance of λ equal to the mean distance between the asperities on the stylus, then each stylus asperity will on the average make one collision, hence the total work done by the stylus over the unit displacement is

$$\mu W = N(1/\lambda)_{\text{stylus}}\epsilon \tag{11.3}$$

where N is the mean number of asperities in contact, and ε the energy loss per collision. Assuming that all the asperities in the apparent area of contact A are in real contact, $N \sim A/d^2$. The loss of elastic energy from the pair of asperities involved in each collision is

$$\epsilon = f_{\text{max}}\delta_{\text{max}} = (Ed/\beta)^2\,\delta_{\text{max}}^2 \tag{11.4}$$

Hence, taking $\lambda = d \sim 5$ nm, assuming the point of contact to be half way up the pyramid ($c = 0.5$), and substituting into Equation (11.3) we find that to obtain a typical value of the friction of 0.05 at a pressure of 20 GPa requires a value of $\delta/d \sim 0.04$. That is, the loss of stored energy is sufficient to account for the friction if the asperities deflect 0.2 nm at the point of contact without, of course, suffering fracture.

11.4 The friction of diamond on diamond

The measurements of the friction as a function of pressure described in Section 11.2.b imply that at least two different mechanisms of friction are involved. Of the three mechanisms discussed in the previous section fracture will account for the damage produced during measurements but not for the magnitude of the friction. We now consider whether the other two mechanisms, the ratchet and elastic loss mechanisms, are consistent with the observed pressure dependence (Samuels and Wilks, 1988). We begin by considering the form of the contact between two rough surfaces.

11.4.a Contact of rough surfaces

The nature of the contact between two rough elastic surfaces with surface asperities is quite complex. Greenwood (1967) considered the simpler but essentially similar case of a flat plane in contact with a rough surface, the height of the asperities being specified by some distribution function. Two limiting conditions are distinguished, corresponding to high and low loads. Under low loads contact is only made between the tips of the highest asperities, and the area of true contact is a small fraction of the apparent contact area. In the regime of high loads all the asperities have come into contact, and the load on each increases as the total load W increases. Rather similar considerations apply for Hertzian type contact (Greenwood and Tripp, 1967) though the analysis is now somewhat more complicated because of the different geometry.

The profiles of the polished diamond flats and styli are considerably more complex than the examples treated above. Three different orders of magnitude are involved in a typical experiment: (i), the bulk depression of the flat is of the order of 1 μm or 2 μm; (ii), both surfaces are modulated by polish lines consisting of hills and valleys with vertical height differences varying from perhaps 10 nm to 100 nm and with a horizontal scale of perhaps 0.5 μm to 5 μm; (iii), the surfaces are further modulated by the basic polishing process which produces a structure with a vertical and horizontal scale of the order of 5 nm. (Note that although the hills and valleys of the polish lines may be clearly visible in the interference microscope their slopes are quite shallow.)

Experiments with hard styli resting on hard materials can produce extremely high loadings of the surface. The maximum apparent mean pressure between the surfaces in the present experiments was about 25 GPa, which is over 10^8 times greater than that due to a cube of diamond of side 5 mm resting on a flat surface and close to the value required to produce gross cracking. These pressures will modify the topographies of the polish lines in the manner discussed by McCool (1983) who considers the elastic deformation of a system of parallel sinusoidal hills and valleys when compressed elastically by a plane surface. His results, although not directly applicable to the topography of the polish lines, imply that the pressures in the present experiments are of the right order of magnitude to ensure contact over most of the valleys as well as the hills. Therefore, we shall assume with Samuels and Wilks that in all their measurements the load was sufficient to cause most of the small-scale asperities on the stylus to be in contact with the flat.

11.4.b Pressure dependence

To discuss the form of the friction associated with the ratchet and elastic loss mechanisms we return to Equation 11.3 which is applicable to both mechanisms. For the ratchet mechanism we take:

$$\epsilon = \alpha h w$$

where w is the mean load on an asperity, h the distance through which the asperity relaxes before the load is carried on a new set of contacts, and α a fraction of order 0.1. Hence, remembering that $W = Nw$, Equation (11.3) gives the coefficient of friction associated with the ratchet mechanism as:

$$\mu_R \simeq \alpha h / \lambda$$

a result of the correct order of magnitude. The dependence of μ_R on pressure will be determined by the behaviour of the relaxation length h which is difficult to calculate, but which will certainly decrease as the surfaces move closer together at the higher pressures.

We now consider how the coefficient μ_E associated with the elastic loss mechanism will vary with the pressure. For a Hertz indentation with a spherical stylus the area of apparent contact $A \propto p^2$ where p is the overall mean contact pressure. The number of asperities in this area is A/d^2 where $d \sim 5\,\mathrm{nm}$ and we assume that the majority are in actual contact, hence

$$N \propto p^2.$$

The mean load on the asperities

$$w = \frac{W}{N} \propto \frac{p^3}{p^2} \propto p$$

and the total elastic energy associated with each encounter is

$$\epsilon_E = \gamma w^2 \propto p^2$$

where γ is a constant including the elastic moduli. Hence, substituting in Equation (11.3) we obtain:

$$\mu_E \propto p.$$

The model used above is of course an approximation for the more complicated real situation where the motion of each asperity, both in and perpendicular to the plane of the surface, is limited by the constraints due to its neighbours. Even so, the above estimates show that both the ratchet and elastic loss mechanisms give coefficients of friction which have the correct order of magnitude and which, respectively, decrease and increase as the mean contact pressure is increased.

Hence, a combination of the ratchet and elastic processes will account for the form of the friction observed in the easy <100> directions on the {001} faces. However, with such a combination of mechanisms we would expect the friction in the hard <110> directions to rise at the higher pressure which it does not. To account for this latter behaviour in the hard direction we note that the elastic loss mechanism will only give greater friction with increasing pressure if the asperities continue to absorb more elastic energy rather than fracture. In fact, we know that the higher loads are near to those sufficient to produce gross cracking. Moreover, the resolved tensile stresses across the {111} cleavage planes due to sliding in the hard direction are about a factor $2^{1/2}$ greater than those produced by sliding in the easy direction. Hence it seems that the friction in the hard direction is limited by the maximum elastic energy which can be stored before fracture.

(The hard <110> direction is the most resistant to abrasion and polish, so it may seem inconsistent to account for the lower friction in this direction by saying that in this direction the asperities are more easily fractured. However, measurements of the abrasion resistance are made on surfaces which are continuously abraded in the course of the experiment. Therefore, apart from an initial start-up condition the surface topography in an abrasion experiment is always characteristic of the direction of polishing. Indeed, it is the difference of surface topography produced by polishing in <100> and <110> directions which is responsible for the different abrasion rates. However, the present friction measurements were made on a surface which had been polished in an easy direction, with a surface topography characteristic of this direction, and on this surface the asperities will be more easily fractured by attack in <110> directions.)

11.4.c Measurements on {011} faces

Samuels and Wilks also made measurements on polished {011} faces which like
{001} faces show considerable variation in abrasion resistance for different
directions of abrasion. The experiments were made in a similar way to those on
the cube face except that passes were made only in two directions because of the
lower symmetry of {011} faces. The easy directions of polish and abrasion are
parallel and anti-parallel to the <100> axes, while the hard directions are parallel
and anti-parallel to the <011> axes. Thus, a {011} face polished in the normal
way in an easy direction presents two principal directions for the friction
measurements, the easy direction parallel to the polish lines and the hard direction
perpendicular to the polish lines, and measurements were made in these two
directions.

As in the previous experiments measurements were only made in conditions free
of ring cracking. However, as sliding in the easy direction produced ring cracks at
pressures of no more than about 8 GPa the pressure range was extended to
somewhat lower pressures by using a stylus of larger radius.

As on the {001} face, the value of the friction in a given direction appears to
be determined primarily by the contact pressure. Therefore it is again possible to
combine the experimental results into one schematic diagram showing the pressure
dependence over the whole pressure range (Figure 11.11). Although the absolute
values of the pressure are different, the form of curves in Figure 11.11 recall similar
features in Figure 11.7. In particular, the coefficient of friction in the easy direction
first falls with rising pressure and then rises, while the coefficient in the hard
direction first falls with rising pressure but then continues to fall at the higher
pressures.

Following the discussion of the previous section we note that the resolved tensile
stresses across the cleavage planes on a polished {011} surface are about ×2
greater for sliding in a <011> hard direction than in a <100> easy direction.
Hence, the asperities will fracture more readily when the stylus moves in a <011>

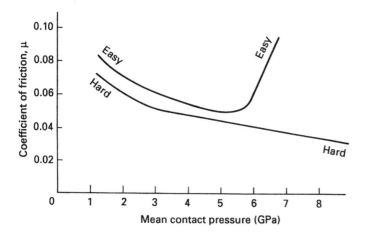

Figure 11.11 Schematic summary of the experimental results for the coefficient of friction for a stylus
sliding on an {011} face as a function of mean contact pressure. The terms easy and hard refer to the
directions of sliding, see text (Samuels and Wilks, 1988)

direction, so we expect that with increasing pressure the effect of fracture will first limit the friction in the hard <011> direction, as seen in Figure 11.11.

11.4.d Summary

The friction measured in the principal directions on both {001} and {011} faces shows a similar dependence on the contact pressure. These and other features are quite well explained, at least qualitatively, in terms of the ratchet and elastic loss mechanisms given above. Various points of detail are further discussed by Samuels and Wilks (1988), mostly in terms of the same mechanisms. We refer here only to one or two of the more significant points.

We first note that the minimum in the value of μ for the soft direction occurs at an appreciably lower pressure on the {011} face, see Figures 11.1 and 11.7. The position of this minimum will be determined by the relative magnitudes of the friction arising from each process. The absolute value of each component will be determined *inter alia* by the detailed topographies of both the polished stylus and the polished flat which are not well specified. However, electron microscope studies show that the topographies of {001} and {011} surfaces are different (Section 9.4.b). Hence we expect the components of the friction to differ from one face to another, with a resulting shift in the position of the minimum.

Another feature of the measurements concerns the difference of μ on {001} faces when sliding parallel and perpendicular to the polish lines. These differences are most marked at pressures below 14 GPa where the ratchet mechanism predominates, so at first sight, it seems that sliding across the hills and valleys of the polishing lines leads to a greater dissipation of energy. However, as discussed above, the pressures involved are probably sufficient to bring both hills and valleys into contact, so the roughness responsible for the irreversible losses will be on a scale of 5 nm rather than the 50 nm depths of the valleys. It therefore seems likely that the polishing process has produced a structure even on the fine 5 nm scale which tends to ease the passage of asperities sliding parallel to the direction of polish. Hence, the friction is least in the direction of polish.

Although the ratchet and elastic loss mechanisms account reasonably well for the form of the results, some details remain unclear. For example, it is not clear why in the high-pressure regime the friction is somewhat greater for sliding parallel to the polish lines. Nor is it clear why the coefficients of friction in all three directions tend to the same value at the lowest pressure (Although in this case the number of contacts between asperities may now be so reduced that statistical variations in their form and distribution tend to blur out directional effects).

11.5   Diamond sliding on other materials

We now consider the friction of single crystal diamond sliding on other materials. Various measurements have been made on a range of materials but under a diversity of experimental conditions.

Several experiments have been made at quite low sliding speeds of the order of 0.1 mms^{-1}. Figure 11.12 shows the friction of rounded diamond styli of different radii sliding over polycrystalline copper under various loads (Steijn, 1964). The coefficient of friction increased steadily with load and was greater for the smaller styli. Steijn also observed that the width of the track in the copper increased with

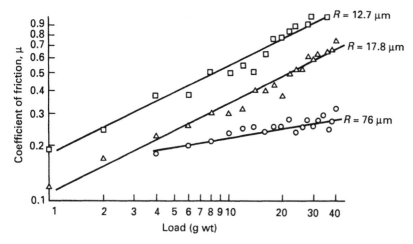

Figure 11.12 Coefficient of friction of diamond styli of different radii R sliding over polycrystalline copper as a function of load (Steijn, 1964)

the load in a comparable manner. Hence it appears that the progressive increase in the friction was associated with a ploughing type of action plastically deforming the copper.

It is clear from the above that the coefficient of friction may depend considerably on the load and the geometry of the stylus but accounts of experiments frequently give little information on these points. However, it is clear that at low loads and low speeds the friction can be of the order of 0.1 as observed by Casey (personal communication, 1973) who slid styli of copper and brass over diamond and by Freeman and Field (1989) who slid diamond styli over hard steel. Even though the friction of a diamond sliding on diamond is quite sensitive to the orientation of the diamond there is hardly any information on the effect of the orientation of the diamond sliding on other materials.

Experiments at speeds typical of machining processes are described by Matsuo, Toyoura and Kita (1986). Diamond grits were pressed by forces of 2.5 N and 4.5 N against discs of hardened steel, cobalt alloy, cemented carbide and sintered alumina spinning to give relative linear speeds of 5 ms^{-1} and 15 ms^{-1}. No details were given of the geometry or orientation of the diamond but a diagram suggests that the tip of a grit was pressed against the wheels. The coefficients of friction for the two metals and the alumina were initially about 0.3 and fell to a steady value of 0.1 or less about a minute after contact was first made. In contrast the friction on the cemented carbide was initially about 0.2 and then rose in about a minute to a fairly constant value of 0.4. Fedetov (1967) appears to have made experiments at comparable speeds to the above and quotes values of μ for sliding on glass and several metals ranging from 0.12 to 0.29, but gives no details of the loads or speeds, or of the form of the diamond.

Finally we mention two other sets of experiments. Miyoshi and Buckley (1980) measured the friction of diamond sliding at slow speed over metals in an ultra-high vacuum under carefully controlled conditions. The results are of interest in connection with the chemistry of clean diamond surfaces as discussed in Section 13.8, but are not directly relevant to typical machining conditions. In a completely

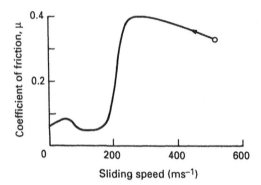

Figure 11.13 Coefficient of friction for a diamond ball sliding on chromium at high speed (Bowden and Freitag, 1958).

different type of experiment Bowden and Freitag (1958) measured the friction of diamond against copper and chromium balls spinning at very high speeds. The results shown in Figure 11.13 give the friction for speeds up to 500 ms^{-1}, very much higher than encountered in machining. As discussed by the authors an increase in μ from about 0.1 to 0.4 at speeds of about 200 ms^{-1} is associated with the surface melting of the metal and its transfer to the diamond, see also similar experiments by Miller (1962).

References

Bowden, F. P. and Brookes, C. A. (1966) *Proceedings of the Royal Society*, **A295**, 244–258

Bowden, F. P. and Freitag, E. H. (1958) *Proceedings of the Royal Society*, **A248**, 350–367

Bowden, F. P. and Hanwell, A. E. (1966) *Proceedings of the Royal Society*, **A295**, 233–243

Bowden, F. P. and Tabor, D. (1950) *The Friction and Lubrication of Solids, Part I*, Clarendon Press, Oxford

Bowden, F.P. and Tabor, D. (1964) *The Friction and Lubrication of Solids, Part II*, Clarendon Press, Oxford

Casey, M. (1972) *D.Phil. Thesis*, Oxford University

Casey, M. and Wilks, J. (1973) *Journal of Physics D*, **6**, 1772–1781

Enomoto, Y. and Tabor, D. (1980) *Nature*, **283**, 51–52

Enomoto, Y. and Tabor, D. (1981) *Proceedings of the Royal Society*, **A373**, 405–417

Fedotov, A. I. (1967) *Russian Engineering Journal*, **47**, (5), 67–69

Freeman, C. J. and Field, J. E. (1989) *Journal of Materials Science*, **24**, 1069–1072

Greenwood, J. A. (1967) *Transactions of the ASME, Journal of Lubrication Technology*, **89**, 81–91

Greenwood, J. A. and Tripp, J. H. (1967) *Transactions of the ASME, Journal of Applied Mechanics*, **34**, 153–159

Hillebrecht, F. U. (1981) *M.Sc. Thesis*. Oxford University

Johnson,K.L. (1985) *Contact Mechanics*. Cambridge University Press, Cambridge

Matsuo, T., Toyoura, S. and Kita, H. (1986) In *Proceedings of the 2nd International Conference on Hard Materials*, (Rhodes, 1984). Institute of Physics Conference Series No.75, (eds E. A. Almond, C. A. Brookes and R. Warren), Adam Hilger, Bristol, pp. 897–905

McCool, J. I. (1983) *Wear*, **86**, 105–118

Miller, D. R. (1962) *Proceedings of the Royal Society*, **A269**, 368–384

Miyoshi, K. and Buckley, D. H. (1980) *Applications of Surface Science*, **6**, 161–172

Rabinowicz, E. (1965) *Friction and Wear of Materials*. John Wiley, New York

Samuels, B. and Wilks, J. (1988) *Journal of Materials Science,* **23**, 2846–2864

Seal, M. (1958) *Proceedings of the Royal Society,* **A248**, 379–393

Seal, M. (1967) In *The Science and Technology of Industrial Diamonds, Vol. 1, Science,* (ed J. Burls), De Beers Industrial Diamond Division, London, pp. 145–159

Seal, M. (1981) *Philosophical Magazine A,* **43**, 587–594

Shaw, M. C. (1984) *Metal Cutting Principles.* Clarendon Press, Oxford

Steijn, R. P. (1964) *Wear,* **7**, 48–66

Tabor, D. (1979) In *The Properties of Diamond*, (ed. J. E. Field), Academic Press, London, pp. 325–350

Chapter 12

Polycrystalline diamond (PCD)

We now review the physical and mechanical properties of the man-made polycrystalline diamond PCD described in Section 1.3. When discussing properties of PCD one must remember that the material is available as ranges of proprietary brands from different manufacturers. Hence, the various products may differ in ways not fully specified, and may be modified in the course of time so that results obtained in the past may not be entirely typical of current production.

12.1 Structure

In order to discuss the structure of PCD we need to know the size of the diamond grains and the nature of the second phase filling the interstices between the grains, often described as the solvent/catalyst. The grain size is readily observed by polishing a specimen and viewing in the optical or electron microscope as in Figure 1.10 (a). The chemical constitution of the second phase may be found using an electron beam microprobe analyser (Section 2.1.b). (Note also that PCD with a cobalt solvent/catalyst is appreciably denser than diamond (Table 12.1), but with a silicon second phase is appreciably lighter than diamond.)

Given a particular specimen of PCD the crucial factor for its satisfactory performance is the strength of the bonding between the grains, and we describe various studies of this bonding. As an introduction, however, we describe some

Table 12.1 Some properties of cobalt based PCD material together with those of diamond and tungsten carbide. The figures give typical approximate values at room temperature, see Section 12.2

	PCD	Diamond	TC
Density (kg m^{-3})	4100	3520	15000
Young modulus (Gpa)	800	1000	600
Velocity longitudinal sound (m s^{-1})	16000	17500	6700
Compressive strength (GPa)	7.4	9	5
Transverse rupture strength (GPa)	1.2		1.7
Fracture toughness (MPa m$^{1/2}$)	9	3.4	11
Knoop hardness (GPa)	50	50–100	50
Thermal conductivity (W m^{-1} K^{-1})	500	1000	100
Thermal expansion (10^{-6} K^{-1})	4.0	2–5	5.4

studies of the condition of the diamond grains after sintering. We will generally be concerned with the most usual type of PCD with a cobalt solvent/catalyst but some reference will be made to other types.

12.1.a Plastic deformation

The high pressures and temperatures used to compact the grains together during the sintering process produce considerable plastic deformation within the grains. Figure 12.1 shows an optical micrograph of a polished surface of PCD where the grains exhibit well marked deformation bands (which are rather similar to those observed on somewhat larger crystals by DeVries (1975). The deformations have been studied in detail by examining thin sections of PCD in the transmission electron microscope (Walmsley and Lang, 1983; Yazu *et al.*; 1983, Lee, DeVries and Koch, 1986). These authors report the common occurrence of twinning, dislocations, slip bands, and forests of dislocations formed by interactions between dislocations, particularly near the points of contact between the grains.

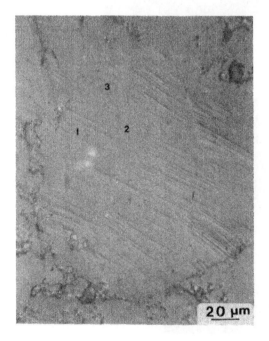

Figure 12.1 Optical micrograph of a polished surface of a heavily deformed diamond particle. 1, 2 and 3 indicate three intersecting sets of deformation bands (Lee, DeVries and Koch, 1986)

Figure 12.2 shows a TEM micrograph of a grain of typical PCD material with a high concentration of dislocations aligned on {111} slip planes. (The dark objects in the lower left of the picture were identified by X-ray analysis as globules of cobalt about 100 nm in diameter.) Figure 12.3 shows a similar micrograph of another grain which is unusually free of dislocations, individual dislocations being clearly visible. Figure 12.4 gives a dark field TEM micrograph of a grain exhibiting at least three primary twin bands running horizontally with a number of secondary twins inside them. Figure 12.5, another TEM micrograph, shows a group of narrower twin bands. (Relatively broad twin bands can be identified by their

Figure 12.2 TEM micrograph showing dislocation slip systems in a crystallite of diamond. The orientation of the surface of the thin specimen is close to {011}. Field width 2 μm. (Walmsley and Lang, 1983)

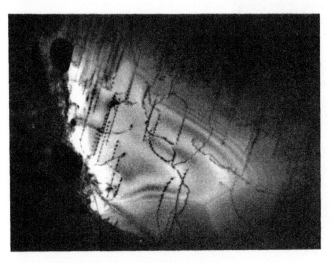

Figure 12.3 TEM micrograph showing a region of low dislocation density in a crystallite of diamond. Field width 3 μm. (Walmsley and Lang, 1983)

characteristic X-ray diffraction patterns. Walmsley and Lang (1983) have used a dark field TEM technique to distinguish between slip bands and twin bands no more than 1.0 nm to 1.5 nm wide.)

The condition of the grains has also been studied by observing the luminescence of PCD samples when excited by the 488 nm and 514.5 nm lines of an argon laser (Davey, Evans and Robertson, 1984; Evans, Davey and Robertson, 1984). Figure

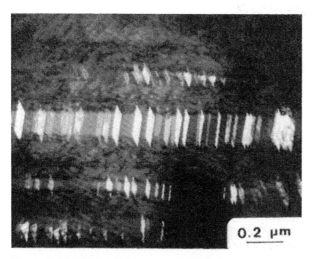

Figure 12.4 TEM dark field micrograph showing primary twin bands (horizontal) with large numbers of secondary twins (vertical) within each band (Lee, DeVries and Koch, 1986)

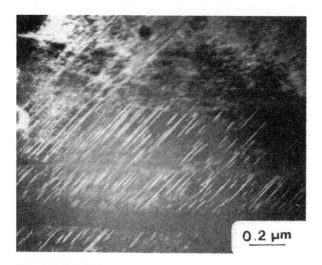

Figure 12.5 TEM dark field micrograph showing a set of narrow secondary twins (Lee, DeVries and Koch, 1986)

12.6 shows the luminescence spectra given by PCD materials produced by sintering 8 μm to 25 μm diamond powder at 900°C at various pressures, together with the spectra of a sample of the original powder. The results are normalized to the first order Raman peak (Section 3.1.c) and show that the luminescence is greater after sintering at higher pressures.

Evans, Davey and Robertson note that the emission spectrum of the unsintered material in Figure 12.6 shows small peaks at 1.945 eV and 2.463 eV, corresponding with the zero-phonon lines of two of the vibronic systems in Table 3.3, and a further line at 575 nm, which is probably formed by combinations of nitrogen atoms and vacancies. The authors suggest that the luminescence is greater in the sintered

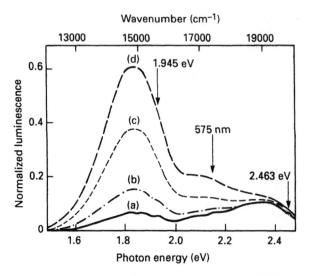

Figure 12.6 Normalized photoluminescence spectra of PCD material fabricated at 900°C under pressures of: (b), 2.6 GPa; (c), 4.6 GPa; (d), 9.9 GPa. Spectrum (a) shows the photoluminescence exhibited by the diamond powder used to fabricate the PCD (Davey, Evans and Robertson, 1984)

specimens because dislocation interactions during sintering produce vacancies which combine with nitrogen atoms to form optical centres. The authors also note that greater deformation causes the zero-phonon lines of the vibronic bands to broaden and then become indistinguishable, apparently because of inhomogenous stresses set up during the sintering process (cf.Section 4.3.b). The presence of internal strains is also reflected in a broadening of the first order Raman line.

Evans, Davey and Robertson (1984) also observed that both the intensity of the luminescence and the broadening of the zero-phonon line were more marked near the points of contact of the grains where the deformation was most intense, and also that greater plastic deformation was present in material of smaller grain size. The ratio of the luminous intensity from the 2.464 eV and 1.945 eV peaks appeared to depend on the pressure and temperature during sintering, so the authors suggest that this ratio could be used to detect pressure gradients in the high pressure cell during sintering. Similar changes to the above have been observed in cathodoluminescence spectra (Collins and Robertson, 1985).

12.1.b Bonding between grains

Some idea of the degree of bonding between the grains may be obtained by leaching away the solvent phase with a suitable solvent such as aqua regia and viewing in the SEM. Figures 1.10 (b) and 12.7 give micrographs of a leached sample at two magnifications and show that the original crystals have grown into an irregular interlocking mass. Details of this intergrowth between grains have been studied by viewing thin sections of PCD in the TEM (Walmsley and Lang, 1988a),

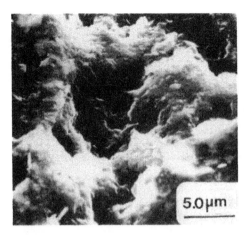

Figure 12.7 SEM micrograph of cobalt based PDC after leaching for 18 hours in aqua regia (Hitchiner, Wilks and Wilks, 1984)

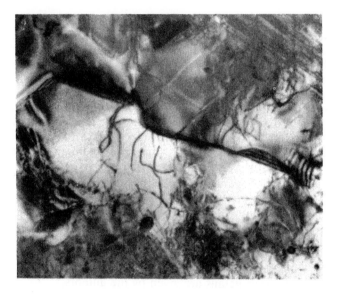

Figure 12.8 TEM micrograph of a region of PCD containing low-dislocation-density regrown diamond, see text. Field width 2.6 μm. (Walmsley and Lang, 1988a)

and Figure 12.8 shows a micrograph which distinguishes between the original grains of diamond and the material grown during the sintering process. The dark regions at the top and bottom of the picture correspond to parts of two adjacent grains which have been heavily deformed during the sintering. The boundaries of these grains are irregular and the gap between them is bridged by two volumes of regrown diamond showing only a small number of dislocations.

We note that the low level of dislocations in the junctions between the grains (Figure 12.8) indicates that during the sintering process (Section 1.3) some of the heavily deformed grains must dissolve in the cobalt and then recrystallize as new

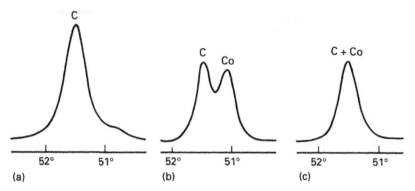

Figure 12.9 The effect of sintering temperature on the shape and position of 111 X-ray diffraction peaks from PCD material; (a), (b) and (c) refer to progressively higher temperatures, see text (Tomlinson and Wedlake, 1983)

diamond relatively free of deformation. Studies in the TEM also show that the regrown volumes of diamond may contain small inclusions of cobalt with preferred orientations relative to the diamond lattice (Walmsley and Lang, 1988a). Less commonly, the regrown material may sometimes have trapped small randomly oriented inclusions of graphite.

The processes of diamond dissolution and regrowth during sintering may be monitored by the observation of X-ray diffraction patterns (Tomlinson and Wedlake, 1983). The first stage of the sintering process (Section 1.3), before the temperature has risen sufficiently to melt the cobalt, is marked by a broadening of the diamond line due to some deformation of the grains, and the appearance of a peak at a somewhat lower angle due to graphite produced on the surface of the grains (Figure 12.9(a)). As the temperature increases the cobalt melts and then dissolves the graphite. Hitherto the cobalt peak has by chance coincided with the diamond peak but the dissolved graphite causes it to shift to a lower angle and it can now be identified (Figure 12.9(b)). At the same time the intensity of the graphite peak decreases. At still higher temperatures all the carbon precipitates from the cobalt as regrown diamond, and the cobalt line again coincides with the diamond line (Figure 12.9(c)). Further details are given by the authors.

12.2 Strength

As mentioned above PCD comes as a range of proprietary materials and different products may differ in ways not fully specified. However, by surveying the available information, particularly that given by General Electric and De Beers Industrial Diamond Division one can form quite a well defined picture of the standard form of PCD material. The results of various measurements described below are summarized in Table 12.1 which gives typical values for a cobalt based PCD material and values for single crystal diamond and tungsten carbide for comparison. These values are not quoted with any precision because they will vary somewhat depending on the particular compositions and grain sizes of the PCD and tungsten carbide. (In addition the values for single crystal diamond depend on the crystallographic orientation of the diamond, so the figures quoted are again representative values.)

12.2.a Elastic constants

The Young modulus has been determined by observing the compression of discs of PCD and by observing the three point bending of a disc as shown schematically in Figure 12.10 (Gigl, 1979; Roberts, 1979). A typical value of the modulus is given in Table 12.1, and also a figure for the velocity of longitudinal sound waves (Roberts, 1979). The Young modulus of PCD is about 20% less than for single crystal diamond, partly of course because of the smaller diamond content in the PCD. The velocity of sound is about 10% less than in the single crystal, as we might expect because the velocity is proportional to the square root of the modulus. Roberts (1979) also measured the Poisson ratio of PCD material in compression tests, using strain gauges to determine the lateral strains. He obtained a value of about 0.22 which together with the Young modulus leads to values of the bulk and shear moduli.

12.2.b Fracture strength

The compression tests described in the previous section were also used to determine the strength of the material, that is the pressure required to produce fracture (Gigl, 1979; Roberts, 1979). The values in Table 12.1 show that the compressive strength of PCD material is of the order of 80% of that of single crystals measured in the same way. (Much higher values of the compressive stresses are observed in experiments with anvils described in Section 7.3.a because anvils are designed to avoid the excess tensile stresses which arise at the edges of the specimen in a simple compression test.)

Gigl (1979) also determined the transverse rupture strength by increasing the loading in three point bending tests to cause failure in tension (Table 12.1). No value is given in Table 12.1 for the transverse rupture strength of single crystal diamond as this parameter is extremely sensitive to particular flaws in the material. In quite different experiments Musikant, Sullivan and Hall (1979) measured the resistance of PCD to the impact of silicon and nylon beads projected at speeds of the order of 5000 ms^{-1}.

Load P

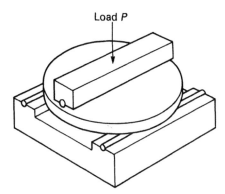

Figure 12.10 Diagram to show method of loading a specimen for a bend test (Roberts, 1979)

Table 12.1 also gives values for the Knoop hardness based on figures given by Roberts (1979) and Brookes and Lambert (1982). We have already referred in Section 8.2.b to the difficulty of interpreting Knoop indentations in single crystal diamond at room temperature, so too much significance should not be read into the exact values of the figures. However, the test indicates that the PCD is somewhat less hard than single crystals.

Notsu, Nakajima and Kawai (1977) and Akaishi *et al.* (1987, 1988) have carried out experiments which show how the hardness may depend on the sintering conditions. In particular by reducing the cobalt content to 2.5% (atomic) and sintering at temperatures up to 2000°C Akaishi *et al.* (1987) obtained Knoop hardnesses as high as 100 GPa to 150 GPa. Values for the Vickers hardness as a function of temperature are given by Lee and Hibbs (1979).

12.2.c. Fracture toughness

The fracture toughness K_c is a measure of the reluctance of a crack to propagate in a material and may be determined by observing how a crack of known geometry extends as a function of the applied stress, see for example Parker (1981). This toughness is closely related to the fracture and cleavage energies described in Section 7.2.b, and it is readily shown that for a simple system of strain (plane strain)

$$K_c^2 = 2E\gamma \tag{12.1}$$

where E is the Young modulus and γ the fracture energy, see for example Parker (1981) or Kelly and Macmillan (1986).

The fracture toughness of single crystal diamond may be found from Equation (12.1) using values of the fracture energy measured by Field and Freeman (1981) who observed the extension of a cone crack under an increasing load (Section 7.5.d), but in an opaque material such as PCD the crack is not visible. Therefore the toughness of PCD has been measured by methods which observe the stress required to fracture specimens of a particular geometry. Values obtained by Lammer (private communication) using a method described by Yarema (1976) and Devin *et al.* (1982) are quoted by Heath and Nicols (1986) and included in Table 12.1.

The values of the elastic constants and the strength of PCD material are somewhat less than those for single crystal diamond, but the fracture toughness of PCD is almost three times that of single crystal diamond and almost equal to that of tungsten carbide. It is this combination of high strength and high toughness which makes PCD such a valuable material. The greater toughness arises primarily because the random orientation of the grains restricts the passage of a fracture from one grain to the next. (Deformation twins produced within the grains during sintering (Section 12.1.a) may also hinder the progress of a cleavage (Lee and Hibbs, 1979).)

Finally we note that PCD is often grown on a relatively thick base of tungsten carbide to which it is firmly bonded. In this case the strength and toughness of the PCD is greatly augmented by the strength and toughness of the carbide.

12.2.d Thermal stability

We now consider the response of PCD material to the high temperatures which may be encountered during the manufacture of a tool and under working conditions. As described in Section 13.3.a, a single crystal diamond begins to oxidize and graphitize when heated in air above 600°C but can be heated to much higher temperatures in a protective atmosphere such as an inert gas, a reducing atmosphere of hydrogen, or in a vacuum. With PCD material the presence of the solvent /catalyst is a further cause of instability. At temperatures above about 700°C the solvent/catalyst in contact with the diamond promotes some graphitization which results first in a reduction in strength and finally to the complete degradation of the material. The solvent/catalyst is also responsible for a second undesirable effect. Because diamond has a very low coefficient of thermal expansion, much lower than that of cobalt, heating the PCD produces differential expansions and internal stresses, additional to those caused by the volume changes associated with any graphitization.

Detailed studies of the thermal deterioration of PCD have been made by Mehan and Hibbs (1989) who used optical microscopy to observe severe deterioration in the structure of the material after heating to temperatures above 800°C, the amount of damage depending on the time the PCD was held at the high temperature. These authors also detected the presence of acoustic emissions while the PCD was being heated, and observed that the number of emissions per unit temperature rise increased greatly as the material began to deteriorate. In addition Mehan and Hibbs detected a low level of acoustic emission when heating a coarse grained (170 μm) PCD at lower temperatures between 400° and 600°C even though no degeneration was visible in the microscope. (However, this last effect was not observed with material of fine grain size (1.5 μm).)

The thermal deterioration of PCD has also been studied by Bex and Shafto (1984) with particular reference to the manufacture of tools involving the brazing of PCD to metal shanks and blades. These authors measured the wear of PCD turning tools which had been held at various temperatures for periods of 1 min, 5 min and 10 min. No loss of performance was observed after heating for 10 min at 700°C or for 1 min at 800°C but the wear increased very considerably for similar periods at higher temperatures. (The acceptable values of time and temperature will of course depend on the particular brand of PCD involved.)

12.2.e Thermally stable PCD

At least two variants of the standard cobalt based PCD have been developed to give greater thermal stability. The first variant is obtained by leaching out the cobalt from standard PCD using a suitable acid solvent so as to leave only the diamond matrix. Figures given by General Electric state that this material is thermally stable up to 1200°C, and that the leaching reduces the elastic modulus and strength by only about 10%. Unfortunately it becomes progressively more difficult and time consuming to leach out the cobalt as the thickness of the PCD samples increases.

Another approach to greater thermal stability is to use silicon rather than cobalt as the second phase (Tomlinson et al., 1985). During the sintering process the silicon reacts with the diamond to form silicon carbide and this carbide acts as a binding agent but without promoting any graphitization. The optical micrograph in Figure 12.11 shows the surface of a specimen of the De Beers product Syndax© 3 after leaching to remove the second phase, and shows that the grains of diamond are clearly distinct from each other. TEM studies by Walmsley and Lang (1988b) indicate the presence of silicon carbide between the grains of diamond and show

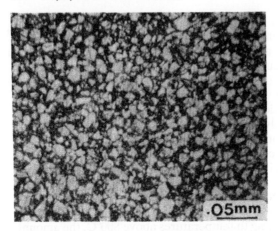

Figure 12.11 Optical interference micrograph of a leached specimen of a silicon based PCD material

no sign of any diamond–diamond intergrowth. It is found that this type of material is stable to about 1200°C. Unfortunately another effect of the use of silicon is that the strength of the PCD is reduced below that of cobalt based material by about 35%. The toughness is also reduced but is still twice that of single crystal diamond. Further details are given by Tomlinson *et al.* (1985).

12.2.f Fatigue

We have already described experiments which demonstrate fatigue processes in single crystal diamond (Section 7.7). That is, repeated stressing of a diamond by loads below the critical fracture load can lead to internal damage and eventually fracture. As PCD materials are essentially masses of crystalline diamond we would expect them to exhibit rather similar fatigue effects.

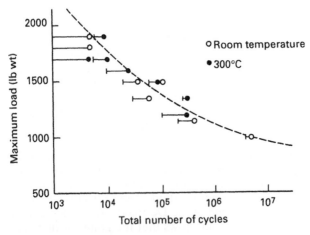

Figure 12.12 Cyclic fatigue test of PCD material giving the number of cycles to failure as a function of the maximum load (Dunn and Lee, 1979)

The most complete study is that of Dunn and Lee (1979). These authors used test specimens 13 mm in diameter consisting of a layer of PCD 0.5 mm thick bonded on to a tungsten carbide substrate about 2.8 mm thick. The specimens were mounted in a dynamic test machine which forced the edge of the disc against a block of tungsten carbide, the geometry of the system being chosen to simulate to some extent the impacts suffered by PCD during rock cutting operations. Each run was continued until the edge was visibly chipped or (in one case) a visible crack appeared on the face of the PCD. The results are given in Figure 12.12 which shows the number of cycles to fracture for a range of loads. Measurements were made at room temperature and 300°C and no significant difference was observed between the two conditions. The authors give micrographs of the fracture surface which show obvious signs of debris generated during the fatigue process.

12.3 Shaping and polishing PCD

PCD is supplied in blanks of various geometries which must be shaped and polished in order to form tools. Generally speaking the work is done in two stages, a rough grinding or polishing to generate the correct shape, and a final polish to give a fine finish. These operations are not entirely straightforward as PCD is extremely resistant to polish. Single crystal diamond may be polished fairly readily by choosing a direction of easy abrasion and avoiding other directions, but the individual grains of diamond in PCD are randomly oriented, so there will always be many grains which present themselves in orientations which are difficult to polish.

12.3.a Mechanism of polish

The first attempts to polish PCD using diamond powder on a cast-iron wheel often produced only very low rates of removal of material. Moreover, these rates often appeared to be almost independent of the speed of the wheel and sometimes even decreased with rising wheel speed. Such results are in complete contrast to the behaviour usually observed when polishing single crystals, and we now describe experiments made to investigate these points further (Hitchiner, Wilks and Wilks, 1984).

Blocks of PCD of grain size of about 10 μm or 25 μm, and of area about 15 mm². were polished on a cast-iron scaife with powder, and the removal rate of material measured as described in Section 9.2.a. The specimen was automatically traversed to and fro across the scaife for a distance of about 4 mm on either side of its mean path in order to avoid grooving the wheel. The mean radius of its track was about 36 mm, and the speed of the wheel could be adjusted from about 50 rev min⁻¹ to 3000 rev min⁻¹. The usual load on the specimen was the weight of the holder or dop moving freely in its slide, equal to 2015 g wt.

It was soon observed that loss of powder from the scaife which is an important factor when polishing single crystal diamond in a hard direction (Section 9.3.b), is particularly important when polishing PCD. In one experiment the cast-iron scaife was periodically recharged with 8 standard size drops of 5 μm–10 μm diamond powder in olive oil. Table 12.2 shows the number of revolutions of the wheel between charges and the rate of removal expressed as the volume removed for each revolution of the wheel. We see that the volume removed per revolution was

Table 12.2 Effect on removal rate of recharging cast-iron scaife when polishing PCD of 25 μm grain size (load 2015 g wt)

Wheel speed (rev min^{-1})	Time between charges (s)	Number of revolutions	Volume removed per revolution ($\times 10^{-6}$ mm^3)
2920	240	11680	0.8
	5	243	15.0
1500	240	6000	0.9
	10	250	10.7
450	240	1800	9.1
	30	225	13.1

Table 12.3 Removal rates of diamond when polished on a cast-iron scaife and on a bonded wheel for: PCD (A75) of 75 μm grain size, PCD (AX4) and PCD (DX1) of 25 μm grain size, an {011} face of a single crystal diamond. (Wheel speed 2932 rev min^{-1}; load 2015 g wt)

Material	Direction	Removal rates ($\times 10^{-4}$ mm^3 s^{-1})	
		Cast-iron wheel	Bonded wheel
PCD (A75)			0.04
PCD (AX4)		12.9	0.04
PCD (DX1)		19.7	
Single	<100>	5.9	3.3
crystal	<110>	0.66	0.14

much greater when the scaife was recharged more frequently. However, provided that recharging took place every 240 rev, the volume removed per second was roughly independent of the speed as we would expect when polishing single crystal diamond. Hence, it appears that charging every 240 rev was sufficient to maintain the supply of powder, and further experiments showed that, once this condition was reached, more frequent recharging made little difference to the removal rate.

Some typical values for the rate of removal of PCD material in the above experiments using a well replenished cast-iron wheel are given in Table 12.3 which also gives the removal rates obtained on single crystal diamond under similar conditions. Before discussing these results we note also that specimens of PCD were also polished on a bonded wheel with a working surface containing ~30 μm diamond powder in a metal matrix (erroneously described as 4–8 μm in Hitchiner, Wilks and Wilks (1984)). In this case the rates of removal were proportional to the speed of the wheel over a wide range of speeds but were much less than the rates obtained with a cast-iron scaife, as shown in Table 12.3.

When samples of polished PCD are examined in the microscope there is virtually no loss of material due to pull-out of the individual crystallites, which appear firmly bonded together. Therefore, polishing must proceed by the abrasion of the individual grains, which no doubt have almost random orientations. The solvent/catalyst phase has a relatively low abrasion resistance but this hardly affects the overall removal rate because the PCD is mainly diamond. Therefore, we would expect the removal rate for the PCD to have some value intermediate between the rates for polishing single crystals in their softest and hardest directions. That is, we might expect the removal rates for PCD on a cast-iron scaife to lie between the values 5.9×10^{-4} and 0.66×10^{-4} mm^3 s^{-1} for single crystal diamond in Table 12.3

but in fact the removal rates given by the cast-iron wheel are appreciably greater than the maximum rate for single crystals.

One possibility to account for the high rate of removal is that the presence of the softer metallic phase may have some significant effect on the polishing process. Therefore the polishing rates for two samples of PCD were measured and the specimens then etched in boiling aqua regia for 18 h until the binder had been removed to a depth of at least 14 μm. The effectiveness of the etch was verified by making abrasion cuts on the etched surface and then viewing the bottom of the cuts in the scanning electron microscope to verify that the binder had been leached out. Polishing rates from the etched surfaces of the two specimens were then measured in the usual way and showed that the removal of the binder had no effect on the polishing rate for one of the specimens and led to a higher rate for the other. Hence, the binder cannot be held responsible for the high rate of wear of PCD relative to that of single crystal diamond.

An explanantion for the anomously high rate of polish of PCD on a cast-iron scaife is suggested by some experiments on the polishing of single crystal diamond. Figure 12.13 is a micrograph of a polished {011} surface of a diamond which had been prepared initially by making three abrasion cuts in order to observe the rate of removal from the surface. Figure 12.13 shows the surface after it had been polished for 8 min at a speed of 450 rev min^{-1} under a load of 2015 g wt, the scaife being recharged in the usual way. The direction of polish was in the easy direction indicated by the arrow in the figure, parallel to the length of the abrasions. The polish on the surface of the diamond is fairly uniform except that beyond the lower end of the abrasions there are regions which are appreciably lower than the rest of the surface by about 1 μm or 2 μm. Material has been removed faster in these areas, almost certainly because the abrasion cut permits the powder on the scaife to abrade the edge as well as the surface of the diamond. Similar effects were observed at other speeds and loads, and when polishing in hard directions of the diamond (Hitchiner, Wilks and Wilks, 1984).

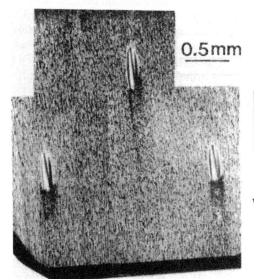

0.5mm

Figure 12.13 Optical interference micrograph of a polished {011} surface of a single crystal diamond, showing additional removal of material at the lower edges of the abrasion cuts. The arrow shows the direction of polish (Hitchiner, Wilks and Wilks, 1984)

The above example shows that the abrasion of a diamond surface proceeds more rapidly if the surface can be attacked from the edge as well as from on top. (It is, of course, well known to polishers that the edge of a diamond is particularly susceptible to damage.) The surface of PCD consists of islands of small diamond crystals surrounded by a much softer metallic phase. Hence the diamond powder has the possibility of sinking into the softer second phase as it moves across the surface, and thus of attacking the diamond crystallites at their edges. Hence, we expect a considerable enhancement of the rate of removal compared with the value obtained on large surfaces of single crystal.

When PCD is polished on a bonded wheel the position is quite different. The diamond powder is held rigidly in the plane of the wheel and has a much reduced possibility of moving down into the binder. Hence, we might expect the removal rate for PCD to have a value intermediate between the rates for the hard and soft directions of single crystals polished on a bonded scaife. In fact, the observed rate is considerably less than either of these values (Table 12.3). This inferior performance has not been investigated in detail but probably arises because diamond powder in the wheel is bonded in a matrix of cobalt which will tend to adhere to the binder in the PCD. This may well result in metal being smeared over the diamond grits and reducing their effectiveness.

Although the rate of removal is less on the bonded wheel, the surface finish produced is much finer. Polishing on the cast-iron scaife produces a typical matt appearance, whereas surfaces polished on the bonded wheel have a shiny appearance with the boundaries of the crystallites more clearly defined. This finer finish almost certainly arises because, as discussed above, the diamond powder in the bonded wheel is less effective in abrading the crystallites of diamond, so that material is removed on a smaller scale.

12.3.b Practical details

As discussed in the previous section much faster rates of polish may be obtained with loose diamond powder on a cast-iron scaife than with diamond bonded wheels. However, the loss of powder from the scaife is so considerable that the method is only practical if the powder is recovered and used again. Therefore most polishing today is done with diamond bonded wheels to avoid this problem. Even so there is now at least one polishing machine on the market which appears to operate by recovering and recirculating loose powder in a suitable fluid (Stähli and Bourquin, 1989).

Although bonded wheels are widely used to polish PCD there is no general agreement on which type of bond, vitreous, resin or metal gives the best results, see for example comments by Heath and Aytacoglu (1984), Metzger (1986) and Werner and Kenter (1989). This variety of opinion arises partly because the choice of a wheel involves various and not necessarily consistent criteria. One operator may desire the maximum removal rate, another the best edge quality, and a third that the wheel needs a minimum of dressing. There are also differences between the various types of PCD including in particular the nature of the solvent/catalyst. Therefore the choice of bond must take account of the details of the particular task to be performed.

Despite some uncertainty regarding the best type of bond, there is general agreement on several principles of operation to be followed when working and shaping PCD. To overcome the relative slow rates of polish the wheels are run at high speed and the specimen is loaded hard against it with a constant force of up to perhaps 500 N (Shimaoka, Tomimori and Kawakita, 1982; Heath and Aytacoglu, 1984; Werner and Kenter, 1989). It is also essential either to oscillate the wheel over the specimen or to use a planetary motion to ensure that the wheel wears uniformly (Metzger, 1986). The large forces and speeds involved tend to produce vibrations which must be avoided by a stiff and massive construction of the grinding machine, see for example Tönshoff, Bussmann and Stanske (1987). The heavy loading of the PCD on the wheel makes it imperative to flood the wheel with an adequate coolant. Finally the wheel must be selected to be free cutting (Section 15.2.c), and any tendency to glaze over countered by frequent dressing.

Using wheels with grits of the order of 50 µm the finish of a shaped surface may be good enough for some purposes but is usually improved by further polishing to give a smoother finish. This is obtained by polishing under gentler conditions, using a wheel with smaller grit size of the order of 8 µm–16 µm, running at a lower speed, under a lower load, on a stiff and well constructed polishing machine. To obtain the smoothest finish some authors recommend the use of still finer powder on a cast-iron scaife (see for example Heath and Aytacoglu (1984)).

12.3.c Electrical discharge machining

Another possible way of shaping a PCD blank is by the use of electrical discharge machining (EDM) often referred to as spark erosion. This is a technique used to deal with any hard material which has a finite electrical conductivity, see for example Barash (1969) and Boothroyd (1975). Using this approach the normal grinding wheel is replaced by a graphite wheel separated from the workpiece by a gap of the order of 10 µm. High voltage pulses applied to the wheel produce sparking across to the workpiece, thus removing material both from the workpiece and to some extent from the graphite wheel (which is rotated to ensure that it wears evenly). To produce more complex shapes a wire may be used as the electrode, this wire being pulled along under tension between reels in order to maintain even wear.

Until recently the main application of spark erosion to the working of diamond was the drilling of single crystal diamond to produce wire drawing dies (Section 9.6.b) but the technique is now being used to machine PCD material. A machine with a graphite wheel for the spark grinding of PCD is described by Silveri (1986), and a technique of shaping PCD with a wire electrode by Spur, Puttrus and Wunsch (1988).

Electrical discharge machining has the considerable advantage that because there is no contact between tool and workpiece no high forces are involved. There is no need for the extremely massive type of machine normally used to grind PCD nor for a high power motor. On the other hand spark erosion proceeds by the high energy of a spark causing local evaporation and there is inevitably a residue of damage left in the surface as shown in the micrographs of Spur, Puttrus and Wunsch (1988). These authors also show that the use of higher currents results in faster machining but at the expense of more damage. Therefore shaping by spark erosion will generally be followed by a fine polish on a normal wheel to remove the damaged layer extending perhaps to a depth of 50 µm or 100 µm. Spur, Puttrus and Wunsch also discuss how the rate of erosion depends on the structure and grain size of the PCD and other variables.

12.4 Friction and wear

The friction of PCD sliding on other materials is a more complex process than the sliding of single crystal diamond because the solvent/catalyst may adhere to the other surface and modify the form of the contact. Therefore we expect the value of the friction to depend considerably on the nature of the other material, on the type of PCD involved, as well as on the conditions of sliding such as load, speed and temperature. We describe three rather different sets of experiments.

Mehan and Hayden (1981) observed the friction and wear when blocks of cobalt type PCD about 6 mm wide were held with a normal force of 223 N against a steel ring about 35 mm in diameter rotating at 50 rev min^{-1}. Measurements were made of the friction force and of the wear on both the PCD and the ring, using three samples of PCD of grain size approximately 2 μm, 10 μm and 25 μm, and rings of two types of steel. The PCD surface was either lapped to a surface roughness with CLA values in the range 0.33 μm to 0.73 μm, or ground to CLA values in the range 0.10 μm to 0.22 μm.

The PCD materials of different grain size behaved quite similarly, but there were large differences between lapped and ground materials. On hard steel running dry, the coefficients of friction for lapped surfaces were approximately 0.75 but for ground surfaces only about 0.15. In addition there was appreciable wear of both ring and PCD with the lapped PCD, but almost zero wear with the ground PCD. Hence there were clearly two distinct regimes, one where the rugosities on the rougher PCD surfaces gouged the steel ring and produced visible debris, and the other where the action of the ground surfaces were much less destructive. It was also observed that the presence of a mineral oil reduced the coefficient of friction for both lapped and ground surfaces to around 0.04. Further details are given in the original paper.

In another set of experiments Feng and Haywood (1988) measured the friction force on styli of copper, cobalt and tungsten carbide when sliding over blocks of cobalt based PCD of either 10 μm or 25 μm grain size. These PCD blocks were commercial material in the as received state with a CLA surface roughness of 0.2 μm for the smaller grain size and of 0.5 μm or 1.0 μm for the larger size. The styli had the form of cones with a 90° included angle and a tip which at the start of an experiment was either flat or hemispherical and about 30 μm in diameter.

Measurements were made under a load of 0.5N, at sliding speeds of about 0.1 mm s^{-1}, for either 20 or 100 cycles of reciprocating motion. Micrographs of tungsten and copper styli after 100 cycles show wear flats of diameter 100 μm and 300 μm respectively, both flats being extensively grooved. Hence, the surface conditions appear somewhat similar to those observed with the rougher lapped specimens used by Mehan and Hayden. Figure 12.14 shows results for copper styli sliding on three PCD surfaces of different roughness as a function of the number of cycles; the friction rising somewhat with the number of cycles.

Another set of experiments used similar apparatus to that of Feng and Hayward (1988) but with the styli of PCD in the form of Knoop indenters sliding over flat surfaces of hard and annealed D3A steel. (Freeman, 1984; Freeman and Field, 1989). Some results are shown in Figure 12.15 and we see that they are quite complex. (Note, however, that we do not expect the friction of a PCD stylus sliding on steel to be the same as that of a steel stylus sliding on PCD because the materials are of quite different hardness and this will much affect the geometry and behaviour of the contact.) The authors suggest that the increases of friction in

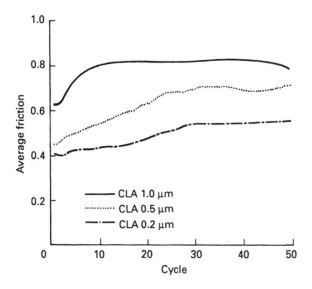

Figure 12.14 The average coefficient of friction for copper cones sliding on specimens of PCD material of different roughness as a function of the number of cycles of sliding (Feng and Hayward, 1988)

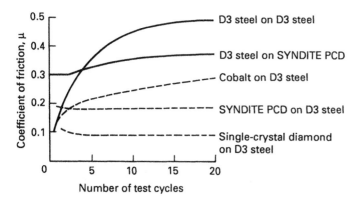

Figure 12.15 The coefficient of friction for various pairs of material as a function of the number of test cycles (Feng and Hayward, see Lammer, 1988)

Figure 12.14 arise from a transfer of copper to the PCD which results in relatively strong adhesion between copper and copper. On the other hand the largest rise in friction in Figure 12.15 occurs for a steel stylus sliding over a flat of similar material. To interpret these results we probably need more information on the nature of any transferred material and of any debris produced. For example Freeman and Field (1989) report that in the experiments on sliding on steel there was significant transfer of steel to a PCD stylus but not to a single crystal diamond.

12.5 Electrical and thermal conductivity

The metallic second phase in PCD material gives rise to an appreciable electrical conductivity corresponding to a resistivity of the order of 10^{-4} Ω m^{-1} (McLachlan, 1984). The metallic phase forms a continuous net, so the magnitude of the conductivity is determined primarily by the size of the finer channels between the crytallites of diamond. For a discussion of how the resistance is expected to depend on the relative fractions of catalyst and diamond see McLachlan (1984). As noted in Section 12.2.b Akaishi *et al.* (1987, 1988) have produced PCD material with an enhanced Knoop hardness by using a low concentration of cobalt and sintering at temperatures up to 1200°C. This material also exhibited a much reduced electrical conductivity. The authors quote a value for the resistance of about 10^8 Ω m^{-1}, compared with typical values of 10^{-4} for conventional cobalt based PCD, and with values of 10^3 to 10^5 for Type IIb diamond and of more than 10^{18} for Type I and Type IIa diamond.

We also note that if the second phase in the PCD is ferromagnetic then the PCD will exhibit some form of ferromagnetism. Detailed observations of this magnetism have been made by Notsu, Nakajima and Kawai (1979) on PCD specimens sintered with cobalt. These authors observed the remanent magnetism left in the specimens when applied magnetic fields of up to 400 Oe (5 A m^{-1}) were reduced to zero. The degree of remanent magnetism appeared to depend on the particular conditions of sintering and was greater in specimens exhibiting greater values of the Knoop hardness.

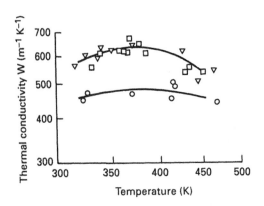

Figure 12.16 The thermal conductivity of three specimens of PCD material of different grain size; ($\square$), 100 μm; (∇), 50 μm; ($\bigcirc$), 10 μm (Burgemeister, 1980)

The thermal conductivity of a cobalt type PCD has been measured by Burgemeister (1980) at temperatures in the range 320 K to 450 K and his results for three specimens of different grain size are shown in Figure 12.16. As might be expected the conductivity is appreciably less in the specimens with the smaller grains. The values near room temperature are only about half those of the single crystal diamonds given in Figure 6.30. In other experiments Burgemeister and Rosenberg (1981) observed that leaching out the cobalt reduced the conductivity by only about 5% to 10%, showing that the heat flow in PCD is carried mainly by the diamond. Hence the lower thermal conductivity of PCD compared with diamond must arise primarily from the reductions of cross-sectional area in the regions of contact between the grains.

Finally we note that other forms of PCD material with different solvent/catalysts may have little or no electrical conductivity, as for example material produced with silicon. As described in Section 12.2.d, the grains of diamond in this latter material are bonded together by the growth of silicon carbide. This carbide produces an appreciable resistance to the flow of heat so the thermal conductivity of the PCD at room temperature is only of the order of 100 W m^{-1} K^{-1} (Tomlinson *et al.*, 1985).

12.6 Thin films

We now consider the mechanical properties of the thin films of polycrystalline diamond now being grown by vapour deposition methods (Section 1.4). One of the principal reasons for the considerable interest in these films lies in the possibility of depositing thin layers of diamond on components of other materials to give wear resistant coatings which would greatly increase the life of the components. Despite

Table 12.4 Depths of successive cuts made to compare the abrasion resistance of a polycrystalline diamond film with: (a), that of an easy direction of abrasion on a polished {001} face; (b), that of an easy direction on a natural {111} face of single crystal diamond (in μm)

1	2	3	4	5	6
{001} / <100>		Film		{001} / <100>	
3.3	3.2	1.7	1.7	2.7	2.7
{111} / <11$\bar{2}$>		Film		{111} / <11$\bar{2}$>	
1.0	0.95	1.7	1.8	1.1	1.05

this interest not much detailed information is yet available on the mechanical behaviour of these thin films.

The abrasion resistance of these films may certainly be comparable to that of single crystal diamond. For example, Table 12.4 shows the result of comparing the abrasion resistance of a polycrystalline film grown on a silicon substrate with that of a single crystal diamond. Using a method similar to that described in Section 10.2.b we compared the resistance of the film with that of the easy direction of abrasion on a polished {001} and on a natural {111} face of a single crystal diamond. The table shows that the film was more abrasion resistant than the easy direction on the {001} face but less resistant than the easy direction on the {111} face. We also note that the abrasion resistance of a film will depend on the conditions of deposition and perhaps on the nature of the substrate, but no information is available on these points.

Although a polycrystalline diamond film may be very wear resistant it may also be very abrasive because its mode of growth results in a surface which may be rough on a scale of microns or less, see Figure 1.16. Some effects of this roughness were observed by Samuels (1987) who measured the friction of a spherically tipped diamond stylus sliding under various loads on as-grown diamond films and on the polished surfaces of single crystal diamonds. After sliding on the polished surface no damage could be seen in the optical microscope either on the surface or on the stylus, but the stylus produced very visible tracks on the surface of the film (Figure 12.17). It was also seen that the effect of repeated passes or a heavier load was to

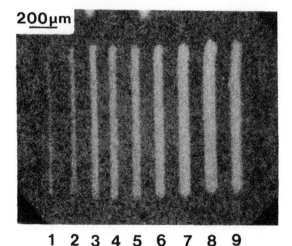

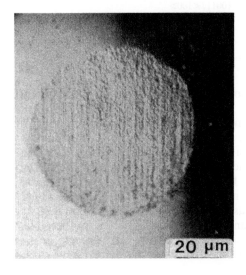

Figure 12.17 Optical interference micrograph of tracks made by a diamond stylus of radius 490 μm sliding on a polycrystalline diamond film (Samuels, 1987)

Track no	1	2	3	4	5	6	7	8	9
Load (g wt)	141	141	141	141	239	389	625	1024	1024
No. passes	1	1	5	10	5	5	5	5	1

Figure 12.18 Optical interference micrograph of the tip of a stylus of radius 490 μm after sliding over a diamond film (Samuels, 1987)

produce some smoothing of the track to give a more reflecting surface. At the same time a considerable flat area was worn on the stylus after only a few passes (Figure 12.18). This wear was also reflected in the value of the coefficient of friction which dropped considerably with repeated passes along the same track (Figure 12.19).

The above results show that for many applications it will be necessary to polish the film to obtain a smooth surface. In this case two precautions are necessary. Because the layer may be only a few microns thick, it must be carefully aligned

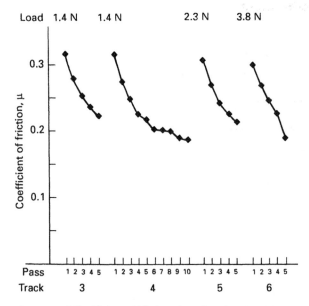

Figure 12.19 Coefficient of friction of a stylus of radius 490 μm sliding on a diamond film under different loads as a function of the number of passes on the same track (Samuels, 1987)

parallel to the polishing wheel to avoid polishing it away. In addition, the polishing must proceed gently as the film may have a tendency to come away, or delaminate, from the substrate. Examples of delamination were seen when thin plates of silicon carrying diamond films were divided by scratching the silicon and bending along the line of the scratch. Sometimes both the silicon and the diamond divided cleanly but on other occasions only the silicon divided along the line of the scratch, while part of the film delaminated and broke elsewhere.

We have also observed delamination while measuring the friction of films. Figure 12.20 shows an optical micrograph of a series of passes of a diamond stylus over a diamond film under increasing loads. The first five tracks are similar to those in Figure 12.17 but the last two show evidence of delamination, the light areas with curved boundaries show where the film has broken away completely and the curved greyish regions where the layer is beginning to come away.

The problem of delamination arises essentially because the arrangement of the atoms in the diamond lattice will not in general match that of the atoms in the substrate, so growth cannot be epitaxial. Not much has been published on these problems no doubt partly because they are crucial to the exploitation of the films. There is not even much information on the strength of the bonding which can be obtained. For example Lewus et al. (1989) refer to scratch tests and wear measurements on coated tools which did not cause any delamination, but without giving any details.

Finally we mention that one of the apparently most detailed studies of the strength of a film is described by O'Hern et al. (1989). These authors used a nano-indenting machine to obtain load–displacement curves when a diamond indenter was pressed against the film to give displacements of up to 300 nm. The load displacement curves for the film and single crystal were fairly similar except that

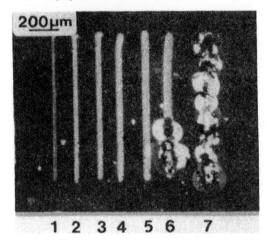

Figure 12.20 Optical interference micrograph of tracks made by a diamond stylus of radius 58 μm sliding on a polycrystalline diamond film (Samuels 1987)

Track no.	1	2	3	4	5	6	7
Load g wt	141	141	239	389	625	1024	1024
No. passes	1	5	5	5	5	2	2

while the displacements on the single crystal were entirely reversible it was observed that on unloading the indenter on the film permanent displacements of about 30 nm were observed. The authors' discussion of the hardness of single crystals when the load–displacement curve was completely reversible is obscure, but the experimental method appears to offer a useful way of assessing the mechanical behaviour of the films. (We also note that the response of thin films to indentation may be much influenced by the mechanical properties of the substrate, see for example Van der Zwaag and Field (1982).)

References

Akaishi, M., Yamaoka, S., Tanaka, J. *et al* (1987) *Communications of the American Ceramic Society,* **70**, C237–C239

Akaishi, M., Yamaoka, S., Tanaka, J. *et al.* (1988) *Materials Science and Engineering,* **A105/106**, 517–523

Barash, M. M. (1969) In *Modern Workshop Technology,* Part II, (ed. H. Wright Baker), Macmillan, London, pp.301–348

Bex, P. A. and Shafto, G. R. (1984) In *Ultrahard Materials Application Technology,* Vol. 3, (ed. P. Daniel), De Beers Industrial Diamond Division, London, pp.22–33, and in *Industrial Diamond Review,* **44**, 128–132

Boothroyd, G. (1975) *Fundamentals of Metal Machining and Machine Tools.* McGraw–Hill Kogakusha, Tokyo

Brookes, C. A. and Lambert, W. A. (1982) In *Ultrahard Materials Application Technology,* Vol. 1, (ed. P. Daniel), De Beers Industrial Diamond Division, London, pp.128–136

Burgemeister, E. A. (1980) *Industrial Diamond Review,* **40**, 87–89

Burgemeister, E. A. and Rosenberg, H. M. (1981) *Journal of Materials Science Letters,* **16**, 1730–1731

Collins, A. T. and Robertson, S. H. (1985) *Journal of Materials Science Letters,* **4**, 681–684

Davey, S. T., Evans, T. and Robertson, S. H. (1984) *Journal of Materials Science Letters,* **3**, 1090–1092

Devin, L. N., Maistrenko, A. L., Simkin, E. S. *et al.* (1982) Translated from *Poroshkovaya Metallurgiya*, No. 5 (233) 88–93, Plenum, New York, pp.419–423

DeVries,R.C. (1975) *Materials Research Bulletin*, **10**, 1193–1200

Dunn, K. J. and Lee, M. (1979) *Journal of Materials Science*, **14**, 882–890

Evans, T., Davey, S. T. and Robertson, S. H. (1984) *Journal of Materials Science*, **19**, 2405–2414

Feng, Z. and Hayward, I. P. (1988) In *Ultrahard Materials Application Technology* Vol 4, (ed. C. Barrett), De Beers Industrial Diamond Division, London, pp.36–46

Field, J. E. and Freeman, C. J. (1981) *Philosophical Magazine*, **43**, 595–618

Freeman, C. J. (1984) *Ph.D. Thesis*. Cambridge University

Freeman, C.J. and Field, J. E. (1989) *Journal of Materials Science*, **24**, 1069–1072

Gigl, P. D. (1979) In *High Pressure Science and Technology*, Vol. 1, (eds K. D. Timmerhaus and M. S. Barber), Plenum, New York, pp.914–922

Heath, P. J. and Aytacoglu, M. E. (1984) *Industrial Diamond Review*, **44**, 133–139

Heath, P. J. and Nicolls, M. O. (1986) In *Proceedings of the Seminar on the Finishing of Surface Coatings*, (Crest Hotel, Coventry 1986), pp.1–27. The Welding Institute, Cambridge

Hitchiner, M. P., Wilks, E. M. and Wilks, J. (1984) *Wear*, **94**, 103–120

Kelly, A. and Macmillan, N. H. (1986) *Strong Solids*, (3rd edn) Clarendon Press, Oxford

Lammer, A. (1988) *Industrial Diamond Review*, **48**, 179–182

Lee, M., DeVries, R. C. and Koch, E. F. (1986) In *Proceedings of the Second International Conference on the Science of Hard Materials (Rhodes, 1984)*, Institute of Physics Conference Series No. 75. (eds E. A. Almond, C. A. Brookes and R. Warren), Adam Hilger, Bristol, pp.221–232

Lee, M and Hibbs, Jr. L. E. (1979) In *Wear of Materials, 1979*, (eds K. C. Ludema, W. A. Glaeser and S. K. Rhee), American Society of Mechanical Engineers, New York, pp. 485–491

Lewus, M. O., Eason, J., Doty, F. P. and Jesser, W. A. (1989) In *Extended Abstracts, Technology Update on Diamond Films*, (eds R. P. H. Chang, D. Nelson and A. Hiraki), Materials Research Society, Pittsburgh, pp.127–130

McLachlan, D. S. (1984) In *Ultrahard Materials Application Technology*, Vol.3, (ed. P. Daniel), DeBeers Industrial Diamond Division, London, pp.34–40

Mehan, R. L. and Hayden, S. C. (1981) *Wear*, **74**, 195–212

Mehan, R. L. and Hibbs, Jr. L. E. (1989) *Journal of Materials Science*, **24**, 942–950

Metzger, J. L. (1986) *Superabrasive Grinding*. Butterworth, London

Musikant, S., Sullivan, R. J. and Hall, K. J. (1979) In *High Pressure Science and Technology*, Vol. 2, (eds K . D. Timmerhaus and M. S. Barber), Plenum, New York, pp.549–558

Notsu, Y., Nakajima, T. and Kawai, N. (1977) *Materials Research Bulletin*, **12**, 1079–1085

Notsu, Y., Nakajima, T. and Kawai, N. (1979) *Materials Research Bulletin*, **14**, 1065–1068

O'Hern, M. E., McHargue, C. J., Clausing, R. E. *et al.* (1989) In *Extended Abstracts, Technology Update on Diamond Films*, (eds R. P. H. Chang, D. Nelson and A. Hiraki), Materials Research Society, Pittsburgh, pp.131–137

Parker, A. P. (1981) *The Mechanics of Fracture and Fatigue*. E. and F. N. Spon, London

Roberts, D. C. (1979) *Industrial Diamond Review*, **39**, 237–241

Samuels, B. (1987) *D.Phil. Thesis*. Oxford University

Shimaoka, H., Tomimori, H. and Kawakita, T. (1982) *Industrial Diamond Review*, **42**, 155–160

Silveri, P. (1986) *Industrial Diamond Review*, **46**, 108–109

Spur, G., Puttrus, M. and Wunsch, U. W. (1988) *Industrial Diamond Review*, **48**, 264–266

Stähli, A. W. and Bourquin, P. (1989) *Industrial Diamond Review*, **49**, 7–8

Tomlinson, P. N., Pipkin, N. J., Lammer, A. and Burnand, R. P. (1985) *Industrial Diamond Review*, **45**, 299–304

Tomlinson, P. N. and Wedlake, R. J. (1983) In *Proceedings of the International Conference on Recent Developments in Speciality Steels and Hard Materials* (Pretoria, 1982), (eds N. R. Comins and J. B. Clark), Pergamon Press, Oxford, pp.173–184

Tönshoff, H. K., Bussmann, W. and Stanske, C. (1987) In *Proceedings of the 26th International Machine Tool Design and Research Conference* (Manchester, 1986), (ed. B. J. Davies), Macmillan, London, pp.349–357

Van der Zwaag, S. and Field, J. E. (1982) *Philosophical Magazine A*, **46**, 133–150

Walmsley, J. C. and Lang, A. R. (1983) *Journal of Materials Science Letters*, **2**, 785–788

Walsmley, J. C. and Lang, A. R. (1988a) *Journal of Materials Science,* **23**, 1829–1834

Walmsley, J. C. and Lang, A. R. (1988b) In *Ultrahard Materials Application Technology,* Vol. 4, (ed. C. Barrett), De Beers Industrial Diamond Division, London, pp.61–75

Werner, G. and Kenter, M. (1989) *Industrial Diamond Review,* **49**, 15–19

Yarema, S. Ya. (1976) *Soviet Materials Science,* **12**, 361–374

Yazu, S., Nishikawa, T., Nakai, T. and Doi, Y. (1983) In *Proceedings of the International Conference on Recent Developments in Speciality Steels and Hard Materials* (Pretoria, 1982), (eds N. R. Comins and J. B. Clark), Pergamon Press, Oxford, pp.449–456

Part III

Applications and wear of diamond

Chapter 13

Wear and surface characteristics

The strength and hardness of diamonds makes them invaluable as tools in a wide variety of machining processes. However, because diamonds are costly their use is only justified in a particular process if their rates of wear are low enough to ensure an adequate working life. Therefore, before discussing the use of diamond tools we consider some general aspects of the wear of diamond and other materials.

13.1 Types of wear

The relative motion of any two surfaces in contact leads to wear, and the nature of this wear depends very much on the conditions of the contact. We will be concerned with at least four main types of wear: fracture, mechanical attrition, thermal degradation, and chemical wear. *Fracture* processes take a variety of forms. Gross cracking may result in the complete loss of a single crystal tool; the fracture of diamond grits in a grinding wheel is an important part of the grinding process; fine microchipping on the edge of a tool may greatly reduce the quality of a machined surface.

Diamond stands at the top of the well known Mohs scale of hardness which is based on a scratch test, a harder material by definition being one that scratches a softer one, see for example Tabor (1954). Diamond heads the scale because only diamond will scratch diamond. However, even if a single attempt to scratch a diamond makes no visible effect, there remains the possibility of some damage having been caused at a microscopic level. In fact if diamond is rubbed for a sufficiently long period, even by materials much softer, measurable amounts of material are removed by processes described as *mechanical attrition*.

Mechanical attrition always occurs when any two materials rub against each other, although it may be obscured by other forms of wear. Attritious wear processes fall into two main types which may be readily observed if a stylus of the harder material is dragged across a surface of the softer material. On brittle materials the stylus may produce damage by *cracking* while on plastic material the damage takes the form of grooving or *ploughing*. Both processes lead to removal of material; progressive cracking isolates small volumes of material from a brittle surface, while the flow of material during ploughing also produces isolated fragments. Quite often both fracture and plastic processes contribute to the wear, as for example in experiments on a range of materials described by Moore and King (1980).

Figure 13.1 Profilometer trace of a brass rod turned in a lathe

When discussing any form of wear it is important to distinguish between the apparent area of contact and the actual or true area of contact. No surface is truly smooth but exhibits some roughness with high and low points as for example in Figure 13.1 which shows a profilometer trace of the surface of a brass rod turned in a lathe. It follows that when any two surfaces are placed together contact will only occur over a limited area, at the peaks of the high points or asperities. The details of this contact will vary greatly depending on the nature of the materials and the preparation of the surfaces, but the initial area of true contact will always be much smaller than the apparent area. The geometric form of this contact is obviously very relevant to processes of wear and friction and has been discussed by several authors, see for example Johnson (1985). Even so, the details of the role played by the asperities are often obscure. (The particular case of the contact of polished diamond surfaces was discussed in detail in Sections 11.3 and 11.4.)

Two further points should be noted regarding wear by attrition. Because the areas of real contact are small the pressures on them may be sufficient to cause the two materials to adhere, thus leading both to the removal of material and to the transfer of material from one surface to the other. The second point is that debris produced by the wear may remain trapped between the rubbing surfaces and promote further wear. Good general accounts of mechanical attrition are given by Bowden and Tabor (1964, 1965) and Rabinowicz (1965), and by Samuels (1971) who stresses the often predominant role of cracking processes in brittle materials.

Wear processes are generally accompanied by friction, which sometimes produces temperature rises sufficient to reduce the strength and wear properties of one or both of the rubbing surfaces, these effects we describe as *thermal degradation*. We discuss these temperature rises in various Sections below but note that, under some conditions with some materials, they can be of the order of several hundred degrees. In this case there is the possibility of atoms diffusing from one surface to the other, and eventually of surface melting. If the temperature rises sufficiently, the materials may burn or otherwise interact with the atmosphere.

The process of burning or oxidation, when material is removed atom by atom, is an example of *chemically induced wear*. This type of wear is also observed during machining processes when tool and workpiece have a chemical affinity for each other, for example very high rates of wear are observed if diamond tools are used to turn ferrous metals (Section 13.5). Chemical reactions may also influence wear processes indirectly. For example, Eiss and Fabiniak (1966) observed that when spheres of single crystal alumina (Al_2O_3) were rubbed on steel the surface became covered with the spinel compound $FeO.Al_2O_3$. Examples involving the interaction of metal and lubricants are given by Hsu and Klaus (1979), and the interaction of atmospheric oxygen with freshly formed metal chips by Duwell, Hong and McDonald (1969). For further references to such chemical reactions, including corrosion and stress-corrosion, see reviews by Shaw (1958/9), Rabinowicz (1965), Evans, U. R. (1981), Tabor (1977) and Shaw (1984).

13.2 Mechanical attrition

We describe some detailed experiments by Crompton, Hirst and Howes (1973) on the wear of diamond rubbing on a variety of materials. The experiments were made using relatively low loads and quite low speeds so any temperature rise induced by the rubbing would have a negligible effect. (The effect of temperature rises in other situations is discussed in Section 13.3.)

13.2.a Time effects

Crompton, Hirst and Howes pressed a stationary diamond against the curved rim of a rotating disc of the second material. The disk was 21 mm in diameter with a rim rounded to a radius of about 0.4 mm. The load on the diamond was usually about 2N, and the disc rotated at 80 rev min^{-1} giving a linear sliding speed of 88 mm s^{-1}. The surface of the diamond was polished to coincide quite closely with {001} and the diamonds oriented so that any misorientations away from {001} had the least effect on the wear (Section 9.3.c). After rubbing for an appropriate time the size of the wear scar was determined by optical interferometry, and the volume of the scar calculated. Hence wear rates were obtained for a range of materials of different hardness.

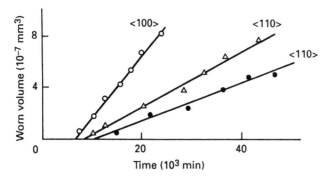

Figure 13.2 The wear of a polished {001} diamond surface rubbed by copper in <100> and <110> directions under different mean pressures; (O), 0.88 GPa; (△), 0.93 GPa; (●), 0.72 GPa (Crompton, Hirst and Howes, 1973)

Two features common to all the results are seen in Figure 13.2 which plots the volume of diamond worn away as a function of the time a diamond was rubbed against a copper disc with a Vickers hardness of 90 VHN. The figure shows that there is an initial induction period during which no wear is observed, after which the volume of the scar rises linearly with the time of rubbing. The implication of these results is that wear does not begin until the rubbing has caused sufficient damage on the surface to allow the removal of material.

Figure 13.2 also shows that the wear of the diamond is less when rubbed in a <110> direction than in a <100> direction. This result is of course consistent with the fact that the diamond shows a greater resistance to polish in this direction (Section 9.3.b). We therefore expect the induction period to be longer in the more abrasion resistant direction as is observed.

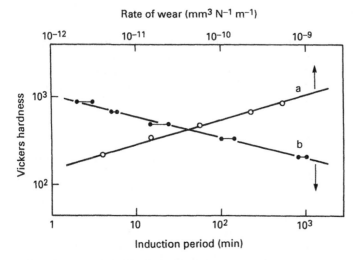

Figure 13.3 Relation between the Vickers hardness of the steel and (a), the rate of wear; (b), the induction period for a polished {001} diamond surface rubbed by silver steel (after Crompton, Hirst and Howes, 1973)

Both the rates of wear and the length of the induction periods vary greatly with the hardness of the rubbing material. Thus Figure 13.3 shows results for five specimens of silver steel which were heat treated to produce hardnesses ranging from 220 to 895 VHN and then rubbed on the diamond in <100> directions. We see that the wear rates vary by a factor of over 10^2 and the induction periods by a factor of almost 10^3.

13.2.b Rubbing on various materials

Crompton, Hirst and Howes (1973) made experiments with diamond rubbing on copper, silver, gold, aluminium, duralumin, tool steel, carborundum and a diamond grinding wheel, in both <100> and <110> directions on the diamond. Figure 13.4 shows their values for the wear rates when rubbing in a <100> direction as a function of the Vickers hardness of the rubbing material for all the materials except gold where the results varied somewhat as they were only able to use specimens in the form of thin foil. (The measured values of hardness and wear rate are plotted directly in Figures 13.3 and 13.4 rather than the derived quantities given by Crompton, Hirst and Howes.)

We see in Figure 13.4 that the wear rates produced by the diamond wheel and carborundum lie approximately on the same line as the points for the steels even though the wear rates vary by a factor of over 10^5. The wear rate of $1.7 \ 10^{-6}$ mm^3 N^{-1} m^{-1} quoted for the wear produced by a diamond bonded wheel is quite comparable to the values given by Hitchiner, Wilks and Wilks (1984) for the removal rate of material when polishing diamond with a bonded wheel. Thus it appears probable that the processes responsible for the wear produced by the steel, carborundum, and the diamond wheel are essentially similar. However, the position is rather different for the other metals.

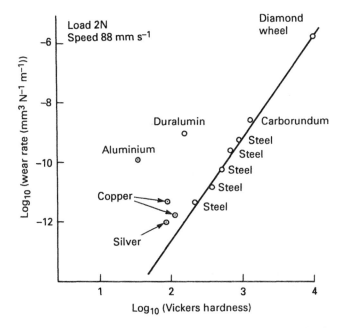

Figure 13.4 Wear rates produced by various materials rubbing an {001} face of diamond in a <100> direction as a function of their Vickers hardness (after Crompton, Hirst and Howes, 1973)

The wear rates for aluminium and duralumin are factors of 10^3 or so above the values given by the straight line in Figure 13.4, so some different mechanism must be involved besides that responsible for the wear due to tool steels. The authors state that during the rubbing the aluminium discs became visibly oxidized and therefore concluded that the high wear rates were due to the presence of aluminium oxide which has a high Vickers hardness. (They also observed another induction period associated with changes in the state of the aluminium discs, which is presumably related to the time required to form an appreciable quantity of oxide.)

The wear rates for copper and silver are low so there may be some uncertainty in the measured values. Even so, the three wear rates given are from 3 to 10 times greater than those given by the extrapolated line in Figure 13.4. It thus seems probable that this wear also is being produced by some additional mechanism. As already noted the wear rates on silver steel depend on the direction of the rubbing. For steel of 895 VHN the wear rates in <100> and <110> directions differed by a factor of about 40 but for steel of 220 VHN the difference was only a factor 5. This result also suggests that there may be another contribution to the wear which is less dependent on the direction of rubbing and which is only significant at low levels of wear.

13.2.c Mechanisms of attrition

The results shown in Figure 13.4 suggest that the wear produced by rubbing with steel and carborundum arises in a somewhat similar way to the wear of diamond by diamond in the polishing process. That is, by a form of fine scale mechanical fracture more readily produced by the harder materials. We described in Section 7.7 how diamond may suffer internal damage or fatigue when subjected to loads insufficient to produce immediate fracture. It appears that the induction periods observed in the present rubbing experiments represent the time required for fatigue damage to build up sufficiently to permit the removal of material. With softer materials the damage to the diamond is less and the induction periods correspondingly longer.

Several details of these wear processes are not well understood. Crompton, Hirst and Howes estimated that materials of hardness greater than 720 VHN should produce immediate fracture of the diamond in agreement with their observation that the induction period is then very small. However, this estimate is based on the magnitude of the load required to produce a ring crack by indenting with a ball 1 mm in diameter rather than by considering the contact of the much smaller asperities responsible for the damage. The authors also discuss the likely size of the wear particles.

Crompton, Hirst and Howes also considered why the results for copper and silver do not fall on the line in Figure 13.4. When rubbing with softer materials the stress fields are smaller and do not penetrate so deeply into the diamond. They therefore suggested that a change in behaviour occurs when the scale of the stress field becomes comparable with the dimensions of the microcracks in the diamond surface left by the polishing process. However, no evidence was given in support of this hypothesis. Indeed, the relatively high values of the wear might equally well be the result of adhesive interactions between the asperities on the metal and the diamond. (As already noted the high wear rates produced by rubbing with aluminium and duralumin are probably adequately explained by the production of hard Al_2O_3.)

13.3 Effects of temperature

13.3.a Oxidation and graphitization

When diamond is heated in air it reacts with the oxygen and burns away. The details of this process have been studied by Loparev et al. (1984) who used thermal differential analysis and gravimetric techniques (see for example Bassett et al. (1978)) to observe the oxidation of small grains of diamond. Figure 13.5.(a) shows the weight loss in arbitrary units and indicates the absorption or evolution of heat when the specimens were heated in air at a rate of 7.5 K min^{-1}. Both curves show that the rate of reaction begins to increase appreciably at temperatures above about 800 K (500°C).

The burning of diamond in air is a more complicated process than might appear at first sight, because graphitization as well as oxidation may be involved. Therefore before discussing burning we first refer to some experiments in which diamond was heated in the absence of oxygen, in a vacuum of 10^{-6} torr or better. Under these conditions diamond can be heated up to 1700 K without undergoing any change. However, after 12 h at 1700 K (Howes, 1962) or 45 min at 1900 K (Evans, T. and James, 1964) the diamond was covered with a grey film, and diffraction patterns obtained in the electron microscope showed that the film consisted of small crystallites of graphite of the order of 10 nm in size. Hence some of the surface of the diamond had been converted to graphite.

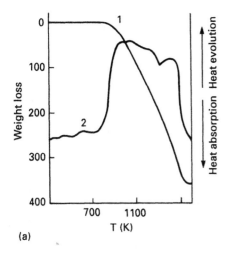

(a)

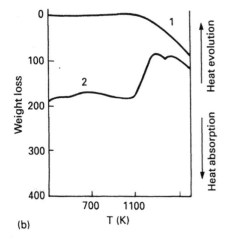

(b)

Figure 13.5 Thermograms indicating weight loss (1) and heat evolution (2) during experiments on heating: (a), typical diamond crystals; (b), crystals containing 5 wt % boron; in arbitary units (Loparev et al., 1984)

In other studies of graphitization Davies and Evans (1972) measured the rate of transformation of diamond between 2150 K and 2300 K. (The mass of graphite produced was determined by boiling the diamond in acid and observing the loss in weight.) They thus found that once graphitization had spread over the whole surface of the diamond the rate of etch could be represented by a relationship of the form:

$$\frac{\mathrm{d}x}{\mathrm{d}t} = A\mathrm{e}^{-\Delta U/RT} \tag{13.1}$$

where ΔU is regarded as the activation energy of the process and R is the gas constant. By choosing specimens with predominately {011} or {111} faces the authors obtained the values:

$\Delta U\{011\} = 730 \pm 50 \text{ kJ mol}^{-1}$

$\Delta U\{111\} = 1060 \pm 80 \text{ kJ mol}^{-1}.$

The authors explained the different behaviour on the two faces by the fact that on {011} faces all the atoms in each atomic layer are joined to the layer below by two carbon–carbon bonds, and that the energy of breaking these bonds is quite close to the observed activation energy of 730 kJ mol⁻¹. However, the bonding between successive {111} layers alternates between a situation in which each atom in a layer is joined to the layer below by one bond and a situation where each atom is joined to the lower layer by three bonds, see Figures 1.18 and 6.1. Hence the rate of graphitization on a {111} face will be controlled by the breaking of three bonds, and in fact the measured activation energy approximates to this value.

The rate of graphitization is considerable reduced by pressures of the order of 5 GPa (Bundy et al., 1961; Davies and Evans, 1972), as might be expected because graphite has a larger molar volume than diamond. The latter authors also discuss the detailed way in which the atoms of diamond rearrange themselves to form graphite but the position is still unclear. Other measurements of rates of graphitization have been made by Russian workers, see for example references given by Fedoseev et al. (1986). The latter authors measured graphitization rates over a wider temperature range than Davies and Evans and observed a considerable lower activation energy below 1900 K. However, they comment that their results may have been affected by the presence of oxygen. Also some of their measurements, not clearly identified, were made on synthetic diamonds which probably contained inclusions of solvent/catalyst which could have influenced the rate of graphitization (see below).

Bearing in mind the above behaviour of diamond when heated in a vacuum we now return to its behaviour when heated in the presence of oxygen. Evans T. (1979) describes experiments by Evans and Phaal (1962) who etched diamonds by passing a flow of low pressure oxygen over {001}, {011} and {111} surfaces of the diamond at temperatures between 920 K and 1620 K. The rate of etch, inferred from the loss in weight, was found to be proportional to the pressure of the oxygen, and rates for an oxygen pressure of 0.4 torr are shown in Figure 13.6 plotted in

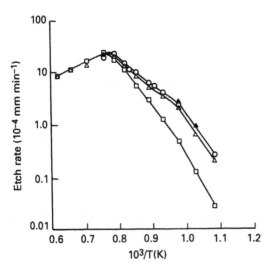

Figure 13.6 Arrhenius plot of the etch rate in 0.4 torr of oxygen of three faces of diamond against the reciprocal of the absolute temperature: (○), {111}; (△), {110}; (□), {100}. (After Evans, 1979)

the form of Equation (13.1). After oxidation at the higher temperatures, the faces were covered with a grey layer which electron diffraction measurements identified as crystallites of graphite less than 5 nm across. This layer was seen on {011} and {111} faces at temperatures of 970 K and above, and on {001} faces at temperatures of 1120 K and above. Hence it appears that small amounts of oxygen will greatly reduce the temperature required to cause graphitization. The slopes of the lines in Figure 13.6 at temperatures below 970 K when no graphite was left on the surface imply an activation energy for oxidation of about 230 kJ mol^{-1} for all three surfaces. There is at present no detailed account of the changes in slope above 970 K and of the decrease in all the reaction rates above 1200 K but it seems likely that the graphite layer tends to protect the diamond against oxidation.

Finally we note that the graphitization of diamond may be much encouraged if the diamond is in contact with certain metals. Gorodetskii *et al.* (1971) observed diffraction patterns of graphite after depositing thin layers of nickel about 50 nm thick on the {111} surfaces of diamond and heating to 400°C. Ikawa and Tanaka (1971) produced graphitization by heating diamonds with iron powder for 30 min at 900°C in a vacuum of 2×10^{-3} torr. Gorodetskii *et al.* (1972) describe the etching of diamond by heating in contact with nickel and iron films; see also Vishnevskii *et al.* (1975). These effects are, of course, not unexpected as nickel and iron act as solvents for both diamond and graphite (Section 1.2.a).

13.3.b Bulk and flash temperatures

In the experiments on oxidation and graphitization described above the whole diamond was heated uniformly in some form of constant temperature enclosure. However, when tools become hot during machining the heat is generated close to the cutting edge and produces a non-uniform temperature distribution as it flows away. Hence we must distinguish between the mean temperature of the tip of the tool and the mean temperature of the whole diamond. These temperatures are determined by the work done by the friction forces which create the heat, and by the thermal resistances of the paths taken by the heat as it flows away through tool and workpiece. Hence by measuring the friction forces and the speed of the workpiece one can estimate the rises in the mean temperatures of tip and tool using standard methods of heat flow analysis.

It is also necessary to take into account the fact that contact between two surfaces generally occurs over a much smaller area than the nominal area, see Section 13.1. Real contact only occurs at the tips of the highest asperities so all the heat generated by the friction forces is produced in these small localized regions. Hence, the heat flowing away from these constricted areas will give rise to temperature differences much greater than the mean temperature rise of the surface. In addition, the contact between any pair of asperities is only transitory, the load being carried by different pairs of asperities as the two surfaces slide over each other. Hence the high temperatures are not only localized in space but are transitory in time and are thus known as *flash temperatures*.

To obtain an estimate of the likely magnitude of these flash temperatures we use a treatment given by Bowden and Tabor (1950), based on a calculation by Jaeger (1942) which gives the temperature rise when an applied load W is carried by just one area of true contact of radius a as:

$$\Delta T = \frac{\mu W V}{4aJ} \frac{1}{k_1 + k_2} \tag{13.2}$$

where μ is the effective coefficient of friction, V the velocity of sliding, J the mechanical equivalent of heat, and k_1 and k_2 the thermal conductivities of the two surfaces. Bowden and Tabor assume that as contact is made at the tips of the asperities the softer material will plastically deform and increase the area of contact until the pressure falls to the yield pressure p_m of the material. Then if we are dealing with one area of contact $W = \pi a^2 p_m$. Then substituting into Equation (13.2) we obtain:

$$\Delta T = \frac{\mu V \sqrt{W p_m \pi}}{4J(k_1 + k_2)} \tag{13.3}$$

Values of ΔT calculated from Equation (13.3) are only an approximation because Equation (13.2) assumes that the whole load is carried by only one area of true contact. In fact there will be many areas of contact so the effective value of W in Equation (13.3) will be reduced. On the other hand it is also necessary to take account of the mutual influence of adjacent areas of contact, but no detailed treatments are available. There is also the further point that for sliding at high speeds we must take into account the fact that the moving surface is continually bringing up colder material. This effect reduces the rise of temperature, and at sufficiently high speeds the rise ΔT is no longer proportional to the speed V but to $\sqrt{V}$ (Jaeger, 1942).

The presence of flash temperatures is well established and various experiments on a range of materials have shown excess flash temperatures of the order of several hundred degrees K, see for example Bowden and Thomas (1954) who rubbed metals against a revolving glass plate and observed flashes of light through the glass, and also verified the above dependence of ΔT on $V\sqrt{W}$. See also work on the rubbing of steel by Uetz and Sommer (1977) and by Quinn and Winer (1987), and also a review by Rabinowicz (1965). Because of the very high thermal conductivity of diamond (Section 6.3.c) we may expect the excess flash temperatures to be relatively low on diamond surfaces.

13.3.c Wear during rubbing

The rubbing experiments described in Section 13.2 were made under light loads at slow speeds so any temperature rises created by the rubbing would be negligible. However, on increasing the load and the speed of rubbing the rise in temperature at the interface must eventually play a significant part in the wear process. For example, Table 13.1 shows wear rates of {001} surfaces of diamond rubbed against

Table 13.1 Wear of a polished {001} face of diamond sliding on glass in <100> and <110> directions under a load of 1 kg in air. (Bowden and Scott, 1958)

Sliding speed m s^{-1}	Wear rate	
	(10^{-3} mm^3 h^{-1})	
	<110>	<100>
4.2	nil[a]	0.29
8.4	0.25	18.3
17	2.52	97.3
35	50.7	325

[a]After 6 h

a rotating glass disc under a load of 1 kg at various speeds (Bowden and Scott, 1958). We see that for sliding in both <100> and <110> directions the wear rate increases very rapidly as the speed increases.

Micrographs of the wear tracks on the glass disc in the above experiments showed that the nature of the surface damage changed markedly between sliding speeds of 5 and 15 m s^{-1}. At the lower speed the wear on the glass appeared as series of scratch lines whereas at 15 m s^{-1} the glass 'showed characteristic markings suggestive of flow of a viscous liquid'. It was also observed that this flow only occurred above a certain critical value of the product $V\sqrt{W}$ where V is the sliding velocity and W the load, and according to Equation (13.2) this condition corresponds to a particular value of flash temperature. Hence the authors concluded that by speeds of 15 m s^{-1} the surface temperature had risen sufficiently to cause melting of the glass. It is therefore not surprising that the rate of wear of the diamond also changed.

Another study of the effect of surface temperatures generated by high sliding speeds is that by Miller (1962) who measured the friction of a steel ball spinning at extremely high speeds and constrained by four diamond surfaces pressed against it with forces of 200 g wt. At speeds between 100 m s^{-1} and 200 m s^{-1} the coefficient of friction increased from about 0.01 to about 0.1, and at speeds above 150 m s^{-1} there was considerable transfer of steel onto the diamond. These effects are similar to those observed in the friction of various combinations of materials when the temperature of one of them approaches its melting point (Rabinowicz, 1965, p.93). Of course the speeds used by Miller, up to 500 m s^{-1}, are much higher than those normally used in machining processes.

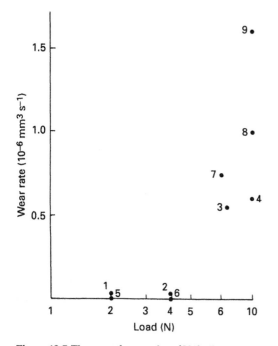

Figure 13.7 The rate of wear of an {001} diamond surface against a clean cast-iron scaife, free of powder, as a function of the load. The figures indicate the order of taking the measurements

Rubbing experiments somewhat similar to those of Crompton, Hirst and Howes (1973) (Section 13.2) have been made by rubbing a polished diamond surface on a clean rotating scaife of either steel or cast-iron . Figure 13.7 shows values for the wear of a diamond surface rubbed against a cast-iron scaife at a linear sliding speed of about 1.5 m s^{-1}, as a function of the load (A. G. Thornton, private communication). The wear rises steeply as the load approaches a value of 10 N suggesting that a further mechanism of wear is now involved over and above that described in Section 13.2. The results for the clean scaife in Figure 9.29 were obtained with rather similar loads and speeds and show the mean temperature of the diamond to have risen by the order of 100°C, so it is quite possible that the increase in wear in Figure 13.7 is due to a rise in temperature. Finally we emphasize that any rates for the removal of material by rubbing in this type of experiment are very specific to the conditions of the particular experiment in question. They will for example depend on the state of the scaife including the initial skimming, the amount of smoothing of the scaife produced by the rubbing, and the presence of any debris.

13.4 Effects of atmosphere

The strength, friction and wear of many materials may be affected by their gaseous or liquid environment. Thus the strength of glass against fracture is considerably reduced when measured under water and other liquids, see for example Culf (1957). The wear and friction of graphite is greatly reduced by the presence of water or oxygen (Bowden and Tabor, 1964, p. 191). It is also well known that the action of lubricants in machining operations may involve not only the physical presence of the film but also chemical reactions with tool and workpiece, see for example Bowden and Tabor (1964) and Shaw (1984). Various examples of wear due to the effect of environment are given by Westwood (1974). We now describe three sets of experiments in which the wear of diamond is modified by its environment.

13.4.a Polishing diamond

Figure 13.8 shows some results on the polishing of an {011} diamond surface on a cast-iron scaife charged with 5 μm–10 μm diamond powder. The equipment was similar to that described in Section 9.2.a but it was modified so that the scaife and polishing head were completely enclosed by a chamber which could be evacuated down to 0.1 torr (Hitchiner, Wilks and Wilks, 1984). The measurements were made under a load of 1.57 kg, with the scaife rotating at 60 rev min^{-1} corresponding to a linear speed of about 1.2 mm s^{-1}, the speed being limited by the need to avoid undue heating of the vacuum seals. Figure 13.8 shows that when polishing an {011} face in a hard <0$\bar{1}$1> direction of abrasion the effect of reducing the air pressure from 760 torr to 0.1 torr was to reduce the polishing rate considerably, and that a period of time was necessary for the rate to adjust to its new equilibrium value. Other (unpublished) experiments confirmed the above results and also showed that replacing 760 torr of air with a similar pressure of argon had the same effect as reducing the pressure.

    The above experiments were made after it was noticed that the rates of polish in air at quite low speeds were appreciably greater than would be expected from

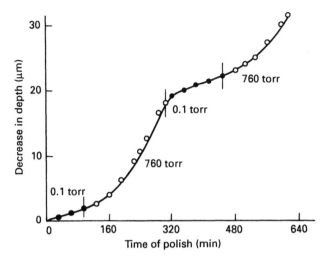

Figure 13.8 Effect of the pressure of the surrounding air on the rate of removal of material when polishing an {011} surface of single crystal diamond in a hard direction at 60 rev min^{-1} (Hitchiner, Wilks and Wilks, 1984)

the normal high speed value by applying the usual relationship that the rate is proportional to the speed of the wheel (Section 9.4.c). Table 13.2 summarizes some results on the effect of varying both the speed of the wheel and the pressure of the surrounding air. These show that in a normal atmosphere reducing the speed of the scaife from 2780 rev min^{-1} by a factor of 46 caused the rate of polishing to fall by only a factor of 6, but when the air pressure was then reduced to 0.1 torr the rate fell to a value which was a factor of 33 down on its original value. A similar but less marked effect was observed when polishing in a <100> direction: reducing the speed by a factor of 46 reduced the polishing rate by a factor of 26, and then reducing the pressure to 0.1 torr further reduced the rate to a value 45 times less than its original value.

To discuss the above results we first note that Table 13.2 shows that at a speed of 60 rev min^{-1} the removal rate is considerably greater in 760 torr of air, but that polishing still proceeds at an appreciable rate when the pressure is reduced to as little as 0.1 torr. Hence somewhat different mechanisms of polishing must be involved at the two pressures, this conclusion is also implied by the different

Table 13.2 Effect of air pressure on removal rate when polishing an {011} face of single crystal diamond on a cast-iron scaife in the <110> direction (load 1567 g wt). (Hitchiner, Wilks and Wilks, 1984)

Speed rev min^{-1}	Air pressure, torr	Removal rate, 10^{-4} mm^3 s^{-1}
2780	760	0.59
60	760	0.098
60	0.1	0.017
		0.019

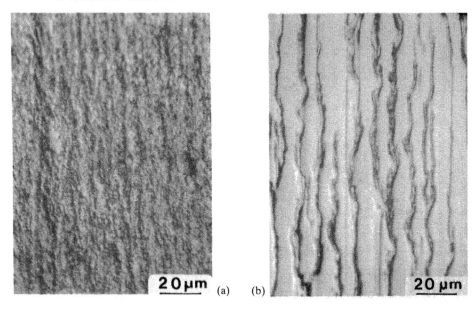

Figure 13.9 Nomarski interference micrographs of {011} surfaces of diamond polished under similar conditions: (a), in 760 torr and (b), 0.16 torr of air

appearances of the polished surfaces (Figure 13.9). Although it was not possible to measure the friction force between the diamond and the scaife at low pressures, measurements in atmospheres of air and argon showed that the force was about 20% smaller in argon.

It is well known that the propagation of cracks in brittle solids may be affected by the atmosphere, see for example studies of cone cracks in glass (Langitan and Lawn, 1970; Mikosza and Lawn, 1971), of the abrasion of silicon (Danyluk and Clark, 1985; Lim and Danyluk, 1985), and general reviews by Lawn (1973) and Westwood (1974). These experiments show that the propagation of the tip of a crack can be affected by the presence of a particular chemical species, as discussed by Lawn (1974, 1975). Hence there is the possibility that when diamond is polished at low speed an atmosphere of air is able to penetrate to the tips of the surface cracks and promote the breaking of the atomic bonds necessary for an extension of the crack. Therefore the polishing rate only falls off proportionately to the speed of the scaife either when the air pressure is reduced or when the air is replaced by the more chemically inert argon.

The changes just described, of reducing the air pressure and of replacing the air by argon, had the effect of reducing the rates of polish. However, it is quite possible that some other atmospheres might bring about greater rates of polish than air. Figure 13.10 shows the results of some recent unpublished experiments on the polishing rates of diamond on a bonded scaife when covered with a layer of water, (A bonded scaife was used because the liquid would have displaced the powder from a cast-iron scaife.) Large increases in the rate of polish were produced by the presence of water, and similar effects were produced by ethanol. Further studies of these effects would be of interest.

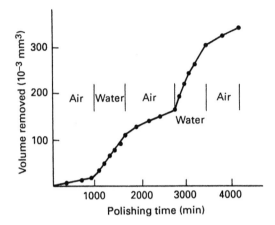

Figure 13.10 Rates of wear of an {011} face of diamond polished at low speed on a bonded scaife and on the same scaife covered with water

13.4.b PCD sliding on steel

Section 12.4 described measurements of the coefficient of friction when PCD materials were rubbed on steel rings (Mehan and Hayden, 1981). The same authors also observed that the friction on some steels was much influenced by the surrounding atmosphere. On replacing the usual atmosphere of air by one of nitrogen the coefficient of friction for the rougher PCD rubbing on 4620 steel was reduced from about 0.7 to about 0.1, while at the same time the wear of the steel fell to virtually zero. The authors made X-ray analyses of the wear debris, and found that in an atmosphere of air it consisted mainly of Fe_2O_3 with small amounts of Fe_3O_4. Hence metal fragments produced by the rubbing must have reacted with the atmosphere of air to give the much harder oxides which then produced wear on the steel ring and a high coefficient of friction.

13.4.c Diamond turning mild steel

In Section 13.5 we discuss the very high wear of diamond tools turning mild steel. However, it is convenient to note here that this wear may be much modified if the atmosphere of the air is replaced by either an inert atmosphere or an atmosphere of a more chemically active species. Figure 13.11 plots the area of the wear flat on

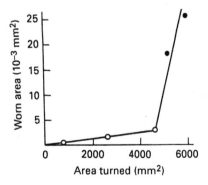

Figure 13.11 Wear of a diamond tool versus area of mild steel turned at 0.15 m s^{-1} in air (○) and in methane (●) (Hitchiner and Wilks, 1984)

a tool turning mild steel against the area of metal turned, first in an atmosphere of air and then in an atmosphere of methane. We see that the rate of wear is considerably greater in methane. Other examples of increased wear in atmospheres of hydrogen and methane are given by Hitchiner and Wilks (1984).

13.5 Wear on mild steel

We now describe some laboratory experiments on the wear of diamond tools turning mild steel under carefully controlled conditions (Thornton and Wilks, 1979). Although the rates of wear are so high that it is not practical to use diamond tools to turn ferrous metals in a factory the types of chemical reactions involved are of some general interest. In particular such reactions may be relevant to the wear of diamond in processes such as grinding and drilling.

13.5.a Measurement of tool wear

To obtain reproducible measurements of tool wear it is necessary to remember that the wear of any cutting tool depends on a large number of parameters. For a turning tool these will include the geometric form of the tool, its positioning with respect to the workpiece, the depth of cut, the linear velocity of the workpiece, and the rate of traverse of the tool. Wear rates may also vary greatly with changes in the constitution of either tool or workpiece. Therefore in order to study the effect of varying one parameter it is essential to maintain all other parameters unchanged.

Measurements were made with turning tools having a nose radius of about 450 μm and a flank clearance angle of 10°. The tools were mounted on a diamond turning lathe in a holder designed to permit the rake face (Figure 13.12(a)) to be

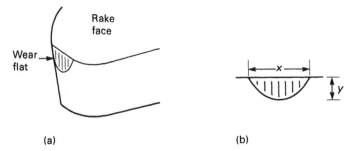

Figure 13.12 Schematic diagram of a wear flat on a diamond tool turning mild steel (Thornton and Wilks, 1979)

set accurately horizontal by optical observation. The rake face was set at the same height as the centre line of the lathe, and 50 μm cuts were made dry on standard billets of mild steel. Although some wear occurred on the rake face of the tool, most was on the flank face in contact with the rotating billet, where the steel quickly wears a vertical flat as shown in Figure 13.12(a). This wear flat, viewed at a relatively low magnification, has a smooth appearance on which is imposed a pattern of fluting with a spacing equal to the feed rate of the lathe (Figure 13.13).

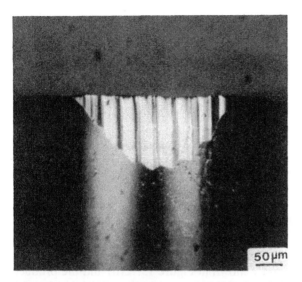

Figure 13.13 Optical micrograph of a typical wear flat produced by mild steel (Thornton and Wilks, 1979)

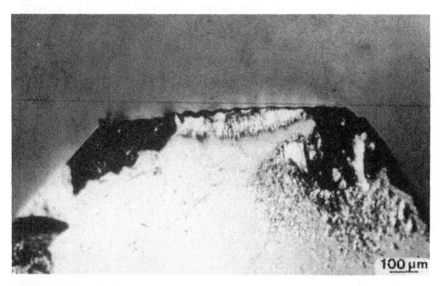

Figure 13.14 Optical micrograph of the rake face of a heavily worn tool. The line is parallel to the axis of the lathe (Thornton and Wilks, 1979)

This fluting appears to be a secondary effect often seen in wear patterns but not relevant to our present measurements. (A discussion of how this fluting arises has been given by Pekelharing (1959) and Lambert (1961/2).) The geometry of the wear is shown in Figure 13.14 which gives a plan view of the rake face of a similar tool which has suffered a large amount of wear. The line on the micrograph has been drawn parallel to the edge of the turned billet and indicates that the wear has taken place on the leading (left hand) edge of the tool as would be expected.

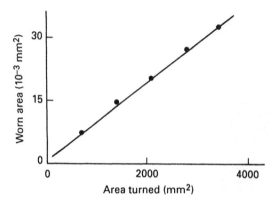

Figure 13.15 Area of wear flat versus metal turned in air at 10 m s^{-1} (Thornton and Wilks, 1979)

To obtain a quantitative measure of the rate of wear, the size of the wear was characterized by the dimensions x and y shown in Figure 13.12(b) and the product xy taken as a measure of the area. Then a plot of xy against the total area of metal traversed during the cuts was found to give an almost straight line, as shown for one particular tool in Figure 13.15. This linear relation appears to be generally valid provided the product xy does not exceed a value of about 30×10^{-3} mm^2. Hence by working in this range we can define a rate of wear specified by the gradient of the line, which we quote in units of 10^{-6} (mm^2 mm^{-2}).

Subsidiary experiments showed that the rate of wear rose by about 30% as the depth of cut increased from 25 μm to 100 μm. As the depth in each experiment could be set to within 3 μm, any variations in the wear rates due to inaccuracies in this setting were negligible. The feed rate of 25 μm rev^{-1} was determined precisely by the gearing of the lathe. As the steel rods decreased in diameter during the wear runs, the speed of the lathe spindle was adjusted to maintain the cutting speed. As a result of these adjustments the speed was not held exactly constant but was kept within about ± 10% of its nominal value. Although the wear rates change substantially with large changes in the velocity of cutting, any differences due to the above changes in the size of the rods were negligibly small.

The rate of wear of the diamond depends on the constitution of the steel. Rods made from a standard EN1A mild steel, from a free cutting EN1A mild steel containing lead, and from EN34 a nickel–molybdenum steel gave wear rates differing by a factor of about 2. Similar observations on the wear of diamond grit grinding different steels have been reported by Ikawa and Tanaka (1971), Tanaka and Ikawa (1973), and Graham and Nee (1974). Therefore, the measurements were generally made on one type of steel, a standard ENIA mild steel. The wear rate was usually so high that one pass over a 10 mm length of a 75 mm diameter rod was sufficient to produce measurable wear.

13.5.b Differences between diamonds

During the above measurements it became clear that though repeated measurements gave similar results the wear rates of different diamonds tended to differ, sometimes appreciably. To review this effect Thornton and Wilks (1979, 1980) measured the wear of some 40 similar diamond tools turning mild steel, the tools being either purchased from commercial suppliers or fabricated from drill stones. It was found that the wear rates of these tools turning in air at a cutting speed of approximately 11 m s^{-1} differed considerably. The values were usually between 1 and 15 units of 10^{-6} mm^2 mm^{-2} but other smaller and larger values ranged from less than 0.3 to 24 units.

In considering the above differences in wear rates between different diamonds the authors began by ruling out certain possibilities. To estimate the reproducibility and consistency of measurements they compared the results given initially by a tool with those given after the diamond had been completely repolished. Nine sets of measurements were made on tools before and after a repolish and in each case the wear rates were similar. The small differences observed, which reflect the accuracy of the measurements and possibly some inhomogeneity in the diamonds were insignificant compared to the wide range of the wear rates.

The crystallographic orientation of the diamonds may also affect the wear but it is extremely difficult to make a definitive experiment on this effect. One can prepare a tool with the diamond in a particular crystallographic orientation and measure the wear, but to change the crystallographic orientation of the diamond involves a complete reshaping of the tool. The new cutting edge will then inevitably be in a different region of the stone, perhaps with different properties. However, the authors describe an experiment on two pairs of tools each pair being fabricated from one high quality gem stone. The diamonds in each pair had different orientations and the wear rates of the two tools of each pair differed by a factor of about 2. This result is consistent with the observations of Tanaka and Ikawa (1973) that the rate of wear of a diamond grit machining different samples of steel varied with the crystallographic orientation of the grit by factors between 1.3 and 2.1 depending on the type of steel.

In addition Thornton and Wilks (1980) observed that there was no correlation between widely differing wear rates and either the bulk temperatures of the diamonds or the forces on them. Hence it appears that the differences in the wear rates are due to differences in the constitutions of the diamonds. We discuss such differences further in Section 13.6.d, but note here that it is obviously most important when comparing wear rates under different conditions, say at different turning speeds, to make experiments using the same tool.

13.5.c Effects of atmosphere

The tool and workpiece in the above experiments could be surrounded by a chamber which allowed the atmosphere of air to be reduced to a pressure of about 0.1 torr or to be replaced by an atmosphere of other gases. It was then found that the rate of wear often depended appreciably on the presence of air or other gas. Figure 13.16 shows the wear rates observed in an experiment turning at a low speed (0.13 m s^{-1}) in which all other parameters were held constant and the pressure of the air varied. The wear rate changes considerably over a narrow range of pressure, but at other pressures is apparently independent of the pressure, and this pattern of behaviour appeared to be fairly general.

The wear with which we are concerned occurs on the flank face of the tool which is pressed into the steel billet (Figure 13.17). Hence the wear flat on the tool is protected from the surrounding atmosphere by the presence of the freshly cut metal. However, the fact that changing the air pressures affects the rate of wear

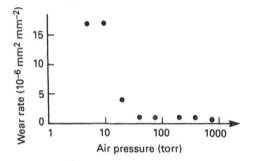

Figure 13.16 Wear rate of a diamond tool turning mild steel at 0.13 m s^{-1} as a function of the surrounding air pressure (Thornton and Wilks, 1979)

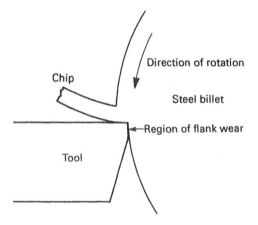

Figure 13.17 Schematic diagram showing the position of the wear flat with respect to the turned metal

implies that at the higher pressures, at least, molecules of air are arriving at the interface. As the wear above and below the transition is largely independent of pressure it suggests that so far as the wear reactions are concerned there are essentially two regimes, in which air molecules are either reaching or not reaching the interface.

The pressure at which the above change in the rate of wear occurs will depend on the several factors which determine the access of air molecules to the interface, such as the cutting speed, the shape of the tool, the quality of the polish on the tool and the presence of any cracks near the tool edge. However, the magnitudes of the pressures and speeds at which the transitions occur in the present experiments are of the same general order as those reported in other machining experiments; see, for example, Rowe and Smart (1966–67) and Williams and Tabor (1977).

The dependence of the wear rate on cutting speed and atmosphere was quite complex. Table 13.3 shows the wear rates of all the tools measured by Thornton and Wilks (1979, 1980) which had values equal or greater than 7 units (10^{-6} mm^2 mm^{-2}) when measured at a cutting speed of approximately 11 m s^{-1}. The table shows that when cutting at 11 m s^{-1} in air a reduction of the pressure to a nominal

Table 13.3 Rates of wear of a number of similar diamond tools turning mild steel in air pressures of 760 and 0.1 torr, at cutting speeds of 11 and 0.16 m s⁻¹. Wear in units of 10^{-6} mm² mm⁻². R = repolished, R2 = repolished twice, etc. (Thornton and Wilks, 1980)

Tool	760 torr		0.1 torr	
	11 m s⁻¹	0.16 m s⁻¹	11 m s⁻¹	0.16 m s⁻¹
DS11R	10	0	4	14
DS12R	7	4	7	15
DS12R2	8	6	5	17
DS19R	12	–	12	–
DS20R	9	0	5	9
T25R	8	3	3	13
T27R	9	4	4	22
T27R2	11	3	7	14
T36R2	10	2	4	22
T37R	15	4	–	–
T37R3	16	–	9	–
T38R	8	3	3	13
T39R	8	2	4	20
T39R2	8	3	5	17
T40R	24	6	–	–
T42	9	7	6	–
T42R	11	4	6	26
T42R2	11	5	7	30

vacuum of about 0.1 torr produced a reduction in the wear rate of all the tools except two which showed no change, the reduction being by various amounts up to 60%. Other experiments showed that replacing a full atmosphere of air by one of argon produced similar reductions in the wear.

Measurements were also made at the much lower cutting speed of 0.16 m s⁻¹. Table 13.3 shows that the wear rates in a full atmosphere of air were now appreciably less, sometimes much less, than at 11 ms⁻¹, as might perhaps have been expected because any temperature rises produced by the turning will be less at the lower speed. However, on reducing the air pressure to 0.1 torr the wear rate invariably increased considerably, generally to values greater than those observed at 11 m s⁻¹ in air. Thus the effect of the air is quite different at high and low cutting speeds, and we discuss this point further in the following section. (The wear rates of the tools which wore more slowly in air at 11 m s⁻¹ showed the same general trends with speed and atmosphere but with greater variation in the performance of individual stones perhaps because of a greater uncertainty in the measured values of the smaller wear rates.)

13.6 Chemical surface reactions

We begin with a further discussion of the experiments on the wear of diamond on steel described in the previous section.

13.6.a Wear on steel at high speeds

At cutting speeds of about 11 m s⁻¹ the wear rate has a value of up to about 15 mm² mm⁻² and depends only weakly on the speed. However, on increasing the speed the wear suddenly rises rapidly so that at $30\,m\,s^{-1}$ it is about 50 times greater and very short passes were necessary to avoid wearing out the tool during the measurements. A characteristic feature of this high rate of wear, not observed at lower speeds, was a shower of white sparks, presumably hot debris from the turning process. The vacuum chamber was then modified to take the extra large billet used to obtain the high cutting speeds and the experiment was repeated in a nominal vacuum of about 0.2 torr. There was now no sign of hot debris but the wear rates were the same as in a full atmosphere of air.

It is clear that if the speed of cutting is continually increased, the surface temperature of the diamond must eventually rise to very high values and cause either oxidation or graphitization. As described in Section 1.2.a iron and nickel act as catalysts which promote the diamond–graphite transformation. Therefore, bearing in mind (i), that the very high wear rates are the same in 0.1 torr and 760 torr of air; (ii), that the high cutting speeds generate temperatures sufficient to ignite the debris in air and (iii), that graphitization must eventually occur with increasing speed, it seems likely that the extremely fast wear processes observed at speeds above about $30\,m\,s^{-1}$ arise from graphitization of the diamond in the presence of iron which acts as a catalyst.

Figure 13.18 Nomarski interference micrograph of the wear flat on a tool after turning mild steel at 30 m s^{-1} (Thornton and Wilks, 1979)

Figure 13.18 shows an optical micrograph of the wear flat on a diamond tool which had worn on steel at a rate of approximately 400 units. Because the magnification is high and the wear band not absolutely flat, only the central part is fully in focus, but the micrograph shows not only vertical grooving caused by the downward flow of metal over the tool but also a structure of horizontal features lying almost perpendicular to the velocity of the moving metal. These horizontal features are inconsistent with abrasive wear but are not unlike those associated with the graphitization of diamond, which tend to develop along preferred crystallographic directions. As the wear flat on the tool of Figure 13.18 had no simple crystallographic orientation, the precise orientation of these features is not significant, and in any case the geometry of graphitization depends in a complex way both on the orientation of the face and on the particular conditions and chemical environment. Not much information is available on the geometry of graphitization but the general form of the features observed is not dissimilar to that observed in other studies (Evans and Sauter, 1961; Davies and Evans, 1972). Hence all the results obtained at $30 \, \text{m s}^{-1}$ are consistent with the wear arising from graphitization.

The wear of diamond on steel, even at only $11 \, \text{m s}^{-1}$, is very high compared with that produced by most other materials of comparable hardness. Figure 13.19 shows a micrograph of the wear flat of a tool, T25R, used at $11 \, \text{m s}^{-1}$ in 760 torr of air. Besides the vertical grooving observed in Figure 13.18 there are linear features inclined at an angle of about 30° to the vertical similar to the horizontal lines in Figure 13.18. The mode of wear appears to be similar in both cases, although the surface features are more marked in Figure 13.18, presumably because the much higher rate of reaction produces a deeper etch pattern. The different orientations of the linear features in Figures 13.18 and 13.19 are due to different crystallographic orientations of the diamonds in the tools. An inspection of the tool in Figure 13.19 in the cathodoluminescent mode of the scanning electron microscope revealed the position of {111} growth layers and suggested that the line features were approximately parallel to these layers.

The inference from the above results is that the wear at $11 \, \text{m s}^{-1}$ occurs by a rather similar mechanism to that at $30 \, \text{m s}^{-1}$ (Figure 13.18). As described in Section

Figure 13.19 Wear flat on a tool after turning at 11 m s^{-1} in 760 torr of air (Thornton and Wilks, 1979)

13.3.a diamond alone does not begin to graphitize until the temperature is raised to at least 1800 K but in the presence of iron graphitization may occur at much lower temperatures. We also know that even a small pressure of oxygen begins to etch a diamond surface at temperatures of the order of 1000 K. Hence it appears possible that the graphitization of a diamond when turning at $11\,\mathrm{m\,s^{-1}}$ may be encouraged by oxygen.

13.6.b Wear on steel at low speeds

We have described above how the wear at $30\,\mathrm{m\,s^{-1}}$ appears to be the result of direct graphitization caused by high temperatures, and the wear at $11\,\mathrm{m\,s^{-1}}$ in air appears to be the result of graphitization promoted by some rise in temperature together with the presence of oxygen. We now consider the wear observed under other machining conditions set out in Table 13.4. We see that starting at a cutting speed of $11\,\mathrm{m\,s^{-1}}$ in air the wear is reduced by reducing the speed or the air pressure as might be expected. However, if both speed and pressure are reduced the wear rate is substantially *increased*.

Figure 13.20 shows a micrograph of a worn tool after turning at low speed and low pressure and we see the appearance of the wear is quite similar to that in Figure 13.19 for a tool turning at $11\,\mathrm{m\,s^{-1}}$ in air. Note in particular the well developed line features inclined to the vertical as in Figures 13.18 and 13.19. In addition, the surfaces of a steel rod turned by this tool at $11\,\mathrm{m\,s^{-1}}$ in 760 torr of air and at $0.16\,\mathrm{m\,s^{-1}}$ in 0.1 torr were of quite similar appearance, as were the surfaces of the swarf when viewed in the SEM. These various results suggest that the mechanism responsible for the wear is rather similar at both speeds.

Figure 13.20 Wear flat on a tool after turning at 0.16 m s^{-1} in 0.1 torr of air (Thornton and Wilks, 1979)

In the previous section we concluded that the wear in air at $30\,\mathrm{m\,s^{-1}}$ and $11\,\mathrm{m\,s^{-1}}$ was produced by graphitization promoted by the excess temperatures generated at the interface. At a cutting speed of $0.16\,\mathrm{m\,s^{-1}}$ the excess temperatures will be greatly reduced. A rough but very conservative estimate is obtained by assuming that the flash temperatures vary as the square root of the velocity (Section 13.3.b). Hence, assuming that the excess temperatures at $11\,\mathrm{m\,s^{-1}}$ are of the order of 800°C, the excess temperatures at $0.16\,\mathrm{m\,s^{-1}}$ will certainly be less than 100°C. It follows that the high rates of wear at $0.16\,\mathrm{m\,s^{-1}}$ in 0.1 torr of air are not caused by any excess temperatures.

Two other experiments were made to look for some dependence of the wear rate at $0.16\,\mathrm{m\,s^{-1}}$ on the temperatures at the interface. In the first the wear was measured in 0.1 torr of air at the much lower speed of $0.02\,\mathrm{m\,s^{-1}}$. The flash temperatures in the surface were thus further reduced but the rate of wear was actually *greater* as shown in Figure 13.21. In the second experiment the wear was measured under the same conditions except that the mild steel rod was heated to 220°C by an induction heater. The area of wear on the diamond increased at the same rate whether the steel was heated or not, that is an excess temperature of 200°C produced no increase in the rate of wear.

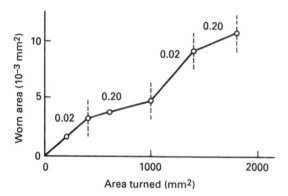

Figure 13.21 Area of wear flat versus area of metal turned at speeds of $0.20\,\mathrm{m\,s^{-1}}$ and $0.02\,\mathrm{m\,s^{-1}}$ in 0.1 torr of air (Thornton and Wilks, 1979)

The above results imply that even at low speeds when the effect of any flash temperatures must be negligible the diamond is wearing by some form of graphitization. Yet a diamond placed on a block of steel at room temperature does not graphitize. It therefore appears that the clean and uncontaminated surfaces generated by the machining have a greatly enhanced chemical activity.

A particularly clear example of such activity has been given by Grunberg (1953) who reported the generation of hydrogen peroxide when zinc, aluminium, magnesium and nickel were machined under water, the volume of gas produced corresponding to each atom on the new surface being chemically active. In the case of diamond and steel the activity is clearly sufficient to produce surface graphitization. The presence of air contaminates the clean surfaces by a covering layer and so reduces the activity and the wear. An atmosphere of argon at low speeds produces a rather similar effect as an atmosphere of air. We discuss these clean surface reactions in the following sections but first note a further point regarding the wear at low pressures.

The results shown in Figure 13.21 for the wear at 0.1 torr and low speeds pose the question of why the wear is greater at the lower speed. As discussed above, both speeds are so low that thermal effects will be negligible. Possibly the lower speed allows more air to reach the interface but the effect of increasing the air pressure at low speeds is to decrease the wear (Table 13.3). One possibility might be that a decrease of cutting speed makes a significant increase in the time available for a reaction to occur. However, at a speed of $0.1\,\mathrm{m\,s^{-1}}$ a steel and a carbon atom will be in close proximity for about 3 ns and this should be ample time to permit the reaction to dissipate excess energy to the thermal motion of the lattice. (The relevant relaxation time is of the order of a period of the atomic vibrations in the lattice, i.e. 10^{-12} s.)

The most likely explanation for the decrease of wear with speed appears to be as follows. The surface of a polished diamond is not flat but as discussed in Chapter 9 has a rough topography of asperities of the order of 5 nm high with a linear scale of about the same order. As a diamond is loaded against a steel surface under static conditions, plastic flow occurs until a sufficient area of steel is in contact with the asperities on the diamond to support the load without exceeding the yield stress of the metal. This will take a small but finite time, so the effective area of contact during machining may fall with rising speed and in this case the rate of reaction will also fall.

It is difficult to estimate the likely magnitude of the above effect, particularly for a metal such as steel where delayed yielding makes the position quite complicated. However, even at as low a speed as $0.1\,\mathrm{m\,s^{-1}}$ the metal will remain in contact with one particular asperity for the order of only 50 ns, while the delays for yielding are commonly much longer, see for example Suh (1967) and Harding (1971). We also note the experiments of Maan and Broese van Groenou (1977) who observed the scratching of mild steel by a diamond stylus under a constant load driven at various constant low speeds. As the scratching speed was raised from $0.3\,\mu\mathrm{m\,s^{-1}}$ to $0.3\,\mathrm{mm\,s^{-1}}$ the area of the cross section of the groove decreased by a factor of about 3 despite the very low magnitude of all the speeds involved.

13.6.c Clean surface reactions

The surfaces of materials exposed to the atmosphere are normally covered by a film of adsorbed gas which tends to reduce the chemical activity of the surface. The experiments described above suggest that by reducing the pressure of the gaseous atmosphere surrounding the diamond tools it is possible to obtain almost complete exclusion of the atmosphere from the wear surfaces during machining. Our usual perception of the reactivity of materials is generally based on their behaviour in a contaminated state. The conditions during machining may be quite different because the tool is cutting into the interior of bulk metal which is initially uncontaminated. In addition, if the tool is wearing at an appreciable rate, clean surface is also being produced at the wear land.

The experiments on diamond and steel suggested that similar chemical activity may be observed in other materials during machining. Therefore rather similar experiments have been made using tools and workpieces prepared from various materials (Thornton and Wilks, 1979; Hitchiner and Wilks, 1984). In one set of measurements tools of copper and nickel were used to turn carefully prepared billets of solid sulphur. The tools were all of similar hardness and all much harder than the sulphur but even so the wear of the copper tools was over ×100 greater than that of the nickel tools.

Other experiments were made turning billets of graphite with much harder tools made of tungsten. At cutting speeds in the range $0.1\,\text{m s}^{-1}$ to $10\,\text{m s}^{-1}$, either in 0.1 torr of air or 760 torr of argon, the wear was much less than 1×10^{-6} mm mm^{-2} but on increasing the air pressure to 760 torr the wear rate increased to a value of about 40×10^{-6} mm mm^{-2}. Thus it appears that the presence of air was responsible for some chemical reaction at the interface but there is no obvious explanation of the very considerable wear of the tungsten tool in terms of the chemistry of bulk materials. For example, if oxygen is the critical constituent of the air, it might be supposed that it would react with the graphite rather than the tungsten, yet it is the tungsten which wears.

Several metals produce an unusually high rate of wear on a diamond tool similar to that produced by steel. We consider these metals further in Section 14.4.d but note here the very different wear of diamond when turning molybdenum. After a very short time fragments of diamond are torn out of the cutting edge and the tool becomes useless (Figure 14.10(d)). The behaviour of a diamond tool turning nickel at low speeds appears similar to that of molybdenum but at higher speeds the wear is smoother and more similar to that on steel.

Chemical reactions are usually discussed in terms of the thermodynamic functions of the two phases, which are determined by the bond strengths in the bulk solids. However, the wear processes in the present experiments, particularly where the temperature remains virtually ambient, are essentially surface phenomena. We may begin to analyse this situation by considering the diamond–steel reaction. The diamond is certainly wearing away, so carbon atoms are being removed because of some linkage to the iron (or other) atoms in the steel. At least five types of bonds are involved in the wear process, as shown schematically:

   1 1 2 3 4 5 5
   Fe–Fe–Fe–Fe–C–C–C–C.

Bonds 1 and 5 are characteristic of the bulk material, 3 is the bond formed across the interface, and bonds 2 and 4 are those linking the uppermost iron and carbon atoms to the next lower layer under the conditions at this particular interface. Any chemical reaction at the interface is controlled by the form and strength of bonds 2, 3 and 4, and these are not fully determined by the behaviour of the bulk material.

The wear of diamond turning molybdenum gives an interesting contrast to the wear on steel. With steel the diamond wears steadily and relatively smoothly; this implies that in the schematic arrangement of bonds shown above, bond 4 is appreciably weaker than the others. With molybdenum there is again a strong interaction with the diamond but instead of a smooth wear process fracture occurs in the bulk of the diamond, and tearing also occurs in the bulk of the metal (Sections 14.3.a and 14.5.e). Setting out the bonds schematically:

   1 1 2 3 4 5 5
   Mo–Mo–Mo–Mo–C–C–C–C

we see that these experimental results imply that all five types of bonds are stronger than the inherent mechanical strengths of both the metal and the diamond, which of course are determined by structural defects rather than bond strengths.

13.6.d Factors affecting surface reactions

Surface reactions may be modified by the presence of foreign atoms in either the diamond or the workpiece and by the presence of chemically active gases. The effect of an atmosphere of air on the wear of diamond turning steel was described in Section 13.5.c. Other experiments show that atmospheres of hydrogen and methane may increase the wear of the diamond by a factor which varies from diamond to diamond and which may be greater than 10 (Figure 13.11). (Preliminary observations of the effect of carbon dioxide and methane reported by Casstevens (1983) are difficult to interpret because the machining conditions are not well defined and because of the presence of a mineral oil as a cutting fluid.)

An example of the action of foreign atoms in diamond is given by Simons and Cannon (1966) who observed that doping with boron greatly reduced the rate at which diamonds burnt in air. They also observed that incompletely burnt diamonds were covered by a continuous layer of boric oxide which appeared to inhibit oxidation. Subsequent experiments on the rates of oxidation of diamonds synthesized with 2 wt % and 5 wt % of boron are described by Loparev et al. (1984). Results for the 5% sample in Figure 13.5(b) show that the onset of oxidation occurs at about 1100 K compared with about 700 K for pure diamond. (The authors ascribe an initial small rise in curve 1 of Figure 13.5(b) to oxidation of the boron.) Rather similar observations are described and discussed by Ogorodnik, Pugach and Postolova (1985). In Section 14.4.d we describe the large changes in the wear of diamond associated with different concentrations of phosphorous in workpieces of electroless nickel.

A further example of the large effects which may be produced by small changes in the constitution of the tool or the workpiece is given by a series of laboratory experiments made with copper tools turning billets of carefully prepared sulphur (Hitchiner and Wilks, 1984). As mentioned in Section 13.6.c the copper tool wore rapidly even though the copper was much harder than the sulphur. However, with a copper tool of similar hardness containing no more than 1.0 wt % of nickel the wear was reduced by a factor of 7. For a discussion of the mechanism involved, see the original paper.

We have already described the wide variations in the wear of different diamond tools on steel (Section 13.5). In the light of the various experiments just described it appears possible that these differences between diamonds arise from the presence of chemically significant impurities. To speculate further we note that many diamonds contain appreciable concentrations of hydrogen (Section 2.1.e), and that atmospheres of hydrogen and methane encourage the wear of diamond, so the differences in wear might arise because of different concentrations of hydrogen.

Additional information on surface reactions is given by studies of lubrication. Several authors have discussed how the machining of metal is affected by the presence of air, oxygen and carbon tetrachloride vapour (Rowe and Smart, 1967; Williams, 1977; Williams and Tabor, 1977). However, nearly all these studies are primarily concerned with the *forces* involved and it is therefore worth stressing a distinction between measurements of friction and wear. Let us consider, for example, a monatomic material A machined by a tool of a monatomic material T. We represent the bonding perpendicular to the surface very schematically as:

A–A–A–A–T–T–T–T

where the outer A–A and T–T bonds are typical of the bulk material while the inner bonds are modified by their proximity to the interface. The centre A–T bond gives rise to adhesion but the force between workpiece and tool is determined by the strength of the weakest bond. If all the bonds are strong, the forces on the moving tool will be high, but even so if the eventual severance occurs either at the A–T or at an A–A bond, the wear will still be low. In fact, the forces on similar diamond tools turning different materials may be quite similar while the wear rates differ by orders of magnitude. Hence, the amount of the tool wear is determined *inter alia* by the relative strengths of the various bonds involved.

13.7 The diamond surface

We now consider the nature of a clean diamond surface. Machining produces vast areas of clean surface but they are very rapidly contaminated by the atmosphere. However, clean surfaces may be studied at leisure by a range of techniques in ultra-high vacuum chambers.

13.7.a Diffraction studies

Atoms on the surface of a solid have fewer neighbours than atoms in the bulk material, so their positions will probably not coincide with the lattice points of the bulk crystal. Hence one of the first items of study when considering the behaviour of a surface is the spatial arrangement of the surface atoms. One of the principal techniques used for this purpose is low energy electron diffraction (LEED) shown schematically in Figure 13.22. An electron beam of energy between 10 eV and 1000 eV strikes the surface and is diffracted into a series of beams each producing a spot on the screen. The essence of the technique is that low energy electrons are easily scattered and therefore only penetrate the crystal for a very short distance. Therefore, to a first approximation, the diffraction pattern is produced by only the top layer of atoms and is rather similar to that observed when a beam of light is incident on a reflection grating ruled with two mutually perpendicular sets of lines. Further general details are given by Woodruff and Delchar (1986).

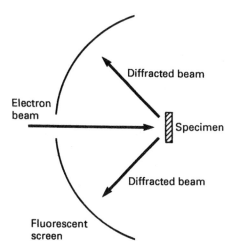

Figure 13.22 Schematic diagram of LEED optics for displaying diffractions patterns (after Prutton, 1984)

The preparation of clean surfaces of diamond is somewhat difficult. A natural surface may have suffered some contamination in the past. Cleavage in ultra-high vacuum (UHV) will produce two clean surfaces, provided no gas is liberated from the diamond (Section 2.2.b), but diamond is a difficult material to cleave in the confines of a vacuum chamber. Attempts to obtain clean surfaces by ion bombardment using argon and other ions have shown that the surface is modified by the bombardment (Evans, S. and Thomas, 1977; Evans, S., 1978, 1980). Therefore most workers have settled for polishing in air using conventional polishing techniques, generally on approximately {111} faces, and finally heating in UHV to drive off any foreign atoms.

Experiments by Lander and Morrison (1966) and Lurie and Wilson (1977) showed that if the temperature during the cleaning process rose to 1000°C significant changes occurred in the surface structure of the diamond. This change is well shown in the more recent experiments by Derry, Smit and Van der Veen (1986). They prepared surfaces by polishing a {111} face with diamond powder and then cleaning in an ultrasonic bath of detergent before placing in the vacuum chamber. Figure 13.23(a) shows the LEED pattern obtained after cleaning by heating in UHV to 800°C and it corresponds quite closely to what we would expect from the first layer of atoms in an undistorted lattice. However, when the diamond was cleaned by heating to 950°C a quite different pattern was obtained (Figure 13.23(b)). This pattern has a different symmetry and corresponds to a new arrangement of the surface atoms known as the *reconstructed* state. This result poses questions about the nature of the initial state, the reconstructed state, and the transition between them.

13.7.b Reconstruction and surface hydrogen

To consider the surface structure of diamond let us suppose that we create two clean surfaces by cleaving a diamond on a {111} plane. The bonds crossing this plane, see Figures 1.18 and 6.1, will be broken and the electrons forming the bonds will rearrange themselves into new configurations of minimum energy. According to a calculation by Ihm, Louie and Cohen (1978) this rearrangement should produce new electronic energy levels located in the band gap and an attempt to observe these electronic states was made by Himpsel et al. (1979). They used a natural {111} surface cleaned by washing ultrasonically and then heating to 150°C. The surface was then irradiated with monochromatic ultraviolet light and the energy spectrum of the emitted electrons analysed to determine the electron states in the diamond. No states in the gap were found and this result was confirmed by Pate et al. (1980) who made measurements on a polished surface. However, Pate et al. (1981) then heated the surface to 950°C and observed a reconstructed LEED pattern as would be expected, and in addition new surface states. Figure 13.24 shows the energy spectra from the photoemission experiments before and after the reconstruction, the curve for the reconstructed surface shows a well marked peak corresponding to an energy 2.5 eV below the Fermi level.

Figure 13.23 LEED patterns from a {111} surface of diamond: (a), after polishing and annealing to 800°C; (b) after heating to 950°C, showing reconstruction (Derry, Smit and Van der Veen, 1986)

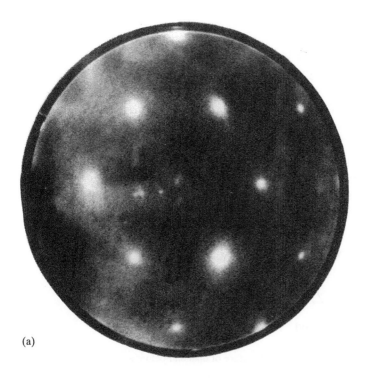

(a)

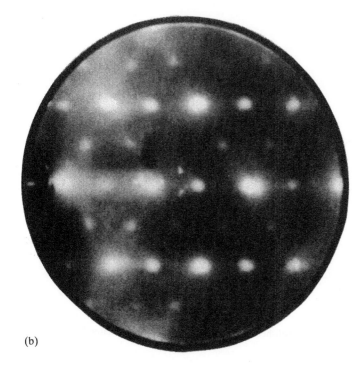

(b)

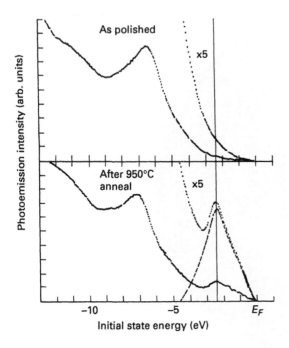

Figure 13.24 Photoemission from a polished {111} surface of diamond before and after reconstruction (after Pate *et al.*, 1981)

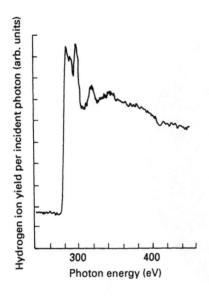

Figure 13.25 The hydrogen ion yield of a polished {111} surface of diamond irradiated by photons (after Pate *et al.*, 1982)

The key to the above results is provided by experiments which show that during reconstruction hydrogen is lost from the surface of the diamond. In one of these experiments Pate *et al.* (1982) observed the yield of hydrogen ions desorbed from the surface by photons of different energy. Figure 13.25 gives a plot of some of their results and shows a large increase in the number of hydrogen ions desorbed when the photon energy exceeds about 280 eV (which is about the energy required to ionize one of the inner core electrons of the carbon atoms). A prior anneal of the surface at 600°C had little effect on the yield of hydrogen but after annealing at 1000°C to produce reconstruction no more hydrogen was emitted. On the other hand a further treatment with atomic hydrogen restored the yield to its original value.

In other experiments Waclawski *et al.* (1982) irradiated a {111} surface with 5 eV electrons incident at 60° to the normal and observed the energy spectrum of the electrons reflected specularly. The upper curve in Figure 13.26 shows a spectrum for an unreconstructed surface plotted against the energy loss, while the lower shows the spectrum obtained after reconstruction. The authors point out that the additional energy losses observed with the unreconstructed surface coincide with various frequencies of vibration associated with carbon–hydrogen bonds.

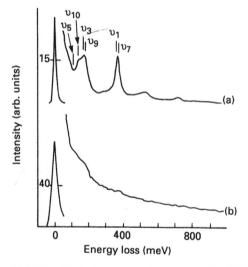

Figure 13.26 Electron energy loss spectra of a {111} surface of diamond: (a), polished and heated to 300°C; (b), after heating to 1000°C. The vertical lines on the upper trace indicate the energies of the C–H stretch vibrations in the infrared spectrum of –CH$_3$ (Waclawski *et al.*, 1982)

In a third set of experiments Derry, Madiba and Sellschop (1983) measured the concentration of hydrogen on an unreconstructed surface using a nuclear technique which irradiated the surface with high energy fluorine ions to initiate the nuclear reaction ^{1}H(^{19}F,$\alpha\gamma$)^{16}O. Then by observing the yield of γ rays as a function of the energy of the incident ions the authors were able to identify the reaction and hence the presence of hydrogen at concentrations ranging from 0.2 to 2.3 monolayers.

The above experiments show that a normal diamond surface may be covered by about a monolayer of hydrogen. In this case otherwise unsaturated carbon bonds are linked to hydrogen atoms, the positions of the carbon atoms on the surface coincide approximately with the diamond lattice, and additional electron energy levels are observed. It also follows from the above work that annealing at about 950°C drives off the hydrogen and that the unsaturated carbon bonds then rearrange and the surface atoms take up their reconstructed positions.

13.7.c Atomic structure and surface preparation

Various theoretical approaches have been made to determine the configurations taken up by the atoms on a diamond surface. A review of this work giving details of the proposed structures for both reconstructed and unreconstructed surfaces is given by Pate (1986). Comparisons between these predictions and the actual state of the surface are not straightforward because of the limited resolution of the experimental methods, one of the main difficulties being the problem of distinguishing effects due to the surface atoms from those due to atoms immediately below. Even so, various apparent inconsistencies remain to be resolved both between different experiments and between experiment and theory.

Pate (1986) describes experiments in which the characteristic LEED and electronic energy levels were observed on a reconstructed polished {111} face. The surface was then treated with atomic hydrogen, the LEED pattern reverted to its unreconstructed form, and the additional electronic levels disappeared, in line with the results described in the previous section. However, on again heating the surface to produce reconstruction, the electron levels reappeared but the LEED pattern remained in its unreconstructed form. These results are discussed by Pate but remain unresolved.

Other experiments not fully understood include those of Vidali and Frankl (1983) who observed LEED patterns and proton scattering from an unreconstructed polished {111} surface and deduced that the surface was partly disordered. Dayan and Pepper (1984) observed the Auger spectra (Section 13.8.a) from an unreconstructed polished surface and found it apparently inconsistent with the electron levels observed by electron energy loss spectroscopy. A further complication is that a diamond surface may also carry some oxygen (Section 13.8.c). (Pate (1986) stresses that *atomic* hydrogen is more readily taken up by the diamond surface than is the molecular hydrogen used by some investigations.)

Lurie and Wilson (1977) observed that LEED patterns gave no evidence of any reconstruction when polished {011} surfaces were heated to 1000°C. Pate *et al.* (1989) have suggested that this was because the structure of an unreconstructed {011} surface is quite similar to that of a reconstructed {111} face, and is therefore sufficiently stable not to undergo further changes. These authors also measured the dispersion curve of the electronic states, which suggested that some changes had occurred in the {011} surface states after heating even in the absence of reconstruction.

Finally we note that nearly all the experiments on surface structure have been made on polished surfaces while the theoretical calculations of the expected behaviour have considered an ideal plane surface. In fact, as described in Chapter 9, a polished surface of diamond is never ideally plane because the polishing process proceeds by microcleavage. We expect the surface to show a structure on a scale of the order of 5 nm with slopes determined by the lie of the cleavage planes and of the order of 50° (Section 9.4). The scale of this structure is large compared with the wavelength of the electrons used in LEED experiments, so the faces seen by the electrons are quite different than the simple {111} or {011} faces assumed in the calculations. Although this point has been noted by Derry, Madiba and Sellschop (1983) in connection with the bonding of oxygen to a polished surface, it has not been discussed in any of the present accounts of LEED or other surface studies.

13.8 Atomic interactions at the surface

We now summarize various experiments made to investigate the interactions between diamonds and other materials on an atomic scale. One of the first of these experiments was that of Sappok and Boehm (1968) who measured the mass of several gases adsorped on to the diamond,including hydrogen, fluorine, chlorine, bromine and oxygen. Subsequently various more advanced techniques, including XPS, LEED and Auger techniques, have been used to observe the interaction of various atomic species with the diamond surface, see for example a review by Thomas (1979). There is at present only a limited amount of information on these atomic interactions but we briefly outline some of the different types of experiments which are now being made.

13.8.a XPS and Auger techniques

In XPS or X-ray photoelectron spectroscopy the surface is irradiated by monochromatic X-rays which eject electrons from the surface with an energy equal to the difference between that of the incident X-rays and the binding energy of one of the inner or core electrons of the surface atoms. Then an analysis of the energy spectrum of the emitted electrons leads to the binding energy of the electron in the atom. Hence one can identify the atomic species in the surface and by making suitable calibrations estimate their concentrations.

Besides identifying chemical species XPS can provide other important information. Chemical bonds between atoms are formed primarily by interactions between the outer electrons of the atoms but these interactions also have a small but significant effect on the core electrons which results in a measurable change in their binding energy. Hence measurements of the binding energy of the core electrons can give information on the chemical bonding, see for example Roberts (1981).

The first pair of curves in Figure 13.27 show part of the energy spectra of photoelectrons from a {111} diamond surface produced by cleaving in an atmosphere of hydrogen (Morar et al., 1986). In order to differentiate between atoms on the surface and in the bulk material the X-ray beam was incident at a glancing angle of about 5° to the surface and with an energy of either 304.73 eV or 330.73 eV. The peaks observed correspond to the ionization of 1s core electrons in the carbon atoms with a binding energy of 285 eV. The mean free path of these electrons within the diamond is only of the order of nanometers so the spectrum is given by only the first few layers. Hence, because this mean free path decreases with increasing energy we expect the spectrum for the 330 eV beam to be more characteristic of the surface.

The top pair of curves in Figure 13.27 show spectra for a surface cleaved in hydrogen and then cleaned by heating to a temperature of about 800°C, well below the reconstruction temperature. There is no apparent difference between the spectra produced by the two X-ray energies. On heating further to drive off the hydrogen and obtain a clean reconstructed surface the main peaks broadened and could be resolved into two components as seen in the lowest pair of spectra in

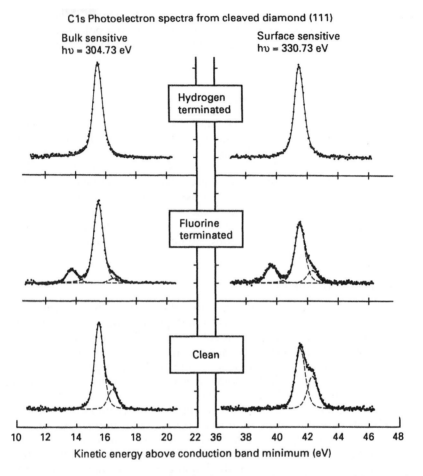

C1s Photoelectron spectra from cleaved diamond (111)

Figure 13.27 XPS spectra from cleaved {111} surface of diamond; see text (Morar *et al.*, 1986)

Figure 13.27, the new smaller component on the right being due to reconstructed surface atoms. Finally the surface was exposed to XeF_2 gas after which the right hand component of the peak decreased and a new peak appeared on the left. It appears that fluorine atoms had bonded to some of the reconstructed carbon atoms to give a new peak, the shift between this peak and the main peak being a measure of the difference between a C–F and a C–H bond. As we would expect, these changes are more marked in the 330 eV spectra which are more sensitive to the surface conditions.

Evans, S. and Ney (1990) have used XPS and other techniques to study the behaviour of films of cobalt evaporated on to {011} diamond surfaces in a high vacuum. The coated surfaces were heated to temperatures above 700°C and it was seen that the cobalt produced a sharp onset of graphitization above about 800°C. The XPS technique was able to distinguish three separate processes, the dissolution of diamond into the cobalt, the diffusion of the dissolved carbon through the cobalt, and the precipitation of this carbon as graphite from the upper face of the cobalt. The authors also concluded that the rate of graphitization was determined by the rate of dissolution of the diamond into the cobalt.

In Auger electron spectroscopy the surface is irradiated by either X-rays or an electron beam thus exciting the core electrons of the carbon atoms. However, in contrast to XPS, these electrons are not ejected with all the excess energy. The excited atoms return to equilibrium by the so-called Auger process in which an electron is emitted while at the same time the core electron falls to a lower excited state. Hence the emitted Auger electrons have an energy spectrum which reflects the energy of the core electrons. The Auger spectroscopy, like XPS, can be used to identify atomic species and give information on the nature of their chemical bonding. For further details of both XPS and Auger techniques see Prutton (1984) and Woodruff and Delchar (1986).

13.8.b Mechanical bonding

If two similar clean metal surfaces are pressed against each other they will generally bond together with a binding strength roughly equal to the tensile strength of the bulk material, see for example Bowden and Tabor (1964). However, when hard solids are pressed together the position is rather different, as shown in a series of experiments by Gane et al. (1974) who loaded two surfaces together and then measured the force required to pull them apart. It was found that the bonding between hard elastic solids such as germanium was extremely small, the authors explained this result in terms of the inevitable microscopic roughness of the two surfaces. In particular any high points among the surface asperities will exert large elastic forces which will tend to rupture any bond when the load is released.

If a hard elastic solid is pressed against a plastic solid the position is again different. Because of the plasticity any high loading at a hard asperity will be spread over a greater area thus reducing the elastic energy which tends to disrupt the bonding. This effect almost certainly accounts for the results shown in Figure 13.28

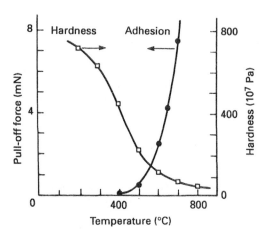

Figure 13.28 Values of the force required to pull apart two clean surfaces of germanium pressed together under a load of 5 mN as a function of temperature (Gane et al., 1974)

for the force required to pull apart two clean surfaces of germanium pressed together under a load of 5 mN. Below 400°C the germanium is hard and brittle and exhibits very little adhesion but above this temperature the materials soften and the adhesion increases considerably. Experiments such as these suggest that there will be very little adhesion between polished diamond surfaces at room temperature. In fact the adhesion measured was so low that it was difficult to obtain precise values (Figure 13.29); the figure also shows the considerably greater adhesion obtained when diamond was brought into contact with a clean copper surface, also in a high vacuum.

The bonding of two diamond surfaces has also been studied by pressing them together under a load and measuring either their static or sliding friction (the static friction being the force needed to initiate relative motion of the surfaces). Pepper

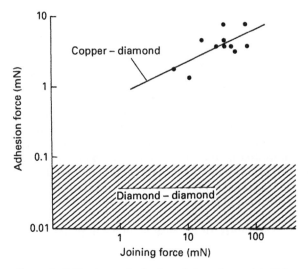

Figure 13.29 The force of adhesion of clean diamond in high vacuum as a function of initial applied load (Tabor, 1979)

(1982) measured the static friction between a copper stylus and a diamond surface in ultra-high vacuum at room temperature after the diamond surface had been subjected to various annealing treatments. A polished diamond surface annealed at 750°C gave a coefficient of friction of about 0.1 but after heating to about 850°C the value of the coefficient increased to about 0.6. Then on exposing the latter surface to atomic hydrogen at about 850°C the friction fell to a value of about 0.3. Hence the high value of 0.6 appears to be associated with the reconstructed diamond surface (Section 13.7), and Pepper discusses the increase in friction in terms of interactions between the electrons in the copper and the extra electron states introduced by the reconstruction.

In experiments on sliding friction Miyoshi and Buckley (1980) studied the variation of the strength of bonding between diamond and a range of the so-called transition metals. The diamond surface was prepared in a high vacuum by ion bombardment and the metals were machined to the form of rounded pins. The values of the coefficients of friction ranged from about 0.4 to 1.1 with the majority falling between 0.4 and 0.5 (Figure 13.30). The authors relate these values to the

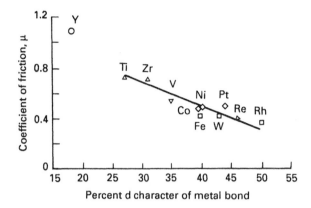

Figure 13.30 Coefficient of friction of a {111} surface of diamond sliding in a <110> direction in a vacuum of 10^{-8} Pa on various transition metals plotted as a function of the percent d character of the metal bond (Miyoshi and Buckley, 1980)

concept of the 'percent d character of the metal bond' using the values for particular metals given by Pauling (1949). However, the only point clearly established is that yttrium, titanium and zirconium which have only two d shell electrons exhibit a higher friction presumably because the greater number of empty d states encourage stronger bonding with the diamond. (The coefficient of both static and sliding friction, will depend on the roughness and topography of the surfaces besides the strength of bonding. Therefore it is probably optimistic to expect detailed correlation between the friction and the nominal percent d character of the metal particularly as no account is taken of the nature of the electron bonding in the diamond.)

13.8.c Oxygen and wettability

Besides carrying a surface layer of hydrogen there is also the possibility that a diamond may sometimes carry a significant amount of oxygen. This oxygen has been detected in experiments by Derry, Madiba and Sellschop (1983) and Hansen *et al.* (1988) who irradiated polished surfaces with 1 MeV protons and observed the energy spectra of Rutherford backscattered atoms. They found no oxygen on surfaces which had been washed with organic solvents, but on surfaces washed with a commercial detergent they identified the presence of ^{16}O and estimated that its concentration was of the order of a monolayer.

The presence of oxygen was confirmed by XPS measurements which gave the concentration as 0.39 monolayers at ambient temperature and 0.22 monolayers after heating to 500 K. (This result suggests that any oxygen will be completely removed during the heating process prior to a reconstruction of the surface). It was also observed that heating in oxygen at 770 K increased the coverage to 1.4 monolayers, while heating in hydrogen at 1020 K reduced the concentration to 0.16 monolayers which was near the limit of detection. As suggested by the authors, it seems likely that the action of the detergents is to replace hydrogen atoms bonded to carbon with –OH hydroxyl groups.

As oxygen bonds are polar they tend to attract water molecules, so the presence of oxygen affects the wettability of diamond, as shown in the experiments of

Hansen *et al.* (1988). These authors assessed the wettability by weighing a parallel-sided slab of diamond half in and half out of water, and after allowing for buoyancy effects arrived at the value of the line force due to the liquid acting vertically down on the diamond. If the diamond is wetted by the water the line force will equal the surface tension of water (72 mN m^{-1}) but the measured force was appreciably less, showing that diamond tends to repel water, and the authors used the value of the force as a meaure of the wettability of diamonds with different surface conditions.

In measurements of the above type Hansen *et al.* (1988) observed that polished surfaces washed with detergent gave tensions of about 40 mN m^{-1}. Heating in oxygen to 700 K increased this tension to about 70 mN m^{-1}, whereas heating in hydrogen to remove the oxygen reduced the tension to about 20 mN m^{-1}. Hence the tension appears to increase with the concentration of oxygen on the surface as expected. The authors also note that the wettability is increased by the presence of ferric Fe^{+++} ions though not by chromic ions. We refer to the wettability of diamond by metals in Section 14.2.d.

References

Bassett, J., Denney, R.C., Jeffrey, G. H. and Mendham, J. (1978) *Vogel's Textbook of Quantitative Inorganic Analysis,* (4th edn). Longman, London

Bowden, F. P. and Scott, H. G. (1958) *Proceedings of the Royal Society,* A248, 368–378

Bowden, F. P. and Tabor, D. (1950) *The Friction and Lubrication of Solids.* Clarendon Press, Oxford

Bowden, F. P. and Tabor, D. (1964) *The Friction and Lubrication of Solids,* Part II. Clarendon Press, Oxford

Bowden, F. P. and Tabor, D. (1965) *In Physical Properties of Diamond,* (ed. R. Berman), Clarendon Press, Oxford, pp. 184–220

Bowden, F. P. and Thomas, P. H. (1954) *Proceedings of the Royal Society,* A223, 29–39

Bundy, F. P., Bovenkerk, H. P., Strong, H. M. and Wentorf, R. H. (1961) *Journal of Chemical Physics,* 35, 383–391

Casstevens, J. M. (1983) *Precision Engineering,* 5, 9–15

Crompton, D., Hirst, W. and Howes, M. G. W. (1973) *Proceedings of the Royal Society,* A333, 435–454

Culf, C. J. (1957) *Journal of the Society of Glass Technology,* 41, 157T–167T

Danyluk, S. and Clark, J. L. (1985) *Wear,* 103, 149–159

Davies, G. and Evans, T. (1972) *Proceedings of the Royal Society,* A328, 413–427

Dayan, M. and Pepper, S. V. (1984) *Surface Science,* 138, 549–560

Derry, T. E., Madiba, C. C. P. and Sellschop, J. P. F. (1983) *Nuclear Instruments and Methods in Physics Research,* 218, 559–562

Derry, T. E., Smit, L. and Van der Veen, J. F. (1986) *Surface Science,* 167, 502–518

Duwell, E. J., Hong, I. S. and McDonald, W. J. (1969) *Transactions of the ASLE,* 12, 86–93

Eiss Jr, N. S. and Fabiniak, R. C. (1966) *Journal of the American Ceramic Society,* 49, 221–226

Evans, S. (1978) *Proceedings of the Royal Society,* A360, 427–443

Evans, S. (1980) *Proceedings of the Royal Society,* A370, 107–129

Evans, S. and Ney, M. (1990) *Journal of Hard Materials,* 1, 169–181

Evans, S. and Thomas, J. M. (1977) *Proceedings of the Royal Society,* A353, 103–120

Evans, T. (1979) In *The Properties of Diamond,* (ed. J. E. Field), Academic Press, London, pp. 403–424

Evans, T. and James, P. F. (1964) *Proceedings of the Royal Society,* A277, 260–269

Evans, T. and Phaal, C. (1962) In *Proceedings of the Fifth Conference on Carbon,* Pennsylvania State University, Pergamon Press, Oxford, pp.147–153

Evans, T. and Sauter, D. H. (1961) *Philosophical Magazine,* 6, 429–440

Evans, U. R. (1981) *An Introduction to Metallic Corrosion,* (3rd edn) Edward Arnold, London

Fedoseev, D. V., Vnukov, S. P., Bukhovets, V. L. and Anikin, B. A. (1986) *Surface and Coatings Technology*, **28**, 207–214

Gane, N., Pfaelzer, P. E. and Tabor, D. (1974) *Proceedings of the Royal Societ*, **A340**, 495–517

Gorodetskii, A. E., Builov, L. L. *et al.* (1971) *Fiziko-Khimicheskiye Problemy Kristallizatsii*, No.2, 62–67

Gorodetskii, A. E., Lukyanovich, V. M. *et al.* (1972) *Smachivayemost i Poverkhnostniye Svoistva Rasplavov i Tverdykh Tel*, 125–127

Graham, W. and Nee, A. Y. C. (1974) *The Production Engineer*, June, 186–191

Grunberg, L. (1953) *Proceedings of The Physical Society*, **66B**, 153–161

Hansen, J. O., Derry, T. E., Harris, P. E. *et al.* (1988) In *Ultrahard Materials Application Technology*, Vol. Four, (ed. C. Barrett), De Beers Industrial Diamond Division, London, pp. 76–87

Harding, J. (1971) *Acta Metallurgica*, **19**, 1177–1188

Himpsel, F. J., Knapp, J. A., Van Vechten, J. A. and Eastman, D. E. (1979) *Physical Review B*, **20**, 624–627

Hitchiner, M. P. and Wilks, J. (1984) *Wear*, **93**, 63–80

Hitchiner, M. P., Wilks, E. M. and Wilks, J. (1984) *Wear*, **94**, 103–120

Howes, V. R. (1962) *Proceedings of the Physical Society*, **80**, 648–662

Hsu, S. M. and Klaus, E. E. (1979) *Transactions of the ASLE*, **22**, 135–145

Ihm, J., Louie, S. G. and Cohen, M. L. (1978) *Physical Review B*, **17**, 769–775

Ikawa, N. and Tanaka, T. (1971) *Annals of the C.I.R.P.*, **XXIV**, 153–157

Jaeger, J. C. (1942) *Journal and Proceedings of the Royal Society of New South Wales*, **76**, 203–224

Johnson, K. L. (1985) *Contact Mechanics*. Cambridge University Press, Cambridge

Lambert, H. J. (1961/62) *C.I.R.P.– Annalen, Bd.X*, 246–255

Lander, J. J. and Morrison, J. (1966) *Surface Science*, **4**, 241–246

Langitan, F. B. and Lawn, B. R. (1970) *Journal of Applied Physics*, **41**, 3357–3365

Lawn, B. R. (1973) In *The Science of Hardness Testing and its Research Applications*, (eds J. H. Westbrook and H. Conrad), American Society of Metals, Metals Park, Ohio, pp. 418–431

Lawn, B. R. (1974) *Materials Science and Engineering*, **13**, 277–283

Lawn, B. R. (1975) *Journal of Materials Science*, **10**, 469–480

Lim, D. S. and Danyluk, S. (1985) *Journal of Materials Science*, **20**, 4084–4090

Loparev, V. V., Veprinstev, V. I., Manukhin, A. V. and Funtikov, E. V. (1984) *Soviet Journal of Superhard Materials*, **6**, 12–16

Lurie, P. G. and Wilson, J. M. (1977) *Surface Science*, **65**, 453–475

Maan, N. and Broese van Groenou, A. (1977) *Wear*, **42**, 365–390

Mehan, R. L. and Hayden, S. C. (1981) *Wear*, **74**, 195–212

Mikosza, A. G. and Lawn, B. R. (1971) *Journal of Applied Physics*, **42**, 5540–5545

Miller, D. R. (1962) *Proceedings of the Royal Society*, **A269**, 368–384

Miyoshi, K. and Buckley, D. H. (1980) *Applications of Surface Science*, **6**, 161–172

Moore, M. A. and King, F. S. (1980) *Wear*, **60**, 123–140

Morar, J. F., Himpsel, F. J., Hollinger, G. *et al.* (1986) *Physical Review B*, **33**, 1340–1345

Ogorodnik, V. V., Pugach, E. A. and Postolova, G. G. (1985) *Thermochemica Acta*, **93**, 705–708

Pate, B. B. (1986) *Surface Science*, **165**, 83–142

Pate, B. B., Hecht, M. H., Binns, C. *et al.* (1982) *Journal of Vacuum Science Technology*, **21(2)**, 364–367

Pate, B. B., Hwang, J. C., Woicik, J. and Lindau, I. (1989) In *Proceedings of the First International Conference on the New Diamond Science and Technology* (Tokyo, October 1988) Japan New Diamond Forum, pp. 174–175

Pate, B. B., Spicer, W. E., Ohta, T. and Lindau, I. (1980) *Journal of Vacuum Science Technology*, **17**, 1087–1093

Pate, B. B., Stefan, P. M., Binns, C. *et al.* (1981) *Journal of Vacuum Science Technology*, **19(3)**, 349–354

Pauling, L. (1949) *Proceedings of the Royal Society*, **A196**, 343–362

Pekelharing, A. J. (1959) *Annals of C.I.R.P.*, **VIII**, 112–120

Pepper, S. V. (1982) *Journal of Vacuum Science Technology*, **20(3)**, 643–646

Prutton, M. (1984) *Electronic Properties of Surfaces*. Adam Hilger, Bristol

Quinn, T. F. J. and Winer, W. O. (1987) *Transactions of the ASME, Journal of Tribology*, **109**, 315–320

Rabinowicz, E. (1965) *Friction and Wear of Materials*. John Wiley, New York

Roberts, M. W. (1981) *Chemistry in Britain*, **17**, 510–514

Rowe, G. W. and Smart, E. F. (1966–67) *Proceedings of the Institution of Mechanical Engineers*, **181**, 48–57

Samuels, L. E. (1971) *Metallographic Polishing by Mechanical Methods*, (2nd edn) Pitman, London

Sappok, R. and Boehm, H. P. (1968) *Carbon*, **6**, 283–295

Shaw, M. C. (1958/9) *Wear*, **2**, 217–227

Shaw, M. C. (1984) *Metal Cutting Principles*. Clarendon Press, Oxford

Simons, E. L. and Cannon, P. (1966) *Nature*, **210**, 90–91

Suh, N. P. (1967) *International Journal of Mechanical Science*, **9**, 415–431

Tabor, D. (1954) *Proceedings of the Physical Society B*, **67**, 249–257

Tabor, D. (1977) *Transactions of the ASME, Journal of Lubrication Technology*, **99**, 387–395

Tabor, D. (1979) In *The Properties of Diamond*, (ed. J. E. Field), Academic Press, London, pp. 325–350

Tanaka, T. and Ikawa, N. (1973) *Bulletin Japanese Society of Precision Engineering*, **7**, 97–101

Thomas, J. M. (1979) In *The Properties of Diamond*, (ed. J. E. Field), Academic Press, London, pp.211–244

Thornton, A. G. and Wilks, J. (1979) *Wear*, **53**, 165–187

Thornton, A. G. and Wilks, J. (1980) *Wear*, **65**, 67–74

Uetz, H. and Sommer, K. (1977) *Wear*, **43**, 375–388

Vidali, G. and Frankl, D. R. (1983) *Physical Review B*, **27**, 2480–2487

Vishnevskii, A. S., Lysenko, A. V., Ositinskaya, T. D. and Delevi, V. G. (1975) *Izvestiya Akademii Nauk SSSR, Neorganicheskie Materialy*, **11**, (9), 1589–1593

Waclawski, B. J., Pierce, D. T., Swanson, N. and Celotta, R. J. (1982) *Journal of Vacuum Science Technology*, **21**(2), 368–370

Westwood, A. R. C. (1974) *Journal of Materials Science*, **9**, 1871–1895

Williams, J. A. (1977) *Journal Mechanical Engineering Science*, **19**, 202–212

Williams, J. A. and Tabor, D. (1977) *Wear*, **43**, 275–292

Woodruff, D. P. and Delchar, T. A. (1986) *Modern Techniques of Surface Science*. Cambridge University Press, Cambridge

Chapter 14

Turning, boring and milling

Diamond can machine hard and difficult workpieces because of its strength and abrasion resistance. Tools made from single crystal diamond can be polished to sharp edges which give extremely fine finishes on suitable materials. This chapter discusses various aspects of turning with diamond but much of the discussion is equally relevant to milling and boring operations which are quite similar cutting processes.

14.1 Single crystal and PCD tools

Table 14.1 gives examples of the types of materials now being worked with diamond cutting tools. It also indicates some of the materials for which diamond tools are unsuitable either because the diamond wears too quickly or fails to produce a satisfactory finish. The most common cause of a poor performance is that the workpiece reacts chemically with the diamond as discussed in Section 14.4.d but it is difficult to predict how a particular material will behave. The only sure way to determine how a diamond tool will behave in a particular situation is to make a trial run. (For example, nickel is listed in Table 14.1 as an unsuitable material, but chemically deposited nickel plating containing about 10% phosphorus behaves quite differently, and is used to coat high precision mirrors before they are turned with diamond, see Section 14.4.d.)

The various applications of diamond tools tend to fall into a spectrum which at one end involve taking shallow cuts to obtain maximum dimensional accuracy and a high standard of surface finish, and at the other end deeper cuts to obtain an

Table 14.1 Typical materials machined with diamond (after Gerchman (1986) and Becker (1988))

Metals
Aluminium alloys, copper alloys, magnesium alloys, zinc alloys, gold, silver, tin, electroless nickel

Plastics
Acrylics, polymethylenemethacrylate, polyvinyl chloride, fibre reinforced plastics

Dielectrics
Ceramics, glass, carbon and graphite, geranium, silicon, gallium arsenide, zinc sulphide, wood, chipboard

Materials unsuitable for diamond machining
Iron, cobalt, nickel, steel, beryllium, molybdenum, titanium, sialon, silicon based glasses and ceramics

Figure 14.1 View of the cutting edge of a PCD tool of 10 μm grain size (De Beers Industrial Diamond Division)

appreciable removal of material with less regard to the surface finish. These different requirements are usually met by using tools made from single crystal diamond for the finer work and tools made from PCD material when deeper cuts and larger forces are involved.

A good single crystal tool can produce high precision finishes because of its extremely fine edge and the low friction of its surfaces (Section 11.5). The use of single crystal diamond is essential to obtain the best finish because the grain size of polycrystalline material sets a relatively coarse limit to the precision of the edge. Thus Figure 14.1 gives a micrograph of a PCD tool of 10 μm grain size which had been carefully polished to give a good edge . At the same magnification a well polished single crystal tool would show no sign of surface structure, but the PCD edge shows marked imperfections which are particularly obvious in the top left part of the figure. In addition, the solvent/catalyst in PCD material will probably increase the friction with the workpiece (Section 12.4) which will also have a bad effect on the finish.

Polycrystalline PCD material comes into its own at the other end of the spectrum. It has a superior toughness, or resistance to chipping and fracture (Section 12.2.c), and material for tools are generally grown on a tungsten carbide base to give additional strength. For example, PCD tools are used in the automobile industry to machine abrasive aluminium alloys and outperform other types of tools by a large factor. In addition PCD material has the considerable advantage that it is isotropic without the complications of orientation which are involved when using single crystal diamond.

14.2 Fabrication of tools

The basic geometry of the diamond in a typical round-nosed turning tool is shown schematically in Figure 14.2 which indicates the *flank* and the *rake face* (or *table*)

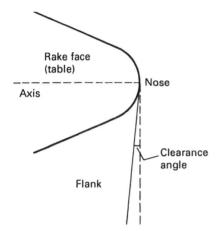

Rake face
(table)

Axis

Nose

Clearance
angle

Flank

Figure 14.2 Sketch showing the diamond in a typical round–nose turning tool

on either side of the cutting edge, and the *clearance angle* between the flank and the normal to the rake face. The latter is kept to no more than a few degrees so as to give as much support to the edge as possible. We now outline the methods of fabricating such tools from either single crystal or PCD material. The use of single crystal tools is much complicated by the fact that both the polishing of the diamond and its wear during use depend greatly on its crystallographic orientation.

14.2.a Single crystal tools

To obtain the best performance from a tool the flanks and the rake face must be polished as smooth as possible to minimize friction and to ensure that the edge where they meet is clean and sharp. We have already seen that the best finishes are generally obtained by polishing in a direction of easy abrasion (Section 9.3.d) but nearly all the published information on the polishing of diamond deals with the preparation of surfaces, whereas the performance of a tool depends on the strength and quality of the cutting edge. There are, however, three well known rules of thumb. The diamond must be positioned so that the lie of the cleavage planes does not encourage the stresses during use to cause chipping by cleavage. When polishing near the cutting edge the direction of polish must be in a direction towards the edge and the diamond and not away from them, to avoid generating tensile stresses. Because it is easier to obtain a good polish on the flat rake face than on the curved flanks, with their varying orientations, the final stage in producing a tool is to give a fine polish to the rake face in order to produce as good an edge as possible.

Besides the above considerations, there is also the requirement that the tool operates with a minimum of wear, this wear may depend considerably on the crystallographic orientation of the diamond. Examples of the effect of orientation when turning aluminium alloys are given in Section 14.4.c but generally speaking there is not much published information on the performance of differently oriented tools on different materials. Therefore the tool maker proceeds on the basis of his experience. The rake face is often oriented as {011} or {001} with the axis of tool and shank parallel to <100>. However, it is possible that an optimum orientation for a particular tool may sometimes be found a few degrees away from these planes and directions. The best orientation is still the subject of experimentation, see for example Hurt and Showman (1986).

The positions of the {001} and {011} planes and the <100> and <110> axes are readily recognized if the diamond is a regular octahedron (Figures 5.9 and 5.10). Alternatively the orientation may be determined by X-ray diffraction as mentioned in Section 5.2.a, see also Weavind, Guykers and Roy (1958), Raal (1963) and Cullity (1978). This procedure is now greatly facilitated by the use of image intensifiers which permit an almost instantaneous determination of the orientation, and thus a ready adjustment of the diamond in its holder.

14.2.b Selection of diamond

A toolmaker will generally prefer to use PCD material if this will give a fine enough finish as it is tougher and presents no problems of orientation. Quality control of the material is also provided by the manufacturer. Hence, apart from the choice of supplier, the toolmaker has only a few points to decide on. Polycrystalline material is generally available with crystallites of various size, perhaps nominally $2\,\mu m$, $10\,\mu m$ and $25\,\mu m$ in diameter. Larger crystallites tend to give a better resistance to wear (Figure 14.3) but cannot take as fine an edge as the $2\,\mu m$ material. Hence a compromise has to be made for the particular task concerned. There is also the possibility of using the PCD variants described in Chapter 12.2.e which are more stable at high temperatures. In another variant the tungsten carbide base which serves as a support for the PCD extends round the outer rim of the PCD to obtain stronger bonding of the PCD to the metal shank.

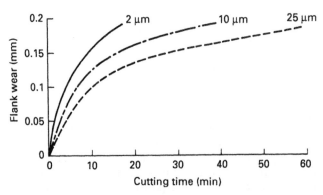

Figure 14.3 Flank wear of three grades of PCD tools of grain size approximately 2, 10, and 25 μm turning a silica-flour-filled epoxy resin: cutting speed $7\,m\,s^{-1}$ (Heath and Aytacoglu, 1984)

The task of selecting single crystal diamonds is more difficult. Their main use is either as single point dressers (Section 17.2) or as high precision cutting tools, and for precision work careful selection is necessary in order to realize their full potential. At present, single crystal diamonds used as turning tools are mainly natural products. These can be of very varied quality so the toolmaker must be able to assess whether a diamond is suitable for his needs. The first step in an assessment is to make a visible inspection of the diamond for the presence of any cracks or inclusions. If the diamond is a regular octahedron with good quality faces the whole of the diamond may be viewed quite readily, but if the form of the diamond is irregular or its faces are not fully transparent then it is essential to polish two opposite parallel windows preferably either {001} or {011} planes to obtain a good view of the interior.

The degree of perfection needed in a diamond is determined by the precision of the required finish. In any case it is desirable to view any diamond at a magnification of at least ×100, preferably by using a stereoscopic optical microscope to reveal any major imperfection in the body of the stone. Typical examples of inclusions and cracks have been given in Sections 2.2 and 10.1. It is essential to avoid any inclusions or cracks anywhere near a cutting edge and preferable to avoid them altogether. Examinations at up to ×500 magnification can also be made but only using special microscope objectives with a long working distance. (The ordinary high power objective has too short a working distance to view the interior of a diamond of any thickness, and this difficulty is accentuated by the high refractive index which increases the effective thickness of the diamond.)

The microscope should also be equipped with polarizing and analysing plates to check that the diamond is free of birefringence and internal strain (Section 2.3.a). An example of the effect produced by internal strain is seen in Figure 14.4(a) which shows a diamond slab which was being polished to obtain a sharp edge on the right hand sloping section. As the polishing proceeded the diamond began to take a nice

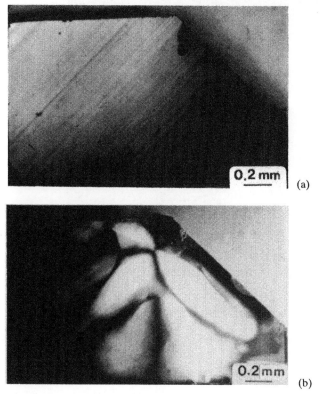

(a)

(b)

Figure 14.4 A polished edge viewed in the optical microscope: (a), with Nomarski technique; (b), in polarized light (Casey and Wilks, 1976b)

edge, but small nicks appeared from time to time and had to be polished out. Finally, just before the edge was finished, a relatively large flake of diamond chipped out as shown in the figure. The stone was then viewed in polarized light, and the birefringence pattern of Figure 14.4(b) was observed. Using this technique a strain-free stone would appear totally dark, whereas areas of light indicate the presence of strains. Areas of light spread out from a point near the chip, but where the flake came away the micrograph is quite dark, indicating that the removal of

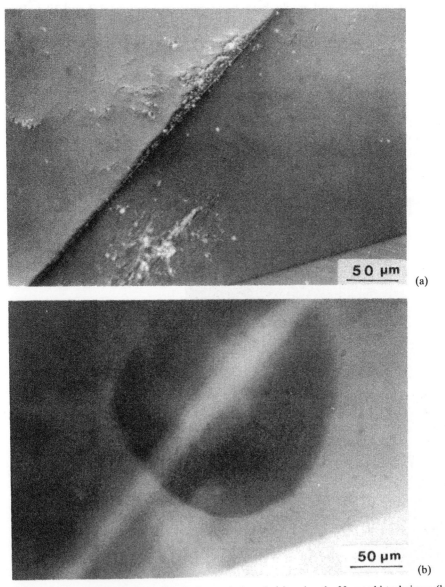

(a)

(b)

Figure 14.5 Shallow edge on a polished diamond viewed: (a), using the Nomarski technique; (b), in polarized light (Casey and Wilks, 1976b)

the chip has released the strain. As another example Figure 14.5(a) shows the edge between two almost parallel facets polished on a diamond. In the centre of the picture the edge is broken as if small pieces were being chipped out by the polishing process. An examination in polarized light, Figure 14.5(b) indicates some defect in the diamond below the surface which is obviously responsible for the chipping of the edge.

Another technique for the inspection of diamonds is the use of dark ground illumination to show up inclusions and defects, see Sections 2.3.c and 6.2.f. In this technique an intense beam of light is passed through two windows polished on the diamond and any scattered light is viewed through a third window on a line perpendicular to the beam (Figure 2.11). Any inclusion large enough to be visible in the optical microscope gives rise to intense scattering when illuminated by the beam. The technique also shows up scattering from smaller inclusions but the interpretation of this lower level of scattering is complicated by the fact that scattering may also arise from the presence of dislocations.

(There is little information on how dislocations may affect the performance of a tool, but diamonds containing high densities of dislocations are generally either brown Type I stones or Type II, and have a somewhat higher resistance to abrasion (Sections 10.2.c). The greater abrasion resistance suggests that these diamonds might be particularly suitable for tools but brown Type I stones usually have a complicated growth structure which leads to marked birefringence and a poor cleavage (Section 10.3). This leaves only the possibility of using Type II diamonds but there is no practical evidence that they give a superior performance. In any case Type II diamonds are already in considerable demand for use as heat sinks (Section 17.6.d).)

Two other techniques of inspecting diamonds have been mentioned in previous chapters: X-ray topography (Section 2.3.b) and the cathodoluminescent mode of the SEM (Section 4.1). The X-ray technique is quite complex but the CL technique offers the possibility of a simple and rapid inspection by viewing the luminescence of a polished face in the SEM. As described in Section 5.2.c CL, patterns can reveal growth layers which may produce irregularities when polishing, and which if complex are usually accompanied by strain and birefringence. These patterns are best viewed on either {001} or {011} faces as the growth planes are then most apparent. Probably the ideal diamond for a tool will be one with a very low level of luminescence and if there is any pattern visible it should be simple and regular. (Despite various remarks in the literature there is little evidence that the presence of platelets correlates with mechanical strength (Section 10.5.c), although a very large platelet located right at the edge of a tool would doubtless cause trouble.)

14.2.c Fabrication

When making a tool the diamond or block of PCD must be secured to a metal shank before shaping and polishing. This is a fairly straightforward operation with PCD material grown on tungsten carbide substrates which can be readily brazed to metal. However, the temperature must be kept below about 700°C in order to avoid damage to the PCD, see Section 12.2.d and Bex (1979).

Single crystal diamond is not so readily brazed and various methods have been used to hold it in position. Perhaps the most common has been to surround the diamond, positioned on the shank, with a sintering powder of a metal such as a copper alloy which is then heated to a temperature of about 700°C. The powder coheres and holds the diamond in place by differential contraction as the tool cools. In order to produce a strong joint the diamond must of course be of such a shape that it can be held firmly by the sintered metal. Note also that care must be taken that the differential contraction does not produce excessive strains on the diamond. Details of other methods of fixing, including purely mechanical mountings, are given by Grodzinski (1953).

Today developments in the brazing of diamond make it possible to obtain strong diamond–metal bonds by using brazes such as those described in the following section. Because these brazes melt at temperatures high enough to cause the diamond to oxidize the brazing must be carried out in either a vacuum or an inert atmosphere. The required temperature may be conveniently achieved by induction heating. As the diamond may tend to move on the shank during brazing it may be convenient to first braze the diamond to a thin sheet of a suitable metal, such as molybdenum, somewhat larger than the diamond. This sheet may then be positioned relatively easily on the steel shank and secured in place with another braze at a somewhat lower temperature.

After the diamond is attached to the shank it is shaped and polished to produce a tool with a cutting edge, flat table and smooth flanks. This polishing presents certain problems. It is difficult to obtain a very smooth polish on curved surfaces because of the varying crystallographic orientation yet any roughness on the curved flanks where they intercept the rake face will produce an imperfect edge. The flanks are usually polished on a radiusing machine and its bearings must be sufficiently massive and well designed to reduce any play or vibration to a minimum. The procedure for shaping a tool of PCD is very similar except that as the PCD is very resistant to polish it is necessary to use greater pressures and therefore more massive polishing machines. For other comments on polishing machines see Sections 9.6.a and 12.3.b.

The details of the final inspection given to a tool will depend on the task which it has to perform, but for precision work the edge should be viewed in the optical microscope with magnifications up to ×500. An optical inspection is often adequate but for the highest precision use is now made of the SEM which gives much higher magnifications. In addition, the great depth of focus of the instrument permits both the edge and the two adjacent surfaces to be viewed at the same time whereas this is not possible in the optical microscope except at low magnifications.

Finally, it is not too difficult an operation to check the crystallographic orientation of a diamond tool by taking a back reflection X-ray diffraction picture, see for example Cullity (1978). This information is often significant because the orientation may affect both the quality of the polish which can be produced during fabrication and the wear of the tool. Checks on the orientations of a set of diamond tools also give information on the ability of the toolmaker to orient the stone within given limits and to maintain consistency of performance.

14.2.d Brazing

We now describe various experiments made to develop suitable brazes for bonding diamonds to metals. The first extensive studies concerned the wetting of diamond and graphite by metals (Naidich and Kolesnichenko, 1961, 1963, 1964, 1966). These authors measured the angles of contact between diamond and a range of molten metals and found that several elements in the transition group gave low angles and good wetting but that outside this group only aluminium and silicon behaved in a similar manner. However, it was also found that copper, silver and tin, alloyed with titanium or chromium, gave good wetting (Naidich and Kolesnichenko, 1964, 1966).

To be effective a braze metal must flow and wet the diamond and must also form a bond with the diamond at least as strong as the strength of the metal. It should not be brittle so that it can accommodate any stresses set up by differential contractions as the tool cools after brazing. Satisfactory brazes may be obtained using the relatively inert metals copper, silver or gold with a few parts per cent of titanium or tantalum. These two transition elements form stable carbides and appear to remove any oxygen or hydrogen adsorped on the diamond surface. According to Paterson and Taylor (1974) good results are obtained using a braze with a constitution by weight of 95% gold, 4.85% tantalum and 0.15% silicon, usually prepared as a thin foil 30 μm to 50 μm thick. References to other suggested brazes are given by Seal (1969). It is also worth noting that at the high temperatures involved (1000°C–1200°C) any oxygen present in the vacuum chamber will combine with the transition metal and reduce its chemical activity (as well as encouraging graphitization of the diamond).

Various experiments have been made to study and optimize the parameters which control the brazing process. Scott, Nicholas and Dewar (1975) measured the contact angles between diamond and drops of molten copper containing up to 10% of chromium or titanium. The specimens were held for 20 min at 1150°C, allowed to cool, and the strength of the bond measured by shearing it in a tensometer. The results with chromium and titanium were quite similar and those for chromium are given in Figures 14.6 and 14.7, the maximum bond strength being given by concentrations of chromium less than those needed to produce maximum wetting. An electron probe analysis of some broken metal–diamond interfaces showed a high concentration of chromium suggesting the presence of a boundary layer consisting primarily of carbide. The authors suggest that the bond strength initially increases with concentration because the surface becomes covered by greater areas of carbide. However, when the whole surface is covered, a further increase in concentration results in a thicker but weaker film. Other rather similar measurements on a range of copper–tin–titanium alloys are described by Evens, Nicholas and Scott (1977).

Dewar, Nicholas and Scott (1976) measured the bond strengths obtained by pressing cylinders of copper or nickel alloyed with chromium, titanium or boron against a diamond at 800°C in a simple vacuum chamber under a pressure of 3.6 kg mm^{-2}, sufficient to plastically deform the metal. The increase and decrease of bond strength with the time of bonding, see Figure 14.8, recall the form of the results in Figure 14.7, but the details of the process are not easy to interpret.

Naidich, Umanski and Lavrinenko (1984) made vapour depositions of molybdenum, tungsten, chromium and titanium on to diamond and measured the strength of bonding by brazing a molybdenum rod to the metal and pulling it off. The bond strength exhibits a maximum value for a particular bonding temperature (Figure 14.9). Other experiments, at these optimum temperatures, showed that the bond strength passed through a maximum value as the time of deposition increased. For bonding conditions corresponding to the rising parts of the strength versus temperature curves, the failure when the coating was pulled off occurred at the diamond–metal interface, but for bonding conditions on the falling part of the curves the failure occurred within the diamond. The authors proposed that the

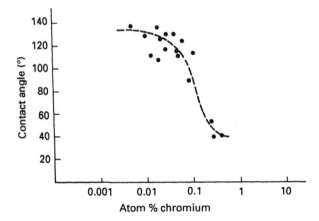

Figure 14.6 Contact angle between different copper–chromium alloys and diamond (Scott, Nicholas and Dewar, 1975)

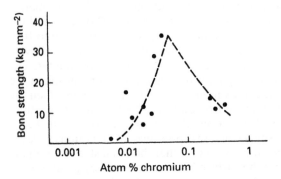

Figure 14.7 The bond strength between solidified sessile drops of copper–chromium alloys and diamond (Scott, Nicholas and Dewar, 1975)

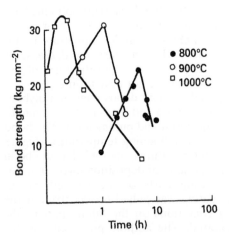

Figure 14.8 The effects of fabrication time and temperature on the bonding of a copper–0.09% boron alloy to diamond (Dewar, Nicholas and Scott, 1976)

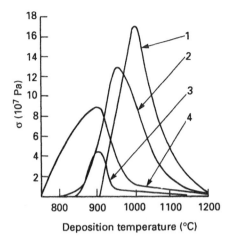

Figure 14.9 The effect of deposition temperature on the bond strength σ between a metal coating and diamond for a deposition time of 1 h: 1 chromium, 2 molybdenum, 3 tungsten, 4 titanium (Naidich, Umanskii and Lavrinenko, 1984)

strength of the bond falls off beyond the maximum for two reasons. First, the carbon diffuses more quickly into the metal than the metal into the diamond, so that if the bond remains too long at high temperatures this diffusion results in the formation of vacancies and voids in the diamond which reduce its strength. Second, the higher temperatures and times encourage graphitization of the diamond.

14.3 Fracture of tools

We now consider how the tool user can best avoid the fracture and chipping which have sometimes been a serious problem. For example, Keen (1974) noted that in one factory turning automotive pistons from aluminium–silicon alloys 60% to 80% of the tools failed because of fracture of the cutting edge. We also know other surveys made at about the same time in factories turning copper commutators, components of vulcanite, and automotive pistons. Provided chipping did not occur tools might produce several thousand components before the edge became blunt and needed repolishing. However, they often failed by chipping, sometimes after producing only a few components, for example one survey showed that 41% of the tools produced only 12% of the output. These problems have now been much reduced, by the use of PCD material, by greater care in the fabrication of single crystal tools, and by greater attention to their proper use in the factory, see for example Hervo (1983).

14.3.a Chipping of single crystals

The production of a good edge resistant to chipping is achieved by care in the selection and fabrication of the diamond as described above. In particular, the diamond must be free of internal stresses and inclusions. Stresses must not be introduced during the fabrication of the tool, and the final polishing must be a fine one to keep any residual cracks as small as possible. Careful attention is also necessary to avoid setting the diamond in orientations that lead to increased tensions across the {111} cleavage planes where most chipping occurs. The orientations generally used are chosen to take account of this point and are probably as good a compromise between achieving both a strong edge and a good polish on face and flanks as can be obtained. (A theoretical analysis of the stresses in a tool is given by Ikawa and Shimada (1982).)

After the toolmaker has completed his work the responsibility for the performance of the tool passes to the user. Diamond tools must only be used in suitable machine tools in good condition. Successful machining involving brittle materials either as tool or workpiece requires a high degree of stiffness in the machine tool. It is vital that the tool be held rigidly because any vibration will produce both an inferior finish and a tendency for the tool to fracture. Yet diamond tools are sometimes mounted on overlong metal shanks with too little stiffness. To ensure the correct positioning of the diamond relative to the workpiece the tool is best mounted on a square shank which fits into a holder permitting only the minimum necessary adjustment in order to avoid incorrect settings. It is also essential that there is a minimum of play in all the components of the lathe or machine tool particularly the bearings and the slide movements.

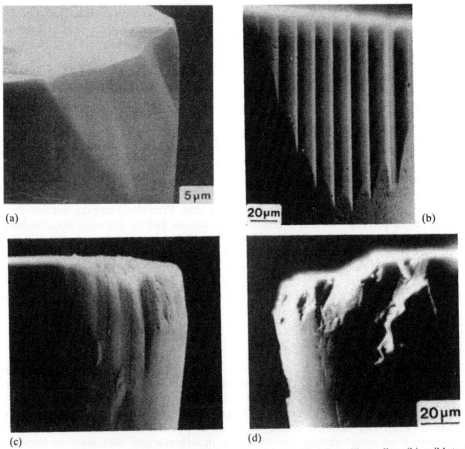

Figure 14.10 Wear lands on diamond tools after turning: (a), an aluminium silicon alloy; (b), mild steel; (c), vulcanite; (d), molybdenum (Wilks, 1980)

The cutting edge of a good diamond tool is relatively robust but like any sharp edge must be protected from accidental impacts particularly with hard metals. When not in use the tip of a tool should at all times be covered with a protective cap to guard against chipping. The tool should be stored in a shock-absorbent material within a small container, so that the container may be dropped without damage to the tool. The machine operator should have a definite place where he can open the box and take out the tool without any risk of dropping it, and finally the cap should only be removed from the tip when the tool is in position on the lathe. Not only do such procedures protect the tools, but they draw the operator's attention to the care needed in handling them.

Chipping and fracture may be much influenced by the nature of the workpiece. Diamond may bond with a particular metal so strongly that turning becomes impossible. For example, if an attempt is made to turn molybdenum, fragments of diamond are soon torn away from the cutting edge as shown in Figure 14.10(d). Chipping may also be caused by imperfections in the material of the workpiece. For example the aluminium silicon alloys used for motor-car components were formerly turned with single crystal tools and now more usually with PCD material. With either type of tool chipping can be caused by an impact of the tool on oversize particles of hard silicon in the alloy. We know of at least one factory where any onset of chipping of single crystal tools was met by the production manager paying a visit to the foundry, fortunately on the same site.

Intermittent cutting with any type of tool will create impact forces and thermal stresses which tend to cause fracture, see for example Shaw (1984), Pekelharing (1980) and Chakraverti, Pandey and Mehta (1984). There is not much information regarding diamond. Under good conditions any chipping due to intermittent cutting will probably be caused by progressive fatigue processes (Section 7.7) rather than a single impact. Therefore, as in all fatigue processes, the probability of failure may be almost negligible if the stresses on the diamond are kept well below the critical fracture stress by the choice of suitable machining conditions. Note of course that a poor polish on a tool will result in a greater number of larger surface flaws which will encourage fatigue type damage.

14.3.b Monitoring performance

Diamond tools may chip while in use because of poor selection of the diamond, poor fabrication, poor machine tools, poor quality of the workpiece material and poor handling of the tools. Proper attention to all these details makes a great deal of difference to the performance of the tool. It is perhaps particularly difficult to specify and maintain the exact conditions of use. A good tool may run for a week or more and turn several thousand components, and during this time be used by different operators making different components from different alloys. On the other hand just one error may greatly shorten the life of a tool for no obvious reason.

Because of the above considerations it is generally very difficult to make completely satisfactory comparisons between tools being used in actual production conditions. Nevertheless much can be done to raise the standard of performance by keeping effective records on each tool as it passes through the factory. These records should include notes on an inspection of each tool in a microscope on arrival, and before use. It should be signed out to the machinist concerned and on return a record made of the number of components produced and its condition checked. Such procedures will both improve performance in the factory and also provide a company with clear indications either of malfunctioning in the factory or of poor quality in the tools bought in from their supplier.

14.4 Attritious wear

The wear of diamond tools due to chipping and fracture can be greatly reduced by the careful selection of single crystal diamonds or by the use of PCD material, as described above. The life of both single crystal and PCD tools will then be determined by some form of attritious or steady wear. There is in fact very little detailed information on the form and characteristics of this type of wear, partly because diamond is used to turn a variety of materials each of which reacts in its own way. However, some quite detailed studies have been made on the wear of single crystal tools turning aluminium alloys, and we begin by describing measurements on the alloy LM13.

14.4.a Wear on LM13

The wear of a tool may be much influenced by parameters such as the cutting speed, depth of cut, feed rate, and the positioning of the tool including any tilt of the rake face. Therefore, if we wish to compare two tools or to study the effect of varying a particular parameter it is essential that all the remaining parameters are held constant. We also note that the use of a machining fluid is often necessary both as a coolant and to improve the finish produced on the workpiece (Section 14.5.e) but that surprisingly little information is available on how the presence of fluids affects the rate of wear of diamond tools. It may be that this lack of information is partly an indication that the choice of fluid is not very critical as far as the wear of the tool is concerned but more information would be of interest.

The wear rates of single crystal diamond turning LM13, an aluminium alloy containing 12% silicon, have been studied by Casey and Wilks (1972) who observed

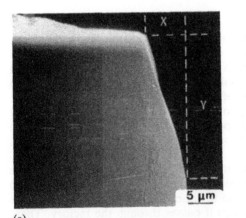

(a)

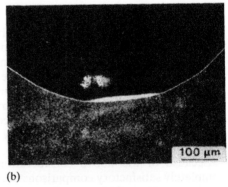

(b)

Figure 14.11 Wear land of a tool after turning an aluminium silicon alloy viewed: (a), in profile; (b), normal to the rake face (which appears black against the out of focus mounting stub) (Casey and Wilks, 1976a)

Table 14.2 Crystallographic orientation of the rake face, axis, and a 45° wear land on tools used to turn aluminium silicon alloys, see text

	Rake face	Axis	45° land
A	{011}	<1$\bar{1}$0>	{001}
B	{001}	<100>	{011}
C	{011}	<100>	10° off {111}

wear lands with the form of sloping facets (Figure 14.10(a)), quite different from the vertical facets produced by turning mild steel (Figure 14.10(b)). To obtain a measure of the wear the dimension X in Figure 14.11(a) was obtained by viewing the rake face normally in an SEM as in Figure 14.11(b). Experiments were made with six tools fabricated from good quality single crystal diamonds so as to be geometrically similar but with three of the diamonds mounted in a crystallographic orientation A, as defined in Table 14.2, and the other three in a different orientation B. The tools were used to turn similar billets of LM13 initially about 100 mm in diameter and 200 mm long at a cutting speed of about 5 m s^{-1} under dry conditions with a 25 μm depth of cut and a feed rate of 25 μm rev^{-1}.

Figure 14.12 shows the width of the wear lands plotted against the number of passes over the billet multiplied by the diameter of the billet, a figure proportional to the amount of metal removed. We see that the widths of the lands increase linearly with the amount of metal removed, and that the three tools in each group behave in a similar way, but that the wear lands on the tools with the A orientation are about ×7 wider than those on the B tools. (Another experiment with a B tool showed that the linear relationship continued for at least 60 passes over the billet when the width of the land had increased to 15 μm.)

Other experiments were made to observe the wear of a tool when the lathe was running at 600 rev min^{-1} and 2000 rev min^{-1} (Casey and Wilks, 1976a). As shown in Figure 14.13 there was no appreciable difference in the slopes of the two wear curves. Thus, although the wear naturally depends on the total number of

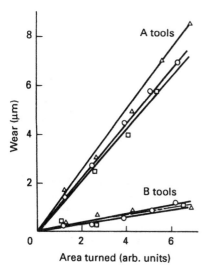

Figure 14.12 The wear of six similar round-nose tools, with two different crystallographic orientations A and B, when turning an aluminium silicon alloy (Casey and Wilks, 1972)

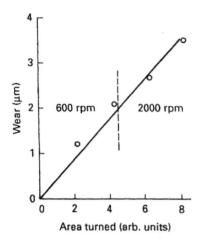

Figure 14.13 Dependence of the wear rate of a diamond tool turning an aluminium silicon alloy on the cutting speed (Casey and Wilks, 1976a)

revolutions of the lathe, it does not depend here on the speed of cutting. Therefore as any excess temperatures generated by the machining will rise with cutting speed we conclude that the wear is not due primarily to thermally activated processes.

The above result together with the marked dependence of the wear on the orientation of the diamond suggest that the wear produced by the LM13 is the result of a mechanical abrasive process probably somewhat similar to that observed by Crompton, Hirst and Howes (1973) in their rubbing experiments described in Section 13.2. Hence we might expect some correlation between the orientation effects when machining LM13 and when polishing diamond on a scaife. To pursue this point further we note that the wear lands have the form of a sloping facet (Figure 14.10.a) and that this geometry is quite complex because the land is not plane but curves round the front of the tool. (It is also not centred on the tip of the tool but is displaced towards the side where the tool first meets the metal.) To give some measure of the geometry involved Table 14.2 includes the approximate crystallographic orientation of a wear land which at the tip of the tool is inclined at 45° to the rake face. Then using the results given in Section 9.3.c we see that the metal moves over the facet on an A tool in a direction of easy abrasion, and on a B tool in a hard direction. Hence we would expect the wear to be greater on the A tools as is observed.

14.4.b Wear lands and forces

The growth of a wear land results in a progressive and undesirable increase in the forces on the diamond. Bex (1975) used a dynamometer to measure the radial or thrust forces while machining aluminium silicon alloys with three sets of tools with the orientations A, B and C set out in Table 14.2. Figure 14.14 shows some of his results for a cutting speed of 10 m s^{-1}, a depth of cut of 50 μm and a feed of 60 μm rev^{-1}, these conditions and the work performed being comparable to those for Figure 14.12 apart from the use of a machining fluid. The radial force on the tools increases almost linearly with the time of machining and the rate of increase is about ×7 greater for a tool with the A type orientation, so the increase of force on a tool provides a convenient way of monitoring the progress of wear (though care must be taken that the dynamometer does not appreciably reduce the stiffness

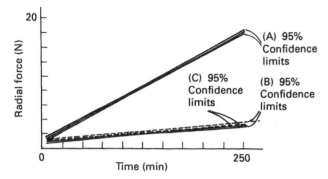

Figure 14.14 Thrust or radial force plotted against time when turning an aluminium silicon alloy, with three diamond tools with the orientations A, B, and C given in Table 14.2 (Bex, 1975)

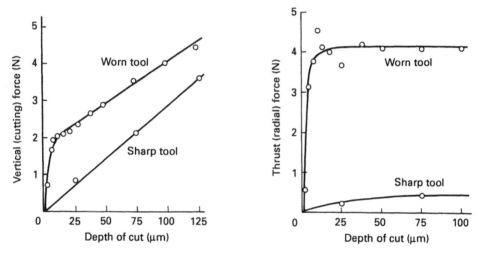

Figure 14.15 Vertical (cutting) and thrust (radial) forces on a sharp new tool and a worn tool turning an aluminium silicon alloy (Casey and Wilks, 1976a)

of the tool mounting). Hence, the results imply that the wear of the tool with the C orientation was comparable to that with the B orientation.

The relationship between the forces on a tool and the form of the wear land is shown in an experiment with two similar single crystal tools, one with a new sharp edge and the other with a wear land about 12 μm wide (Casey and Wilks, 1976a). Figure 14.15 shows the forces observed when machining LM13 under similar conditions to those mentioned above save that the depth of cut was varied. With a sharp tool the cutting force increases almost linearly with the depth of cut, as might be expected because the total amount of metal removed as plastically deformed chips is proportional to the depth of cut. However, with the blunt tool there is an additional contribution to the cutting force at all depths. Figure 14.15 also shows that the thrust or radial force is much greater with a worn tool, the initial rapid increase of the force at low depths of cut arising because the tool tends to rub rather than cut.

Finally we note that there is always a third component of the force on a tool which acts parallel to the direction of the feed but this is generally an order of magnitude smaller than the other two components.

14.4.c Abrasive wear

There is not much quantitative information on the wear rates of single crystal tools on other materials commonly machined with diamond (Table 14.1). However, all these materials generally cause the initially sharp edge of the tool to deteriorate into a sloping or rounded facet (Figures 14.10(a) and 14.10(c)) quite different from the wear lands produced by mild steel (Figure 14.10(b)). Therefore most of this wear probably arises from mechanical abrasive processes like those observed in the rubbing experiments of Crompton, Hirst and Howes (1973) described in Section 13.2.

It would obviously be of interest to correlate the wear of diamond turning tools with the wear rates observed by Crompton, Hirst and Howes. However, the results of these rubbing experiments are not directly applicable to wear during machining because of the different geometries and conditions. Crompton, Hirst and Howes rubbed the diamond under a constant load whereas during machining both the cutting and radial forces increase with changes in the size and shape of the wear land. In addition the wear produced by rubbing was usually determined by the hardness of the rubbing material but the wear produced by aluminium was two orders of magnitude higher than expected. The authors accounted for this result by noting that the surface of the aluminium became visibly oxidized during the experiment and that the oxide is much harder than the metal. However, the wear during machining is produced by the rubbing of freshly cut metal which has probably not yet oxidized.

There are also other points which remain to be resolved. For example, it is sometimes suggested that the wear caused by aluminium silicon alloys is primarily due to the much harder silicon phase of the alloy, but this is by no means certain in view of measurements made of the wear produced by an aluminium 4% magnesium alloy N8 of comparable hardness to LM13 (Wilks, J., unpublished). In a typical experiment the tool, oriented in an abrasion resistant direction, turned the inside of a cylinder about 500 mm in diameter and 600 mm long, the depth of cut being 10 μm, the cutting speed about 5 m s^{-1}, and the feed rate 20 μm rev^{-1}. The wear land was viewed in the SEM and compared with the wear produced when turning a billet of LM13 under comparable conditions, and it appeared that the wear produced by the N8 was about a factor 2 greater than that produced by the LM13 even though it contained no silicon.

The above results are in line with the results of Keen (1974) who observed that approximately the same amount of wear was produced by aluminium silicon alloys containing 0.6% and 11–12% silicon. On the other hand Casey and Wilks (1976a) measured the wear of a tool turning two billets of LM13 heat treated to give a bulk Vickers hardness of 80 VHN and 140 VHN respectively. In spite of this considerable difference the wear rates on the two billets were very similar as if the abrasiveness of the LM13 was due primarily to the much harder silicon phase. These various results suggest that the wear of the alloys is not determined solely by the hardness of the material and that some other factor is involved. More information would be of interest.

Finally, we note that very low wear rates of diamond have been reported when turning brass, much lower than produced by the aluminium alloys LM13 and N8, even though brass is of comparable hardness (Thornton and Wilks, 1979). This behaviour is not understood but may arise because of a very long induction time before the wear becomes appreciable, as in the rubbing experiments described in Section 13.2.

14.4.d Chemical wear

The wear of diamond on mild steel is extremely high (Section 13.5). In the experiments on the aluminium alloy N8 mentioned above the wear land on a diamond tool after turning for a total path length of about 40 km was still appreciably smaller than that produced by turning mild steel under comparable conditions for a path length of only 0.1 km. The rate of wear on mild steel is greater by a factor of the order of 10^3. As discussed in Section 13.6 these high wear rates arise from chemical affinities between the diamond and the steel.

Table 14.3 based on unpublished results of Thornton and Wilks gives the order of magnitude of some high wear rates on various other materials measured in air under similar conditions to those used in the experiments on steel. There is no obvious direct correlation between these wear rates and the chemical properties of the workpiece but iron, nickel and cobalt have the ability to dissolve diamond and promote graphitization. In addition all the metals in Table 14.3 are classified as transition elements lying in the same part of the periodic table and with chemical activities associated with an inner shell of d electrons. Tungsten is also a transition metal but the high wear produced by tungsten carbide is probably partly due to its much greater hardness.

(One might hope to relate the different wear rates of the transition elements in Table 14.3 with measurements by Miyoshi and Buckley (1980) of the strength of the static bonding of clean diamond to various transition elements described in Section 13.8.b. However, their results for cobalt, iron and nickel lie close together (Figure 13.30) while the wear rates on these three metals differ greatly (Table 14.3). These different results are not necessarily inconsistent because a rate of wear is not determined solely by the strength of the bond between the diamond and the metal, see Section 13.6.c).

The difficulty of predicting the magnitude of chemical wear is underlined by observations of the wear of diamond turning electroless nickel, that is nickel

Table 14.3 Order of magnitude values for the wear of diamond single crystal tools turning various materials causing high rates of wear. Wear in units of $10^{-6}\,mm^2\,mm^{-2}$, as described in Section 13.5.a

Cobalt	<1
Cast-iron	1
Zirconium	
Mild steel	10
Titanium	
Tungsten carbide	
Nickel	100

deposited chemically on a metal surface from suitable solutions, see for example Dennis and Such (1972). Electroless nickel plating produced in this way has various desirable properties, it is readily machineable, and is used *inter alia* to produce surfaces on various high precision components which are then machined with diamond. This last application is at first sight surprising for, in order to obtain the exact dimensions which are required, it is essential for the wear on the diamond tool to be very low, whereas the wear of diamond turning nickel is extremely high (Table 14.3). However, electroless-nickel plating can be turned without undue wear because the chemical deposition process introduces an appreciable amount of phosphorus which greatly modifies the reactivity of the nickel. Further discussion of this behaviour is given by Hitchiner and Wilks (1984) and by Taylor *et al.* (1986) who describe how the wear depends both on the phosphorus content of the plating and on subsequent annealing treatments (Figure 14.16).

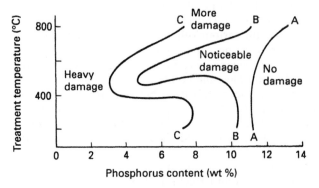

Figure 14.16 Dependence of the wear and damage on the rake face of a diamond tool turning electroless nickel on the phosphorus content and the temperature of the heat-treatment of the metal (Taylor *et al.*, 1986)

It is not well understood how the presence of phosphorus in electroless nickel so greatly reduces the activity of the nickel atoms. We note, however, that the changes in the electronic state of the metal produced by the higher concentrations of phosphorus result in the metal ceasing to be ferromagnetic, and that the ferromagnetism is associated with 3d electrons that are probably responsible for the chemical activity. These changes in electronic behaviour are also accompanied by changes in the crystal structure of the metal which have been observed in X-ray diffraction studies (Figure 14.17). With increasing phosphorus content the film passes from a crystalline to an essentially amorphous form (Graham, Lindsay and Read, 1965). Hence, if the film is truly amorphous, it may be that its chemical inertness arises from the absence of geometric defects, as has been suggested to explain the high corrosion resistance of some metals in the amorphous state (Polk and Giessen, 1978).

(Chemical wear produced by pure nickel is so high that any accompanying abrasive wear will be negligible in comparison. However, it is quite possible that the wear produced by electroless nickel with a much lower chemical activity may have both chemical and abrasive components. Syn, Taylor and Donaldson (1986) studied the edge profiles of diamond tools during the course of turning long paths of electroless nickel, and Figure 14.18 shows a set of these profiles obtained in one

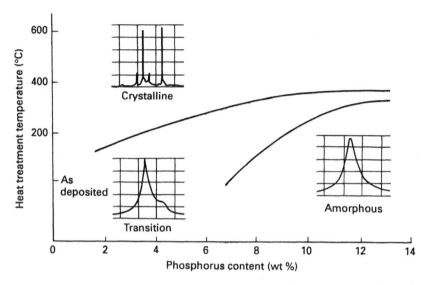

Figure 14.17 Map showing types of X-ray diffraction spectra of electroless nickel according to phosphorus content and temperature of a heat treatment (Taylor *et al.*, 1986)

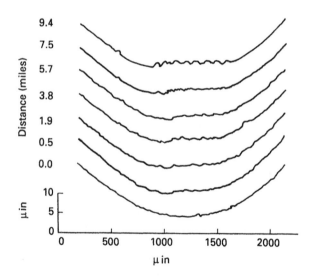

Figure 14.18 Profilometer traces of grooves made by plunge cuts with a diamond tool after various cutting distances (Syn, Taylor and Donaldson, 1986)

run. The rougher parts of the edge show quite abrupt changes of direction suggesting that the edge has suffered mechanical abrasion. On the other hand the profiles show various features associated with the feed rate of the lathe (100 μin per revolution) which suggest that wear was also taking place on a much finer scale either mechanically or chemically.)

14.4.e Wear of PCD tools

In an experiment to investigate the behaviour of PCD tools when drilling rock Hibbs and Lee (1978) used a block of PCD to turn a cylinder of sandstone and took micrographs of the worn PCD surfaces. They observed that there was very little pull out of individual crystallites and that the mode of wear appeared quite similar to the attritious wear of single crystal diamond. In addition, however, PCD material has a much greater toughness and is less likely to chip or crack (Section 12.2.c). As an example of this difference Figure 14.19 shows the wear flat on a PCD tool after turning a small length of molybdenum while Figure 14.10(d) shows the gross damage which resulted to a single crystal tool attempting to turn the same metal.

20 µm

Figure 14.19 Wear on a PCD tool after turning molybdenum (Casey and Wilks, 1976b)

There is not much information on the rates of attritious wear of PCD tools. We have already noted in Section 14.2.b and Figure 14.3 that the larger grain material may give a somewhat longer tool life, though not so fine a finish. The results of a factory survey by Keen (1974) and of some laboratory tests by Casey and Wilks (1976b) on PCD tools turning the aluminium alloy LM 13 suggest that the rates of wear are comparable to those on good single crystal tools. A brief report of wear rates when turning zirconium oxide, alumina and silicon nitride is given by Kiso *et al.* (1987). According to Gerchman (1986) silicon-based glasses and ceramics do not machine well with diamond. We have observed an extremely high rate of wear on a PCD tool, comparable to that produced on diamond by nickel, when turning a billet of Sialon©, a proprietary material containing silicon, nitrogen and oxygen which has a high hardness of about 1800 VHN. Possibly this very high wear may arise from chemical as well as abrasive processes, as the clean surface conditions produced by turning may permit the oxygen and silicon in the Sialon to react with the diamond.

Besides showing an enhanced toughness, PCD material has other characteristics which must be borne in mind when used as a tool. The lower thermal conductivity

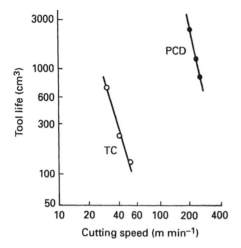

Figure 14.20 Tool life (volume cut) as a function of cutting speed for (○) tungsten carbide and (●) PCD tools turning a glass fibre reinforced epoxy resin (Spur and Wunsch, 1988)

of PCD material (Section 12.5) will result in considerably higher flash temperatures at the machining interface. The presence of the second phase in the PCD usually reduces its thermal stability (Section 12.2.d), so temperatures above about 600°C must be avoided. Figure 14.20 gives a plot of tool life against cutting speed when turning an epoxy resin reinforced with glass fibres. This material is hard and abrasion resistant and has a low thermal conductivity so is particularly liable to

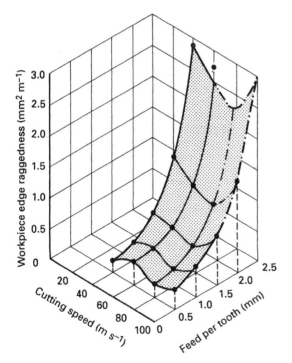

Figure 14.21 Curves showing the effect of the feed per tooth and cutting speed on the raggedness of the milled edge of plastic coated chipboard (Saljé and Stühmeier, 1988)

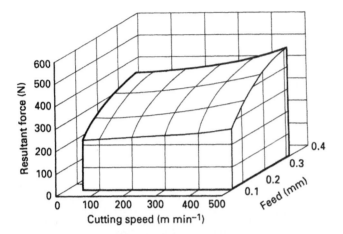

Figure 14.25 Micrograph of a copper disc machined by: (a), face turning; (b), a plunge cut using the same tool (King and (Wilks, 1976)

thermal degeneration, and the wear increases rapidly as the speed increases (Spur and Wunsch, 1988). (The diagram also shows the much inferior performance of a tungsten carbide tool.) There is also the possibility with PCD material that the solvent/catalyst may be smeared across tool and workpiece and perhaps react chemically with the workpiece.

To obtain the best results when turning with PCD it will generally be necessary, as with most tools, to vary the machining conditions to optimize performance. Good examples of the type of experiments which should be made are given by Saljé and Stühmeier (1988) and by Spur and Wunsch (1985). The former authors used PCD tools to mill laminated chipboard containing a central layer or core of denser and abrasive material and were particularly concerned to reduce any raggedness at the edges. Figure 14.21 shows how this raggedness varied with both the cutting speed and the feed rate. Results were also obtained showing how changing the speed and feed affected the cutting forces. Spur and Wunsch (1985) describe studies of the effect of cutting speed on the cutting force when turning the resistant fibre reinforced plastics mentioned above (Figure 14.22).

14.5 Surface finish

We now consider the quality of the finish produced by diamond turning. We deal primarily with the performance of simple crystal tools as these are invariably preferred to PCD tools when a fine finish is required. The use of a machining fluid such as a light oil generally produces beneficial effects so virtually all high quality machining is performed wet. We discuss the action of these fluids in Section 14.5.e.

14.5.a Factors affecting finish

The production of a good finish requires a well made tool with a good cutting action operating in a machine-tool which holds diamond and workpiece accurately

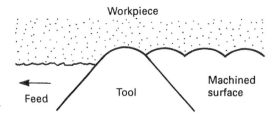

Workpiece

Feed   Tool   Machined surface

Figure 14.23 Sketch showing the ideal geometric finish produced by a round-nose turning tool

and rigidly in position. Then under optimum conditions the topography of the surface should be determined solely by the geometry of the tool and the feed rate. Thus Figure 14.23 shows the finished surface left by a round nosed tool turning a cylindrical billet, the height variations in this surface being equal to $f^2/8R$ where f is the feed rate per revolution and R is the radius of the nose. For the ideal geometry produced by tools of other shapes see for example Chisholm (1969) and Shaw (1984).

In practice the topography of a machined surface may differ greatly from the ideal. Figure 14.24 shows an optical interference micrograph of the central area of a copper disc which had been faced turned (with the feed direction perpendicular to the axis of the lathe). When viewed by eye the surface looked bright and highly polished with only a suggestion of some speckled effect, but the micrograph reveals at least five different kinds of imperfection: (i), a central pip with ribs radiating

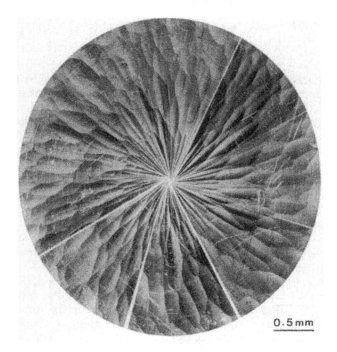

0.5 mm

Figure 14.24 Composite view of the central area of a turned copper disc using the Nomarski interference technique (King and Wilks, 1976)

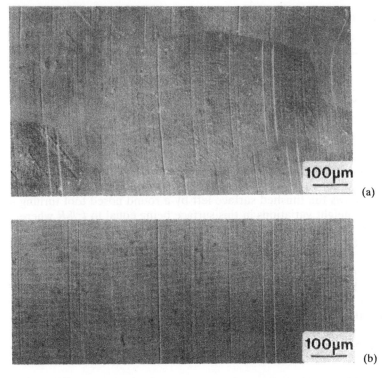

(a)

(b)

Figure 14.25 Micrograph of a copper disc machined by: (a), face turning; (b), a plunge cut using the same tool (King and Wilks, 1976)

outward; (ii), a series of elongated shaped patches; (iii), a system of intermittent concentric lines; (iv), a number of irregular scratch marks; and (v), fine structure within each individual patch (hardly visible in the reproduction).

Figure 14.25(a) shows a detail of Figure 14.24, and Figure 25(b) the surface made by a plunge cut (without any traverse) with the same tool. An analysis of these and similar figures (King and Wilks, 1976) shows that the patches are depressions whose cross-sections in the radial direction correspond with the curved edge of the tool. The almost vertical lines in Figure 14.25(b) produced by the plunge cut are ridges produced by nicks in the edge of the tool, which also give rise to the curved ridges in Figure 14.25(a). These ridges all lie on one spiral, as might be expected, but only cover about 10% of the total length of the spiral and are interrupted at the boundaries of the patches. Hence the tool cannot have been in steady contact with the disc throughout the machining process. The same pattern of spiral lines appears in the same position on each patch, so that the finish of all the patches is produced by the same section of the edge of the tool. These results imply a relative motion of tool and workpiece in a direction parallel to the axis of the lathe, and that a patch is the impress of the shape of the tool when it is closest to the disc. Subsequent measurements showed an axial play on the lathe bearing of about 1 μm, in good agreement with the depth of the patches.

The centre pip feature in Figure 14.24 though striking is essentially trivial and arises because the tool was stopped in its traverse just before reaching the centre

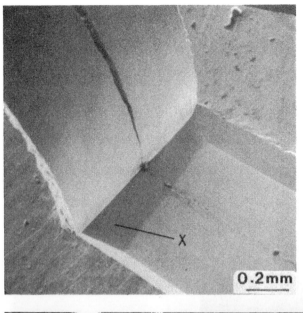

Figure 14.26 (a), SEM micrograph of the root of an arrested chip; (b), section of the brass specimen and chip, after an etching process which causes deformed metal to appear white in the micrograph, see text (Casey and Wilks, 1976a)

of the disc, the curvature of the sides of the pip corresponding with the curvature of the tool. The irregular scratches are due either to swarf scraping the disc during turning or possibly to careless handling after turning. Finally, within each patch, see Figure 14.25(a), there are series of fine horizontal lines which appear to be chatter marks arising from vibrations during the turning, see for example Lewendon (1980).

We also note that turning is a complex process in which both chip and workpiece may suffer considerable plastic deformation as shown by the example in Figure 14.26. A simple straight-edge diamond chisel was polished to give a large negative-rake wear land, and mounted as a planing tool in a quick-stop device. It was then used to machine a block of brass, divided into two so that it could subsequently be viewed in section. Figure 14.26(a) gives a general view of the arrested chip taken in the scanning electron microscope, and shows the sloping edge of metal X which had been in contact with the wear land. As the tool moves forward, the metal below this sloping edge must be pushed away, so heavy deformation results. Figure 14.26(b) shows a section through the metal after an etching process which reveals heavy plastic deformation both in the chip and below the newly formed machined surface.

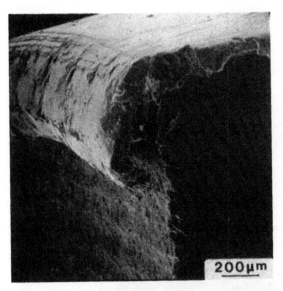

Figure 14.27 SEM micrograph of the end of a turned billet of an aluminium silicon alloy showing metal spreading over the edge (Casey and Wilks, 1976a)

Plastic deformation during turning may sometimes displace appreciable quantities of metal over the surface of the workpiece. For example, Figure 14.27 shows an SEM micrograph of one end of a circular slotted billet of the aluminium alloy LM13 after making 50 passes taking a cut of 25 μm along its length. The aluminium rich phase of the alloy is relatively soft and the micrograph shows metal which has been pushed forward (to the left of the picture) until it finally overhangs the edge of the billet.

14.5.b Structure of workpiece

Variations in the structure of the workpiece material will tend to produce imperfections of finish. Figure 14.28(a) shows an SEM micrograph of a disc of an

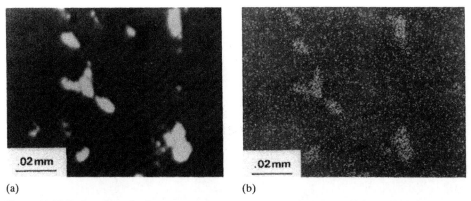

(a)   (b)

Figure 14.28 Surface imperfections on face-turned aluminium alloy viewed: (a), in SEM; (b), in microprobe set to record copper (Wilks, 1980)

aluminium alloy which had been face turned to give a high quality finish but which showed obvious imperfections when viewed by eye. An analysis of the micrograph showed that the defects appearing white were small protuberances rising above the general level of the surface which reflected the electrons more strongly. Figure 14.28(b) shows the same area of the disc viewed in a microprobe analyser set to detect the copper component of the aluminium alloy. We see that the copper is not uniformly dispersed but is concentrated in the regions of the protuberances which obviously result from a change in mechanical behaviour produced by the copper. Impurities, precipitates and any voids in the material will produce irregularities in a similar way.

Figure 14.29 shows other imperfections of surface finish arising partly from the two phase nature of the workpiece, an aluminium 12% silicon alloy in which one phase consists mainly of aluminium and the other almost entirely of silicon. The surface in Figure 14.29(a) was turned by a tool with a sharp but chipped edge. The tool cuts fairly cleanly and reveals areas of the aluminium phase which reflect brightly and areas of the silicon phase which appear dark, while the many vertical lines are due to nicks in the edge of the tool. The surface in Figure 14.29(b) was produced by a good tool which had been withdrawn after a long life because it had developed a pronounced wear land. We see that the softer aluminium phase has been pushed over the surface so that much less silicon is visible.

Even if the workpiece material consists only of one pure phase, the form of the machined surface may still be affected by the crystallinity and other structure of the material. For example, Figure 14.30(a) shows an interference micrograph of a surface produced by a plunge cut made into the face of a copper disc. The copper is polycrystalline and has been cut sufficiently sharply to reveal the outlines of all the grains in the surface, see also Stadler, Freisleben and Heubeck (1987). In addition the grains in polycrystalline materials are generally randomly oriented so differences in orientation may affect the amount of plastic deformation caused by machining and this gives rise to variations in the surface finish. For example Figure 14.31 shows the surfaces produced by two similar cuts on two single crystals of aluminium with different crystallographic orientations; surface (a) has a relatively smooth finish whereas (b) is much rougher because its orientation has encouraged considerably more plastic deformation and cross slip (Ohmori and Takada, 1982).

(a)   (b)

Figure 14.29 Nomarski micrographs of surfaces of the aluminium silicon alloy LM13 turned: (a), by a new but chipped tool and; (b), by a worn but unchipped tool (King and Wilks, 1976)

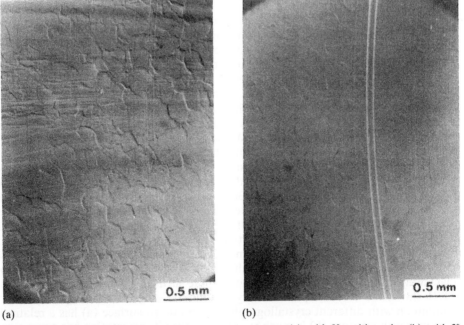

(a)   (b)

Figure 14.30 Nomarski micrographs of plunge cuts on copper: (a), with 3° positive rake; (b), with 3° negative rake (King and Wilks, 1976)

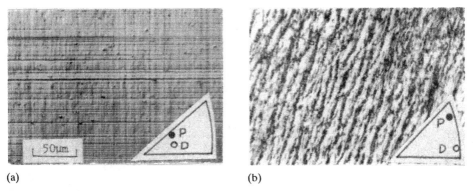

(a)   (b)

Figure 14.31 Micrographs of surfaces of two single crystals of aluminium machined in different crystallographic directions (Ohmori and Takada, 1982)

14.5.c Quality of tool

The effect of the quality of the tool edge is well shown by some observations of the finish produced by single crystal diamond tools turning aluminium alloy LM13 pistons for automobile engines. (Although single crystal tools have largely been replaced by PCD tools for this type of operation the results are relevant to any form of turning.) Figure 14.32 shows profilometer traces of the surfaces of three pistons,the first profile shows the satisfactory performance given by a new tool with a good edge. The second profile was produced by a relatively new tool 1802 which had deteriorated rapidly because of many small nicks chipped on the cutting edge. (Figure 14.33).The third profile was obtained with another tool 1806 which had worn smoothly (Figure 14.33) and had only been withdrawn from use after a long life when the height variations in the surface exceeded the allowed tolerances.

Figure 14.34 shows SEM micrographs of the surfaces produced by the above three tools. The new tool produces a satisfactory finish for the required component. The relatively new but chipped tool gives an obviously inferior finish because metal is picked up by the nicks in the edge and pulled as well as cut. In contrast the worn tool with the wear land has produced a finish which is generally much smoother than either of the other two. It appears that the wear land has pushed the softer aluminium phase over the whole surface to give a smooth finish except for some

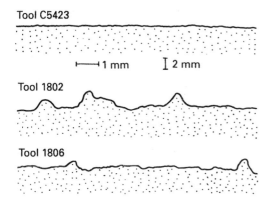

Tool C5423

⊢——⊣1 mm   I 2 mm

Tool 1802

Tool 1806

Figure 14.32 Profilometer traces of aluminium–silicon pistons turned by three different diamond tools, see text. (King and Wilks, 1976)

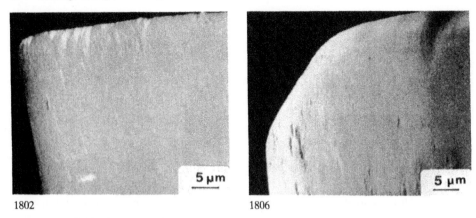

1802   1806

Figure 14.33 SEM micrographs showing side views of tools 1802 and 1806 (King and Wilks, 1976)

C5423

Figure 14.34 SEM micrographs of the surface finish produced by a new tool C5423 and two worn tools, 1802 and 1806 (King and Wilks, 1976)

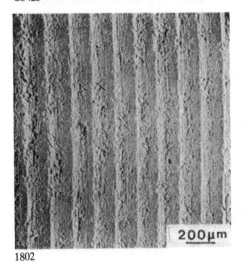

1802

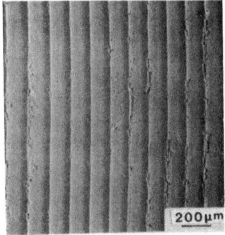

1806

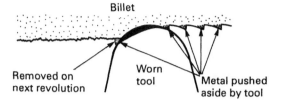

Figure 14.35 Sketch to illustrate the pushing of metal by a worn tool (King, 1976)

flakes of metal left at the edge of each band of machining. As well as these flakes, responsible for the rejection of the tool, there is a marked asymmetry in the cross sections of the machining bands not present in those produced by the new tool. This asymmetry arises because the metal pushed by the wear land flows to either side of the cutting edge (Figure 14.35). The metal pushed to the leading edge of the tool is removed on the next revolution of the workpiece but that on the trailing side remains to produce the asymmetry.

Another example of the poor finish resulting from a chipped edge is given in Figure 14.36 which shows the surface of a piston where the finish deteriorated when the tool was damaged by an inclusion in the metal. We note in particular the presence of flakes of metal which greatly increase the surface roughness (King and Wilks, 1976). In this example the damage to the tool was particularly severe, but even minor damage to the edge will produce the same type of effects on a smaller scale.

The finish produced by a tool may also become unsatisfactory because of a tendency for metal to adhere to the rake face, particularly for deep cuts involving high forces. This may cause an accumulation of metal, or built-up edge (BUE), on the face as in Figure 14.37 which shows a large BUE on a poorly polished diamond.

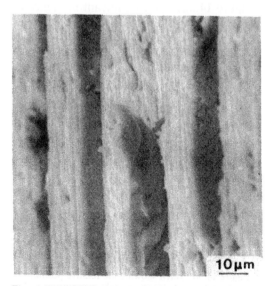

Figure 14.36 SEM micrograph showing detail of the unsatisfactory finish produced by a chipped tool (King and Wilks, 1976)

Figure 14.37 Large built-up edge of an aluminium alloy on a poorly polished diamond

Such built-up edges generally increase the friction between tool and chip and lead to a deterioration in the quality of the finish. These effects are best kept to a minimum by ensuring that the rake face is polished as smooth as possible and by the use of a machining fluid.

14.5.d Tool settings

The rake angle at which a tool is set (Figure 14.38) has a significant influence on the quality of the finish. Figure 14.30(a) shows a surface produced by making a plunge cut into a copper disc with a large radius tool with a positive rake angle of 3° (and a clearance angle of about 1.5°), and Figure 14.30(b) the surface produced by the same tool tilted to give a negative rake of 3°. A comparison of the micrographs reveals two main differences. The curved lines produced by nicks in the edge of the tool are much more marked for the negative rake position, while the structure of the crystallites stands out much more clearly with the positive rake. Thus positive rake causes the tool to cut more cleanly, whereas negative rake acts

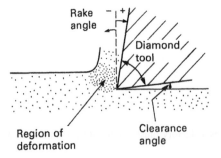

Figure 14.38 Sketch showing the relative positions of cutting edge and clearance angle on a diamond turning tool

in a similar way to a sloping wear land which pushes the metal and tends to smooth over irregularities. Thus it might appear that provided the edge of the tool is free from nicks it would be advantageous to use some negative rake. On the other hand negative rake leads to higher cutting and radial forces which may result in vibration or chatter marks, as in Figure 14.25. Therefore the optimum position is best found by experimenting with different angles in machine tools which are as stiff as possible.

Provided the clearance angle of a tool (Figure 14.38) is of the order of 5° its actual value has little effect on the surface finish, but with appreciably smaller angles the newly machined workpiece may be rubbed by the flank of the tool thus giving rise to additional forces and possibly vibration. On the other hand Weinz (1968) describes the use of both negative rake and a low clearance angle to give a high gloss surface by a form of burnishing process. In practice it is usually not easy to separate the effects of rake and clearance angles, and for any particular task the choice of angles is usually an empirical compromise determined by the state of the machine tool, the machining conditions, and the properties of the workpiece.

The best finishes are usually obtained when the forces on the tool are small and this generally implies a small depth of cut. However, the cutting conditions should also be as constant and uniform as possible. Figure 14.38 shows a tool with a sharp edge but in practice any edge will be blunted or rounded to some extent, and if the effective radius of the nose begins to approach the thickness of the chips the geometry of the cutting process becomes more complex, see for example Boothroyd (1975). The cutting conditions begin to fluctuate and if the cut becomes too shallow the tool may not cut at all but only rub the workpiece. These considerations have generally set a limit to the depth of cut of about $2\,\mu m$, see for example Furukawa and Moronuki (1988), but Ikawa, Shimada and Morooka (1987) have used a very sharp tool to make much finer cuts (Section 14.6.d).

14.5.e Machining fluids and surface interactions

Most diamond machining is carried out wet, with a machining fluid applied either as a liquid or as a spray mist to act as both lubricant and coolant. The use of a fluid is essential under heavy machining conditions, such as deep cuts at high speed, where the heat generated would otherwise cause damage to either tool or workpiece. However, even if no cooling is necessary, fluids acting as lubricants generally have a beneficial effect on the surface finish, although a good finish may sometimes be obtained when turning dry.

A useful discussion of machining fluids and their mode of action when used with conventional (non-diamond) tools is given by Shaw (1984). He describes measurements of the cutting force when five different metals were machined both dry and with eleven different fluids. The forces were reduced below the dry values by between 10% to 90%, and a particular feature of these results was that each fluid was not equally effective with each metal. Thus the action of a fluid may be very specific to the constitution of the workpiece, implying that this action is at least partly chemical.

Figure 14.39 shows part of a disc similar to that in Figure 14.24 which was turned starting from the right with no fluid. The initial finish was poor with very pronounced chatter marks (Section 14.5.a) and dark patches where pieces of metal had been torn out. Then after turning a short length, a few drops of oil were placed on the surface and the left hand side of the micrograph shows the final finish

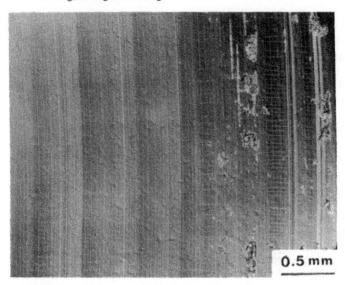

0.5 mm

Figure 14.39 Nomarski micrograph of a copper disc turned with varying amounts of lubrication, see text (King and Wilks, 1976)

obtained after the oil had spread. Only a small amount of oil was used in the above experiment so it would almost certainly have acted as a lubricant rather than a coolant, reducing the friction between tool, chip and workpiece.

The most important factor determining finish is generally the rubbing of the workpiece against the tool. However, the effect of this rubbing depends on a large number of variables besides the presence or absence of a lubricant, such as the area of rubbing, the thrust and friction forces, the temperature, and the hardness and work-hardening properties of the workpiece. Therefore it is not surprising that different sets of machining operations sometimes produce quite different finishes, for example good results have been obtained turning copper without a lubricant in conditions apparently similar to those of Figure 14.39, see Casey and Wilks (1976b).

It is unlikely that any temperature rises in the above experiments were sufficient to affect the finish but this is not always the case. In another experiment a billet of copper was turned by a single crystal tool and a good reflecting finish was obtained both when using a fluid and when turning dry. However, the machining was then repeated using a PCD tool of similar geometry (Casey and Wilks, 1976b). A good finish was obtained when using a fluid but without a fluid the surface was extremely rough (Figure 14.40). Closer inspection showed that the roughness was due primarily to chips and swarf adhering to the surface, and that on some of the chips one side appeared smooth and bright showing that melting had occurred. This very different performance between the PCD and single crystal tools must almost certainly have been due to the higher friction of the PCD and its lower thermal conductivity. The friction forces may be about 5 times greater than with single crystals (Section 12.4), and the thermal conductivity a factor 2 lower (Section 12.5), hence we may expect any temperature rises to be considerably greater with PCD material.

Figure 14.40 Surface of a copper billet turned dry using a PCD tool (Casey and Wilks, 1976b)

We also note that chemical reactions between diamond tools and workpieces such as described in Section 14.4.d generally result in imperfections of the finish. We have already remarked on the strong interaction between diamond and molybdenum (Section 13.6.c) and Figure 14.41 shows profilometer traces of brass and molybdenum surfaces produced by the same round-nosed single crystal tool turning dry. The brass surface approximates to the ideal geometric pattern but there is no sign of this type of pattern on the molybdenum, and the height variations are about 30 times greater.

The above behaviour of molybdenum is a rather extreme example of bonding between a diamond tool and the workpiece, and one may hope to inhibit such bonding by introducing a machining fluid between the surfaces. Nevertheless there

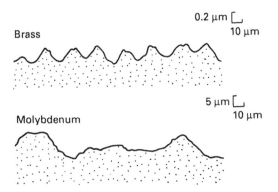

Figure 14.41 Profilometer traces of the surface finish produced on brass and molybdenum by the same round-nose tool under similar conditions. Note the difference in the vertical scales. (Casey and Wilks, 1976b)

Figure 14.42 SEM micrograph showing holes in a turned surface of platinum

are quite a number of materials on which it is difficult to produce a high quality finish. For example all the metals listed in Table 14.3 have high rates of wear and generally give rather poor finishes. As another example Figure 14.42 shows a surface turned on platinum which gives rise to appreciable wear and tends to give a poor finish. Although much of the surface is smooth, in some areas material has been pulled away leaving irregular cavities. (The behaviour of platinum may be particularly complex because it exhibits a high rate of strain hardening, see Rushforth (1978).)

As already mentioned it is difficult to predict the strength of clean surface reactions, and to determine the extent to which a machining fluid penetrates to the interacting surfaces. Therefore the only certain way to determine whether a good finish can be obtained on a particular material is by experiment. Some improvement can usually be effected by varying the machining conditions but the results will seldom be entirely satisfactory if the workpiece material is unsuitable, see for example Allen, Hauschildt and Bryan (1975) who describe detailed experiments made in unsuccessful attempts to obtain a good finish on beryllium.

14.6 High precision turning

14.6.a Precision engineering

Modern technology uses many devices with components of very small size and many devices with components machined to very close tolerances. The need to store as much information as possible on a semiconductor chip has produced a variety of constructional techniques. Beams of ions and electrons are used to machine and shape components and to 'write' fine patterns for photo-etching techniques with a positional accuracy equal in principle to the resolving power of the SEM. Hence precision engineering is now moving towards the fabrication of components on the scale of 10 nm and less, see for example a review by Taniguchi (1983).

Besides requirements for very small scale engineering there is also an increasing demand for greater precision and dimensional accuracy in some of the more conventional types of items produced by turning and grinding. A useful account of these developments is given by McKeown (1987). The efficiency of compressors in aircraft engines is being steadily increased by machining the fan blades with greater precision. The gearboxes to transmit the power in the next generation of aircraft are being designed to give less wear and noise by machining the teeth of the gears more exactly to the best theoretical shape.

Diamonds play an increasingly important part in high precision engineering because they are able to machine components to very close tolerances. Components now produced by diamond turning include drums for copiers, polygon mirrors for laser scanners, large precision mirrors in high power lasers, infrared lens shaped from germanium for use in thermal imaging systems, and large metal lenses to focus X-rays. In response to these various requirements diamond turning techniques have been steadily refined and can now produce smooth surfaces with a particular geometric configuration specified to within 0.1 μm, and with a surface roughness of no more than a few nm.

The above developments in machining have been accompanied, and indeed overtaken, by developments in the measurement of dimensions and of surface profiles and roughness. New methods of surface assessment which aim to produce height resolutions of 1 nm or less are described in a review by Vorburger (1987). These methods include extra high sensitivity profilometers of the traditional type, and a new family of optical interference microscopes which use various techniques to assess the phase of the light in the image. These techniques generally use laser light and record the image on some form of photodiode detector so that the information in the image can be processed electronically. Information on height differences is then obtained by displacing the object vertically through extremely small distances on its mounting of a piezo-electric cell, and processing the observed changes in intensity, see for example Perry, Moran and Robinson (1985).

Although these optical methods can detect very small height differences of the order of nanometers their lateral resolution is limited to that of the optical microscope. However, it is now becoming possible to obtain both very high lateral and vertical resolution with either the scanning tunnelling microscope described by Binnig and Rohrer (1982) and Golovchenko(1986) or the force microscope described by Binnig, Quate and Gerber (1986).

14.6.b The machine tools

The performance of a lathe or similar machine tool is determined by the quality of the spindle bearing and of the slides carrying the cutting tool. Diamond tools only give their best performance in machines with a maximum rigidity and a minimum of play in the bearings. For a detailed analysis of possible imperfections in the various motions of a lathe see for example Bryan, Clouser and Holland (1967). Today an increasing number of high quality lathes have been developed using hydrostatic air or oil bearings. Various authors, Bryan (1979), Miller *et al.* (1979) and McKeown *et al.* (1986) describe lathes capable of turning workpieces of diameters up to 2100 mm, 1400 mm and 1000 mm respectively. Other machines designed for smaller workpieces are described by Wada (1982), Sumiya, Ueda and Tsukada (1982), Falter and Dow (1987), Eda, Kishi and Ueno (1987) and Becker (1988). Much work has also been done to ensure the exact positioning of the tool particularly by determining its position either by interference methods using laser beams, see for example Bryan and Carter (1979), or by observing Moire fringes formed by two diffraction gratings, see for example McKeown and Bent (1979).

One of the most significant factors in the development of high precision machining is of course the use of computer control of the machine tool. For example computer control of a lathe fitted with X and Y slides makes possible the generation of surfaces of arbitrary shape to the precision of which the lathe is capable, see for example Gijsbers (1980). Of the many other possibilities we mention only two. When a tool machines a non-spherical surface the point of contact of the diamond with the surface will move round the edge of the diamond and if the radius of the tool is not exactly constant this will result in some dimensional inaccuracy. However, this can be corrected by assessing the curvature of the tool and modifying the program accordingly. Computer control can also compensate for residual play in the motion of the spindle. For example Patterson and Magrab (1985) describe a system in which the exact position of the spindle is sensed by a capacity detector which operates a fast servo system to make appropriate adjustments to the tool mounted on a piezo-electric actuator.

14.6.c The diamond tool

The developments in machine tools described above have been accompanied by improvements in the polishing and production of the diamond cutting tools. There have been no major changes in design but rather a continuing attention to detail in order to obtain a finer edge free of defects and shaped to closer tolerances. For example, Dillow (1987) refers to commercial tools with tolerances on the absolute values of the radius of about $10 \mu m$ and on the roundness of only $2 \mu m$. He also states that using suitable equipment the angles between faces can be polished to tolerances of only 2 min of arc.

The quality of the cutting edge of a tool is largely determined by the quality of the polish on the adjacent surfaces (Section 14.2.a). Hence there is a continuing trend to develop fine finishes on high quality tools, either by the use of extra fine polishing powder on very stiff specially designed polishing machines or sometimes by suitable chemical polishes as described in Section 9.5.d. Experiments are also being made to use the techniques of ion milling (Section 9.6.c) to generate sharp edges with radii of curvature no greater than 50 nm (Asai et al., 1988). However, the method has so far only been applied to straight edges or styli and not to the production of tools.

A particularly significant trend in the development of high quality tools is the increasing use of the scanning electron microscope. The great depth of focus of this instrument permits views of both the edge and adjoining surfaces of a tool and the form of the wear lands and other possible damage, as in the micrographs in Figure 14.10. As another example, Figure 14.43 shows a micrograph of a straight edge tool, the wear being revealed by comparing the edge with the white reference line (Becker, 1988). Other SEM micrographs of wear on high precision tools are given by Syn, Taylor and Donaldson (1986), Hurt and Showman (1986) and Nishimura et al. (1988). We have in previous chapters given many examples of the use of the CL mode of the SEM to reveal impurities and imperfections in the diamond, and this technique is likely to become increasingly important in the selection of diamonds for high quality tools, as in a study by Nishimura et al. (1988).

Figure 14.43 SEM micrograph of a single crystal diamond cutting edge after prolonged use. The white line is drawn as a reference level. (Becker, 1988).

The SEM may also be used to assess the sharpness of the edge of a tool or the roundness of a stylus. Suganuma (1985) describes a technique in which two detectors for secondary electrons are placed symmetrically about the beam axis on either side of the specimen. The signals from these detectors are processed by taking their sum and their difference. The sum signal is used to produce a normal SEM micrograph while the difference term gives information on the inclination of the specimen surface to the axis of symmetry. Hence by using a microprocessor to compare and analyse the signals the author claims to map the profile of a surface with a resolution of 2 nm normal to the surface and of 10 nm in the plane of the surface. The technique has been used by Asai *et al.* (1988), to measure the radii of tool edges in the range 20 nm to 60 nm.

14.6.d Workpiece and finish

The increasing demand for finer surfaces focuses attention on the homogeneity of the workpiece material. Any polycrystalline material is inhomogeneous because the individual crystallites are randomly oriented and their elastic and plastic behaviour depend on their crystallographic orientation. Thus, Furukawa and Moronuki (1988) in an experiment to analyse the fluctuations of the forces on a tool cutting at a speed of 0.1 m s^{-1} observed changes in the cutting force of about 50% which they were able to correlate with the boundaries of particular crystallites. Other examples of effects produced by grain boundaries are given by Eda, Kishi and Ueno (1987) and Nishiguchi *et al.* (1988).

Techniques are now available to machine surfaces with roughnesses of no more than 10 nm rms which can be measured to the order of a nm, see for example Syn, Taylor and Donaldson (1986). Various experiments are being made to achieve even better performances. Small depths of cut are generally desirable because the forces on a tool are then smaller and produce less distortion of the workpiece. Cuts are commonly made with depths in the range 5 μm to 20 μm the lower limit being set by the fact that with too small a cut the cutting process becomes unstable and the tool rubs rather than cuts. However, by using a tool with a very sharp edge, Ikawa, Shimada and Morooka (1987) apparently obtained stable cutting condition for depths of cut down to 50 nm. On the other hand, with a worn tool which was less

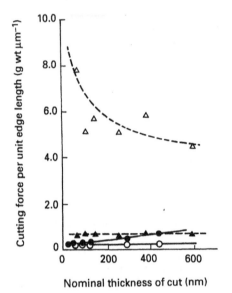

Nominal thickness of cut (nm)

Figure 14.44 Relation between the nominal thickness of cut and the forces per unit length of cutting edge when machining copper at a speed of $10\,\mathrm{m\,s^{-1}}$. ($\bullet$), ($\bigcirc$) show the tangential and normal forces on a sharp tool and ($\blacktriangle$), ($\triangle$) the corresponding forces on a worn tool respectively (Ikawa, Shimada and Morooka, 1987)

sharp the radial force on the tool became very large for depths of cut below 600 nm (Figure 14.44), and at the same time the chips became rougher and less uniform than those produced by the sharp tool. In another set of experiments Nishiguchi *et al.* (1988) studied how the surface finish and the condition of the chips was affected by the angle which the length of the straight edge of a chisel type tool made with the surface. The roughness declined as this angle was reduced to 0.1° but did not decrease further on reducing the angle to less than 0.01°, apparently because of burrs produced by the turning.

Other experiments have been concerned with the residual stress which may be produced in a machined surface. This stress may sometimes give a useful hardening effect but in some materials may lead to undesirable changes over a period of time. The surface hardness of samples of machined copper and aluminium was measured by Evans *et al.* (1987) using a microhardness indenter. Provided the depth of the indents was greater than about 0.5 μm the hardness was independent of this depth, but measurements with lower penetrations gave hardnesses up to three times greater, thus indicating considerable work hardening in a layer of the order of 100 nm deep. A preliminary report of other similar measurements of microhardness is given by Decker (1986). The residual stress may also be determined by X-ray diffractions measurements, see for example Cullity (1978) and James and Cohen (1980). Sugano *et al.* (1987) give results showing an increase in stress as the tool cuts a greater distance, presumably because the tool is blunting.

Finally we note that the use of metal coatings to permit high quality finishes to be obtained on a component is now a quite common technique. Several considerations besides the requirement of a good finish may be involved in selecting the material for a particular component. These considerations may include strength and stability, electrical and thermal conductivity, the reflection coefficient for electromagnetic radiation, and in the case of dielectric materials transparency to radiation. These various criteria can often best be met by machining the bulk of the component from one material to give strength and stability and then depositing a thin layer of another material which has the required reflectivity or another characteristic and which can be machined to give a good finish. Various metals have been used as platings including gold, silver, copper and electroless nickel, see for example Gerchman (1986). (The preparation of films of electroless nickel was discussed in Section 14.4.d.)

References

Allen, D. K., Hauschildt, H. W. and Bryan, J. B. (1975) *UCRL–51916*. National Technical Information Service, US Department of Commerce

Asai, S., Taguchi, Y., Kasai, T. and Kobayashi, A. (1988) Program and Abstracts *First International Conference on the New Diamond Science and Technology*, pp. 152–153. Tokyo: Japan New Diamond Forum

Becker, K. (1988) *Industrial Diamond Review*, **48**, 69–72

Bex, P. A. (1975) *Industrial Diamond Review*, **35**, 11–18

Bex, P. A. (1979) *Industrial Diamond Review*, **39**, 277–283

Binnig, G., Quate, C.F. and Gerber, Ch. (1986) *Physical Review Letters*, **56**, 930–933

Binnig, G. and Rohrer, H. (1982) *Helvetica Physica Acta*, **55**, 726–735

Boothroyd, G. (1975) *Fundamentals of Metal Machining and Machine Tools*. McGraw–Hill Kogakusha, Tokyo

Bryan, J. B. (1979) *Precision Engineering*, **1**, 13–17

Bryan, J. B. and Carter, D. L. (1979) *Precision Engineering*, **1**, 125–128

Bryan, J., Clouser R. and Holland, E. (1967) *American Machinist*, Dec 4, 149–164

Casey, M. and Wilks, J. (1972) In *Diamond Research 1972*, supplement to *Industrial Diamond Review*, 11–13

Casey, M. and Wilks, J. (1976a) *International Journal of MachineTool Design and Research*, **16**, 13–22

Casey, M. and Wilks, J. (1976b) In *Proceedings of the 16th International Conference on Machine Tool Design and Research 1975*, (eds F. Konigsberger and S. A. Tobias), Macmillan, London, pp.553–563

Chakraverti, G., Pandey, P. C. and Mehta, N. K. (1984) *Precision Engineering*, **6**, 99–105

Chisholm, A. W. J. (1969) In *Modern Workshop Technology: Part II Machine Tools and Manufacturing Processes*, (3rd edn), (ed. H. Wright Baker), Macmillan, London, pp. 1–42

Crompton, D., Hirst, W. and Howes, M. G. W (1973) *Proceedings of the Royal Society*, A **333**, 435–454

Cullity, B. D. (1978) *Elements of X-ray Diffraction* (2nd edn) Addison Wesley, Reading, Mass

Decker, D. L. (1986) *Proceedings of the SPIE, Ultraprecision Machining and Automated Fabrication of Optics*, **676**, 2–7

Dennis, J. K. and Such, T. E. (1972) *Nickel and Chromium Plating*, Newnes-Butterworth, London, pp.279–300

Dewar, B., Nicholas, M. and Scott, P. M. (1976) *Journal of Materials Science*, **11**, 1083–1090

Dillow, H. R. (1987) *Proceedings of the SPIE, Micromachining of Elements with Optical and other Submicrometer Dimensional and Surface Specifications*, **803**, 82–86

Eda, H., Kishi, K. and Ueno, H. (1987) *Precision Engineering*, **9**, 115–122

Evans, C., Polvani, R., Postek, M. and Rhorer, R. (1987) *Proceedings of the SPIE, In-Process Optical Metrology for Precision Machining*, **802**, 52–66

Evens, D., Nicholas, M. and Scott, P. M. (1977) *Industrial Diamond Review*, **37**, 306–309

Falter, P. J. and Dow, T. A. (1987) *Precision Engineering*, **9**, 185–190

Furukawa, Y. and Moronuki, N. (1988) *Annals of the CIRP*, **37**, 113–116

Gerchman, M. C. (1986) *Proceedings of the SPIE, Optical Component Specifications for Laser-Based Systems and other Modern Optical Systems*, **607**, 36–45

Gijsbers, T. G. (1980) *Proceedings of the SPIE, Aspheric Optics: Design, Manufacture, Testing*, (Sira), **235**, 44–49

Golovchenko, J.A. (1986) *Science*, **232**, 48–53

Graham, A. H., Lindsay, R. W. and Read, H. J. (1965) *Journal of Electrochemical Society,* **112**, 401–413

Grodzinski, P. (1953) *Diamond Technology.* (2nd edn.) N.A.G. Press, London

Heath, P. J. and Aytacoglu, M. E. (1984). *Industrial Diamond Review,* **44**, 133–139

Hervo, P. (1983) *Industrial Diamond Review,* **43**, 152–153

Hibbs, L.E. and Lee, M. (1978) *Wear,* **46**, 141–147

Hitchiner, M. P. and Wilks, J. (1984) *Wear,* **93**, 63–80

Hurt, H. H. and Showman, G. A. (1986) *Proceedings of the SPIE, Ultraprecision Machining and Automated Fabrication of Optics,* **676**, 116–126

Ikawa, N. and Shimada, S. (1982) *Annals of the CIRP,* **31**, 71–74

Ikawa, N., Shimada, S. and Morooka, H. (1987) *Bulletin Japan Society of Precision Engineering,* **21**, 233–238

James, M.R.and Cohen, J.B. (1980) In *Treatise on Materials Science and Technology,* Vol. 19, (ed. H. Herman), Academic Press, New York, pp.2–55

Keen, D. (1974) *Wear,* **28**, 319–330

King, A.G. (1976) *D.Phil Thesis.* Oxford University

King, A. G. and Wilks, J. (1976) *International Journal of Machine Tool Design and Research,* **16**, 95–113

Kiso, H., Taguchi, T., Fukuhara, M and Kimura, T. (1987) *Bulletin Japan Society of Precision Engineering,* **21**, 142–143

Lewendon, B. N. (1980) *Proceedings of the SPIE, Aspheric Optics: Design, Manufacture, Testing,* (Sira), **235**, 50–56

McKeown, P. A. (1987) *Annals of the CIRP,* **36**, 495–501

McKeown, P. A. and Bent, R. G. (1979) *Precision Engineering,* **1**, 19–23

McKeown, P. A., Wills-Moren, W. J., Read, R. F. J. and Modjarrad, H. (1986) *Advanced Manufacturing Processes,* 1, (1), 133–157

Miller, D. M., Hauver, G. H., Culverhouse, J. N. and Greenwell, E. N. (1979) *Proceedings of the SPIE, Advances in Optical Production Technology,* (Sira), **163**, 55–66

Miyoshi, K. and Buckley, D. H. (1980) *Applications of Surface Science,* **6**, 161–172

Naidich, Yu. V. and Kolesnichenko, G. A. (1961) *Poroshkovaya Metallurgiya, 55–60*

Naidich, Yu. V. and Kolesnichenko, G. A. (1963) *Poroshkovaya Metallurgiya, 49–53*

Naidich, Yu. V. and Kolesnichenko, G. A. (1964) *Soviet Powder Metallurgy, 191–195*

Naidich, Yu. V. and Kolesnichenko, G. A. (1966) *Soviet Powder Metallurgy, 156–158*

Naidich, Yu. V., Umanskii, V. P. and Lavrinenko, I. A. (1984) *Industrial Diamond Review,* **44**, 327–331

Nishiguchi, T., Maeda, Y., Masuda, M. and Sawa, M. (1988) *Annals of the CIRP,* **37**, 117–120

Nishimura, K., Chujo, T., Yoshinaga, H. *et al.* (1988) In *Program and Abstracts of First International Conference on the New Diamond Science and Technology,* (Tokyo, 1988), Japan New Diamond Forum pp. 154–155

Ohmori, G. and Takada, S. (1982) *Bulletin Japan Society of Precision Engineering,* **16**, 3–7

Paterson, D. and Taylor, P. (1974) *Mullard Research Laboratories, Report 2902,* April 1974

Patterson, S. R. and Magrab, E. B. (1985) *Precision Engineering,* **7**, 123–128

Pekelharing, A. J. (1980) *Wear,* **62**, 37–48

Perry, D.M., Moran, P.J. and Robinson, G.M. (1985) *Journal of the Institution of Electronic and Radio Engineers,* **55**, 145–150

Polk, D.E. and Giessen, B.C. (1978) In *Metallic Glasses,* (eds J. J. Gilman and H. J. Leamy), American Society for Metals, Metals Park, Ohio, pp.1–35

Raal, F. A. (1963) In *Proceedings of the First International Congress on Diamonds in Industry* (Paris 1962), (ed P. Greene), De Beers Industrial Diamond Division, London, pp.13–19

Rushforth, R. W. E. (1978) *Platinum Metals Review,* **22**, 2–12

Saljé, E. and Stühmeier, W. (1988) *Industrial Diamond Review,* **48**, 171–178

Scott, P. M., Nicholas, M. and Dewar, B. (1975) *Journal of Materials Science,* **10**, 1833–1840

Seal, M. (1969) *Industrial Diamond Review,* **29**, 408–412

Shaw, M. C. (1984) *Metal Cutting Principles.* Clarendon Press, Oxford

Spur, G. and Wunsch, U. E. (1985) *Industrial Diamond Review,* **45**, 195–199

Spur, G. and Wunsch, U. E. (1988) In *Ultra Hard Materials Application Technology*, Vol 4, (ed. C. Barrett), De Beers Industrial Diamond Division, London, pp.88–101

Stadler, H. J., Freisleben, B. and Heubeck, C. (1987) *Proceedings of the SPIE, In-Process Optical Metrology for Precision Engineering*, **802**, 67–69

Sugano, T., Takeuchi, K., Goto, T. and Yoshida, Y. (1987) *Annals of the CIRP*, **36**, 17–20

Suganuma, T. (1985) *Journal of Electron Microscopy*, **34**, 328–337

Sumiya, M., Ueda, K. and Tsukada, T. (1982) *Bulletin Japan Society of Precision Engineering*, **16**, 16–22

Syn, C. K., Taylor, J. S. and Donaldson, R. R. (1986) *Proceedings of the SPIE, Ultraprecision Machining and Automated Fabrication of Optics*, **676**, 128–140

Taniguchi, N. (1983) *Annals of the CIRP*, **32**, 573–582

Taylor, J. S., Syn, C. K., Saito, T. T. and Donaldson, R. R. (1986) *Optical Engineering*, **25**, 1013–1020

Thornton, A. G. and Wilks, J. (1979) *Wear*, **53**, 165–187

Vorburger, T. V. (1987) *Annals of the CIRP*, **36**, 503–509

Wada, R. (1982) *Bulletin Japan Society of Precision Engineering*, **16**, 8–15

Weavind, R. G., Guykers, C. J. and Roy, A. R. (1958) *American Society of Tool Engineers*, **58**, paper 103

Weinz, E. A. (1968) Pamphlet, *Gloss Turning of Metals and Plastics with Diamond Tools*. Ernst Fr Weinz WEKA GmbH Idar-Oberstein, Germany

Wilks, J. (1980) *Precision Engineering*, **2**, 57–72

Chapter 15
Grinding

Most materials can be ground by abrasives such as alumina or silicon carbide, but very hard materials cannot be dealt with effectively in this way. Diamond grit was first employed on a large scale to grind tools of tungsten carbide and is now widely used to grind cemented carbides, glass and hard ceramics. (Diamond grit has also been used to grind hard steels but this application is now superceded by cubic boron nitride abrasives which are not as hard but have much less chemical affinity for ferrous metals, see Section 15.4.c.)

15.1. The grinding process

A conventional grinding wheel with alumina or silicon carbide as the abrasive consists almost wholly of a mass of alumina or carbide sintered together with some small volume of bonding agent. However, the design of diamond wheels is quite different with the diamond grit located in a relatively thin abrasive layer on the perimeter or face of the wheel. Moreover this layer consists mainly of some bonding material in which the diamond grit is embedded, forming probably no more than 25% of the total mass of the layer. Hence the details of the mechanisms involved when grinding with diamond wheels are appreciably different from those which obtain when using conventional abrasives and in this chapter we consider only grinding with diamond.

Grinding is a complicated process, and not easy to study because of the difficulty of observing the interface between the wheel and the workpiece. However, there is now an increasing amount of published information on the nature of the process and how it is affected by the properties of the diamond grit. We now go on to review this information. Further discussions of various grinding processes are given in books by Metzger (1986), Holz and Sauren (1988), and Krar and Ratterman (1990): see also a review article on the grinding of glass by Khodakov, Korovkin and Al'tshuller (1980).

15.1.a The complexity of grinding

We start by considering a simple grinding operation in which one side of a rectangular bar is ground by the perimeter of a wheel rotating at high speed. This operation is shown schematically in Figure 15.1 where the shading on the wheel indicates an abrasive layer of diamond grit embedded in a suitable bonding

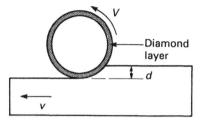

Figure 15.1 Schematic diagram indicating the depth of cut *d*, the direction of the table velocity *v*, and the peripheral velocity of the wheel *V*.

material. As the workpiece is moved forward against the wheel the grits in the rim remove material either as chips formed plastically or as fragments cracked out of brittle material. Hence as the bar is traversed along its length a layer of material is removed leaving a flat ground surface.

Clearly, if the bond material is not strong enough to retain the diamond, the grits will be pulled away bodily and the workpiece will be rubbed only by the bond. On the other hand, if the bond holds the grit very firmly, the edges and points of the grit will eventually be rubbed flat or broken off, and the wheel becomes ineffective. Grinding can only proceed as a steady and continuous process if the wear of the grits is accompanied by some wear of the bond at an appropriate rate, so that worn grits fall out and new unworn grits are uncovered to take their place.

The above situation is reflected in a series of experiments by Reinhart (1967) on the grinding of a cemented carbide. Figure 15.2 shows the volume of carbide removed as a function of the time of grinding by a bronze bonded wheels working under different loads to give three different pressures at the interface. At the lowest pressure the rate of removal of material, given by the slope of the lowest curve, decreases steadily with time. Under the highest pressure the removal rate is

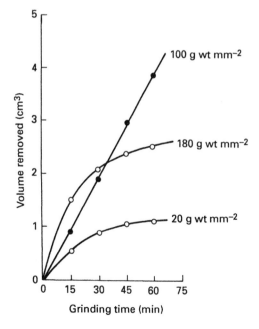

Figure 15.2 Volume of carbide removed by a bronze-bonded wheel for various surface pressures. 100 μm grit, 50 concentration, wet grinding (Reinhart, 1967)

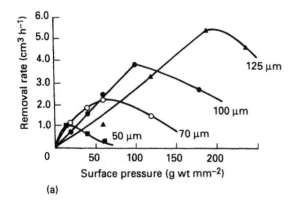

(a)

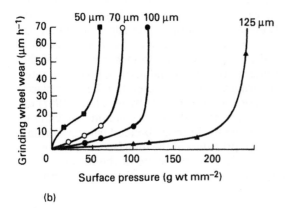

(b)

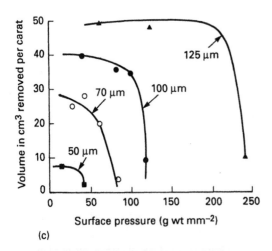

(c)

Figure 15.3 Effect of surface pressure on (a), the hourly removal rate; (b), the rate of wheel wear; (c), the specific efficiency, of bronze-bonded wheels with different grit sizes, 50 concentration, wet grinding (Reinhart, 1967)

considerably greater but also decreases with time. In contrast, the removal rate at the intermediate pressure remains unchanged as grinding proceeds and at the end of the run exceeds that produced by the higher pressure. Hence these results show that there is an optimum pressure for this operation.

In another experiment Reinhart used four wheels with similar concentrations of grit (ct cm^{-3}) but with four different sizes of grit ranging from about $50\,\mu$m to $125\,\mu$m. Figure 15.3(a) gives the total volume removed after grinding for one hour with each wheel as a function of the pressure. For each of the four wheels there is an optimum pressure which gives the greatest removal rate, and the value of this optimum rate increases with increasing grit size. (Note, however, that at other pressures increasing the grit size may reduce the removal rate.) Figure 15.3(b) shows that the wear of the wheels increases monotonically with pressure but only relatively slowly until the pressure reaches the optimum value after which the wear increases rapidly. Figure 15.3(c) shows how the volume of carbide removed per carat of diamond depends on the pressure; we see that the ratio remains approximately the same until the optimum pressure is reached and then falls off rapidly.

To discuss these results Reinhart gives micrographs of the surface of a wheel after grinding for periods of three hours under the pressures shown in Figure 15.2. Figure 15.4 shows that the appearance of the surface of the wheel after grinding depends considerably on the pressure. In Figure 15.4(a) many of the grits seem little changed by the grinding. Hence it appears that this lowest pressure was insufficient to force the grit into the carbide, with the result that after some blunting of the edges of the grits the wheel tended to rub rather than cut. After grinding at the optimum pressure (100 g wt mm^{-2}) the grits appear quite sharp (Figure 15.4(b))

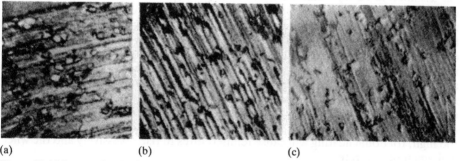

(a)   (b)   (c)

Figure 15.4 Micrographs of a bronze-bonded wheel, $125\,\mu$m grit, 50 concentration, after grinding for 3 hours at pressures of: (a), 20 g wt mm^{-2}; (b), 180 g wt mm^{-2}; (c), 240 g wt mm^{-2} (Reinhart, 1967)

presumably because the pressure is now sufficient to cause enough fracture of the grit to renew the cutting edges. At the same time the debris from the grinding slowly erodes the bond material with two beneficial effects. More clearance is provided for the machining fluid to remove debris, and new grits emerge from the surface to replace worn ones. At higher pressure the position is different again because the forces are now sufficient to tear the grits out of the bond (Figure 15.4(c)), so the workpiece is rubbed only by the bond and the performance of the wheel falls off rapidly.

Although carbides are now generally ground by wheels with resinoid rather than metallic bonds the general form of the behaviour described above is fairly typical of most forms of grinding. In particular the results stress the importance of achieving a balance between the wear of the bond and the wear of the diamond so that the performance of the wheel remains unchanged as the rim wears down. However, it is difficult to design a grinding wheel because of the large number of factors which determine its performance. The interaction between the grit and the workpiece depends on the strength and plasticity of the workpiece. Chips from the workpiece and debris from the wheel interact with both wheel and workpiece. Temperature rises produced by the frictional forces during grinding may affect the behaviour of the wheel. Although a machining fluid is generally used as a coolant there is always some uncertainty about how far it is able to penetrate to the actual working surface, and in addition some grinding processes are influenced by the chemical constitution of the machining fluid.

Finally we note that the interdependence of all the various factors influencing a grinding process makes it difficult to set up a controlled experiment where the effect of varying one parameter may be clearly isolated. For example, the effect of a change in the size of the grit may be studied by comparing the performance of two wheels containing grits of different size but useful information will only be obtained if the different sized grits are of the same quality, and are equally well bonded in the wheel.

15.1.b Geometry of grinding

In the simple grinding operation shown in Figure 15.1 the volume of material removed from the workpiece per second is equal to $v \times d \times w$ where v is the velocity of the traverse table, d the depth of cut or infeed, and w the width of the bar assumed to be less than the width of the wheel. This relationship implies that the removal rate does not depend on the peripheral speed of the wheel. As this may appear surprising, let us suppose that grinding is proceeding satisfactorily and that we then reduce the speed of the wheel considerably. The wheel will be less effective in removing material but the grinding machine is set to maintain the feed rate and the table speed at constant values. Hence provided sufficient power is available to maintain the feed and table speed against rising forces the removal rate remains unchanged. However, as the wheel speed is reduced the forces will eventually become so high that the wheel is damaged both by mechanical action and frictional heating. At the other end of the scale if the speed of the wheel is too high, frictional heating will cause temperatures to rise excessively and the wheel will again deteriorate.

The grinding arrangement shown in Figure 15.1 where the abrasive layer is formed on the rim of the wheel, and the axis of the wheel is parallel to the ground surface, is known as *peripheral grinding*. In this arrangement wheel and workpiece come together in an almost line contact, and the grits attack the workpiece in one direction only. Another common type of grinding geometry known as *face grinding* generally uses a cup-shaped wheel (Figure 15.5) mounted with its axis perpendicular to the surface being ground as in Figure 15.6. The contact area is now much larger and the workpiece is abraded by grits moving in all directions during the traverse of the wheel, so face grinding offers the possibility of a better surface finish but the method also presents problems. The greater area of contact generally results in lower grinding pressures and hence a reduction of the forces on each grit which may slow down the regeneration of the wheel. Also, because of the greater width of the area of contact, perhaps up to 10 mm, it is more difficult for machining

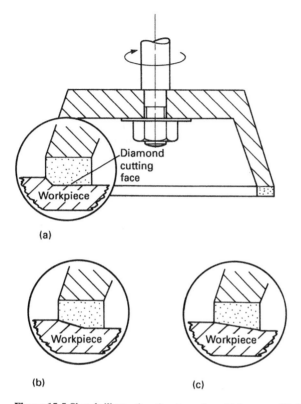

(a)

(b)   (c)

Figure 15.5 Sketch illustrating the stages by which an equilibrium wear face develops on a cup wheel (Dyer, 1963)

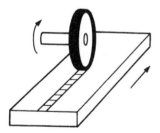

Peripheral grinding

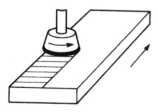

Face grinding

Figure 15.6 Sketch showing the geometry of peripheral and face grinding (after Metzger, 1986)

fluids to reach the actual grinding interface and for chips and debris to be removed. Therefore, wheels for face grinding usually contain a lower concentration of grit, to help maintain the force on each grit and encourage the entry of the fluid.

There is also a difference in the geometry of the wear of cup and peripheral wheels. If a new wheel of each type is used to machine a flat surface as in Figure 15.6, the working edge of the peripheral wheel remains parallel to the spindle as it wears down, but the surface of the cup wheel does not remain plane. There is intense wear at the edge of the wheel as it comes against the workpiece, and eventually the surface takes up an equilibrium condition as shown in Figure 15.5(c).

The infeed or depth of cut in diamond grinding is commonly of the order of 0.05 mm to 0.10 mm, greater depths of material being removed by traversing the worktable back and forth, this technique is known as *reciprocating grinding*. It is also possible to take deeper cuts of up to a few mm depth provided that the traverse speed is reduced from about 50 mm s^{-1} used in reciprocating grinding to values of about 0.5mm s^{-1}. This procedure is known variously as *deep, plunge* or *creep grinding* and offers several advantages, see Section 15.3.d.

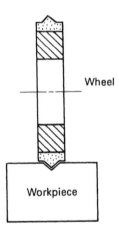

Figure 15.7 Profile grinding (after Metzger, 1986)

Besides the above types of grinding geometry there are several other arrangements of which we mention only three. In *external grinding* a cylindrical workpiece is rotated about its axis while the external surface is ground by the rotating grinding wheel. In *internal grinding* the bore of a cylinder is ground in a similar way. *Form* or *profile grinding* is a type of peripheral grinding where the edge of the wheel is shaped to produce a certain geometry as in Figure 15.7. Practical details of these and other arrangements are given by Metzger (1986).

15.1.c Surface finish

So far no reference has been made to the quality of the surface produced by the grinding process. However, there will generally be a requirement to produce a surface with a limited surface roughness, with a certain dimensional accuracy, and with a minimum of surface damage. Despite the complexity of the grinding process the methods of obtaining a good surface finish are relatively clear and straightforward. The essential principle is to ensure that the cuts made by the individual grits should be as small and fine as possible. These conditions are most obviously met by the use of fine grit and low forces on the wheel. They are also achieved by using high wheel speeds so that the workpiece only advances a short distance during the time of contact of each grit, with the result that the chips are thinner and the finish correspondingly better.

The use of fine grit, low forces and a slow traverse helps to produce a fine finish but all these factors tend to reduce the rate of removal of material from the workpiece. Therefore if an appreciable amount of material has to be ground away it is often advisable to remove most of the material with a wheel of large grit and to finish off with a finer wheel. In the case of brittle materials the grinding process may produce surface cracks deeper than the height of the surface roughness, which will have a deleterious effect on the strength and other properties of the component. Therefore it is sometimes necessary to remove an appreciable amount of material by fine grinding in order to obtain the highest quality surface. (If a large amount of material has to be removed in this way it may be better to grind down in stages using three or more wheels of different grit size, see for example Broese van Groenou (1979).)

Quite apart from the performance of the wheel the quality of a ground surface will be determined by the performance of the grinding machine. Large forces may be set up during grinding so it is essential that the machines are of massive construction designed to avoid vibration. In particular, any vibration may lead to regenerative chatter and unwanted patterns on the surface of the workpiece, see for example Sexton (1982). The design of grinding machines to avoid such vibrations is discussed by various authors, see for example Low, Nakano and Kato (1988). A type of diamond wheel designed to reduce chatter, with the diamond layer mounted more flexibly on the wheel, is described by Sexton (1982).

15.2 Diamond wheels

Diamond is used to grind various materials which exhibit wide ranges of hardness and other properties, so grinding wheels must be designed to meet particular requirements. We now indicate the types of diamond grits and bonding materials which are used by the toolmaker to produce a range of wheels with different characteristics.

15.2.a Types of grit

We have already seen in Section 15.1 that the size of the grit in a wheel determines both the rate of removal and the surface finish. Also, in order to produce wheels which wear down steadily and maintain their cutting action on different materials it is necessary to use grits with differing resistance to fracture. Several manufacturers provide synthetic grits in a range of sizes and with different strengths or friability.

Grits are generally graded for size by sieving through a series of meshes. Each size, or more accurately each group of sizes, is characterized either by the average linear particle size in μm or by the US mesh size which is the number of meshes per linear inch in a sieve which just passes the particular grit. Note that the mesh nomenclature characterizes smaller grits by larger numbers, and it is not always

Figure 15.8 SEM micrograph of a good quality blocky grit (De Beers Industrial Diamond Division)

Figure 15.9 SEM micrograph of a friable grit (De Beers Industrial Diamond Division)

clear in the literature whether a figure refers to μm or mesh! The grits used in grinding wheels commonly lie in the range 50 μm to 350 μm.

Figure 15.8 shows a typical, good quality, synthetic, single crystal with a blocky shape and sharp edges, obtained by growing under carefully controlled conditions at a relatively slow rate. The crystals of highest strength are those most free from internal defects, particularly metallic inclusions remaining from the synthesis. Most manufacturers offer a range of two or more grades of this blocky type of grit which give some differences in strength and morphology, as described in their literature.

Figure 15.9 shows another type of grit which appears quite different from the single crystal diamond just discussed. This type of crystal is synthesized under conditions which promote faster growth, so that the diamonds grow from many nuclei and form 'grossly imperfect' crystals (Wedlake, 1979). There is not much detailed information on the structure of these crystals but their general irregularity is seen in Figure 15.9, and according to Wedlake they have 'a mosaic or polycrystalline structure, are full of inclusions, and have a very rough surface and no characteristic diamond-like morphology.' Whatever their exact structure the essential point is that they are less strong and more likely to fracture than the single crystals in Figure 15.8. It is possible, with suitable conditions of synthesis to produce these mosaic type crystals in forms which break down in a steady and relatively controlled way. This mosaic type grit is often referred to as *polycrystalline* or *friable*.

Most polycrystalline grits are available with a metal coat, generally of nickel or copper put down by chemical deposition (Figure 15.10). This coating usually has a mass of about 50% that of the diamonds, and acts in several ways. It provides a

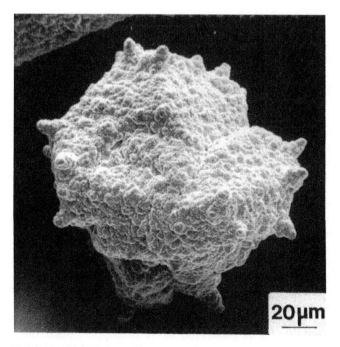

Figure 15.10 SEM micrograph of a metal coated grit (De Beers Industrial Diamond Division)

larger and rougher surface area which gives better bonding in a wheel, and its thermal capacity and conductivity help to take the heat generated during grinding away from the actual cutting edge. In addition the coat forms a protective shell which tends to prevent fragments escaping prematurely (Tomlinson, 1976).

Besides the above main types of grit a variety of other types are either available or have been proposed (often in patent applications). For example, Tomlinson, Notter and Penny (1978) describe a grit of size about 300 μm, termed multigrain, which consists of a mass of much smaller friable crystals connected together by a metal matrix. Also, Tomlinson (1978) has described grits grown to have a high ratio of length to width which have obvious advantages if the grits can be oriented with their long axis perpendicular to the wheel. Various suggestions have also been made for roughening the surface of grits to increase the surface area and thus obtain stronger bonding in the wheel, see for example Borse (1987).

Thus far we have referred only to synthetic grits which have now largely displaced grits produced from natural diamonds because being synthesized under controlled conditions they can offer a reliable product with specific properties. Even so grit from natural diamond is still produced by crushing diamonds unsuitable for other purposes and then sorting the fragments for size and shape. (The sorting for shape is made using a sloping table which is vibrated as the particles move down, thus directing the particles into a row of collecting boxes (Bruton, 1981).) This natural grit has two useful characteristics. Firstly, because the crushed particles are produced by fracture they tend to have sharp edges and points. Second, because the crushing process will have produced fracture at flaws and lines of weakness the final product should have a relatively high strength and low friability.

15.2.b Types of bond

The main features of the different types of bonds have been described by various authors, including Metzger (1988) and Holz and Sauren (1988). The two most common types of bond are the resinoid and the metallic. *Resinoid* bonds are usually based on phenolic or polyimide resins together with a filler material which can act in two ways. It may improve the flow characteristics of the heterogeneous mixture of grit and resin during the sintering process, and may also improve the resistance of the bond to the abrasive action of the workpiece and the grinding debris. Silicon carbide is commonly used as the filler, either alone or with some solid lubricant such as graphite to reduce friction and therefore the temperature of the wheel. (One of the few accounts of the effects of different fillers is given by Mastyugin (1987).)

Metallic bonds are produced by sintering a mixture of grit and powder metal, generally a bronze alloy although steels and carbides are sometimes used. The main characteristic of these bonds is that they are stronger than resinoid bonds (Table 15.1) and hold the diamonds more firmly. A graphite filler mixed with a metallic powder gives the possibility of varying the strength of the bond by introducing some porosity into the metal matrix. Because higher temperatures are required for sintering than for curing resinoid bonds, care must be taken that the diamond grits are not damaged by the heat, see for example Bullen (1975).

Metzger (1986) describes a recent development in which a resinoid bond is produced with a metallic filler such as powdered silver or copper. Such fillers function in two ways. They may act to some extent as a binder and thus help to

Table 15.1 Mechanical properties of typical samples of three major types of bond (after Metzger, 1986)

	Resin bond	Vitrified bond	Metallic bond
Brinell hardness, kg wt mm^{-2}	230	380	280
Rupture strength, kg wt cm^{-2}	1 050	1 240	2 070
Modulus of elasticity, kg wt cm^{-2}	174 000	600 000	790 000

hold the grit more firmly. Then because of their high thermal conductivity they facilitate heat flow and help reduce temperature rises at the grinding surface. Modifications of this type present the possibility of making wheels with a wide range of performance.

Besides the above two main types of bond there are several other possibilities. *Vitrified* bonds are manufactured from fusible powdered glasses together with fillers such as graphite or copper. These are not as strong as metal bonds but are considerably harder (Table 15.1) and offer a lower friction and a more free-cutting action. However, they are brittle and therefore more susceptible to mechanical damage through careless handling, and their thermal conductivity is lower so they are more susceptible to thermal damage. They are also relatively expensive.

In a wheel with an *electroplated* bond a single layer of diamond grit is held in place by a thin layer of nickel electroplated on the rim of the wheel, the thickness of the nickel being about 1 to 1.5 times that of the grit. This type of bond is relatively inexpensive and is convenient for fabricating wheels in a variety of shapes and sizes. However, the life of the wheel is restricted by the single layer of grits, and there is little scope in the plating process to vary the strength of the bond. Practical details of the plating process are described by Lindenbeck and McAlonan (1974).

Much ingenuity has been given to producing new variants of bonds to give improved performance. There are many references to these attempts in journals and patent abstracts. For example, the possibilities of a cast-iron binder are discussed by Nakagawa, Suzuki and Uematsu (1986) and Chikaoka and Watanabe (1988), and experiments to study the effect of various fillers in resinoid wheels by Mastyugin (1987). One of the more important of these developments is the so-called crushable bond for wheels used for form grinding, see Section 15.2.e.

15.2.c Specifications

Before going on to discuss the design of grinding wheels it will be useful to summarize some of the terms used to describe the form of a wheel. The maker's specification of a wheel besides giving its size and shape will include the size of the diamond grit, the type of bond, and the concentration of the grit. This concentration is a measure of the total volume of grit per unit volume of the grinding layer, and is often quoted on a scale such that a concentration of 100 indicates a layer containing 25% of diamond by volume. Concentrations on this scale typically range from 50 to 125 (Holz and Sauren, 1988). The performance of a wheel is generally summarized by quoting the rate of removal of material achieved and the accompanying rate of wear of the wheel. This rate of wear is usually given as the so-called G ratio, which is obtained by dividing the volume of material removed in a particular operation by the volume of the wheel which is worn away.

Some of the other terms used to describe the performance of grinding wheels are much less precise in meaning than the parameters mentioned above. Thus the words *hard* and *soft* applied to a wheel do not refer to the indentation hardness of the abrasive layer. A hard wheel is one in which the diamond is firmly held by a strong bond thus permitting the use of higher forces on the grit. If the grits are held less firmly by a weaker bond the wheel is described as soft. The term *free cutting* is frequently used to describe a situation where the wheel continues to remove material steadily using relatively low power and generating only low noise and low temperature rises (Metzger, 1986). Although a recognized condition in practice, free cutting remains a qualitative term not readily quantified.

15.2.d Design of wheel

We now consider the problem of designing or selecting the most suitable wheel for a particular operation. As will be described in Section 15.3 the magnitude of both the rate of removal of material and the wear of the wheel depend considerably on the materials being machined and on the machining conditions. To give some idea of these variations Table 15.2 shows typical values quoted by Metzger (1986) for

Table 15.2 Examples of G ratios and removal rates obtained in various grinding operations (after Metzger, 1986)

Workpiece	Operation	Grit	Bond	G ratio $mm^3\ mm^{-3}$	Removal rate $cm^3\ min^{-1}$
Cemented carbide	Deep feed cup wheels	Ni-clad	Resin	150 to 500	0.1 to 1.5
Structural ceramics	Surface grinding	Ni-clad	Resin	1000 to 5000	1.2
Photo-chromic glass	Lens generating	Mono-crystal	Porous metal	15000 to 30000	50 to 70

the removal rates and G ratios obtained when grinding different materials. Because of these large differences it is useful to begin by dividing the workpiece materials into two groups the harder group including tungsten carbide and other cemented carbides, and a second less hard group including ceramics and glasses.

We have already seen that the design of a wheel to produce a good removal rate with low wear involves balancing the wear of the grit against the wear of the bond for the particular workpiece in question. The carbide group of materials is so hard and tough that the points and edges of the grits soon become blunted, so it is necessary to use a friable grit which breaks down by fracture and provides more sharp edges. In addition, the bond must not be too strong, so that worn out grits are removed and replaced by new grits revealed by the erosion of the bond. Therefore the standard practice for grinding carbides is to use a friable grit in a resinoid bond.

Glasses are much less hard than carbides and therefore more easily abraded by the grits which do not blunt so rapidly. Hence it is possible to use stronger grits embedded in a stronger bond to obtain much higher removal rates (Table 15.2). Therefore the wheels used for this group of materials generally contain blocky single crystal grit in a metallic bond, see for example Wapler and Juchem (1987). Ceramic materials are intermediate in strength and hardness between glasses and carbides and show a varied range of properties and structure. Therefore various combinations of grit and bond, including vitrified bonds, have been recommended for different ceramics, see for example Juchem and Wapler (1979) and Juchem (1988); some experimentation may be needed to make the best choice from the various options.

The best design of wheel is often not obvious because of the large number of parameters involved and of the need to meet particular requirements. Tight tolerances are more readily obtained with a wheel which wears slowly. Face and plunge grinding involve a larger area of contact between wheel and workpiece so the wheel should provide for more clearance between bond and workpiece to allow the coolant to flush out the debris. Although a machining fluid is generally used as a coolant some processes require the work to be ground dry and here resin bonded wheels which generate less heat are usually preferred. There is also the question of cost, the metallic and vitrified wheels are more expensive, and also more liable to damage by mishandling, particularly the more brittle vitrified wheels. Hence the choice of the best type of wheel can be quite difficult and similar operations in different factories may quite possibly be performed by different types of wheels. The best way of making a choice is probably to consult one of the larger grit or wheel manufacturers who have much information based on practical experience. Books by Metzger (1986) and Holz and Sauren (1988) and journals such as the *Industrial Diamond Review* give many examples of the use of diamond wheels in a wide range of applications.

Finally we note that electroplated bonds give a convenient method of making wheels for form grinding (Section 15.1.b). Complex profiles are readily produced by machining a metal wheel to the appropriate shape and then depositing a layer of electroplating and a single layer of grit. Electroplated wheels tend to have a relatively high number of grits per mm^2 of wheel, giving less load per particle and less wear. Also, because the grits stand quite proud of the plating, there is greater clearance for the removal of debris. On the other hand this type of wheel tends to give a rougher surface finish (Holz and Sauren, 1988). Details of several design variations of the electroplated layer and its supporting substrate are given by Prudnikov *et al.* (1986).

15.2.e Truing and dressing

The final stages in the manufacture of any abrasive wheel consist of truing and dressing. It is essential for successful grinding that the wheel has the correct geometry. In peripheral grinding the working surface must have exact circular symmetry about the driving spindle, and in face grinding must be plane perpendicular to the spindle . This geometry is achieved by the process of *truing* which generally consists of running the wheel against rollers of a hard abrasive such as silicon carbide which are braked so that their surface slips over that of the wheel, for details see Metzger (1986) and Holz and Sauren (1988). Another method of truing, by running the wheel against a block of mild steel, is described by Notter and Shafto (1979) and by Holz and Sauren (1988).

The truing process tends to leave the grits flush with the bond so before the wheel is ready for use it must be *dressed* so that the grits stand sufficiently proud of the bond to do their work. This dressing is usually done by holding an abrasive stick of a conventional abrasive such as alumina or silicon carbide against the rotating wheel with a light pressure. In this way bond material is removed preferentially leaving the grits clearly exposed, for further details see Metzger (1986) and Holz and Sauren (1988).

As described in Section 15.1.a the tool maker aims to produce a wheel whose properties remain constant throughout its life as the abrasive layer wears down. However, if the wheel is not fully matched to the workpiece or if the machining conditions are not ideal the surface condition will deteriorate, the grits wear down to the level of the bond, and the wheel gives a correspondingly lower performance. It is then necessary to dress the wheel, as described above, to restore the surface condition.

A particular form of truing is used to produce wheels for profile grinding (Section 15.1.b). These shaped wheels were formerly produced on a lathe by turning them with a single point diamond tool controlled by a pantograph following a template of the required form, but this method has now been largely superceded by the use of diamond wheels with so-called crushable bonds. These bonds, either metallic or vitrified, are designed to have a controlled brittleness so that the desired edge geometry may be imposed on the wheel by forcing it against a carbide or steel roller with a profile which is the inverse of the required shape. (For further details see Schwämmle and Lowin (1983), Metzger (1986), Holz and Sauren (1988) and Barnard (1989).)

15.3 Performance and wear of wheels

We now give examples of how the performance of diamond wheels depends on the parameters of the grinding process. In considering these examples we must remember that it is often difficult to predict the outcome of a particular grinding operation from observations of grinding other materials with different wheels. Even so, the examples below have been chosen to indicate a variety of trends which are commonly observed.

15.3.a Measurements of wear

The wear of a wheel is often quite small so measurements must be made with sensitive micrometers or dial gauges and care taken to see that the wheel is running true and wearing uniformly. Bailey and Sexton (1981) describe an arrangement to measure the diameter of a wheel *in situ* by monitoring the pressure of air flowing out through an orifice in a plate positioned close to the wheel. This arrangement also has the advantage that it shows up any non-uniformity in wheel or grinding machine and allows an analysis of unwanted vibration.

A second point arises because to produce even a small amount of wear on a wheel it may be necessary to machine large volumes of material. Hence the usefulness of any conclusions drawn from such measurements will depend on the uniformity of the workpieces. A good account of experiments made to check on the variability of workpieces is given by Weavind (1959). It is also necessary to ensure that the conditions of machining do not change during the experiment. While this should not present problems in the laboratory it may often prove difficult under production conditions in a factory.

A further consideration is how to deal with any scatter in the measurements. Statistical variations will occur to a greater or lesser extent in the results of any measurement but they are often quite pronounced in measurements on grinding.

The fact that the grinding process involves the wear and loss of both bond and grit, and perhaps progressive changes in the surface state of the wheel, inevitably leads to some variation between results obtained from apparently similar measurements. Therefore to compare, for example, the performance of two wheels it is not enough to measure the wear of each wheel in one run. One must also make experiments to determine the scale of the scatter when the same measurements are repeated, because any differences in performances are only significant if they are greater than this scatter. For examples of the treatment of scatter in measurements see Metzger (1988, 1989), and for a general discussion of the design of experiments on grinding wheels see Metzger (1986).

Finally it is important to remember that both wheels and workpieces are generally proprietary products. Hence although they may be supplied to some nominal specification, they may vary from batch to batch to a greater or lesser extent.

15.3.b Wheel and table speed

We now discuss how the grinding process is affected by the speed of the wheel and the workpiece. Although relatively few experimental results have been published there is general agreement that the speed of the wheel should be as high as possible. The upper limit is set by two considerations. The speed must not exceed the safe limit set by the strength of the wheel. High speeds may also generate sufficient heat to damage either wheel or workpiece. The linear speed of the grits is often around $20\,\mathrm{m\,s^{-1}}$ or $30\,\mathrm{m\,s^{-1}}$.

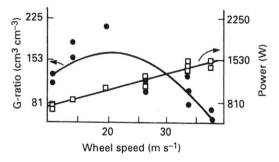

Figure 15.11 Dependence of G ratio and power requirement on wheel speed for a resin bonded diamond wheel grinding carbide (Metzger, 1981a)

It is worth stressing that increasing the speed of the wheel does not in itself change the time required to carry out a particular piece of machining which is determined only by the infeed and the speed of the table. However, as explained in Section 15.1.c, a higher wheel speed results in smaller forces on each grit and therefore a better finish and a higher G ratio, see for example Holz and Sauren (1988). Even so, higher speeds will eventually produce a greater loss of grit because of heavier impacts on the workpiece, and temperature rises which reduce the strength of the wheel, so the performance of the wheel finally falls with increasing speed (Figure 15.11). (More complex variations of G ratio with speed observed when grinding various types of alumina are described by Caveney and Thiel (1972).)

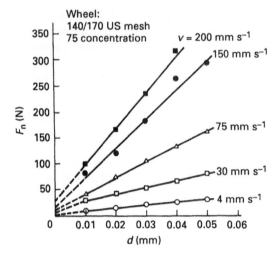

Figure 15.12 Normal force F_n on a resin bonded diamond wheel grinding carbide as a function of infeed d and table velocity V (Zelwer and Malkin, 1980)

During grinding the wheel experiences forces which depend on the magnitude of the infeed and the table speed. Figure 15.12 shows how the normal force on a wheel grinding carbide varies with the infeed or depth of cut, and with the table speed (Zelwer and Malkin, 1980). We see that the force is roughly proportional to the depth of cut as might be expected. Other measurements showed that the tangential force on the rim, opposing the rotation of the wheel, behaved quite similarly being roughly proportional to the depth of cut, although its magnitude was only about 30% that of the normal force. Hence, as the tangential force determines the work done by the wheel it follows that the *specific grinding energy*,

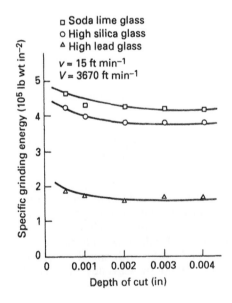

Figure 15.13 Specific grinding energy versus depth of cut for a metal bonded wheel grinding three different glasses (Huerta and Malkin, 1976)

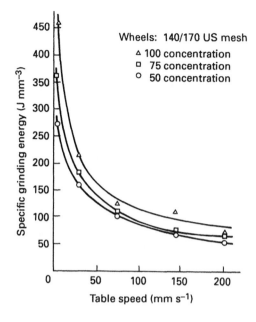

Figure 15.14 Specific grinding energy versus table speed for a resin bonded diamond wheel grinding carbide (Zelwer and Malkin, 1980)

the energy required to remove unit volume of workpiece, is almost independent of the depth of cut, as shown in Figure 15.13 which give results for three different glasses. (The glasses giving the higher energies are those with higher melting points, as might perhaps be expected (Huerta and Malkin, 1976).)

Figure 15.14 shows the specific grinding energy as a function of table speed when grinding carbide with resin bonded wheels (Zelwer and Malkin, 1980), and Figure 15.15 shows a similar type of curve for the grinding of glass with a metal bonded wheel (Huerta and Malkin, 1976). In both cases the energy increases at low table speeds. This effect arises because as discussed in Section 15.4.a the removal of material always involves some plastic deformation of workpiece and chips, and much of the energy of grinding goes to produce this deformation. In addition, the

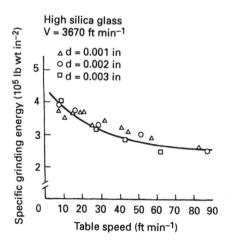

Figure 15.15 Specific grinding energy versus table speed and infeed d for a metal bonded diamond wheel grinding glass (Huerta and Malkin, 1976)

average plastic strain in a thin chip is generally more than in a thick one because the strain is always greatest closest to the rake face of the tool. Hence if all the workpiece material is removed as thin chips it will suffer greater plastic deformation. Therefore, as low table speeds lead to thinner chips we may expect greater specific energies, as are observed. The effect is more marked with carbide (Figure 15.14) than with glass (Figure 15.15) possibly because glass is more brittle, so that plastic deformation makes a smaller contribution to the grinding energy.

15.3.c Grit concentration and size

The main effect of varying the concentration of grit in a wheel is on the G ratio. Figure 15.16 shows the G ratio of a set of similar resin bonded wheels with different concentrations of grit when grinding tungsten carbide. As might be expected, increasing the concentration reduces the wear of the wheel because the force on each grit is reduced. Eventually, however, the G values fall off at the highest concentrations presumably because the greater number of grits cannot be so effectively bonded in the wheel and there is more loss of grit.

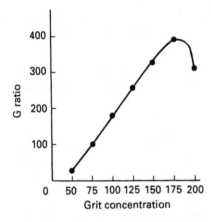

Figure 15.16 G ratios observed when grinding carbide with resin bonded diamond wheels of similar grit size but different concentrations (Tavernier, 1966)

Changes in grit concentration also affect the specific grinding energy. Thus the results for grinding carbide in Figure 15.14 show that this energy is somewhat larger at the higher concentrations. It seems that a higher concentration results in a lower loading of each grit and therefore finer chips and a greater total amount of plastic deformation as described in the previous section. Rather larger effects while grinding glass are reported by Huerta and Malkin (1976) but these results are less instructive as the grain size was varied as well as the concentration.

Figure 15.17 summarizes experiments by Pahlitzsch (1967) on the removal rates of carbide by resinoid wheels with the same volume concentration of grit but with different sized grits. Since the concentration remains constant the number of grits decreases rapidly with increasing grain size but even so the wheels with the larger grits are much more effective, presumably because of the greater loading on each particle. In addition the wear of the wheels was much less with the larger grits (Figure 15.18) presumably because larger grits can be held more securely by the bond. Apart from these curves surprisingly little detailed information has been published. However, it is generally agreed that larger grit sizes generally lead to

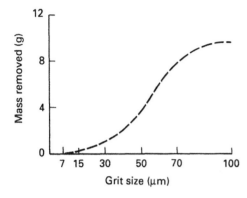

Figure 15.17 Mass of carbide removed in 1 h by resin bonded diamond wheels with grit of similar concentration as a function of grit size (Pahlitzsch, 1967)

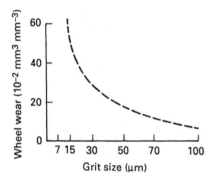

Figure 15.18 Specific wear rates of resin bonded diamond wheels of similar concentration grinding carbide as a function of grit size (Pahlitzsch, 1967)

greater removal rates and higher G ratios, and of course to a rougher finish, see for example Holz and Sauren (1988).

Finally, as grinding is a complex process it is not surprising to find exceptions to general trends. For example, we describe in the next section how the use of larger infeeds in a particular case of creep feed grinding may lead to G ratios which decrease with rising grit size in contrast to the above results (Juchem and Cooley, 1984). Hence, the only sure way to determine the best parameters for machining a particular workpiece is to make trial runs varying the appropriate parameters. Good examples of this approach are given by Dyer (1968) and Busch and Thiel (1971) who describe experiments to determine the best conditions for the difficult task of using diamond wheels to grind hard steels. (Hard steels are nowadays generally ground with CBN abrasives, see Section 15.4.b.)

15.3.d Creep feed grinding

Creep feed grinding uses a much larger infeed and a much smaller table speed thus completing the work in one or two passes. The method, which is described in detail by Metzger (1986) and Holz and Sauren (1988), offers certain advantages. In any form of grinding considerable wear occurs at the start of each pass as the wheel makes contact with the workpiece, as shown in an experiment by Metzger (1981b). This author ground a bar of carbide and then a similar bar divided into five separate shorter bars so that the wheel made five times as many entries into the material, and observed that the G ratios obtained when grinding the divided specimens were less by the order of 50%. In creep grinding the number of passes is reduced to a minimum so we expect a corresponding reduction in the wear of the wheel.

A further advantage of creep grinding is that the reduced table speed results in finer cuts leading to a smoother finish, see for example Geisweid and Gärtner (1978). This, together with reduced wheel wear, makes possible the machining of components to closer tolerances with a better finish. However, these advantages are not obtained without employing suitable grinding machines.

The forces in grinding are often roughly proportional to the infeed and the velocity of the table (Section 15.3.b). This suggests that if the table speed is reduced proportionally to the increase in the infeed then the force on the wheel should remain approximately constant. However, we saw in Section 15.3.b that at low table speeds when the depth of the cut made by each grit becomes small the specific grinding energy increases considerably (Figures 15.14 and 15.15). Therefore when table speeds are shifted towards creep grinding conditions the grinding energy and therefore the forces may rise considerably even if the table speed is reduced proportionally to the change in infeed, see Geisweid and Gärtner (1978). Hence, it is necessary to use more massive and stiffer grinding machines to take advantage of the benefits of creep grinding, as emphasized by Metzger (1986).

The larger forces involved in creep grinding produce correspondingly more heat in wheel and workpiece. On the other hand the area of contact between wheel and workpiece is also larger and this tends to reduce the temperature rises produced by the heat. However, the overall effect is that creep grinding tends to produce higher temperatures, see for example Geisweid and Gärtner (1978), so an adequate supply of coolant is particularly necessary for this type of grinding.

The details of the creep grinding process are not yet fully understood. Metzger (1981a) observed G ratios when grinding carbide which decreased with increasing grit size, which is the opposite dependence to that usually observed in reciprocal grinding (Figure 15.17). Another point is that the relative grindability of different workpieces may depend on the method of grinding. For example, Metzger (1981a, 1982) observed that the G ratio for the reciprocal grinding of a carbide containing 36% TiC/TaC was an order of magnitude less than that for a conventional carbide, yet for creep grinding the difference was only a factor 2. To explain this behaviour it is necessary to take account of the detailed differences between creep and reciprocating grinding. In creep feed grinding the chips are finer, and the area of contact much larger, making it more difficult for debris to leave and for grinding fluid to enter, so the conditions responsible for the wear of the grit and the bond are appreciably different. The geometry of the chip forming process also changes as the depth of cut increases, and with deep cuts will be significantly different if the grits enter at the upper rather than the lower end of the interface (Figure 15.1). In fact grinding rates and G values may vary by about 10 to 20% according to the direction of revolution of the wheel.

15.3.e Grinding fluids

In any type of grinding it is generally advantageous to use a suitable grinding fluid, see for example Springborn (1967). The fluid acts in several ways. It cools the wheel and workpiece and thus helps to avoid thermal damage, particularly to workpieces of low thermal conductivity. It may also act as a lubricant which reduces the friction between wheel and workpiece, and hence reduces both the power required and the production of heat. In addition a quite different but equally important function is to flush out and remove the grinding debris from the face of the wheel so that it does not build up round the grits and reduce their cutting action. There are also two further possibilities. In grinding processes which give rise to chemical interactions between wheel and workpiece, such as those described by Coes (1971), the fluid may act as a barrier blocking off the reaction. On the other hand a fluid may itself react directly with wheel or workpiece.

The liquids used as grinding fluids fall into three main groups: (a), neat (undiluted) mineral oils with or without additives; (b), emulsions consisting of a relatively small concentration of mineral oil in water together with emulsifiers and additives to protect against corrosion; and (c), water solutions with inhibitors to prevent corrosion and perhaps other additives. Neat oils are generally the best lubricants but water has a greater thermal capacity and is more effective as a coolant. In addition, neat oils may give rise to fire and health hazards. Hence a compromise of water based emulsions is often preferred.

When grinding with diamond wheels, as with other types of wheels, a change of grinding fluid often makes a difference to the rates of removal and wheel wear, and to the surface finish. Chalkley (1968) describes experiments in which metal bonded wheels were used to grind specimens of carbide and alumina using 15 different neat oils, 10 soluble oil emulsions, and water and aqueous solutions. The results for alumina are summarized in Figure 15.19 and we see that all the removal rates decrease steadily with time. This is often a characteristic feature of experiments measuring removal rates and presumably arises because of a progressive deterioration of the wheel which was not regenerated by dressing during the run. It was also found that the removal rates varied considerably from fluid to fluid and from workpiece to workpiece for no very obvious reasons. For example, water gave a poor performance with the alumina but acted quite well with the carbide.

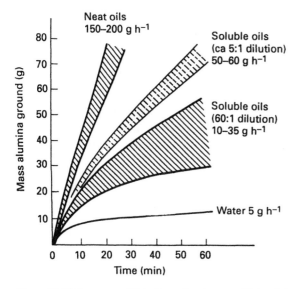

Figure 15.19 Summary of the effect of various machining fluids on the mass of alumina removed during grinding. Figures give removal rates after 60 min. (Chalkley, 1968)

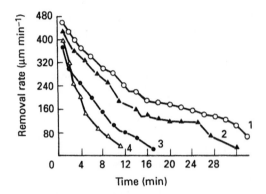

Figure 15.20 Removal rates when grinding glass with: 1 water, 30% glycerine; 2 water, 1% triethanol amine, 1.5% nickel sulphate; 3 water, 1% triethanol amine; 4 water. (Al'tshuller, Ashkerov and Korovkin, 1983)

Further examples of the effect of different fluids on the G value when grinding carbide are given by Oates and Willmington (1975). There are a number of papers on the use of different fluids when grinding glass, see Kerkhof and Schinker (1972), Rupp (1974), Al'tshuller, Ashkerov and Korovkin (1983), and Edwards and Hed (1987). Figure 15.20 shows some typical results for removal rates given by Al'tshuller, Ashkerov and Korovkin (1983). A review article is given by Korovkin and Al'tshuller (1987) (but see Section 16.3.b for comments on the experiments by Westwood *et al.* mentioned in the article.) The influence of fluids on the surface finish of glass has been studied by Korovkin, Al'tshuller and Ashkerov (1985) and Hed and Edwards (1987) whose results show that changes of fluid leading to increased removal rates may be accompanied by either an increase or a decrease in the surface roughness. Examples of the effect of different fluids on the G values when grinding steel are given by Hughes (1968), Busch and Thiel (1971), and Oates and Willmington (1975).

The action of these fluids is not well understood, partly at least because a change of fluid may affect more than one parameter of the grinding process. A fluid with a greater lubricating power may reduce friction and heating with beneficial effects but may also reduce the traction forces responsible for the removal of material, see for example Broese van Groenou, Maan and Veldkamp (1979). It is therefore not easy to predict the most suitable fluid, and the optimum machining conditions. Useful empirical information is given in the trade literature, see also Metzger (1986) and Holz and Sauren (1988). We return to the mode of action of machining fluids in Section 16.3.

To produce the maximum effect a grinding fluid must come as close as possible to the actual working surfaces. It is therefore important that the fluid be directed in sufficient volume in such a direction that it penetrates to the grinding interface. Details of suitable arrangements are given by Metzger (1986) and Holz and Sauren (1988) who also discuss the arrangements necessary to maintain the fluids free of debris and contamination.

Finally, the use of grinding fluids generally involves quite complex arrangements to circulate the fluid, to contain it as it is flung off the wheel, and to keep it clean and free of debris. In addition, the fluid mist generated by the wheel makes it difficult or impossible to view the progress of the grinding. Therefore dry grinding is sometimes preferred either for reasons of simplicity and convenience or to be able to view the workpiece, as when sharpening tools by grinding. In this case resinoid wheels with friable grits are generally preferred because they generate smaller forces and less heat and therefore may be used dry provided that the working conditions are chosen to avoid excessive heating.

15.3.f Condition of ground surfaces

We now consider how the grinding process determines the roughness and strength of the ground surface. We have already remarked that smoother surfaces are obtained by using higher wheel speeds and slower table speeds thus producing finer chips. Figure 15.21 gives an example of how the roughness of a surface depends on the table speed and the concentration of the grit. With increasing concentration the load on the wheel is carried by a greater number of grits so each grit makes a lighter cut thus giving a finer finish. Figure 15.22 shows how the roughness may also depend on the choice of grit. The SNDMB is a natural diamond grit with sharp edges which cut cleanly. The other two are synthetic grits with a blocky form and less sharp edges and therefore likely to produce more roughness. We also note

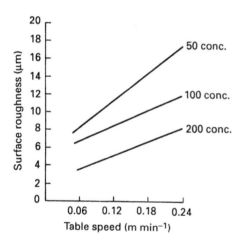

Figure 15.21 Finish produced on glass by diamond wheels with different concentrations of 150 μm grit as a function of table speed (Wapler and Juchem, 1987)

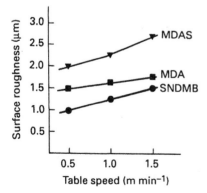

Figure 15.22 Finish produced on glass by diamond wheels of similar specification but with different grit types as a function of table speed; see text (Wapler and Juchem, 1987)

that the surface roughness is affected to only a small extent by the magnitude of the infeed, see for example Chandrasekar, Shaw and Bhushan (1987). This happens because the action of the wheel which produces the final finish is not much affected by how much material has been removed previously, see Figure 15.1.

In Section 15.4.a we describe single point scratching experiments which show that during grinding material may be removed by plastic flow or brittle fracture, the relative importance of the two processes being determined by the nature of the workpiece, the load on the wheel, and other conditions. Each of these processes leaves some damage in the ground surface, in the form of either cracks or plastic deformation. Thus a ground surface may be weakened by the presence of cracks and may contain internal stresses set up either by plastic deformation or by the failure of cracks to close completely after the grinding (Section 7.7.d).

Example of changes in strength produced by grinding are given by Huerta and Malkin (1976) who used four point bending tests to measure the strength of glass beams after grinding with different wheels containing two concentrations and two sizes of grit. Some of their results are shown in Figure 15.23 as a Weibull plot which

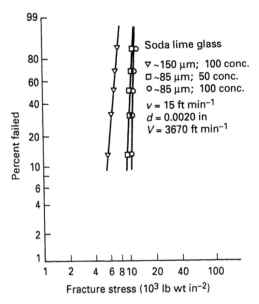

Figure 15.23 Weibull plots of strength of glass ground by wheels of differing grit size and concentration (after Huerta and Malkin, 1976)

displays the spread in the values of the measured strengths which arise from the range of crack sizes produced by each particular wheel, for details of the Weibull plot see Atkins and Mai (1985). Increasing the concentration of the 85 μm grits from 50 to 100 led to a small increase in strength presumably because each grit was then less heavily loaded. The specimens ground with 150 μm grits are somewhat less strong because with the same volume concentration the number of large grits is much less and the loading on each correspondingly greater. Similar results are quoted by Inasaki (1987). Also, if the workpiece is not isotropic, as is often the case, the structure and damage in the surface will depend to some extent on the direction of abrasion.

The depths of the surface cracks produced in a ground surface may be estimated in several ways. Hed and Edwards (1987) used a taper section technique described for example by Bowden and Tabor (1950). Chandrasekar, Shaw and Bhushan (1987) etched the surface at a known rate and used an SEM to observe obvious changes in the surface as the cracks were removed. Kayaba and Fujisawa (1986) found the depths of cracks by observing the increases in strength as the layer of cracked material was polished away in small stages. It turns out that the depth of these cracks are generally quite appreciable and it has been suggested that the depths are often of the order of ×3 the peak to valley surface roughness (Hed and Edwards, 1987).

The internal stresses in the surface of a specimen produced by grinding may be observed by standard X-ray techniques as described for example by Cullity (1978) and James and Cohen (1980). Another technique used by Chandrasekar, Shaw and Bhushan (1987) is to observe the curvature of a thin flat specimen as the strained material is etched away in stages. In ceramics and similar materials the stresses are generally compressive and so tend to close up any cracks and may have a beneficial effect on the strength of the surface, see for example Lange, James and Green (1983) and Chandrasekar, Shaw and Bhushan (1987).

A detailed and instructive comparison of ground surfaces of steel and a nickel–zinc ferrite was made by Chandrashekar, Shaw and Bhushan (1987) who gave micrographs which showed deformation by plastic flow and by cracking. In particular they observed that the grinding of the ferrite had produced both plastic flow and brittle fracture while, of course, the steel was ground primarily by plastic action. The results also showed that the performance of the ferrite when used for magnetic recording was adversely affected by any residual stresses. An example of the effect of surface finish on the electrical quality and stability of quartz crystals used as oscillators is given by Schmitt (1980).

15.3.g Cost effectiveness

The use of diamond grinding wheels can greatly reduce the cost of many machining processes. However, these results are only obtained if the wheels are used on properly designed grinding machines and operated under the appropriate machining conditions. There is also a further point of considerable importance. As described in the previous sections there is now a large amount of information available on the effect of working conditions on removal rates and wheel wear. Therefore, as wheels are costly items it might seem that the working conditions should be selected to give the least wear and greatest G ratios but this is quite the wrong procedure.

The cost of a grinding operation is very specific to the particular operation in question but in all cases the cost can be divided into two parts, the cost of the wheel and all the other costs including labour, overheads, amortization of the grinding machines etc. Provided the wheel is being used under reasonably efficient conditions, higher removal rates will certainly cause more wheel wear and therefore a steady increase in the cost per unit volume of material removed. However, the other costs decrease with increasing removal rate because the operation takes less time. Therefore a balance must be struck between the two components of the cost, and the removal rate chosen to give the lowest *total* cost.

15.4 Studies of the grinding process

We now describe experiments which have been made to study various details of the grinding process.

15.4.a Single point experiments

A number of authors have studied the cutting action of diamond grits by observing the action of a diamond indenter drawn across the surface of a workpiece material under a controlled load. To give some background to these experiments Figure 15.24 summarizes the behaviour of a typical brittle material when indented by a pointed indenter. When the load is first applied the only effect is to produce a small volume of plastic deformation around the point of contact (Figure 15.24(a)). Then as the load is increased a so-called *median* crack is formed going down into the material (Figure 15.24(b)), and both the crack and the volume of plastic deformation increase with rising load. If the load is now reduced somewhat the median crack will close (Figure 15.24(d)). Then on reducing the load further the stress which has been induced in the material by the plastic deformation produces the so-called lateral cracks which spread out as in Figure 15.24(e). Finally as the load is reduced to zero the lateral cracks expand and may run out to the surface. Note that the lateral cracks only form as the load is reduced.

When an indenter is slid over a surface in a scratch type of experiment it creates rather similar damage to the above all along its track. Several authors have observed and discussed this type of damage on various glassy and ceramic materials (Broese van Groenou, Maan and Veldkamp (1975); Veldkamp, Hattu and Snijders (1978); Swain (1979); Molloy, Schinker and Döll (1987); Zhang, Tokura and

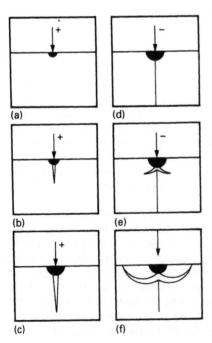

Figure 15.24 Schematic diagram showing formation of a vent crack under point indentation. Median vent forms during loading (+) half cycle, lateral vents during unloading (−) half cycle. Fracture initiates from the plastically deformed zone shown dark (Lawn and Swain 1975)

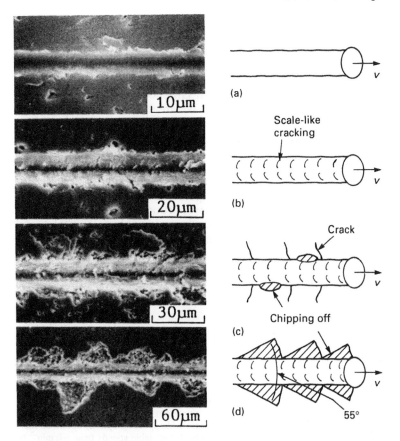

Figure 15.25 Micrographs of scratches on alumina made by a diamond stylus of 1.6 μm radius under progressivly greater loading (Zhang, Tokura and Yoshikawa, 1988)

Yoshikawa (1988)). Figure 15.25 due to Zhang, Tokura and Yoshikawa (1988) shows the effect of progressively higher loads on a diamond indenter scratching alumina. With relatively light loads the only damage is a plastic groove, as in Figure 15.24(a). On increasing the load the plastic grooving becomes more pronounced and cracks appear in the surface (Figure 15.24(b)). At still higher loads lateral cracks break out into the surface behind the indenter (Figure 15.24(c)) producing fragments which either fall out immediately or are removed on subsequent passes. Rather similar scratches have been observed on silicon by Puttick and Shahid (1977), and on Sialon by Nakajima, Uno and Fujiwara (1989).

There is rather less information on the behaviour of cemented carbides during grinding possibly because the presence of the binder phase introduces greater complexity. Optical studies by Luyckx (1968) of the surface of tungsten carbide fractured in bending tests showed evidence of both fracture and plastic flow. High speed single point scratching experiments have been made by Zelwer and Malkin (1980) who conclude that material is removed during grinding in the form of chips not greatly dissimilar from those produced when grinding metals.

Various authors have measured the forces acting on a single grit in scratching experiments. In some experiments the grit was slid, under a constant load, at low speeds ranging from $10\,\mu m\,s^{-1}$ to $1000\,\mu m\,s^{-1}$ (Broese van Groenou, Maan and Veldkamp, 1975; Veldkamp and Klein Wassink, 1976). Other experiments have been made with the grit attached to the rim of a rotating wheel to make a predetermined depth of cut at linear speeds of up to $30\ ms^{-1}$. In all these experiments the ratio between the tangential and normal forces, F_t and F_n, depends on the shape of the stylus (Maan and Broese van Groenou, 1977), and for cones with semi-vertical angles less than $45°$ F_t may be greater than F_n because of increased penetration. Prins (1971a) and Gielisse and Stanislao (1972) stress that while material is removed primarily by the force F_t, the normal forces F_n generally produce subsurface damage prior to removal.

Measurements of F_t and F_n using a rotating grit have been made on alumina (Prins, 1971a; Gielisse and Stanislao, 1972), Sialon (Nakajima *et al.*, 1989), and steel (Prins, 1971b). Gielisse and Stanislao cutting alumina at speeds up to $30\ ms^{-1}$ observed that the forces on a grit were roughly proportional to the depths of cut, as might be expected, and that the forces decreased considerably with rising speed (Figure 15.26). Even at very high speeds the temperature of the grit rose by only

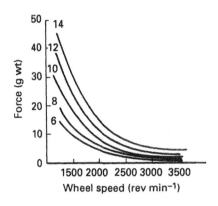

Figure 15.26 Force on a diamond grit scratching aluminia in a single grit experiment as a function of wheel speed and table speed. Depth of cut 0.002 in; diameter of wheel 7 in; table speeds from $6\,ft\,min^{-1}$ to $14\ ft\ min^{-1}$ (Gielisse and Stanislao, 1972)

a relatively small amount, so the authors account for the reduced forces by considering the resilience of the alumina and the energy of the impact between the grit and workpiece. The greater energy of impact at high speeds produces higher stresses and more fracture in the alumina, with a corresponding reduction in the forces on the wheel. Although the details of the processes are not fully resolved, the results clearly indicate the need to avoid low wheel speeds when grinding brittle materials.

The force measurements in scratch experiments provide basic information for an assessment of the specific grinding energy which determines the power required in a particular grinding process. Several authors have discussed the relationship between the forces and the grinding energies for a range of ceramic and glassy materials, see for example Veldkamp and Klein Wassink (1976), Broese van Groenou and Veldkamp (1978/9) and Broese van Groenou and Brehm (1979). These comparisons are instructive but the correlations obtained are generally not too close, no doubt because the exact conditions of abrasion during grinding are much more uncertain than the conditions in scratch experiments. See also a discussion by Kirchner (1984) who stresses that during grinding material may be removed by multiple as well as single passes of the grit.

15.4.b Grit wear

We now describe a study and comparison of the wear of diamond and CBN grits on alumina and hard steels (Hitchiner and Wilks, 1983, 1987). Experiments were made with grits of either diamond or CBN, of size about 125 μm, embedded in nickel electroplated on steel rods of about 2 mm diameter to form grinding wheels or *nibs* which could be easily examined in the optical microscope and the SEM (Figures 15.27 and 15.28). The nibs were used to grind blocks of M2 steel and commercial high density alumina, 2 mm wide and 80 mm long, on a small purpose built grinding machine described by Stokes, Cooley and Juchem (1983) using a plentiful supply of coolant. Repeated passes at an infeed of 50 μm were made over the specimens which were mounted on a dynamometer which measured the normal force at the start of each pass.

1.0 mm

Figure 15.27 Photograph of a grinding nib; see text (Hitchiner and Wilks, 1983)

100 μm

Figure 15.28 SEM micrograph of the grits on a typical new nib (Hitchiner and Wilks, 1983)

Typical values for the force on diamond nibs turning M2 steel are shown in Figure 15.29. As the nibs were relatively large compared to the width of the specimens it was possible to make three separate wear runs on each nib and these are referred to as Bands 1, 2 and 3. The forces rose steadily with the distance ground as shown in the figure, but when they reached some value beyond 1000 g wt the plating began to strip off and the run was concluded.

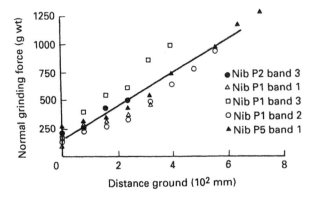

Figure 15.29 Force on MDASE nibs grinding M2 steel (Hitchiner and Wilks, 1983)

Figure 15.30 SEM micrograph of a fractured grit (Hitchiner and Wilks, 1983)

Figure 15.31 Optical micrograph showing wear flats on grits (Hitchiner and Wilks, 1983)

Table 15.3 Performnce of MDASE and EDC diamond grits and ABN300 CBN grits grinding M2 steel (Hitchiner and Wilks, 1983)

Grit	Nib band	Distance ground mm	Wear flat area 10^{-2} mm^2 mm^{-2}	Fracture area		Grinding force	
				Initial 10^{-2} mm^2 mm^{-2}	Final 10^{-2} mm^2 mm^{-2}	Initial g wt	Final g wt
MDASE	P2/3	320	1.1	3.9	7.5	190	600
	P1/1	400	1.8	1.3	7.2	130	600
	P1/2	480	1.3	1.6	6.6	160	1100
	P1/3	640	2.9	1.6	10.7	130	1100
EDC	P2/1	480	1.9	6.5	17.0	160	400
	P3/3	480	1.2	4.4	16.0	120	350
	P2/2	880	1.6	5.6	15.0	80	1100
	P2/3	800	1.8	5.3	14.0	130	1100
ABN300	P2/1	800	0.6	6.1	32	115	220
	P2/2	800	0.7	6.0	35	80	220
	P2/3	16000	2.6	6.9	44	100	500

Microscopic examinations during the wear runs showed that until the final breakdown there was little loss of grit and little damage to the nickel. Hence the principal change was in damage to the grits. Two types of damage were observed. Figure 15.30 shows an SEM micrograph of a typical fracture produced during a run. The other type of damage took the form of smooth wear flats not easily detected in the SEM but which reflected strongly when aligned normally to the axis of an optical microscope as in Figure 15.31. (As the microscope has a small depth of field only the flats themselves are sharp in the figure.) Table 15.3 summarizes the areas of flat and fracture on the diamond and CBN nibs after grinding M2 steel. Two types of diamond grits were used MDASE and EDC, the former being designed to be of a more blocky shape and more resistant to fracture. The table shows that while both grits showed quite similar areas of wear flats the less strong EDC grits showed appreciably more fracture, as might be expected. However, this difference in behaviour is small compared to the difference between the behaviour of the diamond and the CBN nibs. The force on the diamond nibs after a total path of 800 mm rose by 900 g wt whereas after 16000 mm the force on the CBN nib had risen by only 400 g wt. For some reason not understood the force on the CBN nibs rose more rapidly during the first 800 mm and there was also an appreciable increase in the areas of flats and fractures. However, during the subsequent slow rise of the force to 500 g wt there was little change in the fracture area but the area of the flats increased by a factor 4.

Table 15.4 shows the results observed when grinding alumina. The distances ground by both types of nibs were roughly comparable, but while the force on the diamond nibs increased by less than 100 g wt, the force on the CBN nib increased by 400 g wt or more. An inspection of the worn nibs showed that the area of wear flat had increased by an order of magnitude more on the CBN nib. There is also more fracture area on the CBN nib, but this is probably not significant because the forces on the EDC and MDASE nibs are quite similar even though the EDC grits show more fracture. Hence it appears that the life of a nib is determined principally by the rate of production of wear flats on the grits. Diamond grit performs better on alumina because diamond is harder than CBN and more resistant to the abrasive action of the alumina. On the other hand, CBN grit performs better on M2 steel because CBN has less chemical affinity with the steel.

Table 15.4 Performance of MDASE and EDC diamond grits and ABN300 CBN grits grinding alumina (Hitchiner and Wilks, 1983)

Grit	Nib / band	Distance ground mm	Wear flat area 10^{-2} mm^2 mm^{-2}	Fracture area		Grinding force	
				Initial 10^{-2} mm^2 mm^{-2}	Final 10^{-2} mm^2 mm^{-2}	Initial g wt	Final g wt
MDASE	P6/2	3500	0.5		6.7	80	180
	P3/1	7000	0.25	1.3	9.5	60	130
EDC	P6/3	2300	0.1		12.0	80	160
	P4/3	7000	0.4	5.7	24.0	30	100
ABN300	P3/2	2800	3.5	5.9	38.0	60	560
	P6/3	3500	1.7		36.0	70	490

Relatively few studies have been published on the retention of grit in grinding wheels even though loss of grit through erosion or failure of the bond is a crucial element in the performance of most wheels. See, however, Broese van Groenou (1979) who gives micrographs showing the erosion of the binder when grinding a ferrite and discusses how this erosion affects the self sharpening properties of the wheel.

15.4.c Wear on steel

We now briefly consider the reasons for the poor performance of diamond wheels grinding hard steels. Figure 15.32 shows the G ratios obtained in an experiment by Graham and Nee (1974) who ground a range of steels of different hardness using CBN wheels. (Because the steels contained carbide phases the hardnesses are quoted as 'abrasive numbers' which are essentially weighted averages of the Vickers hardness.) We see from the figure that the G ratios appear to be determined largely by the hardness of the steel. The authors also ground the same steels with diamond wheels and found G ratios 10 to 30 times smaller even though diamond is appreciably harder than CB. This result suggests that the high wear of the diamond wheels does not arise from a process of mechanical abrasion but from

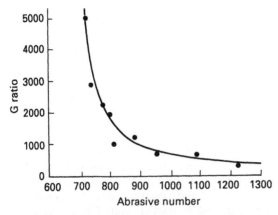

Figure 15.32 G ratios for CBN wheels grinding various steels of different hardness (or abrasive number, see text) (Graham and Nee, 1974)

chemical reactions of the types outlined in Section 13.6; see also Hitchiner and Wilks (1987). In this case there may be a possibility of reducing wheel wear by suitably doping the diamond grits during synthesis.

Evidence of graphitization on a diamond grit is given by an experiment in which a single grit was moved against a steel plate at a speed of about 20 m s^{-1} to simulate conditions during grinding (Komanduri and Shaw, 1975). Figure 15.33 shows a wear flat produced on the tip of a grit which shows groove like markings parallel to the direction of abrasion. A detail of this flat (Figure 15.34) shows that the grooves present a fine sawtooth pattern not readily accounted for by a process of mechanical abrasion, and the authors conclude that it arises by preferential graphitization along {011} planes.

The interaction of the grit with the steel was further investigated by Komanduri and Shaw (1976) who scratched a specimen of pure iron with a diamond grit moving at high speed. Using etching and Auger techniques (Section 13.8.a) they observed a high concentration of carbon on the surface of the iron which penetrated down

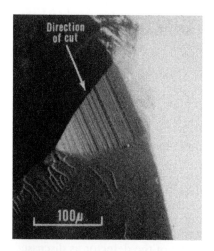

Figure 15.33 SEM micrograph of the tip of a diamond grit after turning steel (Komanduri and Shaw, 1975)

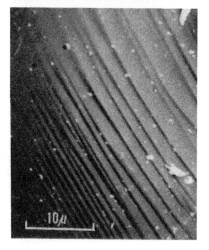

Figure 15.34 Detail of the grooves seen in Figure 15.33 (Komanduri and Shaw, 1975)

into the surface to a depth of about 16 nm. This result clearly demonstrates the motion of carbon atoms from the diamond to the iron, although of course the rate of diffusion of carbon into steel will depend on the presence of carbon already present in carbides and grain boundaries.

In other experiments Ikawa and Tanaka (1971) and Tanaka and Ikawa (1973) observed the graphitization of diamond in static contact with iron and steel, and the accompanying diffusion of carbon into the metal. Their various estimates suggest that in these static experiments at 1000°C the graphitization of the diamond proceeds an order of magnitude faster than the diffusion of carbon into the metal. Hence it appears probable that any graphite produced by the wear of a grit during grinding will be removed primarily by abrasion rather than by diffusion into the steel. Ikawa and Tanaka also attempt to correlate the rates of wear of grits during grinding to the rates of graphitization observed in their static experiments but do not determine the working temperature of the grit.

15.4.d Measurement of temperature

In order to obtain a full understanding of the grinding process we need to know the temperatures generated at the grinding interface. It is obviously difficult to measure temperatures at an interface with relative velocities of the order of $30\,\mathrm{m\,s^{-1}}$, but encouraging progress has been made by Ueda, Hosokawa and Yamamoto (1985, 1986) by observing the infrared radiation emitted by the grits. These experiments were made on the grinding of steel with alumina wheels but give basic information relevant to all types of grinding, so we now describe this work before commenting on earlier measurements on diamond wheels.

The basic arrangement used by Ueda, Hosokawa and Yamamoto (1985) is shown schematically in Figure 15.35. The workpiece is mounted on a table which moves past the wheel as shown. The temperature of the wheel after passing the work is sampled by placing one end of an optical fibre 50 μm in diameter close to the wheel in the position shown. This light pipe collects infrared radiation from the heated grits and passes it to a semiconducting (InAs) detector, so that each grain creates an electrical pulse as it passes the pipe. The grit diameter was about 700 μm and the wheel speed about $30\,\mathrm{m\,s^{-1}}$ so pulses were only about 20 μs long and an essential feature of the experiment was the rapid response time of the detector as discussed by the authors.

Figure 15.36 shows some typical temperatures pulses obtained when the table was traversed past the wheel, each set of pulses being obtained during one revolution of the wheel. The authors identify three pulses a,b,c arising from three grits on different parts of the wheel. The fact that the heights of these pulses are not constant shows that the condition of the grains and their surroundings is changing as grinding proceeds. The authors summarize all their measurements in Figure 15.35 which gives histograms showing the distribution of temperatures recorded when the detector is placed at various angular distances θ from the workpiece. The histograms show how the temperatures of the grits fall after leaving the workpiece, some cooling more rapidly than others. The authors make an extrapolation to estimate the temperature of the grit in contact with the workpiece at θ = 0 and conclude that the mean temperature of grains in the cutting zone was about 1100°–1200°C.

Other experiments were made to determine the temperature of the workpiece (Ueda, Hosokawa and Yamamoto, 1986). A hole was drilled in the back of a steel

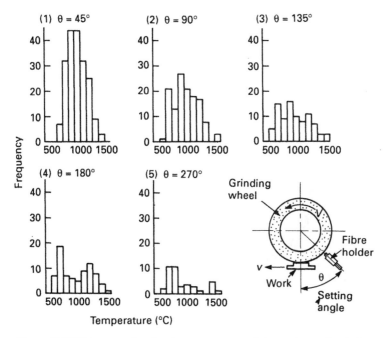

Figure 15.35 Histograms showing the temperatures of the cutting grains during grinding as observed at different distances from the workpiece (Ueda, Hosokawa and Yamamoto, 1986)

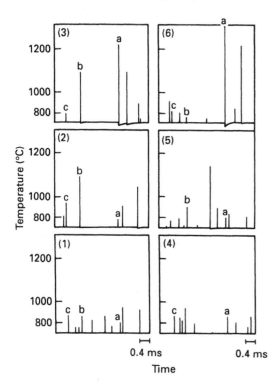

Figure 15.36 Temperature pulses observed during a series of passes of an alumina wheel grinding steel, see text (Ueda, Hosokawa and Yamamoto, 1985)

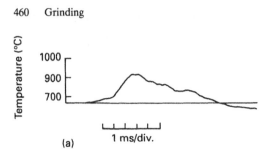

(a)   1 ms/div.

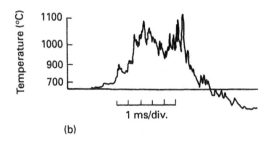

(b)   1 ms/div.

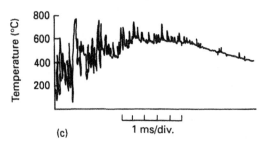

(c)   1 ms/div.

Figure 15.37 Temperature changes observed on the underside of a steel sheet during one pass of an aluminia grinding wheel measured with an InAs detector, thickness of the steel (a), 80 μm; (b), 20 μum. (c), shows the temperature of the top surface measured in a similar experiment using the thermocouple technique of Peklenik (Ueda, Hosokawa and Yamamoto, 1986)

specimen to within 100 μm of the ground surface. An optical fibre was then inserted to view the infrared radiation from the underside of the remaining thin layer of steel as the specimen was ground down. Figure 15.37(a) shows the rise and fall in temperature during one pass when the collector was 80 μm below the surface and Figure 15.37(b) the rise and fall after the surface had been ground away leaving a depth of only 20 μm. We see that in the latter case it is possible to distinguish peaks from individual grains and that maximum temperatures are appreciably higher. Figure 13.38 shows that by extrapolating the results at different depths it is possible to make an estimate of the temperature at the surface. (The figure also shows that the temperatures recorded by a PbS detector with a slower response time were much lower.)

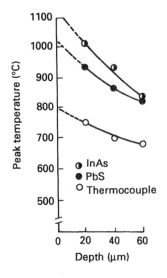

Figure 15.38 Peak temperatures observed with 3 different detectors as a function of the depth of the detector below the ground surface, see text (Ueda, Hosokawa and Yamamoto, 1986)

Ueda, Hosokawa and Yamamoto (1986) also measured the temperature of the workpiece by two thermocouple methods. The first used the simple technique of forming the couple by spot welding a constantan wire to the bottom of a small hole drilled from the underside of the surface and gave only a very broad pulse rising to no more than 700°C. The second method devised by Peklenik (1958) involves drilling a small hole right through the workpiece and then introducing a central wire surrounded by a layer of insulation. As grinding proceeds the production of chips and the plastic deformation of the surface will close a circuit between wire and workpiece thus giving rise to a thermal emf. However, as shown in Figure 15.37(c) the form of the heat pulse is different from that given by the infrared technique and the maximum temperature is less than 800°C, no doubt because the junction is smeared out over a relatively large area with a range of temperatures.

The above infrared techniques have yet to be applied to diamond grinding. The only measurements made at the face of a diamond wheel appear to be those of Primak and Skripko (1970) and Geisweid and Gärtner (1978). Using the Peklenik technique the former observed temperatures up to 400°C when grinding alumina, and the latter temperatures up to 800°C when grinding carbide. Measurements of the temperatures of a single grit traversing a workpiece of alumina at high speed have been made by Gielisse and Stanislao (1973) using a conventional thermocouple technique. In a somewhat similar experiment on steel Sagarda and Khimach (1973) used a semiconducting diamond grit to form a thermocouple with the metal. However, the results described above suggest that all these earlier measurements are of limited accuracy.

15.5 Lapping and polishing

Lapping and polishing are fine grinding processes used to produce exact dimensions and smooth finishes on components already shaped to size by grinding or other machining. Lapping, which precedes polishing, uses a coarser abrasive and is concerned primarily with achieving the correct dimensions of the component. Polishing uses finer abrasives and produces the final surface finish (Note, however, that in the particular case of the polishing of diamond the term polishing generally refers to both the initial shaping process and the final finishing.) General accounts of polishing and lapping, including the use of diamond abrasives, are given by Samuels (1971) and Fynn and Powell (1979). These authors also discuss how these processes may affect the structure of the subsurface layer of the workpiece.

Diamond powder is particularly suitable for lapping and polishing hard materials because its greater hardness and resistance to abrasion ensures that the cutting points remain sharp for much longer than those in other powders. It would be convenient to polish with bonded wheels in which the diamond powder was held in a matrix as in grinding wheels but fine powders cannot be bonded so securely to the matrix because of their very small surface area. Therefore standard practice is to polish with loose powder spread over the surface of a suitable wheel or lap (although the continuing development of metal bonded wheels now presents the possibility of using micron size powder for suitable applications).

The laps for polishing and lapping are made from a variety of materials ranging from steel, cast-iron, brass and copper to wood, fibres and felt. Using the harder materials the diamond powder is pressed more firmly on the workpiece giving a greater removal rate but a less fine finish. Conversely the powder sinks into the softer laps to give a better finish but with a low removal rate. Therefore it is common practice to polish with a range of progressively smaller powders in order to reach the final finish as quickly as possible. The powder is usually introduced as a suspension in a light oil such as olive oil which is smeared evenly over the lap. The use of a suspension helps to prevent the particles agglomerating together, while its viscosity helps to retain the powder on the lap. General accounts of practical details of the lapping and polishing processes are given by Swan (1964) and Davis (1974). Various attempts have also been made to lap and polish with wheels in which the diamond powder is held by electroplating, and with diamonds electroplated to flexible nets which can follow the contours to be polished, see Herbert (1978, 1980).

The quality of the finish achieved by lapping and polishing depends greatly on the careful grading of the diamond powder both for shape and size. Just a few oversize particles may cause much damage and scratching on an otherwise satisfactory surface. Extensive studies have been made on how best to specify the shape and size of particles in a particular sample, see for example Freeman (1963), Dyer and Wedepohl (1963) and Spooner (1976). Experience shows that the diamond particles should be blocky in shape (Figure 15.39) so the particles make about the same depth of cut in all orientations. Powder is produced mainly by synthesis under conditions which give blocky type crystals but some is still obtained from natural diamonds. These are first crushed and then treated by some form of processing similar to ball milling to reduce the variety of shapes prior to sorting for size by sieving, for details see Ranier (1969). The fine powders produced by shock synthesis (Section 1.2.d) which have a form of polycrystalline structure give a somewhat different performance and on certain materials may give either a better finish or a higher removal rate (Bergman, Bailey and Coverly, 1982).

A survey of the parameters influencing the polishing and lapping of tungsten carbide and other materials is given by Davis (1974). Experiments on carbide show that as might be expected the removal rate increases with increasing pressure on

Figure 15.39 Samples of blocky synthetic diamond powder (De Beers Industrial Diamond Division)

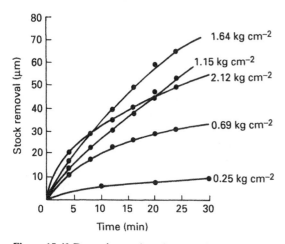

Figure 15.40 Dependence of stock removal on pressure when lapping tungsten carbide with 8–15 μm powder (Davis, 1974)

the lap, but eventually begins to fall with increasing pressure presumably because the grit is breaking down into smaller pieces or is being driven deeper into the lap (Figure 15.40). Other results given by Davis show that with the same mass concentration of diamond in the lapping slurry the removal rate increases roughly in proportion to the grit size, with the surface roughness also increasing. Figure 15.41 shows the finish obtained when an initially rough carbide surface was lapped

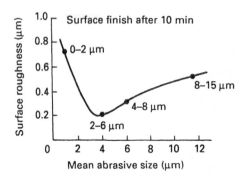

Figure 15.41 Finish produced after lapping carbide for 10 min with grits of different size (Davis, 1974)

for 10 min with different sized powders. Note that the the best finish was not given by the smallest powder because that made too small an effect in the time available.

Finally, diamond lapping on a different scale is used in the stone industry particularly after sawing medium hard stones such as limestones and marbles. The final polish on these stones is generally produced using conventional carbide abrasives but diamond grit is now being used to speed the preliminary lapping process prior to finer polishing (Jennings, 1987). Examples of the use of diamond wheels for this type of work are given by Decroly (1980) and Büttner (1984), and of the use of diamond rollers by Asperti and Ledru (1987).

15.6 Electrolytic grinding

This technique has been described by several authors, Grodzinski (1953), Hughes and Notter (1965), Pahlitzsch (1967), Grodzinskii and Zubatova (1977), Hughes (1978) and Malkin and Levinger (1979). The essence of the technique is that an electrolyte is introduced into the gap between a metal bonded wheel and the workpiece which arises because the grits in the wheel stand proud of the bond. The workpiece is electrically insulated and a low voltage is applied between wheel and workpiece inducing a current of perhaps $0.2\,\mathrm{A\,mm^{-2}}$ to $1.0\,\mathrm{A\,mm^{-2}}$. In this way the rate of removal of material may be increased sometimes by a factor of 2 or 3, but the method has never been widely used because of several disadvantages. Arrangements must be made to maintain a supply of clean electrolyte at the interface, to avoid sparking, and to ensure adequate safety precautions. The finish obtained is often inferior to that given by normal grinding, partly because the current may act preferentially on different constituents in a workpiece.

References

Al'tshuller, V. M. , Ashkerov, Yu. V. and Korovkin, V. P. (1983) *Soviet Journal of Optical Technology,* **50**, 514–516

Asperti, G. and Ledru, P. (1987) *Industrial Diamond Review,* **47**, 210–211

Atkins, A. G. and Mai, Y. W. (1985) *Elastic and Plastic Fracture*, Ellis Horwood, Chichester, p. 505

Bailey, M. W. and Sexton, J. S. (1981) *Industrial Diamond Review,* **41**, 190–191

Barnard, J. M. (1989) *Industrial Diamond Review*, **49**, 31–34

Bergman, O. R. , Bailey, N. F. and Coverly, H. B. (1982) *Metallography*, **15**, 121–139

Borse, D. , Ernst Winter und Sohn (1987) *European Patent Notification* 0 254 940

Bowden, F. P. and Tabor, D. (1950) *The Friction and Lubrication of Solids*, Part I, Clarendon Press, Oxford, p. 9

Broese van Groenou, A. (1979) In *The Science of Ceramic Machining and Surface Finishing II*, (eds B. J. Hockey and R. W. Rice), National Bureau of Standards Special Publication 562, US Government Printing Office, Washington DC, pp. 147–156

Broese van Groenou, A. and Brehm, R. (1979) In *The Science of Ceramic Machining and Surface Finishing II*, (eds B. J. Hockey and R. W. Rice), National Bureau of Standards Special Publication 562, US Government Printing Office, Washington DC, pp. 61–74

Broese van Groenou, A. and Veldkamp, J. D. B. (1978/79) *Philips Technical Review*, **38**, 105–118

Broese van Groenou, A. , Maan, N. and Veldkamp, J. D. B. (1975) *Philips Research Reports*, **30**, 320–359

Broese van Groenou, A. , Maan, N. and Veldkamp, J. D. B. (1979) In *The Science of Ceramic Machining and Surface Finishing II*, (eds B. J. Hockey and R. W. Rice), National Bureau of Standards Special Publication 562. US Government Printing Office, Washington DC, pp. 43–59

Bruton, E. (1981) *Diamonds*, (2nd edn) NAG Press, London

Bullen, G. J. (1975) *Industrial Diamond Review*, **35**, 363–365

Busch, D. M. and Thiel, N. W. (1971) *Industrial Diamond Review*, **31**, 412–423

Büttner, H. (1984) *Industrial Diamond Review*, **44**, 191–194

Caveney, R. J. and Thiel, N. W. (1972) In *The Science of Ceramic Machining and Surface Finishing* (eds S. J. Schneider and R. W. Rice), National Bureau of Standards Special Publication 348, US Government Printing Office, Washington DC, pp. 99–112

Chalkley, J. R. (1968) *Tribology*, **1**, 204–208

Chandrasekar, S., Shaw, M. C. and Bhushan, B. (1987) *Transactions of the ASME, Journal of Engineering for Industry*, **109**, 76–82; 83–86

Chikaoka, Y. and Watanabe, K. (1988) *US Patent No. 4,750,914*, June 14, 1988

Coes, L. (1971) *Abrasives*. Springer-Verlag, New York

Cullity,B. D. (1978) *Elements of X-Ray Diffraction* (2nd edn) Addison Wesley, Reading, Mass

Davis, C. E. (1974) *Industrial Diamond Review*, **34**, 54–61, 94–99

Decroly, J. C. (1980) *Industrial Diamond Review*, **40**, 336–339

Dyer, H. B. (1963) In *Proceedings of the First International Congress on Diamonds in Industry* (Paris, 1962), (ed. P. Greene), De Beers Industrial Diamond Division, London, pp. 147–158

Dyer, H. B. (1968) *Industrial Diamond Review*, **28**, 6–13

Dyer, H. B. and Wedepohl, P. T. (1963) In *Proceedings of the First International Congress on Diamonds in Industry* (Paris, 1962), (ed. P. Greene), De Beers Industrial Diamond Division, London, pp. 65–71

Edwards, D. F. and Hed, P. P. (1987) *Applied Optics*, **26**, 4670–4676

Freeman, G. P. (1963) In *Proceedings of the First International Congress on Diamonds in Industry* (Paris, 1962), (ed. P. Greene), De Beers Industrial Diamond Division, London, pp. 27–39

Fynn, G. W. and Powell, W. J. A. (1979) *The Cutting and Polishing of Electro-Optic Materials*. Adam Hilger, Bristol

Geisweid, G. and Gärtner, W. (1978) *Industrial Diamond Review*, **38**, 285–288

Gielisse, P. J. and Stanislao, J. (1972) In *The Science of Ceramic Machining and Surface Finishing*, (eds S. J. Schneider and R. W. Rice), National Bureau of Standards Special Publication 348, US Government Printing Office, Washington DC, pp. 5–35

Graham, W. and Nee, A. Y. C. (1974) *The Production Engineer*, June, 186–191

Grodzinski, P. (1953) *Diamond Technology*, (2nd edn) NAG Press, London

Grodzinskii, E. Ya. and Zubatova, L. S. (1977) *Machines and Tooling*, **48**, 52–54

Hed, P. P. and Edwards, D. F. (1987) *Applied Optics*, **26**, 2491

Herbert, S. (1978) *Industrial Diamond Review*, **38**, 166–168

Herbert, S. (1980) *Industrial Diamond Review*, **40**, 458–461

Hitchiner, M. P. and Wilks, J. (1983) In *Advances in Ultrahard Materials Application Technology*, Vol. 2. (ed. P. Daniel), De Beers Industrial Diamond Division, London, pp. 100–111

Hitchiner, M. P. and Wilks, J. (1987) *Wear*, **114**, 327–338

Holz, R. and Sauren, J. (1988) *Grinding with diamond and CBN*. Ernst Winter & Sohn, Hamburg

Huerta, M. and Malkin, S. (1976) *Transactions of the ASME, Journal of Engineering for Industry*, **98**, 459–467; 468–473

Hughes, F. (1968) *The Engineer*, **226**, Nov 22nd, 786–787

Hughes, F. (1978) *Diamond Grinding of Metals*. De Beers Industrial Diamond Division, London

Hughes, F. and Notter, A. (1965) *Industrial Diamond Review*, **25**, 476–480, 584–587; (1966) *Industrial Diamond Review*, **26**, 8–13, 231–233, 290–292

Ikawa, N. and Tanaka, T. (1971) *Annals of the CIRP*, **19**, 153–157

Inasaki, I. (1987) *Annals of the CIRP*, **36**, 463–471

James, M. R. and Cohen, J. B. (1980) In *Treatise on Materials Science and Technology*, Vol. 19, (ed. H. Herman), Academic Press, New York, pp. 2–5

Jennings, M. (1987) *Industrial Diamond Review*, **47**, 216–218

Juchem, H. O. (1988) *Industrial Diamond Review*, **48**, 158–161

Juchem, H. O. and Cooley, B. A. (1984) *Industrial Diamond Review*, **44**, 313–319

Juchem, H. O. and Wapler, H. (1979) *Industrial Diamond Review*, **39**, 43–50

Kayaba, N. and Fujisawa, M. (1986) *Proceedings Japan Society of Precision Engineering*, 355–356. In Japanese. Reported by Inasaki, I. (1987)

Kerkhof, F. and Schinker, M. (1972) *Glastechnische Berichte*, June, 228–233

Khodakov, G. S., Korovkin, V. P. and Al'tshuler, V. M. (1980) *Soviet Journal of Optical Technology*, **47**, 552–560

Kirchner, H. P. (1984) *Journal of the American Ceramic Society*, **67**, 347–353

Komanduri, R. and Shaw, M. C. (1975) *Nature*, **255**, 211–213

Komanduri, R. and Shaw, M. C. (1976) *Philosophical Magazine*, **34**, 195–204

Korovkin, V. P., Al'tshuller, V. M. and Ashkerov, Yu. V. (1985) *Soviet Journal of Optical Technology*, **52**, 715–716

Korovkin, V. P. and Al'tshuller, V. M. (1987) *Soviet Journal of Optical Technology*, **54**, 380–386

Krar, S. F. and Ratterman, E. (1990) *Superabrasives Grinding and Machining*. McGraw-Hill, New York

Lange, F. F., James, M. R. and Green, D. J. (1983) *Communications of the American Ceramic Society*, February, C16–C17

Lawn, B. R. and Swain, M. V. (1975) *Journal of Materials Science*, **10**, 113–122

Lindenbeck, D. A. and McAlonan, C. G. (1974) *Industrial Diamond Review*, **34**, 84–88

Low, S. K., Nakano, Y. and Kato, H. (1988) *Journal of Mechanical Working Technology*, **17**, 367–376

Luyckx, S. B. (1968) *Acta Metallurgica*, **16**, 535–544

Maan, N. and Broese van Groenou, A. (1977) *Wear*, **42**, 365–390

Malkin, S. and Levinger, R. (1979) In *The Science of Ceramic Machining and Surface Finishing II*, (eds B. J. Hockey and R. W. Rice), National Bureau of Standards Special Publication 562. US Government Printing Office, Washington DC, pp. 305–315

Mastyugin, L. I. (1987) *Soviet Journal of Optical Technology*, **54**, 116–118

Metzger, J. L. (1981a) *Industrial Diamond Review*, **41**, 304–310

Metzger, J. L. (1981b) *Industrial Diamond Review*, **41**, 192–195

Metzger, J. L. (1982) In *Ultra Hard Materials Application Technology*, Vol. One, (ed. P. Daniel), De Beers Industrial Diamond Division, London, pp. 16–23

Metzger, J. L. (1986) *Superabrasive Grinding*. Butterworth, London

Metzger, J. L. (1988) *Industrial Diamond Review*, **48**, 270–277

Metzger, J. L. (1989) *Industrial Diamond Review*, **49**, 116–117

Molloy, P., Schinker, M. G. and Döll, W. (1987) *Proceedings of the SPIE, In-Process Optical Metrology for Precision Machining*, **802**, 81–88

Nakagawa, T., Suzuki, K. and Uematsu, T. (1986) *Annals of the CIRP*, **35**, 205–210

Nakajima, T., Uno, Y. and Fujiwara, T. (1989) *Precision Engineering*, **11**, 19–25

Notter, A. T. and Shafto, G. R. (1979) *Industrial Diamond Review*, **39**, 203–210

Oates, P. D. and Willmington, H. M. (1975) *Industrial Diamond Review*, **35**, 322–327

Pahlitzsch, G. (1967) *Industrial Diamond Review*, **27**, 340–345, 390–396

Peklenik, J. (1958) *Industrie-Anzeiger*, **80**, 10–17

Primak, L. P. and Skripko, G. F. (1970) *Industrial Diamond Review*, **30**, 318–320

Prins, J. F. (1971a) *Industrial Diamond Review*, **31**, 497–503

Prins, J. F. (1971b) *Industrial Diamond Review*, **31**, 364–370

Prudnikov, E. L. *et al.* (1986) *Soviet Engineering Research*, **6**, 67–68

Puttick, K. E. and Shahid, M. A. (1977) *Industrial Diamond Review*, **37**, 228–233

Rainier, D. M. (1969) *Industrial Diamond Review*, **29**, 1–6

Reinhart, H. (1967) In *Proceedings of the International Industrial Diamond Conference*, (Oxford, 1966), Vol. 2: Technology, (ed. J. Burls), De Beers Industrial Diamond Division, London, pp. 19–35

Rupp, W. J. (1974) *Applied Optics*, **13**, 1264–1269

Sagarda, A. A. and Khimach, O. V. (1973) *Russian Engineering Journal*, **LIII**, (6), 73–76

Samuels, L. E. (1971) *Metalographic Polishing by Mechanical Methods*, (2nd edn) Pitman, London

Schmitt, P. (1980) *Industrial Diamond Review*, **40**, 294–297

Schwämmle, J. and Lowin, R. (1983) *Industrial Diamond Review*, **43**, 20–24

Sexton, J. S. (1982) *Industrial Diamond Review*, **42**, 161–170

Spooner, T. A. (1976) *Industrial Diamond Review*, **36**, 284–294

Springborn, R. K. (1967) ed. *Cutting and Grinding Fluids: Selection and Application*. American Society of Tool and Manufacturing Engineers, Dearborn, Michigan

Stokes, R. J. , Cooley, B. A. and Juchem, H. O. (1983) In *Ultrahard Materials Application Technology*, Vol. 2, (ed. P. Daniel), De Beers Industrial Diamond Division, London, pp. 92–99

Swain, M. V. (1979) *Proceedings of the Royal Society*, **A366**, 575–597

Swan, R. J. (1964) *Industrial Diamond Review*, **24**, 11–17

Tanaka, T. and Ikawa, N. (1973) *Bulletin Japan Society of Precision Engineering*, **7**, 97–101

Tavernier, R. (1966) In *Progress in Industrial Diamond Technology*, (ed. J. Burls), Academic Press, London, pp. 107–122

Tomlinson, P. N. (1976) *Industrial Diamond Review*, **36**, 426–432

Tomlinson, P. N. (1978) *Industrial Diamond Review*, **38**, 123–129

Tomlinson, P. N. , Notter, A. T. and Penny, A. L. (1978) *Industrial Diamond Review*, **38**, 204–211

Ueda, T. , Hosokawa, A. and Yamamoto, A. (1985) *Transactions of the ASME, Journal of Engineering for Industry*, **107**, 127–133

Ueda, T. , Hosokawa, A. and Yamamoto, A. (1986) *Transactions of the ASME, Journal of Engineering for Industry*, **108**, 247–251

Veldkamp, J. D. B. , Hattu, N. and Snijders, V. A. C. (1978) In *Fracture Mechanics of Ceramics*, Vol. 3, (eds R. C. Bradt, D. P. H. Hasselman and F. F. Lange), Plenum, New York, pp. 272–301

Veldkamp, J. D. B. and Klein Wassink, R. J. (1976) *Philips Research Reports*, **31**, 153–189

Wapler, H. and Juchem, H. O. (1987) *Industrial Diamond Review*, **47**, 159–162

Weavind, R. G. (1959) *Industrial Diamond Review*, **19**, 126–127; 146

Wedlake, R. J. (1979) In *The Properties of Diamond*, (ed. J. E. Field), Academic Press, London, pp. 501–535

Zelwer, O. and Malkin, S. (1980) *Industrial Diamond Review*, **40**, 133–139, 173–176

Zhang, B. , Tokura, H. and Yoshikawa, M. (1988) *Journal of Materials Science*, **23**, 3214–3224

Chapter 16

Sawing and drilling

Diamond tools are widely used in the stone industry, in the prospecting and exploiting of oil and mineral resources, in the construction industry, and in a variety of applications on a smaller scale such as the sawing of ceramics for the electronics industry. This chapter considers the two main divisions of these applications, sawing and drilling.

16.1 Sawing

16.1.a Methods of sawing

Stone and rock is sawn with diamond impregnated segments mounted on steel blades, wheels or wires. The diamond grit is embedded in a metal such as cobalt or bronze which acts as a bonding agent holding a dispersed distribution of grit. Thus the surface of a segment has a structure somewhat similar to that of a grinding wheel except that the grits are appreciably larger, ranging from about 150 μm to 1000 μm in size.

The wheels for circular saws are cut from steel sheet to shapes such as those in Figure 16.1, the slots and holes round the perimeter being designed to ensure

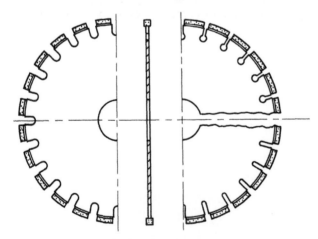

Figure 16.1 Plan and section views of three diamond wheel saws (E. Winter und Sohn)

Figure 16.2 A 2.5 m diameter diamond saw cutting slate (Jennings, 1987)

adequate access of coolant and removal of debris. The wider spacings on one of the saws makes for better access and removal, while a closer spacing tends to cut more precisely and cleanly. Laser technology is now widely used to cut out the wheel and to attach the segments by brazing. This latter technique, described for example by Clauser and Valle (1987), gives very controlled and localized heating,which makes it possible to obtain stronger and more reliable bonds while avoiding damage to the diamond segments.

Wheel saws range in size from diameters of about 200 mm to perhaps 3000 mm. Figure 16.2 shows a 2500 mm wheel cutting a block of stone mounted on a table which is traversed under the saw. To obtain good results one must ensure the accurate alignment and smooth functioning of the sawing machine, and in particular the true running of the saw. With large wheels revolving at high speed large stresses due to centrifugal force are built up in the steel and may cause the wheel to distort. Therefore the steel is usually work-hardened by pre-stressing to increase its resistance to deflection, for details see Büttner and Mummenhoff (1973). The mounting and clamping of the wheel to its shaft so that it runs true and avoids resonances is particularly important.

Smaller mobile saws are used in the construction industry to cut through stone and concrete particularly when making modifications to existing buildings. A diamond saw cuts clean and with a minimum of noise and vibration, an important point if the building is still in use and occupied. Mobile saws are also employed on roads and airport runways both for maintenance work, such as cutting sections for pipes and cables, and for cutting arrays of grooves in the concrete to increase friction and prevent aquaplaning in wet weather, see for example Sulten (1989). Figure 16.3 shows a typical set of wheel saws ganged together for grooving airport runways. The design of these mobile saws including the non-trivial problem of guiding them precisely along the desired track is discussed by Tönshoff *et al.* (1984).

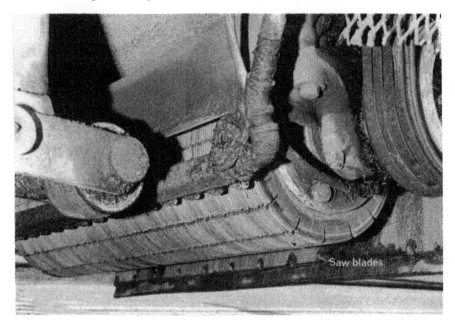

Figure 16.3 A gang of diamond saws for road grooving (Busch, 1979)

Figure 16.4 A set of gang saws sawing marble (Hayes, 1989)

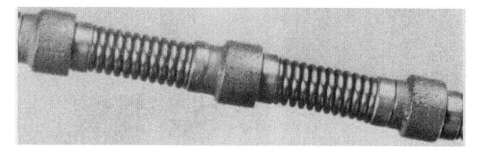

Figure 16.5 Detail of an impregnated diamond wire saw; see text (Diamond Boart. S.A.)

The maximum depth of cut possible with even the largest wheel saws is of the order of 1 m so larger blocks of stones from the quarry are usually first sawn by straight steel blades carrying diamond segments mounted along their length. The blades are either mounted singly or in gangs (Figure 16.4). A general description of both blade and wheel sawing machines is given by Smith (1974).

Blades may have a working length of up to 4 m and an overall length up to 5 m and will therefore tend to curve under load. As any such curvature has highly undesirable effects on the sawing process each blade is stressed in the opposite direction by applying a tension to the ends by clamps mounted so that the tension deflects the blade downwards. Details of arrangements for adjusting the tensions to the appropriate values are given by Miller (1967). A convenient method of monitoring the tension in a blade is given by Wiemann *et al.* (1983) and a brief but useful note on the practical details of sawing with blades by Veglio (1987).

The latest method of sawing, only developed on a large scale since about 1980, makes use of diamond segments or beads mounted on a loop of flexible multistranded steel wire. Figure 16.5 shows an example of an impregnated bead mounted on a wire which is hardly visible because of the springs which in this particular example are used as spacers to hold the beads in place. Figure 16.6 is a sketch showing a typical mode of use in a quarry. The wire has been threaded through holes drilled vertically and horizontally and is now being dragged round by the pulley and is cutting through the block. Accounts of wire sawing in quarries are given by Pinzari (1983), Thoreau (1984) and Daniel (1986a); a view of a typical system is shown in Figure 16.7. Applications to the cutting of blocks of quarried stone are described by Cai and Mancini (1988), and to the sawing of reinforced concrete in buildings by Schaffner and Blaser (1987).

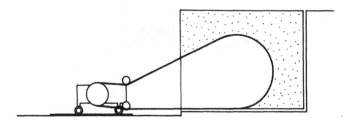

Figure 16.6 Sketch of an arrangement for cutting into a quarry face with diamond wire (Pinzari, 1983)

Figure 16.7 Diamond wire sawing marble from the quarry floor (Trancu, 1980)

We also mention the chain saw and the bump cutter. Both in quarrying and construction work it is often necessary to cut deeper into a face of rock or concrete than a wheel can penetrate. In this case band or chain saws with diamond segments are used to cut into and then along a block. Examples of a 10 m deep cut through a line of terraced houses are given by Weber and Zilm (1987) and of cuts 2 m deep in a quarry face by Daniel (1986b).

The bump cutter is better described as a concrete planing or milling machine and is used to smooth concrete roads and runways by machining off a layer of the order of 5 mm thick. Examples of this work are described by Herbert (1971) and Anon (1972). At first sight the cutter blades appear similar to those on a grooving machine (Figure 16.3) except that the circular blades are stacked closer together. In fact the diamond segments are arranged to lie on a single helix to ensure that their cutting action is spread uniformly over the whole width of the surface. For further details of the cutter and its operation see Perrett (1969).

All the above sawing processes are carried out with a copious supply of coolant, essentially water, to flush away the debris and avoid undue rises of temperature. The filtering and recycling of large volumes of coolant, and the ultimate disposal of slurries and additives, are generally an important part of the cost of any sawing operation. The use of mineral additives in the water, at concentrations between about 1% and 3%, can reduce both the wear of the blade and the power required, see Büttner (1982) and Tönshoff and Schulze (1982) who also give references to earlier observations. The action of additives is often very specific to the stone being sawn and is generally chosen on the basis of experience; some practical recommendations are given by Büttner (1982). We discuss the action of different coolants in Section 16.3 together with comparable effects observed in drilling experiments.

16.1.b Diamond segments

The action of the diamond segments in a sawing process is rather similar to that of a grinding wheel. As sawing proceeds the segments wear down and new diamonds emerge from the matrix and abrade the workpiece. As in grinding, the diamonds suffer wear both by polishing actions which blunt their edges and by fracturing. Then as the bond is worn away by the debris of the sawn stone the grits are torn from the matrix and lost. Examples of these various stages of wear are shown in the SEM micrographs in Figure 16.8.

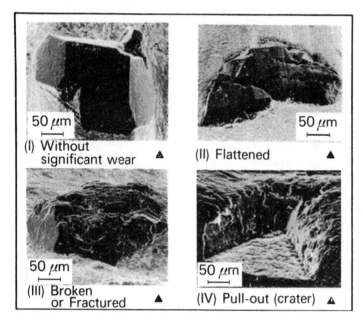

Figure 16.8 Micrographs showing types of grit wear in an impregnated diamond segment (Mamalis, Schulze and Tönshoff, 1979)

If a saw is to perform satisfactorily the diamond segments must maintain the same cutting action as they wear down. Therefore, as in grinding, it is essential to avoid conditions where the diamonds become progressively more polished and less efficient at cutting. Hence the operator must ensure that the sawing conditions are heavy enough to produce enough fracture of the grids to maintain their efficiency. Also, as in grinding, the toolmaker must take into account the mutual interactions between diamonds, bond and workpiece. The workpiece determines the resistance to sawing, and the bond must be chosen to be strong enough to resist premature pull-out, but not so strong and resistant that it prevents the replacement of worn out grits.

A particularly important factor in sawing is the large volume of debris which will not only wear down the bond but will also affect the cutting action of the grits.

For example Garner (1967) made grooves in rock with single grits and observed that up to 50% of the depth of the groove was filled with fine debris. The action of debris is also discussed by Bridwell and Appl (1974) who stress the importance of the grits standing proud enough of the bond to give sufficient clearance for the coolant to flush out the debris. A theoretical discussion of the cutting action of a segment taking account of the presence of debris is given by Aleksandrov, Lévin and Mechnik (1986).

16.1.c Wear measurements

The operating conditions for diamond saws must be chosen to obtain a balance between a range of different criteria. For example, it may be advantageous to accept greater rates of wheel wear if the sawing can be completed more quickly at less cost. The best overall conditions will depend both on the type of sawing equipment available and the labour and overhead costs in the factory, so the costing of each operation must be considered on an individual basis. In this section we consider the more general problem of how the wear of the diamond segments depend on the principal parameters that can be varied in the sawing shed. Most of the available information has been obtained in experiments with wheel saws but similar considerations apply to blade and other types of saws. (For a convenient method of measuring the wear of a wheel *in situ*, see Herbert (1983).)

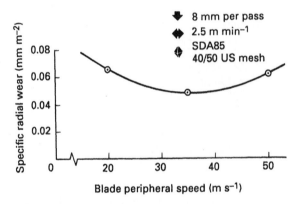

Figure 16.9 Specific radial wear of a wheel saw as a function of peripheral blade speed (Bailey and Bullen, 1979)

Figure 16.9 shows how the rate of radial wear of a particular wheel saw cutting at a constant rate ($m^2 s^{-1}$) depends on the peripheral speed of the wheel (Bailey and Bullen, 1979). We see that in this example the wear has a minimum value at a speed of about 35 m s^{-1}. The authors conclude that the wear arises from two components as shown schematically in Figure 16.10. One component described as mechanical damage arises from the forces responsible for separating chips of material from the workpiece. This term decreases with increasing blade speed because, as in grinding (Section 15.1.c), higher wheel speeds result in finer chips. The other contribution, termed impact damage, arises directly from the collisions of the grits with the workpiece. This term becomes increasingly important at high velocities as appears to be the case in the single point grit experiments of Gielisse

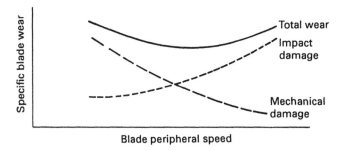

Figure 16.10 Sketch to indicate the contributions of mechanical and impact damage to the wear in Figure 16.9 (Bailey and Bullen, 1979)

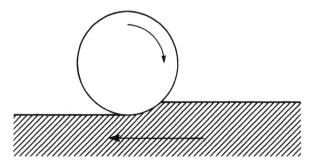

Figure 16.11 Sketch to illustrate sawing in the down position, see text

and Stanislao (1972) mentioned in Section 15.4.a. As shown in Figure 16.10 a combination of these two components accounts for the observed minimum. Other examples of this type of minimum are given by Finnigan (1968).

When discussing the speed of a wheel there are two directional effects. Thus, in Figure 16.11 the diamond grits move downwards as they cut through the workpiece, whereas on reversing the wheel they move upwards. As in grinding (Section 15.3.d) the cutting conditions at the interface are significantly different for up and down cuts if the cuts are relatively deep. In down cuts the grits enter the workpiece cutting at maximum depth, and in up cuts they enter at zero depth. An analysis by Bridwell and Appl (1974) derives a quite different distribution of debris under the segments for up and down cutting. Bailey and Bullen (1979) point out that besides affecting the geometry of the process a change in the direction of wheel or workpiece may sometimes induce vibrations in the sawing machine.

16.1.d Mechanism of sawing

Various experiments have been made to study the detailed mechanisms of the sawing process. Figure 16.12 gives curves showing the power required when sawing red granite as a function of the peripheral speed of the wheel (Finnigan, 1968). The rate of sawing ($m^2 \ s^{-1}$) is the same for curves 16.12(a) and 16.12(b) but the traverse speed for curve 16.12(a) is ×4 greater than for curve 16.12(b) and the depth of cut ×4 less. Even though the sawing rates are the same there are

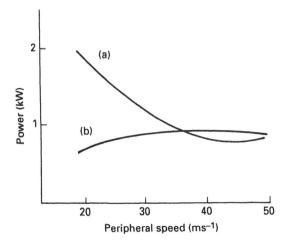

Figure 16.12 Power required when sawing a red granite as a function of peripheral blade speed. The sawing rates were the same for curves (a) and (b), but the table speeds were approximately 120 and 30 mm s⁻¹ respectively, and the depths of cut 2mm and 8 mm respectively (Finnigan, 1968)

Figure 16.13 Micrograph of a diamond impregnated saw segment (Büttner, 1974)

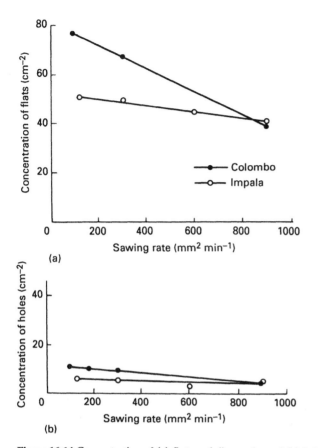

Figure 16.14 Concentration of (a) flattened diamonds, and (b) holes left by pulled out diamonds, as a function of sawing rate when sawing Colombo and Impala granites (Mamalis, Schulze and Tönshoff, 1979)

considerable differences both in the magnitude of the power required and in the form of its dependence on the speed of the wheel. Equally striking differences were observed in the wear of the wheel. Finnigan explained these differences by the presence of vibrations but the measured amplitudes of these vibrations were not greatly different under the two conditions. The main result of the experiment is to underline the complexity of the sawing process.

The grits in a saw are more easily observed than those in a grinding wheel. They are generally larger and used at a lower concentration, and can readily be inspected with an optical microscope (Figure 16.13). Hence it is quite possible to relate the performance of a saw with the behaviour of the grits. In a set of laboratory studies Mamalis, Schulze and Tönshoff (1979) ran a saw through blocks of granite at different sawing rates and then assessed the state of the grits after each cut by classifying them into four groups on the lines set out by the examples in Figure 16.8: no significant wear, flattened, fractured, and pulled out. Figure 16.14 gives the number concentrations of (a) flattened grits and (b) holes left by pulled out grits observed after sawing at different cutting rates. The concentrations of both

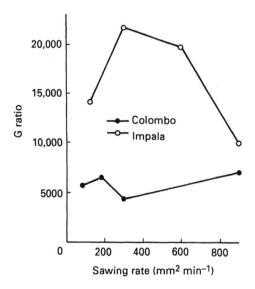

Figure 16.15 G ratios when sawing Colombo and Impala granites as function of sawing rate (Mamalis, Schulze and Tönshoff, 1979)

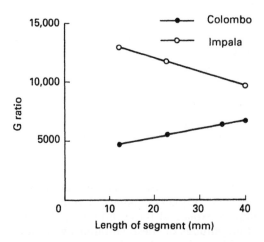

Figure 16.16 G ratios when sawing Colombo and Impala granites as a function of the length of the diamond-impregnated segment (Mamalis, Schulze and Tönshoff, 1979)

holes and flats varied considerably more with cutting speed for the Colombo granite than for the Impala granite.

Mamalis, Schulze and Tönshoff also observed how the G ratios for a saw cutting the above two types of granite depended on the rate of sawing (Figure 16.15) and on the length of the impregnated diamond segments (Figure 16.16). These diagrams show that the dependence on cutting rate and length is quite different for the two types of rock. Hence correlations obtained in one experiment with a particular rock and set of machining conditions may not be of general application.

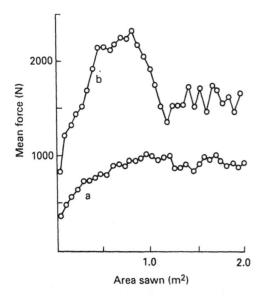

Figure 16.17 Mean cutting force on a saw sawing Colombo granite, depth of cut (a) 10 mm, (b) 30 mm, but rate of sawing same for each curve. (after Ertingshausen, 1985)

A further complication concerning the measurement and discussion of sawing rates arises because even if the sawing conditions remain unchanged the rate of cutting may change with time as the diamond segments come into equilibrium with a new workpiece. For example, Ertingshausen (1985) measured the forces on a workpiece by mounting a dynomometer on the traverse table. In one run with a 10 mm depth of cut the cutting force (in the plane of the table) increased steadily to an equilibrium value. (Figure 16.17). However, in another run with a second saw cutting at the same rate ($m^2 s^{-1}$) but with a 30 mm depth of cut the cutting force initially passed through a maximum before coming to equilibrium. The author suggests that these differences arise because the deeper cut is accompanied by a lower table speed. This lower speed results in thicker chips which produce a higher equilibrium force and also account (by a somewhat involved argument) for the observed maximum. Ertingshausen also observed differences in the wear patterns around the periphery of the saw when cutting in up and down directions.

In other studies Tönshoff and Warnecke (1982) measured the radial force on a saw as a function of infeed and grit size both at a constant volume concentration of grit and for a constant number concentration, and found the form of the results difficult to understand. Effects of grit size are also briefly discussed by Büttner (1974). Tönshoff and Warnecke comment that the 'thoroughly contradictory results' obtained in different investigations probably arose because the sawing conditions were not fully specified and varied from experiment to experiment. Certainly the experiments described above show that the performance of a saw may depend very critically on just one parameter such as the type of rock, rate of infeed, or time of sawing.

It is clear that more information is required for a full understanding of the sawing process. More observations are needed on the wear of the grits perhaps by using detachable segments, on the amount of debris left in the grooving, and on the height of the grits above the matrix which affects the flow of coolant and the removal of debris. One must also remember that rock is a natural product which may vary in texture and homogeneity, even within a single block of stone. It is therefore necessary in any experiment to check that the rock specimens are behaving consistently. Yet accounts of sawing experiments often give no evidence of whether measurements could be repeated to give consistent results.

16.1.e Selection of diamond

We now consider the choice of diamond for use in saws. It is well established by experience that the grits should be strong blocky stones such as the regular cubo-octahedral single crystals which are readily produced by synthesis. The grits in saws are generally larger than those used for grinding lying in a range from about 150 μm to 1000 μm. This size ensures that the grits can be firmly held by the bond and still project sufficiently high above it to allow an adequate flow of fluid to remove the large volume of debris. On the other hand, large grits will tend to take deeper cuts and thus be subject to large and perhaps excessive forces. There is also the consideration that larger grits tend to be less strong (Section 7.3.b), and are also more costly.

Before the synthesis process was capable of producing grits of the required size much effort was made to select and process natural diamonds not suitable for use as gemstones. The best stones for sawing and drilling are of good blocky morphology and as free as possible from inclusions and other defects. The best of them may well be of gem quality apart from a pronounced colour. After selection, diamonds for sawing are generally processed, probably by some form of rotary milling where they are tumbled against each other. This milling has two effects. Any weak stones are fractured and the remnants removed, and unduly sharp edges are smoothed away to produce a blocky shape. Examples of substantial improvements in grit performance after processing are given by Hill (1975). Finally, the grits may be treated to promote better adhesion to the bond either by roughening the surface by etch or other means, or by applying a coating of a metal such as titanium as described by Bailey and Collin (1978).

The production of synthetic diamond offers the possibility of much greater control over the morphology and strength of the grits than is possible with natural diamonds coming from a wide variety of sources. Manufacturers now produce grits with a choice of strengths, the higher strengths being used for more severe conditions and for dealing with harder materials. Bullen and Bailey (1979) give examples of the behaviour of different grits when drilling rock, a process sufficiently similar to sawing for the results to be relevant. Holes were drilled in grey granite using drill bits with two types of good quality crystals SDA100 and SDA100S, the latter being described as the stronger. In each run the bits were loaded to achieve penetration rates of $150\,\text{mm\,min}^{-1}$, $250\,\text{mm\,min}^{-1}$ or $300\,\text{mm\,min}^{-1}$. The two grits gave a comparable performance at $300\,\text{mm\,min}^{-1}$ but at the lower penetration rates the stronger SDA100S performed less well. The authors concluded that this inferior behaviour arose because the loading on the bit at the lower penetration rates was insufficient to produce enough fracture to maintain the cutting efficiency of the stronger grit.

The various results outlined above show that the saw manufacturer has to select a type of grit appropriate to the workpiece, to decide on the size and concentration of the grit, and to match the metal matrix both to the grit and to the material of the workpiece. As mentioned above the grit must be large enough to be firmly retained and give adequate clearance to the coolant, while on the other hand the cost rises rapidly with size. The choice of grit concentration is equally important but there is not much detailed information on the effects of changing this concentration. Most experiments to study changes in grit size have been made keeping the volume concentration of the grit constant. In this case a change of grit size is accompanied by a much larger change in the number of grits, and the results are difficult to analyse. Unfortunately in one of the few experiments to observe the effect of change of grit size while keeping the number of grits constant (Tönshoff and Warnecke, 1982) the results are quite complex (Section 16.1.d). Hence the toolmaker in designing a saw must optimize the parameters involved in the light of experience rather than of theoretical principles. (The possibility of assessing the resistance of a rock to sawing by hardness or abrasion tests is discussed by Wright and Wapler (1986). However, according to Wright (1988) the best test appears to be some simple form of sawing operation.)

16.1.f Sawing ceramics

Besides sawing stone and rock diamonds play an important role in cutting many of the materials used in advanced technologies particularly in the electronics industry. The silicon for computer chips is grown in the form of cylindrical crystals up to 150 mm or 200 mm in diameter and then divided by diamond sawing. In the first stage of sawing, or *wafering*, a wheel saw is used to slice the crystal into wafers no more than 0.5 mm thick. Then in the second stage the wafers are held on a substrate and sawn, or *diced*, to produce small squares having sides perhaps 0.3 mm to 0.5 mm. Other materials sawn by diamond are listed by Cesak (1986) and include various other semiconducting materials such as germanium, gallium arsenide, indium phosphide, etc., as well as ferrites, alumina, etc. A brief survey of the use of diamonds to saw glass is given by Seifarth (1986).

Although all the above materials can be cut fairly readily by a diamond saw, there remains the considerable problem of how to maintain good dimensional accuracy on the thickness of wafers and other components without wasting costly material. If a block of silicon is being cut into wafers $500\,\mu m$ thick and the saw is of equal thickness then half the material is lost. Therefore most of the development work on this type of sawing has been concerned with the design of wheels and sawing machines rather than the details of the diamond layer.

The saws used for wafering are all of the internal diameter, ID, type shown schematically in Figure 16.18. The annular shaped steel wheel is held round the

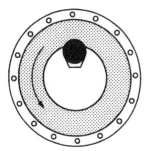

Figure 16.18 Schematic diagram of a cylindrical billet being cut by an ID saw

outer perimeter and the workpiece is fed in against the inner perimeter as indicated. In order to give stability to the steel blade the clamping system round the periphery is designed to apply a strong tension to stiffen the blade, rather in the same way as the skin of a drum is tensioned. Details of various methods of clamping are given by Büttner (1985). Using these techniques it is possible to use wheels no more than about 150 μm thick to cut specimens up to 150 mm or more in diameter with a thickness constant to a few μm, see for example Hayes (1989).

The diamond grits in these saws are held in a layer of nickel electroplated around the inner perimeter of the wheel. A typical profile of a saw showing the abrasive layer is shown in Figure 16.19. Note that the layer is less thick away from the edge

Figure 16.19 Section through the cutting edge of an ID saw blade (Büttner, 1985)

in order to reduce the friction on the workpiece. Cesak (1986) quotes typical grit sizes of 50 μm for the sides and 100 μm for the tip, the larger size giving a greater cutting rate and the smaller size a better finish with less surface damage. Even with a steel blade only 150 μm thick the overall thickness of the wheel will be of the order of 500 μm, comparable to the width of the wafer, with a considerable loss of material. Of course the losses during dicing will be less because only shallow cuts are needed, and these can be sawn with the outer periphery of thinner blades clamped between discs extending almost to the cutting edge.

16.2 Drilling

Diamond drills are used in a wide range of applications in the construction and mineral industries. To give some indication of these applications we begin by considering the drilling of stone and concrete for the construction industry, and then discuss the various types of bits used for deep drilling in the oil industry.

16.2.a Impregnated tube drills

Figure 16.20 shows examples of tube drills of the type used to drill holes of the order of 20 mm to 400 mm diameter and up to perhaps 5 m deep through concrete and other structures with a minimum of disturbance to the surroundings. The drill consists essentially of a steel tube with curved impregnated segments of diamond brazed to its working edge. Thus the drill is rather similar in design and operation to the impregnated saws described in the previous section, and cuts an annular hole leaving an isolated central core of concrete or stone.

Figure 16.20 Range of diamond tube drills
for drilling masonry. (MG utensili
diamantati s.p.a., Torino)

Practical details and examples of the use of tube drills are given by Smith (1974).
The design of the diamond segments is determined by the same considerations as
apply to the design of segments for sawing as discussed above. Some results on
how drilling rates and wear of the segments depend on grit size and concentration,
and on the power applied to the drill, are given by Van Biljon and Swersky (1975).
As in all types of drilling the coolant plays an important role in removing the
drilling debris. This is achieved in tube drills by the impregnated segments being
wider than the thickness of the steel tube so that the coolant may be forced down
the inner annular space between the tube and the rock and then up the other
annular space on the outside of the tube. The effects of different types of additives
in the coolant are discussed in Section 16.3.

16.2.b Surface set diamonds for deep drilling

Oil wells are commonly drilled to depths of over 5000 m using highly developed
techniques with sophisticated methods of monitoring the progress of the work. A
good general view of the general arrangements and problems involved in deep
drilling is given by Jenner (1984) but for our purpose it is sufficient to note that
the holes are bored by a drill bit at the end of a drill rod which is extended length
by length as the hole deepens. The diameter of the bit depends on particular
circumstances but bits are available in sizes ranging perhaps from 20 mm to 400
mm diameter. The bit is forced down on the rock by the weight of the drill rod,
and rotated by a torque from the surface transmitted down the rod or produced
by a turbine mounted above the bit and driven by the flow of coolant.

In order to cool the bit and carry away the drilling debris a drilling fluid is
pumped down the centre of the drill rod to pass through nozzles in the bit and then
back up through the outer annular space between the rod and the walls of the hole.
This fluid may be essentially water or a mud, the latter being a water-based mud
suspension selected to give optimum performance for a particular well, taking into
account the cooling power required, the removal of debris, and other effects such
as possible interactions between the fluid and the walls of the hole.

All bits for deep drilling were formerly made with steel or tungsten carbide but
diamond bits offer two considerable advantages. First, their superior strength and

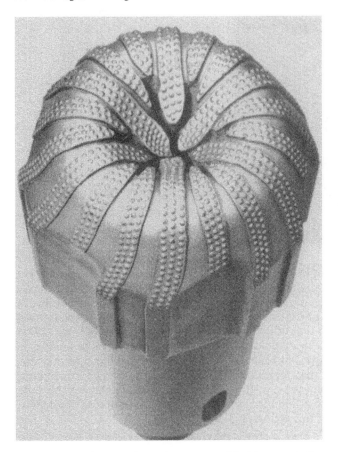

Figure 16.21 Surface–set diamond bit for oil drilling (Anon, 1985)

hardness enables an adequate penetration rate to be achieved in very hard rocks which would otherwise be difficult to drill. Second, diamond bits drilling softer rock may wear much more slowly than other bits because of their high abrasion resistance, and low wear is particularly important because a worn bit can only be replaced by pulling up the drill rod, length by length. Hence the time and cost of changing a bit is considerable, so even though diamond bits are much more expensive they can lead to large savings.

A typical diamond bit for deep drilling is shown in Figure 16.21. In this type of bit the diamonds are located individually on the surface of a matrix such as tungsten carbide, and are described as *surface set*. Note the side bars which contain smaller diamonds and prevent undue wear of the sides of the bit. The centre of the bit is recessed because the diamonds close to the axis of rotation have a much smaller velocity and would suffer damage if they came directly into contact with unbroken rock. Note also the channels or waterways between the diamond pads to permit adequate flow of the coolant. For a general description of variants of this type of bit see Panhorst (1978).

The size of the cutting diamonds in surface set bits ranges from about 4 per carat to 20 per carat depending on the type of rock involved. The action of the bit is first to break up the rock surface by crushing or plastic grooving under the loading of the individual diamonds (as described below). Then the damaged material is removed by the subsequent passage of other diamonds and carried away by the flow of the coolant. The volume of rock removed by a diamond will depend on how high it stands above the general level. Therefore larger diamonds will remove more material but will also experience greater forces. Hence, when drilling very hard rocks where the forces in any case are high it is preferable to use smaller diamonds. The considerations involved in the selection of the single crystal diamonds are similar to those already discussed in connection with sawing in Section 16.1.d. The use of PCD material in surface set bits is described in Section 16.2.c.

Another important type of surface set bit is the so-called core bit. A bit of the type shown in Figure 16.21 breaks up all the rock in the hole, which is then removed as debris. However, when drilling to prospect for oil or minerals it is necessary to obtain samples of the solid rock. This is done by using a core drill with a form somewhat similar to the tube drill described in the previous section. As the bit drills down, a core of solid rock is left in the end of the drill rod and is brought to the surface by some suitable arrangement. Figure 16.22 shows a detail of diamonds set in such a drill.

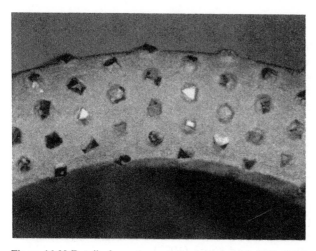

Figure 16.22 Detail of a surface-set diamond core bit (Hill, 1975)

Deep drilling is a complex operation and successful economic results are only achieved by taking into account many considerations, see for example a case history given by Barnett, Jeansonne and Mitchell (1988). Even to analyse the performance of the bit one must take into account the forces on the diamonds, the temperature, the part played by the debris, the action of the coolant, the response of the diamond, the nature of the rock being drilled, the last being a particularly large variable. We now briefly outline some experiments and calculations which have been made on these lines.

Appl and Rowley (1968) considered the cutting action of a single diamond moving under load across the surface of a rock. As the diamonds in a drill are generally of a somewhat rounded shape they were assumed on average to be spherical. A calculation was then made of the volume of material removed and the force on the diamond as the diamond ploughs through the rock, assuming that the material behaves plastically. Proceeding in this way, and taking a value for the resistance to plastic deformation obtained from experiments in which rocks were indented with a wedge shaped tool (Gnirk and Cheatham, 1965), the authors obtained some fairly good correlations between theory and experiment. That is, their values for the cutting or friction force as a function of the normal force and the radius of the diamond were similar to experimental results obtained by Garner (1967) on an Indiana limestone under simulated borehole conditions with pressures up to 300 atm.

The above analysis was extended to cover the performance of a drill bit in which the diamonds are either randomly or uniformly spaced (Rowley and Appl, 1969). These authors proceeded by assuming a certain penetration or volume removal rate, and hence deduced the mean volume of material plastically displaced by each diamond per revolution of the bit. Then, using values of the strengths of various rocks available in the literature, they calculate both the force on the diamonds and the corresponding load on the bit required to produce the assumed penetration rate. They thus obtained curves for the penetration rates as functions of the load on the bit and its rotary speed in reasonable agreement with direct measurements in the literature.

A further analysis was made by Moore, Walker and Appl (1978) to include the effects of wear by assuming that the originally spherical diamonds wore to give plane wear flats. These authors gave curves which show fair agreement with experiments but more extensive measurements of wear and penetrations rates are needed to obtain a conclusive picture. The authors also found it necessary to introduce empirical adjustments to allow for variations of drilling efficiency with the input power, the bit velocity and the type of fluid.

16.2.c PCD cutters for deep drilling

When PCD first became available various attempts were made to replace the single crystal diamonds in drill bits by similarly shaped blocks of PCD. Even though PCD material is more resistant to cleavages and fracture these first attempts were not very successful. (We return to this possibility in Section 16.2.e.) However, it was soon realized that PCD could be used to make a quite different type of bit which gives a good performance with long life in rock of soft to medium hardness, see for example Golis (1983) and Paterson and Shute (1982).

Figure 16.23 shows an example of a PCD bit consisting of a steel body which carries a number of blocks of PCD which act as cutters. Figure 16.24 gives a detail of a typical PCD cutter mounted on a steel stud which fits into the body of the bit. As described in Section 1.3 the actual PCD is generally in the form of a layer about 1 mm thick grown on a thicker base of tungsten carbide which can be readily brazed on to the steel stud. (Sometimes the carbide may also extend round the sides of the PCD to give further strength.) Working in relatively soft rocks the cutters behave quite differently from surface set diamonds, and act rather as cutting tools with a negative rake. As shown schematically in Figure 16.25 these cutters create small chips of rock which may coalesce under the high ambient pressures to form

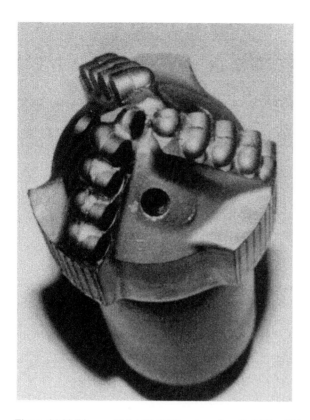

Figure 16.23 Diamond bit with PCD cutters for oil drilling (Dieckmann, 1988)

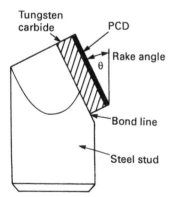

Figure 16.24 Mounting of a typical PCD cutter (after Hoover and Middleton, 1981)

quite lengthy chips, see for example Warren and Armagost (1988). Note also that because the tungsten carbide base wears back faster than the PCD, the cutters exhibit a self-sharpening action which tends to maintain a sharp edge of diamond (Figure 16.26).

The different nature of the rock removal process when using a bit with PCD cutters is particularly evident in an experiment described by Clark (1988) on the

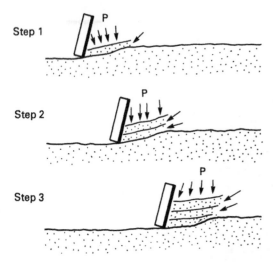

Figure 16.25 Sketch to illustrate chip formation by a negative rake cutter (Warren and Armagost, 1988)

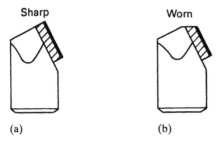

Figure 16.26 Diagram to show the self sharpening effect on PCD cutters (Hoover and Middleton, 1981)

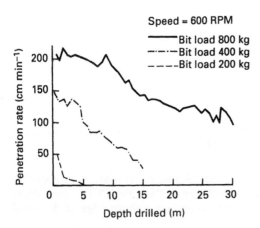

Figure 16.27 Penetration rates of PCD bits drilling Pennant Sandstone as a function of depth drilled and bit load (Clark, 1988)

drilling of a hard and abrasive Pennant sandstone. Figure 16.27 shows the penetration rate versus the depth drilled for three similar bits working under different loads. As is to be expected the higher loads produce the greater penetration rate, and the rate falls off with increasing depth presumably because the bit is becoming blunt. Also the rate of wear of the bit, as shown by the decrease in penetration rate with depth, is less for the higher loads and penetration rates. The result may appear surprising but arises because a greater load on PCD cutters results in deeper cuts which remove a given amount of material in fewer revolutions. Hence, as the wear at the edge of the cutter is much less dependent on the depth of cut than on the number of revolutions, the wear is not so great at the higher loads.

The design of a bit must ensure the efficient removal of the rock debris by the drilling fluid or mud in order to avoid it balling up on the cutter. In addition, the types of rock to be drilled vary greatly in their properties. Therefore there is considerable variety in the geometric form of the drill bits, both in the arrangement of the cutters and the layout of the fluid nozzles, see for example Feenstra (1988a) and Kerr (1988). Examples of the geometries of the cutters themselves and their mounting on the steel studs are given by Dennis and Clark (1987). A useful review of current developments and possible future applications is given by Feenstra (1988b).

16.2.d Analysis of cutter performance

The performance of the PCD cutters in a bit depends on a variety of factors including the hardness of the rock, the cutter design and the wear of the PCD. It is obviously difficult to simulate deep drilling conditions in the laboratory. In particular the ambient pressure is perhaps about 100 atm, the volume of mud flow is very large, and in softer rocks appreciable wear is only produced after long drilling. Hence information on the performance of cutters comes mostly either from reports of performance in the field or from laboratory experiments which may only approximate to real working conditions.

Studies of the effect of varying the rake angle of the cutter (Figure 16.24) have been made by Hibbs and Flom (1978), Hough (1986) and Shafto (1985). However, only the latter author made an actual drilling experiment with a liquid coolant, Hough drilled with 'air flushing', and Hibbs and Flom simulated drilling by turning on a lathe. One of the more realistic sets of laboratory experiments is that of Warren and Armagost (1988) who drilled under a pressure of 8.6 MPa but only to depths of about a metre. However, although these distances are small the penetration rates took up steady values after drilling for only about 0.3 m. Values are given for the penetration rate and the torque on the bit when drilling limestones and shales as a function of the load on the bit for both new and worn cutters, using either oil- or water-based muds. The authors give an informative discussion of their results including the effects of wear and of any balling of the mud around the cutters.

One of the principal limitations to high rates of drilling is the need to avoid overheating the PCD. The standard type of PCD is less stable at high temperatures than single crystal diamond because the solvent/catalyst has two undesirable effects on sintered diamond. Its greater expansion coefficient creates internal stresses on heating and its catalytic activity causes degradation of the diamond (Section 12.2.d). When PCD cutters are used to turn or drill rock at least two types of wear are involved (Lee and Hibbs, 1979; Hibbs and Sogoian, 1983). At low speeds the appearance of the worn surfaces is typical of the type of mechanical abrasive wear seen when the PCD is subjected to a rough polishing, while at high speeds there is a much larger component arising from thermal damage.

It is difficult to quote an exact temperature above which thermal wear becomes important for three rather different reasons. First, the effect of a temperature depends on the time that the PCD is held at that temperature. It is, for example, quite possible to braze a block of PCD to a metal shank in a few seconds and cause little damage even though a similar steady temperature would produce severe deterioration. Second, the temperature stability of PCD depends on the grain size (Section 12.2.d) and other characteristics of the particular proprietary brand of PCD in question. Third, a further complication is shown by an experiment on the wear of PCD turning stone by Lee and Hibbs (1979) who observed that the wear rate rose markedly as the cutting speed increased above a certain value. This rise suggested the onset of thermal wear but micrographs of the worn surfaces showed a structure not much different from that at lower speeds. Hence there may be a temperature region in which the PCD begins to lose strength without obvious signs of disintegration. (Such behaviour is also suggested by the acoustic emissions observed by Mehan and Hibbs (1989) when heating a coarse grain PCD above 400°C as described in Section 12.2.d).

Although the position regarding the onset and nature of thermal damage is somewhat obscure, one can still note some significant temperatures. It is clear that temperatures above about 800°C will cause rapid disintegration by thermal effects alone. It is also clear that considerably lower temperatures may produce a loss of strength. For example Clark and Shafto (1987) show that holding specimens for 30 min at various temperatures above 650°C produced large reductions in their useful life when turning a cylinder of granite, see also Clark (1988). There is, however, not much information on the changes responsible for the increased wear.

Although the temperature may greatly affect the performance of a PCD cutter, it is not easily measured *in situ* at the bottom of a deep well. Therefore estimates of temperatures have generally been made either via simulation experiments or by calculations. Hibbs and Sogoian (1983) made detailed experiments with cutters on a vertical axis lathe making facing cuts on marble and sandstone (using a flood coolant). They observed that thermocouples in the PCD close to the wear flat registered temperatures ranging from about 100°C to 500°C depending on the type of stone, the cutting speed, and the condition of the cutters, increasing greatly as they became worn. (Measurements are also given of the wear and cutting forces without a coolant.)

Attempts have been made to calculate the working temperature of cutters given the load on the bit and its rate of rotation. Using this approach one first calculates the normal force on each cutter and hence the friction force F_μ which is responsible for the generation of heat. Virtually all the work done by the drill appears as heat at the cutters, and flows away from the PCD to either the rock or mud. To estimate these forces and heat flows, experimental values are required for the effective coefficient of friction between the cutter and the rock, for the thermal conductivity of the rock and the cutters, and for the coefficients of heat transfer from the cutters to the mud and in the mud. Hence the calculations involve a formidable programme but estimates of excess temperatures have been given by Ortega and Glowka (1984), who quote values up to the order of 500°C; see also Glowka and Stone (1985). However, because of the difficulty of determining all the various

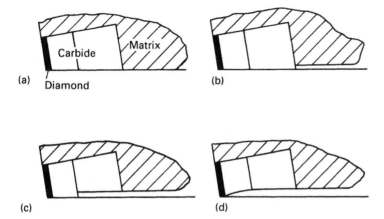

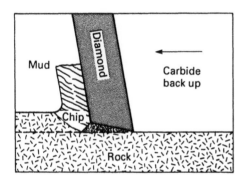

Figure 16.28 Progressive wear of a PCD cutter, see text (Zijsling, 1984).

Figure 16.29 Sketch showing geometry of a cutting edge, see text (Feenstra, 1988a)

parameters and of obtaining values of all the various physical constants, the principal importance of these calculations is not so much in the numerical values but to show how the temperatures are affected by changes in the design factors.

The complexity of the working conditions of a PCD cutter is further underlined by Zijsling (1984) who observed the form of the wear on a cutter in a simulated drill-hole experiment. The PCD layer in this cutter was mounted on a double block of tungsten–carbide set in a matrix-type bit (Figure 16.28(a)). Initially with a new cutter the rock abrades the PCD, the carbide and the matrix, but as the matrix and carbide wear the forces are concentrated on the PCD (Figure 16.28(b), (c) and (d)). A further complication observed by Zijsling was that the edge of the PCD developed a wear land, as shown in Figure 16.29, with effectively a very high angle negative rake, and he argues that the volume under this land is filled with broken down rock or 'rock flour' which affects the temperature distribution. For further details of the various above calculations, see the original papers and comments by Feenstra (1988a).

16.2.e Thermally stable PCD bits

Because of the temperature limitations on the use of PCD bits described in the previous section there is an increasing interest in the more thermally stable types of PCD described in Section 12.2.d. We mentioned in Section 16.2.a that soon after the production of the first PCD material attempts were made to use pieces of PCD in surface set bits in the same way as single crystal diamond. These attempts were not very successful, one of the difficulties being that the hard rocks drilled by surface set bits give rise to high forces and high temperatures which reduced the strength of the PCD. However, the thermally stable variants of PCD now available will operate at temperatures up to about 1200°C and are being used successfully in surface set bits. For example, Figure 16.30 shows a core bit for rock drilling surface set with small oriented cubes of thermally stable PCD. For further details of the design and performance of this type of tool see Clark and Shafto (1987) and Stewart, Falter and Tomlinson (1988).

Figure 16.30 Surface-set diamond core bit using oriented cubes of thermally stable PCD (Stewart, Falter and Tomlinson, 1988)

The advantages of the various types of thermally stable PCD suggest that we may expect its increasing development and use. At present this material is only available in relatively small pieces and therefore cannot replace the large PCD blocks used as drill cutters. However, experiments are now being made with so-called mosaic cutters in which the working surface of a PCD cutter is inset with pieces of thermally stable material (Cerkovnik and Mason, 1988). Note that at present the thermally stable material is not available on a base such as tungsten carbide which is used to give additional strength to normal PCD.

16.3 Action of fluids

As mentioned in Section 16.1.a sawing and drilling is generally carried out with an ample supply of some water based fluid which has the double task of cooling the bit and carrying away the machining debris. Besides acting in these two ways a fluid may also interact more directly with the machining interface to produce changes in cutting and drilling rates and in the wear of the cutters. We now describe some of these effects and then discuss what mechanisms may be involved.

16.3.a Effect of fluids

Figure 16.31 shows some results obtained by Tönshoff and Schulz (1982) in a laboratory experiment in which blocks of granite were sawn by a wheel blade carrying a single diamond-impregnated segment. The figure gives the G ratio

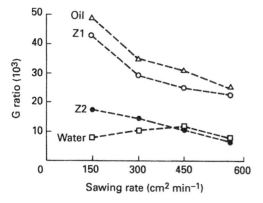

Figure 16.31 G ratio for a diamond impregnated saw cutting Colombo granite as a function of sawing rate for four different fluids: Z1, 2% aqueous solution of organic components; Z2, water plus a wetting agent (Tönshoff and Schulz, 1982)

(Section 15.2.c) for the wear of the segment as a function of the rate of sawing for four different fluids: tap water, water plus a wetting agent, a 2% aqueous solution of organic components, and a low viscosity cutting oil. We see that there are substantial differences between the G values associated with the fluids. The authors also give examples of similar differences when sawing a second type of granite. Other measurements showed that the magnitude of the normal force on the segment also depended considerably on the nature of the fluid. Further examples of the influence of the fluid on blade wear are given by Büttner (1982) together with values of the power requirements with different fluids.

Figure 16.32 gives the results of an experiment by Selim, Schultz and Strebig (1969) to observe penetration rates when drilling quartzite with a 32 mm diameter diamond-impregnated bit flushed with either water or a 12% solution of glycerine. We see that both penetration rates decrease as the distance drilled increases and the drill becomes blunt, but that the decrease is much less rapid when using the solution of glycerine. The authors also give several other examples of the effects of various liquids on the rate of penetration, on the wear of the bit, and on the torque acting on the bit.

Conditions are of course much more complicated when deep drilling in the field with a flushing fluid which is generally a mud with a complex constitution. Figure 16.33 gives an example of the effect of different muds on penetration rate in trials described by Barnett, Jeansonne and Mitchell (1988); other examples are given by

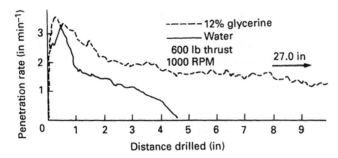

Figure 16.32 Penetration rate of a surface set bit drilling granite as a function of distance drilled with two different fluids (Selim, Schultz and Strebig, 1969)

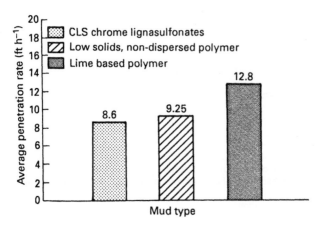

Figure 16.33 Drilling rate of a 7 5/8 in PCD cutter in field trials using 3 different types of mud (Barnett, Jeansonne and Mitchell, 1988)

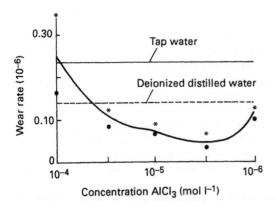

Figure 16.34 Wear rate of a spherical diamond of radius 2.3 mm cutting Georgia granite using cutting fluids of different concentration of aluminium chloride in deionised distilled water. The wear rate quoted is the inverse of the G ratio (Appl, Rao and Walker, 1981)

Holster and Kipp (1983) and Warren and Armagost (1988). Examples of the effect of different muds used as machining fluids when turning rock are given by Hibbs and Sogoian (1983). The action of these muds is particularly complex because the performance of a drill depends not only on the cutting action of the bit and its rate of wear but on the ability of the fluid to clear the cutters of debris and to carry the debris away. It may also be necessary to select a fluid which does not react adversely with the rock wall of the hole. Therefore any discussion of the behaviour of these muds is beyond the scope of this book, and we confine our discussion to various laboratory experiments made under relatively well defined conditions.

Figure 16.34 shows the wear rate of a spherically shaped single crystal diamond used as a tool to turn a cylindrical rod of granite flushed with ionized water containing different low concentrations of aluminium chloride (Appl, Rao and Walker, 1981). The solid circles indicate five points taken first in random order, and the asterisks a second set of points in random order taken as a check on the reproducibility of the results, an important precaution not observed by all authors. The difference between the two points for a concentration of 10^{-4} mol 1^{-1} is rather large but the agreement between other pairs is reasonably satisfactory, particularly as all the later points lie systematically higher presumably because the tool blunted during the experiment. Hence there is a clear indication that the wear rate of the diamond is affected by the concentration of the aluminium chloride.

Finally we refer to experiments by Robinson (1959) who drilled various rocks in a compression cell specially designed so that the pressure on any interstitial fluids in the pores of the rock could be varied independently of the pressure surrounding the rock. These showed that the rock strength increased as the pressure surrounding the rock became greater than the pressure in the pores. Subsequently using the same apparatus Robinson (1967) showed that, quite apart from any pressure effects, the drilling rate varied by up to 40% when using different additives in the water.

16.3.b Analysis of drilling experiments

We now briefly consider how fluids may affect the rates of sawing and drilling and the wear of the diamond. We begin by considering systems where the fluid acts under relatively simple conditions such as those described by Tönshoff and Schulz (Figure 16.31) and by Appl, Rao and Walker (Figure 16.34). Two quite different types of explanation have been put forward to account for such results. One assumes that the fluid modifies the interaction between the diamond and the workpiece, thus changing the friction and hence the penetration rate and the wear of the tool. The other approach assumes that the presence of the fluid directly modifies the strength of the workpiece material.

Before discussing the various sawing and drilling experiments further it is useful to note that there is now a range of experiments which show that the surface strength of a material may be modified by the presence of a fluid, either liquid or gaseous. For example cracking in sapphire is affected by water vapour (Lawn, 1975); in glass by water, toluene and silicon oil (Mikosza and Lawn, 1971), and by solutions of different acidity (Lawn and Wilshaw, 1975); cracking in silicon by ethanol (Lim and Danyluk, 1985). An extensive review of earlier work in this field is given by Rehbinder and Shchukin (1972) who quote a variety of other examples of the effect of fluids on various parameters including the pendulum hardness. (This hardness is essentially a measure of the damping capacity of the surface which carries a hard steel knife edge supporting an oscillating pendulum).The review also notes that these so-called Rehbinder effects might be used to assist the removal of rock during drilling, either by making the rock more brittle to promote fracture, or more plastic to promote removal by shear.

Bearing in mind the above background we now refer to experiments by Westwood and collaborators, reviewed by Westwood (1974). They observed the effect of various fluids in a range of laboratory experiments drilling alumina (Westwood, Macmillan and Kalyoncu, 1973) and quartz and granite (Westwood, Macmillan and Kalyoncu, 1974). They found that a range of alcohols of the form $CH_3(CH_2)_nOH$ with n running from 1 to 12 gave considerably greater penetration rates than either toluene or water. In addition,they observed prominent maxima in the drilling rates for the alcohols $n=5$ and $n=8$, and similar maxima in the values of the pendulum hardness. (For experimental details of the measurement of pendulum hardness see Kuznetsov (1957) and Westwood, Parr and Latanision (1972), and cautionary remarks by Heins and Street (1965).) The authors suggest that these unexpected maxima correlate with zero values of the zeta potential although it is not clear what mechanisms are involved. (The zeta potential between a liquid and a solid is a concept from colloid chemistry involving an emf set up by moving surface charges at an interface, see for example Sennett and Olivier (1965) and Hunter (1981).)

Two other features of the above experiments are hard to understand. In a range of experiments on alumina Swain, Latanision and Westwood (1975) observed similar maxima to those described above. Then in a further experiment the alumina was ground with an impregnated wheel and the removal rates measured in the presence of the different n-alcohols. These rates steadily decreased with increasing n number with no sign of the maxima which might be expected from the behaviour during drilling. In other experiments drilling magnesium oxide (Westwood and Latanision, 1972) and calcite (Westwood, 1974) the authors replaced the diamond bit by a tungsten carbide bit and observed that the drilling rates depended on the n numbers in a different way, but did not propose any satisfactory explanation. The only other experiments of this type are those by Wiederhorn and Roberts (1975), who measured the friction of glass spheres sliding on blocks of silicon carbide and the wear produced on the glass, in the presence of the n alcohols for $n=1$ to $n=12$. They observed that both the friction and wear decreased steadily with increasing n number with no sign of any maxima, and concluded that the action of the alcohols was essentially to modify the friction. It thus appears that further experiments are required to confirm the various maxima described by Westwood *et al.* and their correlation with the zeta potential.

Quite different results are described by Cooper and Berlie (1976, 1978) who drilled marble and granite with diamond impregnated core drills at penetration rates comparable to those used in the field (which are considerably higher than those reported by Westwood (1974)). When drilling marble they found no significant differences in penetration rates over a range of aqueous fluids containing, among other things, $AlCl_3$ and $NaOH$. However, Figure 16.35 shows results obtained drilling granite in an experiment in which the fluid was alternately water and one or other of the n-alcohols with $n=1$ to $n=10$. Using water the rate of penetration falls as drilling proceeds presumably because of the blunting of the bit. The figure also shows that on switching to an alcohol there was no immediate change in the value of the penetration rate, but the rate ceased to fall as drilling continued and sometimes even appeared to rise.

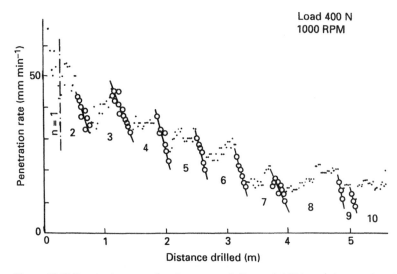

Figure 16.35 Penetration rate of an impregnated diamond drill in granite as a function of the distance drilled using the normal alcohols of the series $CH_3(CH_2)_nOH$ alternating with water as the flushing fluid. The straight lines are drawn on the points obtained with water (Cooper and Berlie, 1976)

Cooper and Berlie noted that there was no immediate change in the penetration rate on switching from water to alcohol. They therefore concluded that the fluids were not producing a direct Rehbinder type of influence on the rock, and hence that the differences in drilling rates arise becuse the fluids have different effects on the wear of the drills. However this argument is not conclusive because it is possible that the full effect of an interaction with a fluid or gaseous environment may take some time to establish itself, as in the experiments on the polishing of diamond described in Section 13.4.a.

There is also the further difficulty that if the effect of the alcohols were simply to reduce the wear of the diamond grits there is no reason why the drilling rates in Figure 16.35 should begin to increase on switching to some alcohols. However, as the authors point out, the wear of any impregnated diamond tool depends (as in the case of grinding) both on the wear and fracture of the diamond and on the wear of the bond and the retention of the diamond. Therefore Cooper (1979) suggests that the observed increases arise because the change of fluid also produced some change in the wear of the bond which produced a shift towards better cutting conditions. (The effect of a machining fluid on the bond material during the grinding of glass is considered by Al'tshuller, Ashkerov and Korovkin (1983).)

In a further discussion of the above experiments Cooper (1979) considers the wear of the diamond and stresses the influence of physico-chemical effects similar to those noted in previous chapters. Thus Figure 16.36 shows the very marked effect of switching from water to solutions of different concentrations of hydrogen-peroxide in an experiment similar to that of Figure 16.35. We see that the hydrogen peroxide greatly increasers the wear of the drill presumably because it is a strong oxidizing agent which reacts with the diamond. Cooper also notes that if two similar new bits are used to drill with different fluids there will probably be little or no difference in performance at first even though large differences develop as drilling

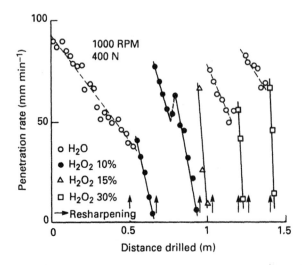

Figure 16.36 Penetration rates in granite of diamond impregnated drills working with water and water containing various concentrations of hydrogen peroxide. The vertical arrows indicate the periodic redressings of the bit (Cooper and Berlie, 1978)

proceeds, as observed for example by Mills and Westwood (1978). Hence it seems likely that no effect of the fluids was seen by Cooper and Berlie (1978) when drilling marble because the rate of wear of the bit was much less, as the experiments were not run for long enough to produce appreciable differences.

References

Aleksandrov, V. A. , Lévin, M. D. and Mechnik, V. A. (1986) *Soviet Journal of Superhard Materials,* **8**, 42–47

Al'tshuller, V. M. , Ashkerov, Yu. V. and Korovkin, V. P. (1983) *Soviet Journal of Optical Technology,* **50**, 514–516

Anon (1972) *Industrial Diamond Review,* **32**, 298–299

Anon (1985) *Industrial Diamond Review,* **45**, 91–92

Appl, F. C. , Rao, B. N. and Walker, B. H. (1981) *Industrial Diamond Review,* **41**, 312–318

Appl, F. C. and Rowley, D. S. (1968) *Society of Petroleum Engineers Journal,* September, 269–280

Bailey, M. W. and Bullen, G. J. (1979) *Industrial Diamond Review,* **39**, 56–60

Bailey, M. W. and Collin, W. D. (1978) *Industrial Diamond Review,* **38**, 8–13

Barnett, K. L. , Jeansonne, J. P. and Mitchell, R. K. (1988) *SPE Drilling Engineering,* March, 7–11

Bridwell, H. C. and Appl, F. C. (1974) *Industrial Diamond Review,* **34**, 51–53

Bullen, E. J. and Bailey, M. W. (1979) *Industrial Diamond Review,* **39**, 352–355

Busch, D. M. (1979) In *The Properties of Diamond,* (ed. J. E. Field), Academic Press, London, pp. 595–618

Büttner, A. (1974) *Industrial Diamond Review,* **34**, 89–93

Büttner, A. (1982) In *Ultrahard Materials Application Technology,* Vol. 1 (ed. P. Daniel), De Beers Industrial Diamond Division, London, pp. 63–71. Also *Industrial Diamond Review* (1980), **40**, 332–335

Büttner, A. (1985) *Industrial Diamond Review,* **45**, 77–79

Büttner, A. and Mummenhoff, H. (1973) *Industrial Diamond Review,* **33**, 376–379

Cai, O. and Mancini, R. (1988) *Industrial Diamond Review,* **48**, 212–214

Cerkovnik, J. and Mason, K. (1988) *Oil and Gas Journal*, May, **82**, 78–84

Cesak, F. (1986) *Industrial Diamond Review*, **46**, 72–75

Clark, I. E. (1988) In *Ultrahard Materials Application Technology*, Vol. 4, (ed. C Barrett), De Beers Industrial Diamond Division, London, pp. 17–35

Clark, I. E. and Shafto, G. R. (1987) *Industrial Diamond Review*, **47**, 169–173

Clauser, G. and Valle, A. (1987) *Industrial Diamond Review*, **47**, 207–208

Cooper, G. A. (1979) In *The Science of Ceramic Machining and Surface Finishing II*, (eds B. J. Hockey and R. W. Rice), N. B. S. Special Publication 562. US Government Printing Office, Washington, pp. 115–137

Cooper, G. A. and Berlie, J. (1976) *Journal of Materials Science*, **11**, 1771–1775

Cooper, G. A. and Berlie, J. (1978) *Journal of Materials Science*, **13**, Letters, 2716–2718

Daniel, P. (1986a) *Industrial Diamond Review*, **46**, 189–194

Daniel, P. (1986b) *Industrial Diamond Review*, **46**, 46–48

Dennis, M. and Clark, D. A. (1987) *Oil and Gas Journal*, Sept 28, 62–68

Dieckmann, M. (1986) *Industrial Diamond Review*, **46**, 199–200

Ertingshausen, W. (1985) *Industrial Diamond Review*, **45**, 254–258

Feenstra, R. (1988a) *Journal of Petroleum Technology*, June, 675–684

Feenstra, R. (1988b) *Journal of Petroleum Technology*, July, 817–821

Finnigan, G. (1968) *Industrial Diamond Review*, **28**, 310–316, 365–367

Garner, N. E. (1967) *Society of Petroleum Engineers Journal*, July, 937–942

Gielesse, P. J. and Stanislao, J. (1972) In *The Science of Ceramic Machining and Surface Finishing*, (eds S. J. Schneider and R. W. Rice), National Bureau of Standards Special Publication No. 348, US Government Printing Office, Washington DC, pp. 5–35

Glowka, D. A. and Stone, C. M. (1985) *Society of Petroleum Engineers Journal*, April, 143–156

Gnirk, P. F. and Cheatham, Jr, J. B. (1965) *Society of Petroleum Engineers Journal*, June, 117–130

Golis, S. W. (1983) *IADC/SPE 1983 Drilling Conference*, (New Orleans, 1983), paper 11391. Society of Petroleum Engineers, Dallas

Hayes, D. (1989) *Industrial Diamond Review,* **49**, 56–58

Heins, R. W. and Street, N. (1965) *Society of Petroleum Engineers Journal*, June, 177–183

Herbert, S. (1971) *Industrial Diamond Review*, **31**, 267–269

Herbert, S. (1983) *Industrial Diamond Review*, **43**, 250–251

Hibbs, Jr, L. E. and Flom, D. G. (1978) *Transactions of the ASME, Journal of Pressure Vessel Technology*, **100**, 406–416

Hibbs, Jr, L. E. and Sogoian, G. C. (1983) *Wear Mechanisms for Polycrystalline Diamond Compacts as utilized for Drilling in Goethermal Environments*. Sandia Laboratories Contract No. 13–9406 SAND82–7213. National Technical Information Service, US Department of Commerce, Springfield, VA

Hill, B. S. (1975) *Industrial Diamond Review*, **35**, 282–288

Holster, J. L. and Kipp, R. J. (1983) *Paper presented at the 58th Annual Technical Conference and Exhibition* (San Francisco, 1983) Society of Petroleum Engineers, Dallas, pp. 1–6

Hoover E. R. and Middleton, J. N. (1981) *Journal of Petroleum Technology*, December, 2316–2321

Hough, Jr, C. L. (1986) *Journal of Energy Resources Technology*, **108**, 305–309

Hunter, R. J. (1981) *Zeta Potential in Colloid Science*. Academic Press, London

Jenner, J. W. (1984) In *Modern Petroleum Technology* (5th edn) Part 1, (ed. G. D. Hobson), John Wiley, Chichester, pp. 123–163

Jennings, M. (1987) *Industrial Diamond Review*, **47**, 216–218

Kerr, C. J. (1988) *Journal of Petroleum Technology*, March, 327–332

Kuznetsov, V. D. (1957) *Surface Energy of Solids*. Department of Scientific and Industrial Research, HMSO, London

Lawn, B. R. (1975) *Journal of Materials Science*, **10**, 469–480

Lawn, B. R. and Wilshaw, R. (1975) *Journal of Materials Science*, **10**, 1049–1081

Lee, M. and Hibbs, Jr. L. E. (1979) In *Wear of Materials, 1979*, (eds K. C. Ludema, W. A. Glaeser and S. K. Rhee), American Society of Mechanical Engineers, New York, pp. 485–491

Lim, D. S. and Danyluk, S. (1985) *Journal of Materials Science*, **20**, 4084–4090

Mamalis, A. G. , Schulze, R. and Tönshoff, H. K. (1979) *Industrial Diamond Review*, **39**, 356–365

Mehan, R. L. and Hibbs, Jr. L. E. (1989) *Journal of Materials Science,* **24**, 942–950

Mikosza, A. G. and Lawn, B. R. (1971) *Journal of Applied Physics,* **42**, 5540–5545

Miller, H. C. (1967) In *Science and Technology of Industrial Diamonds,* Vol. 2 Technology, (ed. J. Burls), De Beers Industrial Diamond Division, London, pp. 205–219

Mills, J. J. and Westwood, A. R. C. (1978) *Journal of Materials Science,* **13**, Letters 2712–2716

Moore, N. B., Walker, B. H. and Appl, F. C. (1978) *Transactions of the ASME, Journal of Pressure Vessel Technology,* **100** Part 2, 164–171

Ortega, A. and Glowka, D. A. (1984) *Society of Petroleum Engineers Journal,* April, 121–128

Panhorst, H-J. (1978) *Industrial Diamond Review,* **38**, 5–7

Paterson, A. W. and Shute, J. P. (1982) *Proceedings of the European Petroleum Conference* (London, 1982) European Petroleum Conference Society of Petroleum Engineers (UK), pp. 575–582

Perrett, C. (1969) *Industrial Diamond Review,* **29**, 236–241

Pinzari, M. (1983) *Industrial Diamond Review,* **43**, 231–236

Rehbinder, P. A. and Shchukin, E. D. (1972) *Progress in Surface Science,* **3**, 97–188

Robinson, L. H. (1959) *Petroleum Transactions of the AIME,* **216**, 26–32

Robinson, L. H. (1967) *Society of Petroleum Engineers Journal,* September, 295–300

Rowley, D. S. and Appl, F. C. (1969) *Society of Petroleum Engineers Journal,* September, 301–310

Schaffner, J. and Blaser, E. (1987) *Industrial Diamond Review,* **47**, 163–164

Seifarth, M. (1986) *Industrial Diamond Review,* **46**, 57–59

Selim, A. A. , Schultz, C. W. and Strebig, K. C. (1969) *Society of Petroleum Engineers Journal,* December, 425–433

Sennett, P. and Olivier, J. P. (1965) *Industrial and Engineering Chemistry,* **57**, 33–60

Shafto, G. R. (1985) *Industrial Diamond Review,* **45**, 237–241

Smith, N. R. (1974) *Users Guide to Industrial Diamonds.* Hutchinson Benham, London

Stewart, A. , Falter, F. X. and Tomlinson, P. N. (1988) *Industrial Diamond Review,* **48**, 117–119

Sulten, P. (1989) *Industrial Diamond Review,* **49**, 246–250

Swain, M. V. , Latanision, R. M. and Westwood, A. R. C. (1975) *Journal of the American Ceramic Society,* **58**, 372–376

Thoreau, B. (1984) *Industrial Diamond Review,* **44**, 94–95

Tönshoff, H. K. , Schmidt, N. , Ertingshausen, W. and Ostertag, A. (1984) *Industrial Diamond Review,* **44**, 4–10

Tönshoff, H. K. and Schulze, R. (1982) In *Ultrahard Materials Application Technology,* Vol. 1, (ed. P. Daniel), De Beers Industrial Diamond Division, London, pp. 50–62

Tönshoff, H. K. and Warnecke, G. (1982) In *Ultrahard Materials Application Technology,* Vol. 1, (ed. P. Daniel), De Beers Industrial Diamond Division, London, pp. 36–49

Trancu, T. C. (1980) *Industrial Diamond Review,* **40**, 329–331

Van Biljon, J. J. and Swersky, G. (1975) *Third Diamond Wheel Manufacturers Institute (DWMI) Technical Symposium,* (Chicago, 1975) Diamond/CBN Abrasives. Paper SC4. DWMI, Cleveland

Veglio, O. (1987) *Dimensional Stone,* Jan/Feb, 45–48

Warren, T. M. and Armagost, W. K. (1988) *SPE Drilling Engineering,* June, 125–135

Weber, U. and Zilm, F. (1987) *Industrial Diamond Review,* **47**, 214–215

Westwood, A. R. C. (1974) *Journal of Materials Science,* **9**, 1871–1895

Westwood, A. R. C. and Latanision, R. M. (1973) In *The Science of Ceramic Machining and Surface Finishing,* (eds S. J. Schneider and R. W. Rice), National Bureau of Standards Special Publication No. 348, US Government Printing Office, Washington DC, p. 141

Westwood, A. R. C. , Macmillan, N. H. and Kalyoncu, R. S. (1973) *Journal of the American Ceramic Society,* **56**, 258–262

Westwood, A. R. C. , Macmillan, N. H. and Kalyoncu, R. S. (1974) *Society of Mining Engineers,* Transactions, **256**, 106–111

Westwood, A. R. C. , Parr, Jr. G. H. and Latanision, R. M. (1972) In *Amorphous Materials,* (eds R. W. Douglas and B. Ellis), Wiley, London, pp. 533–543

Wiederhorn, S. M. and Roberts, D. E. (1975) *Wear,* **32**, 51–72

Wiemann, H. H. , Büttner, A. , Ertingshausen, W. and Schwarz, W. (1983) In *Ultrahard Materials Application Technology,* Vol. 2 (ed. P. Daniel), De Beers Industrial Diamond Division, London, pp. 126–138

Wright, D. N. (1988) In *Ultrahard Materials Application Technology*, Vol. 4, (ed. C Barrett), De Beers Industrial Diamond Division, London, pp. 47–60

Wright, D. N. and Wapler, H. (1986) *Annals of the CIRP,* **35**, 239–244

Zijsling, D. H. (1984) *SPE Technical Conference* (Houston, September 1984), Paper 13260. Society of Petroleum Engineers, Dallas

Chapter 17

Miscellaneous applications

The previous three chapters have described the main applications of diamond as tools for turning, grinding, sawing and drilling. Before reviewing various other applications we first note the great variety of materials and components which are now machined by diamond. Diamond turning tools are used to produce motor-car engine blocks (Herbert, 1987) and contact lenses (Biastoch, 1983; Herbert, 1983). The allied process of milling is used in the woodworking industry (Moitzi, 1985; Jennings, 1989; Steinmetz and Heath, 1989), to give a fine gloss to metal jewellery components (Herbert, 1976), and to recondition aircraft cabin windows (Bühler, 1986). Drilling operations range from oil wells several kilometres deep to the dentist drilling into root canals (Herbert, 1974). Diamond sawing may be at the quarry face, or in an advanced factory dividing silicon crystals to form computer chips, or shaping material for body armour (Herbert, 1986). Diamond grinding is used to shape carbide tools and grinding wheels, to polish the edges of most glass used in motor cars (Heger, 1981), to produce the exact shape for large optical glass mirrors (Herbert, 1984a), and to polish X-ray telescopes to probe the depths of outer space (Reisert, 1989).

17.1 Jewellery

Without doubt the best known application of diamond is as jewellery. A clear regular octahedron is a most attractive crystal but it only takes on the character for which diamond is famous after being cut and polished to shape. The cutter and polisher create a geometric form whereby the diamond captures light and reflects it back in flashes of brilliance and colour. There are a variety of possible geometries but the best known is the so-called brilliant cut with 58 polished facets.

Figure 17.1 shows a cross section of a typical brilliant with the largest facet uppermost, this being the *table* through which the stone is usually viewed. The widest circumference is known as the *girdle*, the part of the stone above the girdle as the *crown*, and the part below the girdle as the *pavilion*. Figure 17.2 shows plan views of the crown and pavilion on a line perpendicular to the table, showing the positions of the table, the small *culet* at the base of the pavilion, and the other 56 sloping facets arranged round them. This shape is chosen so that most of the light entering the diamond through the table is totally reflected by the pavilion facets and emerges again either through the table or the crown facets. Thus the diamond responds to light as a set of highly reflecting mirrors. Hence as a diamond ring

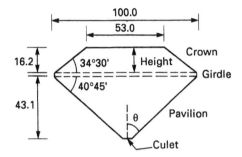

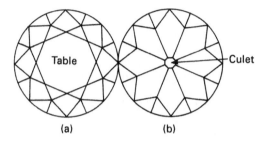

Figure 17.1 Cross section of a typical brilliant cut diamond (Dodson, 1979)

Figure 17.2 Plan view of a brilliant cut diamond viewed: (a), towards the table; (b), towards the culet (Dodson, 1979)

moves on the hand of its owner the diamond reflects many different images of a light so producing the well-known sparkle effect.

It might seem that other gemstones besides diamond could be cut to give similar effects but diamond has a combination of properties which make it unique. It has an extremely low absorption of visible light so that virtually no light is lost in the diamond. It has a particularly high refractive index (2.54) which greatly increases the probability that light entering the table will be reflected back either totally or partly by the other facets. Besides having a high value the refractive index shows considerable dispersion, that is, its value varies appreciably from one end of the visible spectrum to the other. Hence the various components of white light are not refracted to the same extent when passing through the facets so the emerging light is coloured. Finally, quite apart from its optical properties diamond is the hardest of all materials, as is recognized in the well-known Mohs scale of hardness, see for example Tabor (1970). Diamond is extremely abrasion resistant, only diamond can scratch diamond, and this is very important for gemstones because any scratching of the table and crown facets will reduce their transparency and therefore the brilliance of the gem as a whole.

There is a reference to the brilliant cut as early as Jefferies (1753) but a detailed analysis of the optics was not given until much later (Tolkowsky, 1919). However, Tolkowsky's calculations showed that the empirically determined form was in fact quite close to the optimum. The most complete analysis is that of Dodson (1978, 1979) who also summarizes earlier discussions and calculations. Dodson defines the performance of a brilliant in terms of three criteria. The *brilliance* is the fraction of the light entering the stone which is reflected back out of the crown facets. The *sparkliness* is the variation in the intensity of this reflected light (as for example when the diamond is rotated). The *fire* is a measure of the relative intensities of the red, green and blue parts of the spectrum of the reflected light.

Dodson sets up mathematical definitions of sparkliness and fire (in terms of autocorrelation functions) and then determines the values of all three criteria by using a computer program to trace rays through brilliants of various dimensions. He makes calculations for a range of different values of the table width, pavilion angle, and height of the crown, and presents his results in a series of graphs. These results give a basis for estimating the best dimensions to obtain the most pleasing gemstones, although of course the relative importance of the three criteria depends finally on the eye of the beholder.

Although most of the detailed studies have dealt with the brilliant cut there are also a range of other cuts with different shapes, numbers and arrangements of facets (Bruton, 1981; Watermeyer, 1982). Although not as popular these other cuts give variety. They are also useful for dealing with diamonds of irregular shape so as not to lose too much material during cutting. The choice of a different geometry may also help to avoid some inclusion or other imperfection in the diamond.

Finally, the unique character of diamond as a gemstone is underlined by comparison with the various diamond simulants which appear on the market from time to time. For example, crystals of strontium titanate have almost exactly the same refractive index but a dispersion about ×5 greater which results in a rather unpleasing display of colour. Strontium titanate is also much less hard than diamond and so the surface is easily scratched. Other materials which have been put forward as simulants include yttrium aluminium garnet (yag) and cubic zirconium, but both have lower refractive indices than diamond and are less abrasion resistant. Comparative measurements of the abrasion resistance of various gemstones are given by Wilks (1973) and of the indentation hardness by Brookes (1973, 1979), see also Caveney (1976).

17.2 Dressers

We have already referred to the processes of truing and dressing when discussing the use of diamond grinding wheels (Section 15.2.e). Truing ensures that the wheel has the required geometric form, and dressing produces the correct surface texture with the diamond grits standing sufficiently proud of the matrix to give a good cutting action. These two processes are equally necessary when using other types of grinding wheels, and diamond *dressers* are used both to ensure the correct dimensions and surface condition of tungsten carbide, alumina and silicon carbide wheels. One of the oldest and most extensive uses of diamond in industry is as a single point tool to true and dress tungsten carbide grinding wheels.

In the simplest form of truing and dressing a single crystal diamond is set to present one of its natural points to the grinding wheel in a trailing aspect. Practical details given by Freeman (1964) and Selby (1970) include the need to use machines which mount the diamond firmly without vibration, to rotate the diamond periodically to ensure uniform wear, to avoid easy directions of wear, and to use an adequate supply of coolant. Because an appreciable flat is soon formed on the diamond, perhaps up to 1 mm^2 in area, the forces on the stone may be quite considerable. Hence single point dressers are generally relatively large diamonds which can be held firmly in their mounting.

Figure 17.3 Example of a multipoint diamond dresser (E. Winter und Sohn)

An alternative to the single point dresser is the so-called multipoint dresser in which several diamonds are set in one tool piece so that the same rate of dressing may be obtained with less wear and less force on each diamond and therefore the diamonds can be considerably smaller. The concept of a multipoint dresser is further extended in the *impregnated* dresser where large numbers of small grits are embedded in a matrix (Figure 17.3). There are also *rotary* dressers for shaping a grinding wheel prior to use for form grinding. Details of these types of dressers have been given by Cantrell (1964), Smith and Coombs (1964), Selby (1970), and Klocke and Blanke (1987).

So far we have referred only to dressing tools with single crystal diamonds but increasing use is now being made of PCD material, particularly for single point or chisel type dressers, for details see Finnigan (1979). A recent example of the use of PCD is given by Warnecke, Grün and Geis-Drescher (1988) who describe a method of dressing by milling with wheels tipped with blocks of PCD.

17.3 Dies

Diamond plays an important part in the wire drawing industry, see for example Urbanek (1966). Single crystal diamond is in several ways an ideal die material for drawing non-ferrous metals. It is very strong and can be polished to give a smooth surface with a low wear and a low friction. Single crystals have long been used as dies for drawing fine wires, and larger dies made of PCD material are now replacing dies previously made from tungsten carbide.

Figure 17.4 gives a cross-sectional view of a typical die. The wire to be drawn down enters from the top and passes through the narrowing approach section to the uniform bearing section which determines the final diameter. Formerly the drilling and shaping of diamond dies was a rather length process. Today, however, the initial hole for the bore is pierced with a laser beam and then polished by diamond powder on a shaped steel needle vibrated ultrasonically. Details of the manufacture and use of these dies are given by Peter (1981), Eder (1983, 1986) and Niederhauser (1986).

Because, the abrasion resistance of a diamond surface depends on its crystallographic orientation, the wear of a die will depend on the direction chosen for its axis. For example Seal (1967) reports that the length of wire drawn in a die was increased three times by changing the direction of the axis. The best choice of direction is not too obvious because as the wire passes through the die its circumference is abrading the diamond on a whole 360° range of planes, and the rates of wear on these planes will be somewhat different. Hence, the originally circular hole will not only grow larger but will loose its shape. However, <110>

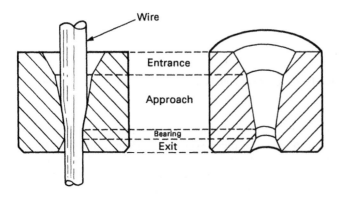

Figure 17.4 Sections of a typical wire drawing die (Eder, 1983)

directions offer the advantage that the wire is abrading the sides of the hole with {001} and {011} orientations in abrasion resistant directions (Seal, 1967; Gane, 1974). It must also be remembered that small changes of crystallographic direction may cause appreciable changes in the pattern of wear (Sections 9.2 and 9.3) but there is not much published information on the optimum orientations.

Dies fabricated from PCD material were first described by the General Electric company (Flom et al., 1975) and offer two considerable advantages. Firstly, the use of a relatively homogenous material avoids the complications involved in orienting single crystals. Second, it is possible to draw wires of much greater diameter. The cost of single crystal diamonds large enough to draw wire of diameter greater than about 2 mm becomes prohibitive and previously dies of tungsten carbide had to be employed. However, PCD blocks are now being manufactured in increasing size and die blanks are presently available up to at least 15 mm in diameter. Descriptions of the manufacture and use of these dies are given by Eder (1983,1986) and Bex (1983). Finally we note that die blanks are now available which are made from the more thermally stable types of PCD described in Section 12.2.d, and although not so strong are more resistant to temperature rises produced by the drawing process.

17.4 Scribers

Perhaps the oldest and best known diamond tool is the glazier's diamond which makes a clean scratch on a glass sheet prior to dividing it by fracture. This tool is no more than the tip of a suitably mounted small diamond which being so much harder than the glass produces a fine groove and hardly wears at all. Diamond has also been used for a long time by the engraver to produce artistic effects on various materials including glass and precious stones, both with a single crystal diamond and with rotating drills and wheels carrying a paste of diamond powder. Details of these traditional techniques are given by Grodzinski (1953). Today metal bonded diamond wheels with vee-shaped edges are used to engrave high quality glass, see for example Herbert (1985).

A more sophisticated version of the glazier's diamond is the diamond tool used to rule optical diffraction gratings. These gratings may be up to 200 mm square and are engraved by specially designed ruling engines (see for example Verrill, 1977) to produce perhaps 100000 lines all accurately parallel and equally spaced. The diffraction grating is a very high precision instrument for the measurement of optical wavelengths but its performance depends critically both on the precision of the ruling engine and on maintaining the same cross-section of the grooves over the whole area of the grating. Hence the diamond must be polished to produce the required form of cross-section for the grating in question and its wear must be negligible over a path length of perhaps 20 km. This condition is met fairly readily by the careful selection of the diamond and by ruling the grating on a film of soft metal deposited on an optically flat glass plate. (Experiments on the wear of diamond under very similar conditions are described in Section 13.2.)

17.5 Wear resistant materials

Besides being an important tool material PCD is now also used as a wear resistant material for items subject to heavy wear. Reported examples include contact surfaces used to grip the workpiece during machining and grinding (Gött, 1986), spindle bearings and facings for slides (Hayes, 1989), components for high pressure homogenizers (Kaphengst, 1989), and for thrust bearings for a down-hole drill turbine (Nagel, 1986). Mehan and Hayden (1981) have compared the friction and wear of PCD materials of different grain size rubbing on steel with that of alumina, silicon carbide, and tungsten carbide under similar conditions, see Sections 12.4 and 13.4.b. Their results show that PCD with a finely ground surface and adequate lubrication has a very low friction and a minimum of wear. (The use of diamond films deposited from the vapour as wear resistant coatings is discussed in Sections 1.4.c and 12.6.)

We also mention various other composite materials designed to be wear resistant in which the diamond is either embedded in some matrix or electroplated on to a surface. Although not as wear resistant as PCD these materials have the advantage that they can be more readily fabricated into components with complex geometries. Sharp (1975) and Zahavi and Hazan (1983) describe composites in which micron size diamond is dispersed in electroplated nickel. Mehan (1982) describes measurements of the wear of composites consisting of diamond grit embedded in a matrix of silicon when rubbed against each other and against steel. Measurements of the wear suffered by a range of materials when rubbing on these composites are described and discussed by Mehan, Hejna and McConnell (1985). At present such composite type materials appear to be mainly in the experimental stage but various possible uses, including applications as handling devices and guide pins in the paper and textile industries have been mentioned by Sharp (1975) and Wapler, Spooner and Balfour (1979).

17.6 Miscellaneous

17.6.a Indenters

Diamonds are widely used as indenters to measure the hardness of materials. In a typical hardness determination some form of pyramidal indenter is pushed into a flat surface of the specimen under a given load to form an indentation by plastic deformation. The force on the indenter divided by the area of the indent (measured in the plane of the surface) then gives a measure of the hardness of the material, see for example Tabor (1970). Diamond being much harder than most other materials is an ideal material for the indenters as it suffers no deformation itself and very little wear or damage.

A diamond indenter may have various geometries, the most common is the four sided pyramid with an apex angle of 136° known as the Vickers indenter. Other shapes may be used for particular purposes. The Knoop indenter has a shallower point and produces an indentation whose length is seven times its width. It is particularly useful for studying the variation of the hardness of anisotropic materials with orientation, see for example Tabor (1970), and tends to give more satisfactory indentations if the specimen is at all brittle. Other types of indenter include the Berkovich which produces indentations with a three-fold symmetry and the pentagonal indenter with a five-fold symmetry. As the symmetry of the pentagonal indenter is different from that of any crystal it will sample a range of directions in the specimen and tends to give a hardness value more or less independent of the relative orientations of indenter and specimen (Brookes and Moxley, 1975).

17.6.b Styli

Diamond styli are widely used as pick-ups in audio disc players. It is obviously important that the stylus follows the variations in the grooving on the disc with a minimum of wear on both disc and stylus, and this is best achieved by using a stylus of diamond with a polished round tip. Figure 17.5 gives a sketch of a typical arrangement with the diamond bonded on to the end of a steel shank, all the dimensions are kept small in order to reduce the inertia of the system. The manufacture of these diamond tips is described by Ebisawa (1972) and Weinz (1975). Small 'logs' of diamond perhaps 1 mm in length and 0.5 mm in diameter are brazed to a steel shank and then ground to the required conical shape prior to polishing the tip. This all important process is generally achieved by tumbling together a large number of tips in a mill until the required radius and polish (Figure 17.6) is obtained.

A considerable effort has been made in the past to develop a system similar to audio discs to produce and play video discs. As there is much more information to be conveyed in a video system the grooves must be much narrower and the

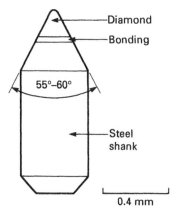

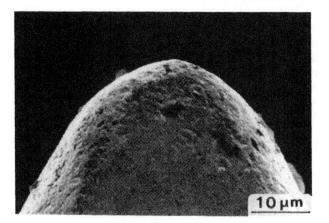

Figure 17.5 Sketch of a diamond record stylus mounted on a steel shank

Figure 17.6 SEM micrograph of the tip of a diamond record stylus (Anon, 1972)

speed of the disc greater, see for example Joschko (1977) and Hackett and Taylor (1981). These systems call for considerable precision in the mounting of the disc, in the design of the pick-up head, and in the design and polishing of the diamond to achieve faithful reproduction and low wear. Even so, good quality video pictures can now be obtained in this way. However, these techniques have been superseded by systems using magnetic tape with their much greater facility for recording.

Diamond styli are also widely used as probes in surface profilimeters, and it is now possible with the technique of ion milling (Section 9.6.c) to manufacture styli with radii of only 100 nm or less. It is of course necessary that the loading of the stylus is kept sufficiently light to cause a minimum of damage to the surface. This is particularly important when using styli of small radii which can generate very high pressures. It must also be remembered when using diamond styli in the latest profilometers working at the level of nanometers that the polished surface of a stylus is itself rough on this scale (Section 9.4.b). This geometry will be particularly important in the atomic force microscope, see for example comments by Albrecht and Quate (1988) and Meyer *et al.* (1988).

17.6.c Knives

We now describe four rather different types of diamond knife which are used to cleave optical fibres, to cut audio discs, and to act as microtome blades and surgeons' scalpels.

Diamond knives or blades are being increasingly used in the telecommunications industry to help join together optical fibres. These fibres are of glassy material and about 100 μm diameter, to effect a satisfactory join two faces must be prepared which are flat, smooth, and perpendicular to the axis of the fibre. This can be achieved by indenting the fibres with a wedged shaped blade and then applying a tension to produce cleavage. A study of this process by Field (1984) shows that the load applied to the blade mut be carefully chosen. Too high a value results in some cracking of the cleaved surface, and too low a load results in a rough surface. There is also an optimum value for the wedge angle of the blade; too great an angle results in a poor cleavage, but if the angle is less than about 45° the knife becomes more liable to damage. Improved results are now being obtained by vibrating the diamond blade at ultrasonic frequencies.

Diamond knives mounted on piezo-electric crystals are used to cut the master moulds for the manufacture of audio discs. Obviously this work calls for particular precision and Figure 17.7 shows a high magnification SEM micrograph of a cutting stylus and the high degree of perfection of the edge (Seal, 1985).

Diamond knives as microtome blades are used for producing sections of biological specimens only 100 nm thick or less, and also sections of harder specimens (Fernandez-Moran, 1956, 1985, 1986). The fine polishing methods which are used to achieve the extremely fine edges on these knives are described in Section 9.5.d. A diamond microtome knife offers three desirable characteristics. Its very sharp edge permits the cutting of very thin sections with very little surface

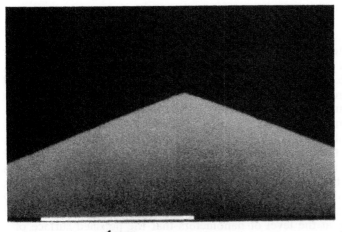

1 μm

Figure 17.7 Micrograph of a diamond stylus for cutting audio disc masters (Seal, 1985)

0.5 mm

Figure 17.8 A standard 45° diamond surgical blade
(Pierse and Ackers, 1987)

damage, see for example Mollenhauer and Bradfute (1987). The edge is strong and therefore reliable, an important point when making serial sections of valuable specimens; its strength also permits the cutting of harder materials. Finally sections of tissue and biological materials slide easily over the knife without wrinkling, partly because of the low friction of a well polished diamond surface and partly because of its water repellant properties (hydrophobicity) described in Section 13.8.c. On the other hand the preparation of a diamond to the high standards required is inevitably costly so diamond knives are used only when essential. (The care required to produce a good knife is underlined by the occasional report of knives whose sections tend to stick to the diamond either because the flanks of the blade are poorly polished or carry some contamination which affects the wettability.)

The sharpness and durability of diamond knives has led to their increasing use as surgeons' scalpels (Figure 17.8), particularly in very delicate operations such as those on the eye. The extreme sharpness of a diamond scalpel has two main benefits. There is a very minimum of damage to the surrounding tissue and also very little displacement of the tissue before cutting begins, in contrast to the quite perceptible displacement and distortion which may be observed when using less sharp knives. Further details of the use of diamond scalpels in eye surgery are given by Herbert (1984b). Other applications are mentioned by Pierse and Ackers (1987) and by Doting and Vincent (1988) who describe an ingenious design of knife to make several different types of cuts in coronary arteries, with just one instrument. Finally, the preparation of these blades calls for a very high degree of precision in their manufacture, including inspection in the SEM (Seal, 1985).

17.6.d Heat sinks

Electronic devices using semiconductors are now capable of generating microwave power and laser energy in devices of remarkably small size. However, although the small size of these devices is very convenient it gives rise to problems associated with the production of the heat which is inevitably associated with the generation of useful power. This flow of heat, at the rate of perhaps 5 W, may be concentrated over an area of perhaps only 10^{-2} mm^2, so the temperature of the device will rise excessively unless the heat can be quickly dissipated into some form of heat sink. An obvious way of removing the heat is to bond the device to a block of high conductivity copper so that the heat spreads out over a much larger area and produces much smaller temperature gradients. An even better method suggested by Swan (1967) is to bond the device to a small block of Type IIa diamond which at room temperature has a thermal conductivity about five times that of copper. Therefore blocks of diamond of the order of $0.5 \times 0.5 \times 0.25$ mm in size are now widely used as heat sinks particularly for laser diodes used in the fibre optical techniques in the telecommunications industry.

In order to obtain good thermal contact with the device the diamond block is generally supplied with some form of coating, perhaps a thin layer of titanium, covered in turn with thin layers of platinum and gold. The titanium bonds well with the diamond, and the gold with the device, while the platinum helps to prevent diffusion effects (Seal, 1978). These various bonds produce some thermal resistance so the overall improvement on changing from a copper to a diamond sink may be about a factor 3 rather than the factor 5 we would expect by considering the conductivity of the diamond alone. The design of heat sinks to give the optimum performance for particular sized devices with particular power outputs is discussed by Berman (1970) and Seal (1988) and by Doting and Molenaar (1988) who have developed computer programs to optimize the mechanical design.

Until recently all diamond for heat sinks was obtained by selecting relatively large good quality stones, either Type II or Type Ia with a low infrared absorption, which have a high thermal conductivity. However, there have recently been two significant developments. First, as mentioned in Section 6.3.c, synthetic Type Ib diamonds are now available, from the Sumitomo company, which appear to give a comparable performance. Second, General Electric has recently produced diamonds with a much reduced concentration of the ^{13}C isotope (Section 1.2.b) which have a thermal conductivity about 50% higher than in any natural diamonds (Section 6.3.c).

17.6.e Anvils and windows

We have already mentioned two applications in previous chapters. Firstly, the use of diamond anvils permits the application of extremely high pressures to materials for experimental study. Diamond is used primarily for this purpose because of its great strength but in addition it is transparent to radiation. Hence it is possible to make optical observations on a specimen under pressure between the anvils. For further details see Section 7.3.a. Second, quite apart from anvils the very low optical absorption of Type II diamond, particularly in the infrared, makes it a very valuable material for optical windows. Because of the strength of diamond, a small window to withstand a pressure of an atmosphere may be only 20 μm thick with virtually negligible absorption over a wide spectrum, and diamond windows have been used on several space probes (Anon, 1979; Seal, 1985).

17.7 Electronics

As described in Section 3.4 some diamonds contain small quantities of boron, have a finite electrical conductivity, and behave as semiconductors. Reviews of this conductivity are given by Collins and Lightowlers (1979) and Collins (1989).

There has long been some interest in using semiconducting diamonds to construct electronic devices which might work at higher temperatures than normal devices based on silicon and be more resistant to hostile environments. However, as discussed by Collins (1989) the semiconducting properties of diamond are by no means ideal. At temperatures below 500°C the number of current carriers varies considerably with the temperature, and at higher temperatures their mobility is much reduced by the thermal motion. In addition, electronic devices based on semiconductors generally depend on the possibility of introducing both n and p type carriers by suitable doping. There are as yet, however, no reports of the successful doping of diamond to produce useful n carriers. (Various attempts to introduce n carriers by ion implantation are summarized by Prins (1983). This author also refers to the possibility of using conducting amorphous layers produced by the implantation of carbon ions, see Section 10.6.c.)

Despite the above limitations of diamond as a semiconductor the production of both large single crystals and thin films of diamond has led to increased interest in this field, particularly as diamond doped with boron may be grown directly from the vapour phase (Poferl, Gardner and Angus, 1973). Geis *et al.* (1987) have described a point contact transistor and a Schottky diode formed on a diamond doped with boron, and Gildenblat *et al.* (1988) describe Schottky diodes formed by depositing gold and aluminium contacts on diamond films. However, it seems unlikely at present that such devices based on diamond can compete with products based on silicon and other materials, particularly as the structure of diamond films presently grown on non-diamond substrates is very imperfect compared to the single crystals generally used for devices. For a further discussion of these points see Collins (1989).

Some natural diamonds exhibit photoconductivity and can therefore be used as radiation detectors because, if a voltage is maintained across the diamond, pulses of current are produced when ultraviolet or nuclear radiation passes through it. (The basic mechanisms of photoconductivity are discussed by Denham, Lightowlers and Dean (1967) and Collins and Lightowlers (1968) but need not concern us here.) The fact that diamond is not toxic and responds to irradiation in somewhat the same way as biological tissue suggests that a small diamond inserted into tissue could form a useful device to measure the radiation received during medical treatment (Cotty, 1956).

Hitherto the main problem in using diamonds as radiation detectors or counters has been that the degree of photoconductivity varies so greatly from one diamond to another that it was not practical to select and calibrate diamonds on a commercial basis. But recent work by Keddy, Nam and Burns (1987, 1988) on specially synthesized diamonds suggests that reliable detectors may soon be available; these authors also give references to earlier work. Keddy, Nam and Burns (1988) also point out that besides measuring the intensity of the radiation, these counting diamonds may be used to measure the total radiation dose received. This is possible because an irradiation produces structural damage in the diamond and if the diamond is subsequently heated to about 300°C some or all of the damage anneals out, and some of the energy associated with the damage is emitted as luminescence. Hence the integrated intensity of this light gives an indication of the total dose of radiation received by the diamond.

References

Albrecht, T.R. and Quate, C.F. (1988) *Journal of Vacuum Science and Technology,* **A6**, 271–274

Anon (1972) *Industrial Diamond Review,* **32**, 69–70

Anon (1979) *Industrial Diamond Review,* **39**, 115–118

Berman, R. (1970) *Diamond Research,* 1970 supplement to *Industrial Diamond Review,* pp.2–4

Bex, P.A. (1983) *Industrial Diamond Review,* **43**, 75–80

Biastoch, G. (1983) *Industrial Diamond Review,* **43**, 66–68

Brookes, C. A. (1973) *Industrial Diamond Review,* **33**, 338–341

Brookes, C. A. (1979) In *The Properties of Diamond,* (ed. J. E. Field), Academic Press, London, pp. 383–402

Brookes, C. A. and Moxley, B. (1975) *Journal of Physics E,* **8**, 456–460

Bruton, E. (1981) *Diamonds,* (2nd edn) revised. NAG Press, London

Bühler, W. (1986) *Industrial Diamond Review,* **46**, 52–54

Cantrell, J. (1964) In *Diamond Abrasives and Tools,* (ed. J. Burls), Pergamon Press, Oxford, pp. 1–15

Caveney, R. J. (1976) *Industrial Diamond Review,* **36**, 45–47

Collins, A. T. (1989) *Semiconductor Science Technology,* **4**, 605–611

Collins, A. T. and Lightowlers, E. C. (1968) *Physical Review,* **171**, 843–855

Collins, A. T. and Lightowlers, E. C. (1979) In *The Properties of Diamond,* (ed. J. E. Field), Academic Press, London, pp. 79–105

Cotty, W. F. (1956) *Nature,* **177**, 1075–1076

Denham, P. , Lightowlers, E. C. and Dean, P. J. (1967) *Physical Review,* **161**, 762–768

Dodson, J. (1978) *Ph.D. Thesis,* University of London

Dodson, J. (1979) In *Diamond Research, 1979,* supplement to *Industrial Diamond Review,* pp. 13–17

Doting, J. and Vincent, J. G. (1988) *Industrial Diamond Review,* **48**, 68

Doting, J. and Molenaar, J. (1988) In *Proceedings of the 4th Annual IEEE Semiconductor Thermal Temperature Measurement Symposium* (San Diego, 1988), pp. 113–117

Ebisawa, T. (1972) *Industrial Diamond Review,* **32**, 150–152

Eder, K. G. (1983) *Industrial Diamond Review,* **43**, 200–204

Eder, K. G. (1986) *Wire Industry,* October, 696–700, 704

Fernandez-Moran, H. (1956) *Industrial Diamond Review,* **16**, 128–133; *Journal of Biophysical and Biochemical Cytology,* **2**, (4) supplement 2, 29–30

Fernandez-Moran, H. (1985) In *The Beginnings of Electron Microscopy* (ed. P. W. Hawkes), Academic Press, Orlando, pp. 178–223

Fernandez-Moran, H. (1986) *Ultramicroscopy,* **20**, 317–328

Field, J. E. (1984) In *Ultrahard Materials Application Technology,* Vol. 3, (ed. P. Daniel), De Beers Industrial Diamond Division, London, pp. 1–10

Finnigan, G. (1979) *Industrial Diamond Review,* **39**, 314–319

Flom, D. G. , Hanneman, R. E. , Rocco, W. A. and Wentorf, Jr, R. H. (1975) *Wire Technology,* January/February, 19–25

Freeman, G. P. (1964) In *Diamond Abrasives and Tools,* (ed. J. Burls), Pergamon Press, Oxford, pp. 17–29

Gane, N. (1974) *Journal of the Australian Institute of Metals,* **19**, 259–265

Geis, M. W. , Rathman, D. D. , Ehrlich, D. J. *et al.* (1987) *IEEE Electron Device Letters,* **EDL-8**, 341–343

Gildenblat, G. Sh. , Grot, S. A. , Wronski, C. R. *et al.* (1988) *Applied Physics Letters,* **53**, 586–588

Gött, W. (1986) *Industrial Diamond Review,* **46**, 265–267

Grodzinski, P. (1953) *Diamond Technology,* (2nd edn) revised. NAG Press, London

Hackett, C. F. and Taylor, B. K. (1981) *RCA Engineer,* Nov/Dec, 26–29

Hayes, D. (1989) *Industrial Diamond Review*, **49**, 118–119

Heger, K. H. (1981) *Industrial Diamond Review*, **41**, 120–125

Herbert, S. (1974) *Industrial Diamond Review*, **34**, 174–176

Herbert, S. (1976) *Industrial Diamond Review*, **36**, 79–81

Herbert, S. (1983) *Industrial Diamond Review*, **43**, 189–193

Herbert, S. (1984a) *Industrial Diamond Review*, **44**, 71–73

Herbert, S. (1984b) *Industrial Diamond Review*, **44**, 249–253

Herbert, S. (1985) *Industrial Diamond Review*, **45**, 171–174

Herbert, S. (1986) *Glass*, July, 249–250

Herbert, S. (1987) *Industrial Diamond Review*, **47**, 100–102

Jefferies, D. (1753) *A Treatise on Diamonds and Pearls*. London

Jennings, M. (1989) *Industrial Diamond Review*, **49**, 1–3

Joschko, G. (1977) *Industrial Diamond Review*, **37**, 117–121

Kaphengst, H. (1989) *Industrial Diamond Review*, **49**, 64–65

Keddy, R. J. , Nam, T. L. and Burns, R. C. (1987) *Physics of Medicine and Biology*, **32**, 751–759

Keddy, R. J. , Nam, T. L. and Burns, R. C. (1988) *Carbon*, **26**, 345–356

Klocke, F. and Blanke, G. (1987) *Industrial Diamond Review*, **47**, 22–25

Mehan, R. L. (1982) *Wear*, **78**, 365–383

Mehan, R. L. and Hayden, S. C. (1981) *Wear*, **74**, 195–212

Mehan, R. L. , Hejna, C. I. and McConnell, M. D. (1985) *Journal of Materials Science*, **20**, 1222–1236

Meyer, E., Heinzelmann, H., Grütter, D. *et al.* (1988) *Journal of Microscopy*, **152**, 269–280

Moitzi, H. (1985) *Industrial Diamond Review*, **45**, 175–181

Mollenhauer, H. H. and Bradfute, O. E. (1987) *Journal of Electron Microscopy Technique*, **6**, 81–85

Nagel, D. D. (1986) *United States Patent No 4, 620, 601*, Nov. 4 1986

Niederhauser, H. R. (1986) *Wire Industry*, October, pp. 709–711

Peter, H. (1981) *Industrial Diamond Review*, **41**, 63–66

Pierse, D. J. and Ackers, R. F. (1987) *Industrial Diamond Review*, **47**, 13–14

Poferl, D. J. , Gardner, N. C. and Angus, J. C. (1973) *Journal of Applied Physics*, **44**, 1428–1434

Prins, J. F. (1983) In *Ultrahard Materials Application Technology*, Vol. 2, (ed. P. Daniel), De Beers Industrial Diamond Division, London, pp. 15–25

Reisert, N. (1989) *Industrial Diamond Review*, **49**, 214–217

Seal, M. (1967) In *Science and Technology of Industrial Diamonds, Vol. 1, Science*, (ed. J. Burls), De Beers Industrial Diamond Division, London, pp. 145–159

Seal, M. (1978) *Industrial Diamond Review*, **38**, 130–134

Seal, M. (1985) *Industrial Diamond Review*, **45**, 58–61

Seal, M. (1988) In *Program and Abstracts First International Conference on the New Diamond Science and Technology* , Japan New Diamond Forum, Tokyo, pp. 162–163

Selby, J. S. (1970) *Tooling*, **24**, No. 3, 35–40

Sharp, W. F. (1975) *Wear*, **32**, 315–325

Smith, N. R. and Coombs, N. J. (1964) In *Diamond Abrasives and Tools* (ed. J. Burls), Pergamon Press, Oxford, pp. 55–80

Steinmetz, K. and Heath, P. J. (1989) *Industrial Diamond Review*, **49**, 164–167

Swan, C. B. (1967) *Proceedings of the Institute of Electrical and Electronic Engineers*, **55**, 451–452

Tabor, D. (1970) *Reviews of Physics in Technology*, **1**, 145–179

Tolkowsky, M. (1919) *Diamond Design*. E. and F. N. Spon, London

Urbanek, J. W. (1966) In *Progress in Industrial Diamond Technology* (ed. J. Burls), Academic Press, London, pp. 37–48

Verrill, J. F. (1977) *Physics Bulletin*, **28**, p. 355

Wapler, H. , Spooner, T. A. and Balfour, A. M. (1979) *Industrial Diamond Review*, **39**, 251–255

Warnecke, G. , Grün, F-J. and Geis-Drescher, W. (1988) *Industrial Diamond Review*, **48**, 106–110

Watermeyer, B. (1982) *Diamond Cutting*, Centuar, Johannesburg

Weinz, E. A. (1975) *Industrial Diamond Review*, **35**, 292–295

Wilks, E. M. (1973) *Industrial Diamond Review*, **33**, 186–189

Zahavi, J. and Hazan, J. (1983) *Plating and Surface Finishing*, **70**, 57–61

Appendix I

Some numerical values

Density of diamond	3520 kg m^{-3}

Distance between adjacent atoms   0.155 nm
Side of the unit cell (Figure 1.19)   0.357 nm

Refractive index (Peter, 1923)

486 nm	2.436
546 nm	2.424
589 nm	2.417
656 nm	2.410

Elastic properties
These properties described in Chapter 7 vary appreciably with the orientation of the diamond. However, for approximate calculations we may use the following values:

Young modulus	1050 GPa
Bulk modulus	500 GPa
Velocity of longitudinal sound	1.8×10^4 ms^{-1}
Velocity of transverse sound	1.2×10^4 ms^{-1}

Reference

Peter, F. (1923) *Zeitschrift für Physic*, **15**, 358–368

Units and conversion factors

Length   The S.I. unit of length is the metre (m)

$1 \text{ m} = 10^3 \text{ mm} = 10^6 \text{ } \mu\text{m} = 10^9 \text{ nm}$

Pressure   The S.I. unit of pressure is the Pascal (Pa)

$1 \text{ bar} = 10^5 \text{ Pa} \simeq 1 \text{ atmosphere}$
$1 \text{ GPa} = 10^9 \text{ Pa} = 10 \text{ kbar}$

Temperature

Temperatures on the Centigrade (°C) and Absolute (K) scales are related as

$T(K) = T(°C) + 273.14$

Energy, wavelength and wavenumber

The energy, wavelength λ and wavenumber $\nu = 1/\lambda$ of a photon are related as

$\text{energy (eV)} = 1240/\lambda(\text{nm}) = 1.240\nu \text{ (cm}^{-1})/10^4$

Index

Although the majority of diamonds used in industry are now produced by synthesis most studies of the properties of diamond have been made on relatively large natural material. Therefore references to the properties of diamond in this index refer primarily to work on natural diamond unless otherwise stated.

Printed and bound by CPI Group (UK) Ltd, Croydon, CR0 4YY

03/10/2024

01040847-0019